Observatory Madras

Results of observations of fixed stars made with the Meridian Circle

Observatory Madras

Results of observations of fixed stars made with the Meridian Circle

ISBN/EAN: 9783337711979

Printed in Europe, USA, Canada, Australia, Japan

Cover: Foto ©berggeist007 / pixelio.de

More available books at **www.hansebooks.com**

RESULTS

OF

OBSERVATIONS OF THE FIXED STARS

MADE WITH THE

MERIDIAN CIRCLE

AT THE

GOVERNMENT OBSERVATORY, MADRAS,

IN THE YEARS

1865, 1866 AND 1867,

UNDER THE DIRECTION OF

NORMAN ROBERT POGSON,

C.I.E., F.R.A.S., & F.M.U.

GOVERNMENT ASTRONOMER AT MADRAS.

PUBLISHED BY ORDER OF THE GOVERNMENT OF MADRAS.

MADRAS.

PRINTED AT THE LAWRENCE ASYLUM PRESS, BY G. W. TAYLOR,

1888.

CONTENTS.

	Page.
Introduction	i
Instrumental Corrections adopted in 1865 ...	iv
Instrumental Corrections adopted in 1866 ...	viii
Instrumental Corrections adopted in 1867	xii
Corrections to the Nautical Almanac Stars in the three years	xvii
Errata	xxi
Separate Results of Observations in 1865	1
Mean Positions of Stars for 1865 January 1st	61
Separate Results of Observations in 1866	125
Mean Positions of Stars for 1866 January 1st ...	181
Separate Results of Observations in 1867	289
Mean Positions of Stars for 1867 January 1st	299
Distribution List of Madras Astronomical Observations	359

INTRODUCTION.

A brief history and description of the Madras Observatory and its appliances, from 1792 to the present time, was given in the volume of *Madras Meridian Circle Observations*, 1862, 1863, 1864. It is therefore sufficient on this occasion to state that the observations of fixed stars during the next three years, the results of which are contained in this volume, were made and reduced upon the same plan as those in the former publication. The only changes introduced are in the arrangement of the Separate Results for each year, which have been printed in double column to save space; and in the flexure correction, which was inadvertently applied twice its proper amount up to the end of the year 1864. In consequence of this oversight the Polar Distances in the last volume will require a small correction to make them comparable with those in the present and future volumes of *Madras Meridian Circle Observations*. The necessary corrections will be furnished with sufficient accuracy by the following table.

From P.D.	Corr.	From P.D.	Corr.	From P.D.	Corr.	From P.D.	Corr.
°	"	°	"	°	"	°	"
0	− 0·9	44	− 0·4	79	+ 0·1	117	+ 0·6
3	− 0·8	52	− 0·3	86	+ 0·2	126	+ 0·7
14	− 0·7	59	− 0·2	93	+ 0·3	137	+ 0·8
25	− 0·6	66	− 0·1	100	+ 0·4	150 & S.P.N.	+ 0·9
35	− 0·5	73	0·0	108	+ 0·5		

The flexure correction finally adopted is,

$$0''\!\cdot\!85 \times \sin \text{Zenith Distance}$$

additive to all Polar Distances under 77° and subtractive from those of greater amount as well as from all observations below pole.

The observations were made throughout the three years by the two chief assistants, C. Ragoonatha Charry and T. Moottoosawmy Pillay, and with occasional exceptions were satisfactory; but it is much to be regretted that the reductions were not completed until both had so deteriorated, mentally and physically, that it has been a labor of unforeseen extent to detect and correct their numerous errors of calculation. The revision of every kind of reduction has to be gone through by the Astronomer personally before publication can be safely ventured upon. With no European assistance and too inadequate a staff of natives even to admit of duplicate calculations, this is not so much a matter of surprise as of disappointment, present delay and too numerous errata. The experience of the two volumes now published will considerably aid the progress and accuracy of future ones, but the task is severe and it is to be hoped peculiar to the Madras Observatory.

The instrumental corrections used during the three years have been exhibited to give a general idea as to the stability of the Meridian Circle when in proper working order. Owing however to a most unaccountable occurrence, such was not the case between March and July 1866, when the index correction will be found subject to most capricious changes, and the inclination correction, though less affected, far from steady. This was very simply explained in July, by finding, what no one could have suspected, that the object glass was unscrewed about one-sixth of a revolution. No clue was obtainable as to who had done the mischief, nor even as to when it was perpetrated, beyond that there was no trace of such defect before February 28th of the same year. Some stranger, who knew too much and too little about instruments, must have tampered with the object glass to gratify idle curiosity, about the end of February, and not having screwed it securely home, left it subject to a tilt on passing the zenith or nadir, which caused a sudden change in the index correction, at first a little under 2". The continual use of the instrument increased the looseness of the object glass, so that by the end of April the effect upon the index error had increased to about 5" and by July it had nearly reached 8". This was immediately rectified when the source of error was detected, on July 11th, and the usual stability of the instrument was at once restored. The foundations are however subject to a slow change, as the ground temperature or perhaps rather the dryness of the soil increases, from the cool season to the hot midsummer; and also to sudden disturbances after heavy rain, instances of which may be seen in the last three months of most years.

It would no doubt have been better to have rejected all the observations taken between March and July 11th and to have had them repeated in another year; but on finding that the observed nadir points always gave either fairly good or unmistakeably erroneous Polar Distances, to avoid such wholesale loss of work done, each observation taken during the time was compared with previous or subsequent determinations and the results accepted or rejected accordingly. Out of 1146 observations made during the time of uncertainty 258 were thus necessarily rejected. The misfortune was not however without a useful result, as it was this unexpected and enforced examination which first realized the general trustlessness of the old native assistants' reductions and the absolute necessity of personal revision by the Astronomer. In the results already printed, these careless mistakes being unforeseen were unfortunately not detected until the separate results had been struck off, causing a most undesirable list of errata and in some cases even involving the reprinting of a few forms. It is hoped however that the scrupulous previous revision now adopted will obviate such annoyance and delay in the publication of the results of future years.

The persistence of the positive sign in the Polar Distance columns of the last table of *Corrections to the Nautical Almanac Stars as given by the Madras Mean Positions*, clearly indicates the necessity of an increase of the assumed latitude. It would appear that the value adopted by Mr. T. G. Taylor in the old Madras Catalogue for 1835, viz., 13° 4′ 9″·1, was nearer the truth than that preferred by his successor, Captain W. S. Jacob, which was 1″ smaller and has been used ever since. This will of course be discussed and allowed for before the final catalogue for 1875 is constructed, but cannot be safely decided at the present time.

The large differences in both R. A. and P. D. of the seven southern stars not observable in Europe, are due to the erroneous positions adopted in the Nautical Almanacs for the respective years; before the improved places furnished by the observations at the Cape Royal Observatory and at Melbourne were available, and when those of the old *British Association Catalogue* were still unavoidably taken. When the Madras Mean Positions of these stars are compared with the more modern values now used in the Nautical Almanac, reduced back to the required years, the corrections are pretty much the same as those for all the other stars in the list.

For α Centauri, the Nautical Almanac gives α^2 as being the principal star, whereas α^1 was the brighter first-magnitude member of the pair, and α^2, the companion, was 7″·5 north following 0s·31 of α^1.

Besides the positions of fixed stars given in this volume, the Moon was also observed with the Meridian Circle on 176 nights, Mars on 33 at the opposition of 1867, and minor Planets on 209 occasions during the three years; but these are deferred until the Star Catalogue is completed, when a volume of planetary and cometary observations will it is hoped follow. My own judgment would have led me to publish this first and the star work afterwards.

Instrumental Corrections adopted in 1865.

Date.		Index.	Run in 5'.	Clock Rate.	Inclination.	Collimation.	Meridian.	Determining Stars.
		"	.	s	s	s	s	
Jan.	4	+ 2·6	− 0·1	− 0·15	− 0·43	− 0·10	+ 0·09	
	6	+ 2·1	− 0·1	− 0·01	− 0·36	− 0·05	+ 0·09	48 R. P. L. and γ[1] Eridani.
	7	+ 1·6	− 0·1	+ 0·09	− 0·39	− 0·08	+ 0·08	
	9	+ 2·1	− 0·1	− 0·01	− 0·37	− 0·06	+ 0·06	
	10	+ 2·0	− 0·1	− 0·04	− 0·34	− 0·06	+ 0·05	35 R. P. L. and ε Urs. Min.
	11	+ 1·1	− 0·1	0·00	− 0·27	0·00	+ 0·05	
	12	+ 1·3	− 0·1	+ 0·04	− 0·33	− 0·02	+ 0·05	
	13	+ 1·1	− 0·1	+ 0·08	− 0·34	− 0·03	+ 0·05	
	14	+ 0·9	− 0·1	+ 0·10	− 0·39	− 0·08	+ 0·04	
	17	+ 0·7	+ 0·2	+ 0·17	− 0·30	+ 0·01	+ 0·04	51 Cephei and δ Urs. Min.
	18	+ 1·2	+ 0·2	+ 0·12	− 0·33	− 0·04	+ 0·06	
	19	+ 1·0	+ 0·2	− 0·06	− 0·38	− 0·01	+ 0·08	
	20	+ 0·6	+ 0·2	− 0·01	− 0·33	0·00	+ 0·10	
	21	+ 0·2	+ 0·2	+ 0·06	− 0·33	0·00	+ 0·12	51 Cephei and λ Urs. Min.
	23	− 0·3	+ 0·2	+ 0·07	− 0·37	− 0·06	+ 0·10	
	24	+ 0·1	+ 0·2	+ 0·07	− 0·33	− 0·02	+ 0·09	
	25	− 0·1	+ 0·2	+ 0·07	− 0·36	− 0·06	+ 0·07	
	26	+ 0·4	+ 0·2	+ 0·06	− 0·34	− 0·05	+ 0·06	
	27	+ 0·5	+ 0·2	+ 0·11	− 0·32	− 0·08	+ 0·05	51 Cephei and δ Urs. Min.
	28	− 0·6	− 0·1	+ 0·14	− 0·39	− 0·09	+ 0·03	
	30	0·0	− 0·1	+ 0·11	− 0·36	− 0·08	− 0·01	
	31	− 0·4	− 0·1	+ 0·09	− 0·41	− 0·09	− 0·02	
Feb.	2	− 0·5	− 0·3	+ 0·25	− 0·30	− 0·04	− 0·06	
	3	− 0·8	− 0·3	+ 0·19	− 0·30	− 0·06	− 0·09	Pollux and δ Urs. Min.
	4	− 1·0	− 0·3	+ 0·08	− 0·30	− 0·05	− 0·07	
	6	− 0·6	− 0·3	+ 0·16	− 0·25	− 0·02	− 0·04	
	8	− 0·9	− 0·3	+ 0·10	− 0·31	− 0·03	− 0·01	48 R. P. L. and ε Urs. Min.
	9	− 0·0	− 0·3	+ 0·05	− 0·30	− 0·06	− 0·01	
	10	+ 0·4	− 0·3	− 0·10	− 0·27	− 0·06	− 0·01	
	11	− 0·3	− 0·3	− 0·09	− 0·14	+ 0·04	− 0·02	
	13	+ 0·4	− 0·3	+ 0·36	− 0·22	− 0·02	− 0·02	
	14	− 0·8	− 0·3	+ 0·16	− 0·27	− 0·06	− 0·06	
	15	− 0·6	+ 0·3	+ 0·01	− 0·17	+ 0·01	− 0·06	
	16	− 0·1	+ 0·1	+ 0·10	− 0·19	− 0·05	− 0·03	51 Cephei and δ Urs. Min.
	17	+ 0·3	+ 0·1	+ 0·11	− 0·17	+ 0·01	− 0·01	
	18	0·0	+ 0·1	+ 0·10	− 0·13	+ 0·04	0·00	
	20	− 0·8	+ 0·1	+ 0·22	− 0·09	+ 0·09	+ 0·03	
	21	+ 0·2	+ 0·1	+ 0·36	− 0·16	+ 0·02	+ 0·04	
	22	− 0·9	+ 0·1	+ 0·21	− 0·17	+ 0·03	+ 0·06	51 Cephei and λ Urs. Min.
	23	− 0·2	+ 0·1	+ 0·31	− 0·14	+ 0·05	+ 0·06	
	24	− 0·4	+ 0·1	+ 0·27	− 0·26	0·00	+ 0·05	
	25	− 0·1	+ 0·1	+ 0·14	− 0·16	+ 0·06	+ 0·04	
	27	0·0	+ 0·1	+ 0·46	− 0·18	+ 0·07	+ 0·08	
	28	− 0·7	+ 0·1	+ 0·37	− 0·28	− 0·04	+ 0·02	
Mar.	1	− 0·3	− 0·3	+ 0·19	− 0·19	− 0·07	+ 0·02	
	2	− 0·6	− 0·3	+ 0·23	− 0·28	− 0·06	+ 0·01	45 and 131 R. P. L.
	3	+ 0·1	− 0·3	+ 0·28	− 0·21	− 0·04	+ 0·01	
	4	− 0·6	− 0·3	+ 0·31	− 0·22	− 0·05	+ 0·02	
	6	− 0·1	− 0·3	+ 0·26	− 0·15	− 0·01	+ 0·02	
	7	0·0	− 0·3	+ 0·20	− 0·15	0·00	+ 0·03	
	9	+ 0·2	− 0·3	+ 0·19	− 0·15	0·00	+ 0·04	
	10	− 0·5	− 0·3	+ 0·17	− 0·09	+ 0·04	+ 0·04	40 R. P. L. and 15 Argûs
	11	− 0·2	− 0·3	+ 0·19	− 0·07	+ 0·03	+ 0·04	
	13	0·0	− 0·3	+ 0·19	− 0·04	+ 0·07	+ 0·04	
	14	− 0·4	− 0·3	+ 0·23	− 0·03	+ 0·00	+ 0·04	
	15	− 0·6	− 0·3	+ 0·22	− 0·11	− 0·05	+ 0·04	
	16	0·0	+ 0·4	+ 0·17	− 0·13	− 0·08	+ 0·04	

Jan. 30.—Index and inclination correction interpolated.

Instrumental Corrections adopted in 1865.

Date	Index.	Run in 5'	Clock Rate.	Inclination.	Collimation.	Meridian.	Determining Stars.
	″	″	s	s	s	s	
Mar. 17	+ 0·4	+ 0·1	+ 0·22	− 0·14	− 0·05	+ 0·04	
18	0·0	+ 0·1	+ 0·26	− 0·10	0·00	+ 0·04	ε Urs. Maj. and λ Urs. Min.
20	+ 0·0	+ 0·1	+ 0·30	− 0·13	− 0·01	+ 0·04	
21	+ 0·2	+ 0·1	+ 0·38	− 0·04	0·00	+ 0·04	
22	+ 0·4	+ 0·1	+ 0·22	− 0·12	− 0·01	+ 0·04	
23	+ 0·1	+ 0·1	+ 0·23	− 0·09	+ 0·08	+ 0·04	70 and 150 R. P. L.
24	− 0·1	+ 0·1	+ 0·40	− 0·12	− 0·01	+ 0·08	
25	+ 0·7	+ 0·1	+ 0·28	− 0·16	− 0·02	+ 0·02	
27	+ 0·6	+ 0·1	+ 0·30	− 0·13	− 0·08	− 0·01	
28	+ 0·4	+ 0·1	+ 0·28	− 0·13	0·00	− 0·02	
29	+ 0·5	+ 0·1	+ 0·30	− 0·10	+ 0·02	− 0·03	
30	+ 0·4	+ 0·1	+ 0·28	− 0·11	+ 0·01	− 0·06	
31	+ 0·6	+ 0·1	+ 0·40	− 0·05	+ 0·06	− 0·06	
Apl. 1	+ 0·7	− 0·2	+ 0·50	− 0·12	− 0·05	− 0·07	70 R. P. L. and δ Crateris.
3	+ 1·0	− 0·2	+ 0·28	− 0·18	− 0·06	− 0·08	
4	+ 1·0	− 0·2	+ 0·30	− 0·09	− 0·01	− 0·01	
5	+ 1·1	− 0·2	+ 0·28	− 0·01	+ 0·04	0·00	
6	+ 1·3	− 0·2	+ 0·26	+ 0·05	+ 0·09	+ 0·02	
7	+ 1·8	− 0·2	+ 0·40	− 0·01	+ 0·03	+ 0·04	
8	+ 1·7	− 0·2	+ 0·37	− 0·01	+ 0·08	+ 0·06	70 R. P. L. and Polaris.
10	+ 1·5	− 0·2	+ 0·28	+ 0·04	+ 0·06	+ 0·06	
11	+ 1·4	− 0·2	+ 0·37	0·00	+ 0·08	+ 0·06	
12	+ 1·3	− 0·2	+ 0·42	− 0·06	− 0·08	+ 0·05	
19	+ 1·7	0·0	+ 0·26	+ 0·01	+ 0·02	+ 0·04	
20	+ 1·6	0·0	+ 0·21	− 0·08	− 0·01	+ 0·04	12 Can. Ven. and Polaris.
21	+ 2·5	0·0	+ 0·23	− 0·02	0·00	+ 0·04	
23	+ 2·2	0·0	+ 0·31	− 0·03	− 0·08	+ 0·04	
24	+ 2·0	0·0	+ 0·10	− 0·01	+ 0·08	+ 0·04	
25	+ 2·0	0·0	+ 0·24	+ 0·10	+ 0·06	+ 0·04	
26	+ 1·8	0·0	+ 0·28	+ 0·01	+ 0·04	+ 0·04	99 R. P. L. and Polaris.
27	+ 1·0	0·0	+ 0·21	+ 0·05	+ 0·05	+ 0·08	
28	+ 1·0	0·0	+ 0·30	+ 0·02	+ 0·04	+ 0·08	
29	+ 2·4	0·0	+ 0·44	+ 0·02	+ 0·02	+ 0·02	
May 1	+ 3·2	− 0·3	+ 0·26	+ 0·04	+ 0·04	+ 0·01	
2	+ 2·9	− 0·3	+ 0·29	+ 0·02	0·00	0·00	99 R. P. L. and ε Corvi.
3	+ 1·5	− 0·3	+ 0·29	+ 0·06	+ 0·02	+ 0·01	
4	+ 1·9	− 0·3	+ 0·25	+ 0·06	+ 0·01	+ 0·01	
5	+ 2·5	− 0·3	+ 0·17	+ 0·12	+ 0·04	+ 0·02	
8	+ 2·2	− 0·3	+ 0·29	+ 0·07	+ 0·02	+ 0·04	92 R. P. L. and Polaris.
9	+ 2·1	− 0·3	+ 0·34	+ 0·06	− 0·02	+ 0·06	
10	+ 2·5	− 0·3	+ 0·38	+ 0·07	+ 0·09	+ 0·08	β Leonis and Polaris.
11	+ 2·1	− 0·3	+ 0·30	+ 0·13	+ 0·08	+ 0·07	
12	+ 1·7	− 0·3	+ 0·30	+ 0·11	0·00	+ 0·07	
13	+ 1·5	− 0·3	+ 0·34	+ 0·10	− 0·02	+ 0·06	
15	+ 1·8	− 0·3	+ 0·27	+ 0·17	0·00	+ 0·06	92, 99 R. P. L. and Polaris.
16	+ 2·1	+ 0·1	+ 0·41	+ 0·12	+ 0·04	+ 0·05	
17	+ 2·3	− 0·3	+ 0·50	+ 0·10	+ 0·05	+ 0·08	
18	+ 2·0	+ 0·1	+ 0·40	+ 0·06	− 0·02	+ 0·06	
20	+ 2·3	+ 0·1	+ 0·42	+ 0·01	− 0·02	+ 0·04	
22	+ 2·1	+ 0·1	+ 0·43	+ 0·05	+ 0·01	+ 0·04	93 R. P. L. and Polaris.
24	+ 2·2	+ 0·1	+ 0·48	+ 0·02	− 0·02	+ 0·01	
25	+ 2·7	+ 0·1	+ 0·44	− 0·04	− 0·04	− 0·03	
26	+ 2·0	+ 0·1	+ 0·39	− 0·01	− 0·06	− 0·06	99 R. P. L. and Polaris.
27	+ 2·1	+ 0·1	+ 0·35	0·00	− 0·01	− 0·06	
29	+ 2·6	+ 0·1	+ 0·48	+ 0·04	+ 0·02	− 0·02	
30	+ 2·4	+ 0·1	+ 0·42	+ 0·02	+ 0·01	− 0·01	
31	+ 2·4	+ 0·1	+ 0·41	+ 0·01	− 0·01	0·00	

Instrumental Corrections adopted in 1865.

Date.	Index.	Run in 5'.	Clock Rate.	Inclination.	Collimation.	Meridian.	Determining Stars.
			s	s	s	s	
June 1	+ 2·3	− 0·2	+ 0·40	+ 0·03	+ 0·08	+ 0·01	
3	+ 1·9	− 0·2	+ 0·44	+ 0·02	− 0·03	+ 0·04	ρ Bootis and Polaris.
5	+ 2·5	− 0·2	+ 0·60	+ 0·14	+ 0·04	+ 0·06	
6	+ 2·5	− 0·2	+ 0·44	+ 0·13	+ 0·02	+ 0·06	
7	+ 1·8	− 0·2	+ 0·51	+ 0·13	0·00	+ 0·06	
8	+ 2·4	− 0·2	+ 0·51	+ 0·10	− 0·01	+ 0·06	
12	+ 1·4	− 0·2	+ 0·47	+ 0·12	− 0·01	+ 0·08	
14	+ 1·8	− 0·2	+ 0·47	+ 0·10	− 0·01	+ 0·09	
19	+ 2·4	+ 0·2	+ 0·68	+ 0·16	+ 0·04	+ 0·11	β U. M., 111 & 88 R. P. L.
20	+ 2·2	+ 0·2	+ 0·64	+ 0·09	0·00	+ 0·11	
23	+ 2·4	+ 0·2	+ 0·80	+ 0·13	+ 0·03	+ 0·10	
24	+ 2·5	+ 0·2	+ 0·60	+ 0·15	+ 0·06	+ 0·09	
29	+ 3·0	+ 0·2	+ 0·78	+ 0·12	+ 0·06	+ 0·07	
July 1	+ 2·4	− 0·3	+ 0·64	+ 0·11	− 0·02	+ 0·07	
3	+ 2·2	− 0·3	+ 0·66	+ 0·15	+ 0·04	+ 0·08	
7	+ 2·4	− 0·3	+ 0·59	+ 0·14	+ 0·06	+ 0·06	
12	+ 2·7	− 0·3	+ 0·57	+ 0·14	+ 0·07	+ 0·08	
14	+ 3·4	− 0·3	+ 0·77	+ 0·04	+ 0·02	+ 0·02	ε Urs. Min. and μ¹ Sagit.
15	+ 3·2	− 0·3	+ 0·81	+ 0·03	+ 0·01	− 0·01	116 R. P. L. and ε Herc.
18	+ 3·0	− 0·3	+ 0·68	+ 0·00	− 0·01	0·00	
21	+ 3·0	− 0·3	+ 0·49	0·00	− 0·03	+ 0·02	
23	+ 3·9	− 0·3	+ 0·51	+ 0·11	+ 0·03	+ 0·02	ε Urs. Min. and μ¹ Sagit.
24	+ 2·7	− 0·3	+ 0·74	+ 0·08	+ 0·03	+ 0·01	
26	+ 3·1	− 0·3	+ 0·76	+ 0·10	+ 0·06	+ 0·01	
27	+ 3·1	− 0·3	+ 0·73	+ 0·07	+ 0·04	0·00	
28	+ 2·6	− 0·3	+ 0·74	+ 0·08	+ 0·06	0·00	
29	+ 3·3	− 0·3	+ 0·78	+ 0·06	+ 0·04	0·00	
31	+ 4·3	− 0·3	+ 0·71	0·00	0·00	− 0·01	
Aug. 2	+ 3·8	+ 0·2	+ 0·66	+ 0·03	+ 0·08	− 0·02	
3	+ 3·5	+ 0·2	+ 0·71	+ 0·04	+ 0·04	− 0·02	
5	+ 3·5	+ 0·2	+ 0·75	+ 0·12	+ 0·06	− 0·03	
8	+ 4·9	+ 0·2	+ 0·68	+ 0·01	+ 0·01	− 0·04	λ Urs. Min. and α Pavonis.
11	+ 4·9	+ 0·2	+ 0·72	+ 0·07	+ 0·06	+ 0·01	δ Urs. Min. and 51 Cephei.
17	+ 5·2	+ 0·2	+ 0·71	+ 0·09	+ 0·02	− 0·08	
18	+ 5·2	+ 0·2	+ 0·73	− 0·08	− 0·01	− 0·04	
23	+ 5·3	+ 0·2	+ 0·70	− 0·02	+ 0·05	− 0·06	
24	+ 4·9	+ 0·2	+ 0·80	+ 0·08	+ 0·04	− 0·06	
25	+ 5·6	+ 0·2	+ 1·02	− 0·07	+ 0·02	− 0·03	
26	+ 5·9	+ 0·2	+ 1·02	− 0·11	+ 0·02	− 0·07	
28	+ 6·0	+ 0·2	+ 0·77	− 0·00	− 0·01	− 0·05	
Sep. 1	+ 6·0	+ 0·2	+ 1·02	− 0·14	+ 0·09	− 0·09	δ Urs. Min. and 51 Cephei.
2	+ 5·3	+ 0·2	+ 1·01	− 0·12	+ 0·04	− 0·08	
5	+ 6·3	− 0·2	+ 0·81	− 0·13	+ 0·02	− 0·08	λ Urs. Min. and α Aquarii.
6	+ 6·4	− 0·2	+ 0·98	− 0·13	+ 0·04	− 0·08	
7	+ 6·5	− 0·2	+ 0·85	− 0·16	+ 0·04	− 0·04	
8	+ 6·4	− 0·2	+ 0·80	− 0·15	+ 0·08	− 0·09	
11	+ 6·1	− 0·2	+ 0·97	− 0·13	+ 0·02	− 0·08	
12	+ 6·1	− 0·2	+ 0·94	− 0·13	+ 0·06	− 0·08	
13	+ 6·1	− 0·2	+ 0·94	− 0·13	+ 0·02	− 0·08	160 and 72 R. P. L.
14	+ 5·7	− 0·2	− 0·20	− 0·11	+ 0·08	− 0·09	
15	+ 5·6	− 0·2	− 0·20	− 0·10	+ 0·04	− 0·10	
16	+ 5·6	0·0	− 0·25	− 0·16	+ 0·03	− 0·10	
21	+ 3·9	0·0	− 0·19	− 0·15	+ 0·03	− 0·14	
22	+ 5·5	0·0	− 0·85	− 0·14	+ 0·01	− 0·15	λ Urs. Min. and α Aquarii.
23	+ 5·0	0·0	− 0·73	− 0·16	+ 0·04	− 0·14	
25	+ 5·3	0·0	− 0·67	− 0·15	+ 0·06	− 0·12	

July 7.—The instrumental corrections interpolated.
Sep. 13.—The transit-clock put forward one minute and its rate adjusted.

Instrumental Corrections adopted in 1865.

Date	Index.	Run in 5′	Clock Rate.	Inclination.	Collimation.	Meridian.	Determining Stars.
	″	″	s	s	s	s	
Sep. 26	+ 5·1	0·0	− 0·69	− 0·12	+ 0·06	− 0·11	
28	+ 4·6	0·0	− 0·70	− 0·10	+ 0·05	− 0·10	
29	+ 4·4	0·0	− 0·79	− 0·15	+ 0·07	− 0·09	
30	+ 4·7	0·0	− 0·80	− 0·10	+ 0·07	− 0·08	λ Urs. Min. and α Gruis
Oct. 2	+ 5·1	− 0·2	− 0·90	− 0·11	+ 0·08	− 0·02	
3	+ 4·1	− 0·2	− 0·85	− 0·12	+ 0·05	+ 0·01	3 Cygni and 70 R. P. L.
4	+ 3·6	− 0·2	− 0·77	− 0·08	+ 0·07	− 0·01	
5	+ 4·3	− 0·2	− 0·76	− 0·17	+ 0·04	− 0·08	
6	+ 4·0	− 0·2	− 0·70	− 0·12	+ 0·06	− 0·06	
7	+ 4·4	− 0·2	− 0·87	− 0·13	+ 0·05	− 0·08	Polaris and 110 R. P. L.
9	+ 4·1	− 0·2	− 0·81	− 0·11	+ 0·04	− 0·06	
10	+ 3·5	− 0·2	− 0·80	− 0·06	+ 0·07	− 0·06	
11	+ 3·3	− 0·2	− 0·50	− 0·07	+ 0·07	− 0·04	
12	+ 3·8	− 0·2	− 0·80	− 0·10	+ 0·02	− 0·03	
13	+ 2·7	− 0·2	− 0·86	− 0·14	+ 0·06	− 0·02	
14	+ 3·8	− 0·2	− 0·91	− 0·10	+ 0·04	− 0·01	160 and 89 R. P. L.
16	+ 2·8	+ 0·1	− 0·98	− 0·10	+ 0·08	− 0·01	
17	+ 3·2	+ 0·1	− 0·89	− 0·13	+ 0·09	− 0·01	
18	+ 3·1	+ 0·1	− 0·87	− 0·11	+ 0·09	− 0·01	
23	+ 4·0	+ 0·1	− 0·80	− 0·13	+ 0·07	0·00	12 R. P. L., Pol. & Fomalhaut
24	+ 4·2	+ 0·1	− 0·78	− 0·12	+ 0·09	− 0·01	
25	+ 3·0	+ 0·1	− 0·70	− 0·14	+ 0·08	− 0·02	
26	+ 3·7	+ 0·1	− 0·78	− 0·14	+ 0·06	− 0·03	
27	+ 3·4	+ 0·1	− 0·78	− 0·15	+ 0·07	− 0·04	
28	+ 3·4	+ 0·1	− 0·82	− 0·15	+ 0·07	− 0·05	
31	+ 5·8	+ 0·1	− 0·84	− 0·24	+ 0·04	− 0·09	Polaris and Achernar.
Nov. 1	+ 5·8	0·0	− 0·84	− 0·25	+ 0·08	− 0·08	
3	+ 6·2	0·0	− 0·75	− 0·28	+ 0·06	− 0·08	Polaris and Achernar.
4	+ 6·6	0·0	− 0·75	− 0·31	+ 0·02	− 0·07	
8	+ 6·4	0·0	− 0·87	− 0·32	+ 0·06	− 0·03	12 R. P. L. and 12 Ceti.
10	+ 7·3	0·0	− 0·96	− 0·46	0·00	− 0·04	
11	+ 7·0	0·0	− 0·97	− 0·41	+ 0·08	− 0·05	
13	+ 6·5	0·0	− 0·88	− 0·45	+ 0·09	− 0·06	
14	+ 6·5	0·0	− 0·90	− 0·49	+ 0·01	− 0·07	Polaris and 114 R. P. L.
15	+ 4·8	0·0	− 0·90	− 0·44	+ 0·08	− 0·09	
16	+ 5·9	− 0·1	+ 0·22	− 0·42	+ 0·06	− 0·10	
17	+ 6·4	− 0·1	+ 0·22	− 0·41	+ 0·04	− 0·13	Polaris and β Ceti.
18	+ 4·2	− 0·1	+ 0·26	− 0·36	+ 0·06	− 0·12	
21	+ 6·9	− 0·1	+ 0·37	− 0·35	+ 0·08	− 0·12	
22	+ 6·4	− 0·1	+ 0·28	− 0·44	+ 0·01	− 0·12	Polaris and Achernar.
27	+ 9·2	− 0·1	+ 0·06	− 0·58	− 0·08	− 0·06	Polaris and 169 R. P. L.
28	+ 9·3	− 0·1	+ 0·26	− 0·47	+ 0·11	− 0·06	
29	+ 9·4	− 0·1	+ 0·44	− 0·58	+ 0·08	− 0·06	
30	+ 8·6	0·0	+ 0·26	− 0·10	0·00	− 0·06	Polaris and Achernar.
Dec. 1	+ 8·5	0·0	+ 0·28	− 0·09	0·00	− 0·05	
2	+ 9·5	0·0	+ 0·15	− 0·01	+ 0·08	− 0·08	
5	+ 7·3	0·0	+ 0·21	− 0·02	− 0·01	+ 0·01	Polaris and β Ceti.
7	+ 7·0	0·0	+ 0·17	− 0·02	− 0·01	+ 0·04	
8	+ 8·4	0·0	+ 0·18	− 0·06	− 0·01	+ 0·06	
9	+ 8·7	0·0	+ 0·22	+ 0·02	− 0·02	+ 0·05	38 R. P. L. and e¹ Eridani.
11	+ 8·3	0·0	+ 0·21	− 0·08	+ 0·06	+ 0·09	Polaris, 38 and 168 R. P. L.
12	+ 7·1	0·0	+ 0·21	− 0·01	0·00	+ 0·09	
13	+ 7·1	0·0	+ 0·26	+ 0·02	+ 0·04	+ 0·10	
14	+ 6·4	0·0	+ 0·22	+ 0·01	+ 0·08	+ 0·10	
15	+ 6·7	+ 0·2	+ 0·20	− 0·04	+ 0·02	+ 0·10	
16	+ 7·1	+ 0·2	+ 0·21	− 0·08	+ 0·01	+ 0·10	

Nov. 15.—The transit clock rate adjusted.
Nov. 21.—The instrument thoroughly examined and cleaned.
Nov. 30.—The inclination error adjusted.

Instrumental Corrections adopted in 1865.

Date.		Index.	Run in 5′	Clock Rate.	Inclination.	Collimation.	Meridian.	Determining Stars.
		″	″	s	s	s	s	
Dec.	18	+ 6·2	+ 0·2	+ 0·21	− 0·08	+ 0·06	+ 0·11	
	19	+ 6·1	+ 0·2	+ 0·21	− 0·08	+ 0·06	+ 0·11	
	20	+ 6·2	+ 0·2	+ 0·27	− 0·01	+ 0·07	+ 0·12	
	21	+ 5·9	+ 0·2	+ 0·33	− 0·07	+ 0·02	+ 0·12	38, 84 and 116 R. P. L
	22	+ 5·8	+ 0·2	+ 0·32	− 0·04	+ 0·08	+ 0·12	
	26	+ 5·2	+ 0·2	+ 0·27	− 0·02	+ 0·04	+ 0·12	
	29	+ 4·6	+ 0·2	+ 0·38	+ 0·04	+ 0·02	+ 0·12	
	30	+ 3·7	0·0	+ 0·31	− 0·01	+ 0·01	+ 0·13	

Instrumental Corrections adopted in 1866.

Date.		Index.	Run in 5′	Clock Rate.	Inclination.	Collimation.	Meridian.	Determining Stars.
		″	″	s	s	s	s	
Jan.	3	+ 3·3	− 0·1	+ 0·37	+ 0·02	+ 0·02	+ 0·08	
	4	+ 3·0	− 0·1	+ 0·35	− 0·01	− 0·01	+ 0·07	
	5	+ 2·5	− 0·1	+ 0·30	+ 0·01	0·00	+ 0·07	51 Cephei and ε Urs. Min.
	6	+ 2·5	− 0·1	+ 0·26	+ 0·04	+ 0·03	+ 0·08	
	8	+ 2·4	− 0·1	+ 0·28	+ 0·05	+ 0·02	+ 0·10	
	10	+ 1·8	− 0·1	+ 0·19	+ 0·08	+ 0·04	+ 0·12	
	11	+ 1·0	− 0·1	+ 0·17	+ 0·06	+ 0·03	+ 0·14	
	12	+ 0·8	− 0·1	+ 0·18	+ 0·01	0·00	+ 0·15	
	13	+ 1·0	− 0·1	+ 0·28	− 0·01	+ 0·08	+ 0·16	
	15	+ 0·0	− 0·1	+ 0·12	+ 0·07	+ 0·04	+ 0·18	51 Cephei and ε Urs. Min.
	16	+ 2·0	− 0·1	+ 0·34	+ 0·07	+ 0·06	+ 0·17	
	17	+ 1·6	− 0·1	+ 0·30	+ 0·02	+ 0·06	+ 0·16	
	18	+ 1·0	− 0·1	+ 0·34	+ 0·08	+ 0·06	+ 0·15	
	19	+ 0·5	− 0·1	+ 0·37	+ 0·04	+ 0·06	+ 0·14	
	20	+ 1·0	− 0·1	+ 0·39	+ 0·08	+ 0·04	+ 0·13	
	22	+ 0·6	− 0·1	+ 0·34	+ 0·08	+ 0·06	+ 0·11	51 Cephei and ε Urs. Min.
	23	+ 1·1	− 0·1	+ 0·35	+ 0·03	+ 0·04	+ 0·11	
	24	+ 0·9	− 0·1	+ 0·37	+ 0·07	+ 0·08	+ 0·12	
	25	+ 0·7	− 0·1	+ 0·40	+ 0·02	+ 0·02	+ 0·12	45 R. P. L. and λ Urs. Min.
	26	+ 0·8	− 0·1	+ 0·36	+ 0·08	+ 0·04	+ 0·12	
	27	+ 0·8	− 0·1	+ 0·31	+ 0·03	+ 0·01	+ 0·11	
	29	+ 0·8	− 0·1	+ 0·32	+ 0·05	+ 0·07	+ 0·11	
	30	+ 1·2	− 0·1	+ 0·32	+ 0·08	+ 0·03	+ 0·10	
	31	+ 0·9	− 0·1	+ 0·38	+ 0·06	+ 0·04	+ 0·10	
Feb	1	+ 0·3 + 1·4 + 1·9	− 0·1	+ 0·36	+ 0·10 + 0·13 + 0·17	+ 0·08 + 0·06 + 0·02	+ 0·10	
	2	+ 1·8	− 0·1	+ 0·38	+ 0·16	+ 0·02	+ 0·09	
	3	+ 0·3	− 0·1	+ 0·41	+ 0·16	+ 0·02	+ 0·09	43 R. P. L. and δ Urs. Min.
	5	+ 0·0	− 0·1	+ 0·16	+ 0·15	+ 0·02	+ 0·08	
	6	+ 1·0	− 0·1	+ 0·29	+ 0·20	+ 0·01	+ 0·07	
	7	+ 0·6	− 0·1	+ 0·32	+ 0·13	− 0·02	+ 0·07	
	8	+ 0·6	− 0·1	+ 0·22	+ 0·22	+ 0·04	+ 0·06	
	9	− 2·3	− 0·1	+ 0·30	+ 0·18	+ 0·02	+ 0·08	
	10	− 1·5	− 0·1	+ 0·27	+ 0·21	+ 0·02	+ 0·06	51 Cephei and δ Urs. Min.
	12	− 4·3	− 0·1	+ 0·25	+ 0·23	+ 0·01	+ 0·07	

Feb. 1 – The index inclination and collimation corrections by three observers; B., P. and M.
Feb. 8 – The [illegible] rooms under repair and being painted inside.
Feb. 9 – The microscopes cleaned and the index error adjusted.

Instrumental Corrections adopted in 1866.

Date.	Index.	Run in 5'	Clock Rate.	Inclination.	Collimation.	Meridian.	Determining Stars.
	.	.	s	s	s	s	
Feb. 13	− 1·5	− 0·1	+ 0·27	+ 0·22	+ 0·08	+ 0·07	
14	− 1·1	− 0·1	+ 0·30	+ 0·27	+ 0·02	+ 0·08	
15	− 1·7	− 0·1	+ 0·27	+ 0·29	+ 0·06	+ 0·09	
16	− 2·6	− 0·1	+ 0·28	+ 0·24	+ 0·05	+ 0·10	
17	− 1·8	− 0·1	+ 0·26	+ 0·22	+ 0·04	+ 0·11	
19	− 1·9	− 0·1	+ 0·24	+ 0·21	+ 0·00	+ 0·12	
20	− 3·5	− 0·1	+ 0·25	+ 0·17	+ 0·05	+ 0·13	46 R. P. L. and λ Urs. Min.
21	− 2·6	− 0·1	+ 0·24	+ 0·16	+ 0·04	+ 0·13	
22	− 2·3	− 0·1	+ 0·36	+ 0·16	+ 0·08	+ 0·13	
23	− 1·6	− 0·1	+ 0·34	+ 0·18	+ 0·01	+ 0·14	
24	− 3·4	− 0·1	+ 0·24	+ 0·17	+ 0·08	+ 0·14	
26	− 3·9	− 0·1	+ 0·25	+ 0·20	+ 0·10	+ 0·14	
27	− 2·6	− 0·1	+ 0·28	+ 0·20	+ 0·09	+ 0·14	
28	− 2·3	− 0·1	+ 0·34	+ 0·10	+ 0·10	+ 0·15	
Mar. 1	− 1·0	− 0·1	+ 0·32	+ 0·18	+ 0·06	+ 0·15	46 R. P. L. and λ Urs. Min.
2	− 2·0	− 0·1	+ 0·30	+ 0·16	+ 0·05	+ 0·14	
3	− 1·8	− 0·1	+ 0·38	+ 0·19	+ 0·07	+ 0·12	
5	− 3·5	− 0·1	+ 0·40	+ 0·19	+ 0·06	+ 0·09	
6	− 1·5	− 0·1	+ 0·39	+ 0·19	+ 0·06	+ 0·08	
7	− 2·7	− 0·1	+ 0·42	+ 0·27	+ 0·11	+ 0·06	
8	− 2·0	− 0·1	+ 0·39	+ 0·22	+ 0·04	+ 0·05	60 R. P. L. and λ Urs. Min.
9	− 2·8	− 0·1	+ 0·39	+ 0·20	+ 0·07	+ 0·05	
10	− 1·6	− 0·1	+ 0·32	+ 0·24	+ 0·04	+ 0·04	
12	− 1·6	− 0·1	+ 0·44	+ 0·31	+ 0·07	+ 0·04	
13	− 2·6	− 0·1	+ 0·48	+ 0·25	+ 0·07	+ 0·04	
14	− 0·9	− 0·1	+ 0·25	+ 0·19	0·00	+ 0·03	
15	− 0·9	− 0·1	+ 0·26	+ 0·26	+ 0·08	+ 0·03	72 R. P. L. and ι Argûs.
16	− 0·3	+ 0·1	+ 0·28	+ 0·26	+ 0·04	+ 0·03	
17	0·0	+ 0·1	+ 0·28	+ 0·31	+ 0·08	+ 0·01	
19	+ 0·3	+ 0·1	+ 0·36	+ 0·32	+ 0·07	0·00	
20	+ 0·1	+ 0·1	+ 0·32	+ 0·31	+ 0·07	− 0·01	
21	+ 0·2	+ 0·1	+ 0·33	+ 0·30	+ 0·05	− 0·02	
22	0·0	+ 0·1	+ 0·35	+ 0·33	+ 0·06	− 0·03	72 R. P. L. and η Argûs.
23	− 0·4	+ 0·1	+ 0·38	+ 0·34	+ 0·07	− 0·03	
24	+ 0·4	+ 0·1	+ 0·30	+ 0·20	+ 0·08	− 0·04	
26	+ 0·3	+ 0·1	+ 0·20	+ 0·36	+ 0·05	− 0·04	
27	+ 0·4	+ 0·1	+ 0·26	+ 0·37	+ 0·07	− 0·05	
28	+ 0·5	+ 0·1	+ 0·30	+ 0·39	+ 0·05	− 0·06	70, 72 R. P. L. and α¹ Crucis.
30	+ 0·4	+ 0·1	+ 0·28	+ 0·38	+ 0·06	− 0·04	
31	+ 0·5	+ 0·1	+ 0·23	+ 0·38	+ 0·07	− 0·04	
Apl. 4	− 2·4	− 0·1	+ 0·15	+ 0·48	+ 0·07	− 0·08	
5	− 0·4	− 0·1	+ 0·14	+ 0·44	+ 0·03	− 0·06	99 and 159 R. P. L.
6	+ 0·2	− 0·1	+ 0·19	+ 0·42	+ 0·04	− 0·06	
7	+ 0·5	− 0·1	+ 0·22	+ 0·34	+ 0·06	− 0·03	
9	+ 1·0	− 0·1	+ 0·38	+ 0·46	+ 0·04	− 0·02	
10	− 1·6	− 0·1	+ 0·31	+ 0·41	+ 0·02	− 0·02	
11	− 1·8	− 0·1	+ 0·19	+ 0·38	+ 0·02	− 0·02	
12	− 2·5	− 0·1	+ 0·21	+ 0·39	+ 0·04	− 0·01	
13	+ 0·6	− 0·1	+ 0·28	+ 0·42	+ 0·02	− 0·01	
14	− 1·6	− 0·1	+ 0·17	+ 0·46	+ 0·05	− 0·01	99 R. P. L. and Polaris.
16	+ 1·9	0·0	+ 0·15	+ 0·51	+ 0·05	− 0·13	
17	+ 2·1	0·0	+ 0·22	+ 0·52	+ 0·07	− 0·20	
18	+ 2·2	0·0	+ 0·25	+ 0·48	+ 0·02	− 0·26	
19	+ 2·0	0·0	+ 0·29	+ 0·45	+ 0·01	− 0·32	108 and 14 R. P. L.
20	− 2·9	0·0	+ 0·19	+ 0·35	+ 0·02	− 0·40	108 and 14 R. P. L.
21	+ 2·6	0·0	+ 0·08	+ 0·58	+ 0·06	− 0·29	
23	+ 1·2	− 0·1	− 0·08	+ 0·07	+ 0·07	− 0·05	99 R. P. L. and Polaris.

Instrumental Corrections adopted in 1866.

Date.	Index.	Run in 5'	Clock Rate.	Inclination.	Collimation.	Meridian.	Determining Stars.
	"	"	s	s	s	s	
Apl. 24	− 2·5	− 0·1	+ 0·00	+ 0·51	+ 0·07	− 0·07	
25	− 4·0	0·0	+ 0·09	+ 0·40	+ 0·08	− 0·10	92, 108 R. P. L. & Polaris.
26	− 3·8	0·0	− 0·10	+ 0.42	+ 0·10	− 0·08	3 Virginis and Polaris.
27	+ 3·0	0·0	− 0·12	+ 0.44	+ 0·08	− 0·11	Arcturus and Polaris.
28	+ 2·5	0·0	− 0·08	+ 0.59	+ 0·09	− 0·04	12 Can. Ven. and Polaris.
30	+ 2·2	0·0	+ 0·16	+ 0·57	+ 0·07	− 0·20	
May 1	+ 2·7	0·0	+ 0·18	+ 0·51	+ 0·06	− 0·27	111 and 84 R. P. L.
2	+ 2·7	0·0	+ 0·23	+ 0·56	+ 0·08	− 0·18	116 and 24 R. P. L.
3	− 1·9	0·0	+ 0·23	+ 0·42	+ 0·02	− 0·14	89 and 12 R. P. L.
4	− 0·4	0·0	+ 0·12	+ 0·56	+ 0·06	− 0·13	89 and 12 R. P. L.
5	+ 2·3	0·0	+ 0·21	+ 0·60	+ 0·10	− 0·09	12 Can. Ven. and Polaris.
7	− 4·1	0·0	+ 0·09	+ 0·55	+ 0·09	− 0·08	
8	+ 3·4	0·0	+ 0·13	+ 0·70	+ 0·12	0·00	99 R. P. L. and Polaris.
9	− 2·2	0·0	+ 0·30	+ 0·65	+ 0·10	− 0·01	
10	+ 1·4	0·0	+ 0·26	+ 0·69	+ 0·07	− 0·08	99 R. P. L. and Polaris.
11	+ 3·2	0·0	+ 0·27	+ 0·69	+ 0·10	+ 0·01	
12	+ 3·4	0·0	+ 0·30	+ 0·65	+ 0·11	+ 0·06	89 R. P. L. and Polaris.
14	+ 3·8	0·0	+ 0·40	+ 0·68	+ 0·11	+ 0·01	
15	+ 3·2	0·0	+ 0·81	+ 0·61	+ 0·07	− 0·01	ρ Bootis and Polaris.
16	+ 2·4	+ 0·1	+ 0·28	+ 0·31	+ 0·13	− 0·04	ε Bootis and Polaris.
17	+ 2·0	+ 0·1	+ 0·45	+ 0·33	+ 0·18	0·00	
18	+ 2·5	+ 0·1	+ 0·44	+ 0·31	+ 0·12	+ 0·04	99 R. P. L. and Polaris.
19	+ 2·3	+ 0·1	+ 0·38	+ 0·29	+ 0·13	+ 0·08	
21	+ 3·3	+ 0·1	+ 0·46	+ 0·31	+ 0·18	+ 0·03	3 Urs. Min. and Polaris.
22	+ 3·8	+ 0·1	+ 0·40	+ 0·31	+ 0·18	+ 0·02	
23	+ 3·6	+ 0·1	+ 0·43	+ 0·33	+ 0·12	+ 0·03	111 R. P. L. and Polaris.
25	+ 4·0	+ 0·1	+ 0·70	+ 0·38	+ 0·16	+ 0·07	12 Can. Ven. and Polaris.
26	− 3·3	+ 0·1	+ 0·09	+ 0·19	+ 0·10	+ 0·04	116 and 36 R. P. L.
28	+ 2·9	+ 0·1	+ 0·57	+ 0·37	+ 0·13	− 0·02	
29	+ 1·5	+ 0·1	+ 0·44	+ 0·27	+ 0·13	− 0·14	η Urs. Maj. and Polaris.
30	+ 2·6	+ 0·1	+ 0·31	+ 0·45	+ 0·14	− 0·11	
31	+ 3·4	+ 0·1	+ 0·42	+ 0·42	+ 0·13	− 0·08	
June 1	− 1·3	+ 0·2	+ 0·48	+ 0·42	+ 0·10	− 0·06	
4	+ 2·3	+ 0·2	+ 0·37	+ 0·50	+ 0·13	+ 0·08	η Urs. Maj. and Polaris.
5	+ 2·7	+ 0·2	+ 0·40	+ 0·51	+ 0·13	+ 0·06	3 Herculis and Polaris.
6	− 4·4	+ 0·2	+ 0·60	+ 0·40	+ 0·17	+ 0·07	
7	+ 3·1	+ 0·2	+ 0·32	+ 0·40	+ 0·13	+ 0·09	ε Bootis and Polaris.
8	− 0·8	+ 0·2	+ 0·65	+ 0·64	+ 0·10	+ 0·10	
9	+ 2·1	+ 0·2	+ 0·71	+ 0·48	+ 0·11	+ 0·10	δ Urs. Min. and Polaris.
13	+ 2·0	+ 0·2	+ 0·44	+ 0·56	+ 0·14	+ 0·08	
16	+ 1·8	+ 0·2	+ 0·48	+ 0·60	+ 0·13	+ 0·06	
19	+ 0·4	+ 0·2	+ 0·50	+ 0·55	+ 0·12	+ 0·04	
22	+ 0·6	+ 0·2	+ 0·54	+ 0·59	+ 0·18	+ 0·02	
25	+ 0·3	+ 0·3	+ 0·56	+ 0·63	+ 0·14	0·00	116 and 40 R. P. L.
26	+ 0·2	+ 0·3	+ 0·66	+ 0·64	+ 0·12	+ 0·01	
27	+ 0·3	+ 0·3	+ 0·68	+ 0·64	+ 0·11	+ 0·02	
29	− 7·0	+ 0·3	+ 0·39	+ 0·63	+ 0·12	+ 0·04	
30	+ 0·9	+ 0·3	+ 0·65	+ 0·67	+ 0·13	+ 0·05	δ Urs. Min. and 51 Cephei.
July 2	+ 0·3	0·0	+ 0·71	+ 0·67	+ 0·12	+ 0·06	
5	− 0·4	0·0	+ 0·73	+ 0·29	+ 0·14	+ 0·06	δ Urs. Min. and 51 Cephei.
6	+ 6·2	0·0	+ 0·75	+ 0·30	+ 0·14	+ 0·17	
7	+ 6·1	0·0	+ 0·78	+ 0·38	+ 0·10	+ 0·30	δ Urs. Min. and 51 Cephei.
13	+ 3·0	0·0	+ 0·73	+ 0·13	− 0·02	+ 0·32	3 Herculis and 51 Cephei.
16	+ 4·0	0·0	+ 0·74	+ 0·19	+ 0·08	+ 0·31	
17	+ 3·2	0·0	+ 0·77	+ 0·13	+ 0·08	+ 0·21	

May 16.—The inclination error adjusted.
June 1.—The microscopes cleaned.
June 19.—The index error adjusted.
July 2.—The inclination error adjusted and the pivots cleaned.
July 5.—The circle clamps repaired and the index error adjusted.
July 12.—The object glass found to be unscrewed about one-sixth of a revolution. This at once explained the unsteadiness of the index and level corrections since last March. It was screwed home and the index error again adjusted.

Instrumental Corrections adopted in 1866.

Date.		Index	Run in 5′	Clock Rate.	Inclination.	Collimation.	Meridian.	Determining Stars
		″	″	s	s	s	s	
July	19	+ 4·2	0·0	+ 0·81	+ 0·12	+ 0·01	+ 0·21	δ Urs. Min. and θ Ophiuchi.
	20	+ 4·7	0·0	+ 0·79	+ 0·17	+ 0·01	+ 0·21	
	23	+ 3·7	0·0	+ 0·83	+ 0·18	− 0·08	+ 0·19	
	24	− 3·9	0·0	+ 0·84	+ 0·18	− 0·01	+ 0·19	
Aug.	3	+ 4·1	0·0	+ 0·78	+ 0·18	+ 0·01	+ 0·14	3 Aquilæ and 51 Cephei.
	7	+ 4·9	0·0	+ 0·85	+ 0·20	0·00	+ 0·13	
	9	+ 5·1	0·0	+ 0·94	+ 0·21	+ 0·02	+ 0·13	δ Urs. Min. and 51 Cephei.
	10	+ 4·3	0·0	+ 0·97	+ 0·24	− 0·02	+ 0·14	
	11	+ 5·2	0·0	− 0·40	+ 0·24	0·00	+ 0·14	
	16	+ 4·2	0·0	− 0·40	+ 0·28	− 0·01	+ 0·16	δ Urs. Min. and δ Aquilæ.
	17	+ 4·5	0·0	− 0·32	+ 0·38	− 0·01	+ 0·11	δ U. M. and μ¹ Sagittarii.
	18	+ 4·5	0·0	− 0·26	+ 0·30	− 0·02	+ 0·11	
	20	+ 4·0	0·0	− 0·28	+ 0·36	− 0·02	+ 0·11	
	21	+ 4·9	0·0	− 0·26	+ 0·28	+ 0·01	+ 0·11	
	22	+ 3·8	0·0	− 0·40	+ 0·32	− 0·01	+ 0·11	δ U. M. and λ² Sagittarii.
	23	+ 4·7	0·0	− 0·51	+ 0·20	− 0·08	+ 0·12	
	25	+ 5·2	0·0	− 0·48	+ 0·26	− 0·08	+ 0·14	
	27	+ 5·3	0·0	− 0·37	+ 0·23	− 0·01	+ 0·17	
	28	+ 6·2	0·0	− 0·45	+ 0·26	− 0·08	+ 0.18	
	29	+ 5·2	0·0	− 0·42	+ 0·31	+ 0·01	+ 0·19	3 Cygni and 70 R. P. L.
	30	+ 5·0	0·0	− 0·28	+ 0·29	+ 0·01	+ 0·17	
Sep.	7	+ 6·0	0·0	− 0·44	+ 0·34	0·00	+ 0·02	λ Urs. Min. and β Aquilæ.
	8	+ 5·3	0·0	− 0·46	+ 0·44	+ 0·04	+ 0·05	λ U. M. and λ² Sagittarii.
	10	+ 6·4	0·0	− 0·38	+ 0·53	+ 0·10	+ 0·08	
	12	+ 6·8	0·0	− 0·46	+ 0·45	+ 0·01	+ 0·02	148 R. P. L. and δ Aquilæ.
	13	+ 6·0	0·0	− 0·47	+ 0·82	+ 0·01	+ 0·08	λ U. M. and ρ Capricorni.
	14	+ 6·9	0·0	− 0·38	+ 0·51	+ 0·06	+ 0·10	
	15	+ 7·1	0·0	− 0·28	+ 0·45	− 0·08	+ 0·11	
	18	+ 7·1	+ 0·1	− 0·37	+ 0·56	+ 0·02	+ 0·13	
	19	+ 7·0	+ 0·1	− 0·40	+ 0·53	+ 0·02	+ 0·13	148 and 69 R. P. L.
	20	+ 7·6	+ 0·1	− 0·41	+ 0·56	− 0·01	+ 0·12	
	22	+ 7·9	+ 0·1	− 0·30	+ 0·50	− 0·01	+ 0·10	
	24	+ 7·4	0·0	− 0·21	+ 0·57	+ 0·02	− 0·08	λ Urs. Min. and β Aquarii.
	25	+ 7·9	0·0	− 0·38	+ 0·53	− 0·06	+ 0·09	
	26	+ 7·8	0·0	− 0·51	+ 0·58	+ 0·01	+ 0·10	
	27	+ 8·4	0·0	− 0·48	+ 0·58	0·00	+ 0·10	
	28	+ 8·8	0·0	− 0·38	+ 0·60	+ 0·01	+ 0·11	
	29	+ 8·5	0·0	− 0·34	+ 0·60	0·00	+ 0·12	
Oct.	1	+ 8·0	0·0	− 0·41	+ 0·50	− 0·02	+ 0·13	
	2	+ 8·7	0·0	− 0·41	+ 0·57	− 0·01	+ 0·14	150 R. P. L. and θ Aquarii.
	5	+ 8·7	0·0	− 0·35	+ 0·56	− 0·05	+ 0·09	
	6	+ 10·1	0·0	− 0·84	+ 0·54	− 0·08	+ 0·07	3 Cygni and 69 R. P. L.
	9	+ 1·4	0·0	− 0·40	+ 0·16	+ 0·02	+ 0·01	
	10	+ 1·1	0·0	− 0·32	+ 0·15	+ 0·01	− 0·01	
	11	+ 2·1	0·0	− 0·35	+ 0·10	0·00	− 0·08	Polaris and 12 Ceti.
	12	+ 2·1	0·0	− 0·68	+ 0·14	0·00	− 0·04	
	15	+ 2·4	0·0	− 0·60	+ 0·16	+ 0·02	− 0·08	
	16	+ 2·0	0·0	− 0·45	+ 0·20	+ 0·07	− 0·09	
	17	+ 3·0	0·0	− 0·82	+ 0·07	− 0·02	− 0·11	
	22	+ 6·2	0·0	− 0·60	− 0·17	− 0·04	− 0·17	Polaris and Fomalhaut.
	24	+ 6·5	0·0	− 0·63	− 0·25	− 0·06	− 0·14	
	25	+ 7·0	− 0·1	− 0·66	− 0·24	− 0·04	− 0·18	Polaris and 101 R. P. L.
	27	+ 6·0	− 0·1	− 0·67	− 0·19	+ 0·06	− 0·13	
	29	+ 6·2	− 0·1	− 0·38	− 0·22	+ 0·08	− 0·14	
	30	+ 5·1	− 0·1	− 0·68	− 0·35	− 0·01	− 0·14	
	31	+ 5·0	− 0·1	− 0·79	− 0·27	+ 0·05	− 0·14	
Nov.	1	+ 4·6	0·0	− 0·68	− 0·30	+ 0·04	− 0·15	
	3	+ 4·2	0·0	− 0·69	− 0·41	− 0·04	− 0·15	Polaris and 101 R. P. L.

Aug. 11.—The transit clock put forward one minute and its rate adjusted.
Oct. 9.—The index and inclination errors adjusted.
Oct. 17 to 22.—Heavy continuous rain; hence the changes in the index and inclination corrections.

Instrumental Corrections adopted in 1866.

Date.		Index.	Run in 5′	Clock Rate.	Inclination.	Collimation.	Meridian.	Determining Stars.
		″	″	s	s	s	s	
Nov.	5	+ 4·5	0·0	− 0·65	− 0·36	− 0·02	− 0·16	
	6	+ 4·7	0·0	− 0·62	− 0·35	− 0·04	− 0·16	
	7	+ 3·4	0·0	− 0·64	− 0·34	− 0·04	− 0·17	158 R. P. L., Pol. & Achernar
	8	+ 3·2	0·0	− 0·70	− 0·32	− 0·02	− 0·16	
	9	+ 3·8	0·0	− 0·60	− 0·28	0·00	− 0·15	
	10	+ 2·7	0·0	− 0·52	− 0·31	− 0·02	− 0·14	
	12	+ 2·6	0·0	− 0·56	− 0·31	0·00	− 0·13	
	13	+ 2·6	0·0	− 0·61	− 0·35	− 0·03	− 0·12	
	14	+ 2·2	0·0	− 0·60	− 0·38	− 0·04	− 0·11	Polaris and 101 R. P. L.
	19	+ 4·9	0·0	− 0·74	− 0·62	− 0·02	− 0·11	Polaris and β Ceti.
	22	+ 4·9	0·0	− 0·79	− 0·72	− 0·03	− 0·09	
	26	+ 5·9	0·0	− 0·96	− 0·88	0·00	− 0·07	Polaris and 101 R. P. L.
	27	+ 6·7	0·0	− 0·97	− 0·94	+ 0·01	− 0·06	
	29	+ 0·5	0·0	− 0·01	+ 0·05	− 0·01	− 0·04	
	30	+ 0·4	0·0	+ 0·01	+ 0·01	− 0·03	− 0·01	
Dec.	6	+ 0·1	+ 0·1	+ 0·32	− 0·10	− 0·03	+ 0·07	35 and 108 R. P. L.
	7	+ 0·5	+ 0·1	+ 0·30	− 0·16	− 0·06	+ 0·07	
	8	− 1·1	+ 0·1	+ 0·20	− 0·18	− 0·03	+ 0·06	
	10	− 0·9	+ 0·1	+ 0·09	− 0·20	− 0·01	+ 0·06	
	12	+ 0·8	+ 0·1	+ 0·08	− 0·24	− 0·04	+ 0·05	
	13	+ 1·0	+ 0·1	+ 0·05	− 0·19	− 0·06	+ 0·06	
	17	+ 0·3	+ 0·4	− 0·01	− 0·12	− 0·04	+ 0·04	
	18	+ 0·4	+ 0·4	− 0·03	− 0·15	− 0·03	+ 0·04	
	19	− 1·6	+ 0·4	− 0·05	− 0·09	− 0·01	+ 0·04	
	20	− 2·2	+ 0·4	− 0·05	− 0·05	+ 0·01	+ 0·03	
	28	− 1·1	+ 0·4	− 0·02	− 0·68	+ 0·01	+ 0·03	35 and 116 R. P. L.

Instrumental Corrections adopted in 1867.

Date.		Index.	Run in 5′	Clock Rate.	Inclination.	Collimation.	Meridian.	Determining Stars.
		″	″	s	s	s	s	
Jan.	3	− 3·6	− 0·1	− 0·32	− 0·02	+ 0·03	+ 0·07	ε Aurigæ and δ Urs. Min.
	4	− 4·1	− 0·1	− 0·11	− 0·05	+ 0·01	+ 0·05	
	5	− 3·9	− 0·1	+ 0·10	− 0·03	0·00	+ 0·05	ε Aurigæ and δ Urs. Min.
	7	− 3·6	− 0·1	+ 0·08	+ 0·08	− 0·02	+ 0·04	51 Cephei and 115 R. P. L.
	8	− 3·4	− 0·1	+ 0·10	+ 0·06	0·00	+ 0·03	
	9	− 4·0	− 0·1	+ 0·17	0·00	− 0·03	+ 0·03	
	10	− 4·7	− 0·1	+ 0·08	+ 0·02	− 0·03	+ 0·01	
	11	− 4·6	− 0·1	+ 0·05	+ 0·08	− 0·01	+ 0·01	
	12	− 5·2	− 0·1	+ 0·10	+ 0·01	− 0·03	0·09	
	14	− 5·4	+ 0·1	− 0·09	+ 0·02	0·00	− 0·03	51 Cephei and ε Urs. Min.
	15	− 5·2	+ 0·1	+ 0·04	+ 0·04	0·00	− 0·03	
	16	− 5·1	+ 0·1	+ 0·01	+ 0·08	+ 0·04	− 0·01	
	17	− 6·0	+ 0·1	− 0·02	+ 0·09	+ 0·06	0·00	
	18	− 5·9	+ 0·1	+ 0·15	+ 0·07	+ 0·04	0·00	
	19	− 6·4	+ 0·1	+ 0·14	+ 0·08	0·00	+ 0·01	
	21	− 7·9	+ 0·1	+ 0·04	+ 0·04	+ 0·02	+ 0·02	
	22	− 6·9	+ 0·1	+ 0·11	+ 0·03	+ 0·02	+ 0·03	
	23	− 7·4	+ 0·1	+ 0·17	0·00	0·00	+ 0·04	51 Cephei and δ Urs. Min.
	24	− 7·8	+ 0·1	+ 0·10	+ 0·05	+ 0·02	+ 0·03	
	25	− 8·1	+ 0·1	− 0·06	0·00	+ 0·01	+ 0·03	

Nov 15.—Heavy rain, followed by change in the corrections.
Nov 28.—The index and inclination corrections and the transit clock rate adjusted.
Dec. 4, 5.—Heavy rain.
Dec 19.—The collimators and microscopes cleaned.
Jan 7.—The index error interpolated.

Instrumental Correction adopted in 1867.

Date		Index.	Run in 5'	Clock Rate.	Inclination.	Collimation.	Meridian.	Determining Stars.
		"	"	s	s	s	s	
Jan.	27	− 8·7	+ 0·1	− 0·21	+ 0·08	+ 0·04	+ 0·01	
	28	− 9·1	+ 0·1	− 0·17	+ 0·10	+ 0·01	− 0·01	
	29	− 9·4	+ 0·1	− 0·01	+ 0·10	+ 0·03	− 0·02	
	30	− 9·8	+ 0·1	+ 0·08	+ 0·09	+ 0·01	− 0·03	β Tauri and δ Urs. Min.
	31	− 9·7	+ 0·1	+ 0·09	+ 0·10	+ 0·02	− 0·01	
Feb.	1	− 10·5	0·0	+ 0·06	+ 0·10	− 0·01	+ 0·01	γ Geminorum & 51 Cephei.
	2	− 11·3	0·0	+ 0·08	+ 0·10	0·00	0·00	
	4	− 10·6	0·0	+ 0·20	+ 0·07	− 0·04	− 0·03	
	5	− 9·9	0·0	+ 0·21	+ 0·13	0·00	− 0·05	
	6	− 10·4	0·0	+ 0·14	+ 0·13	− 0·08	− 0·06	Pollux and δ Urs. Min.
	7	− 10·3	0·0	+ 0·19	+ 0·22	0·00	− 0·06	
	8	− 10·4	0·0	+ 0·28	+ 0·26	+ 0·01	− 0·04	
	9	− 10·3	0·0	+ 0·14	+ 0·10	− 0·08	− 0·02	
	11	− 10·5	0·0	+ 0·16	+ 0·25	+ 0·03	0·00	60 R. P. L. and ε Urs. Min.
	12	− 10·8	0·0	+ 0·22	+ 0·24	0·00	0·00	
	13	− 10·7	0·0	− 0·71	+ 0·19	− 0·04	0·00	
	14	− 11·1	0·0	− 1·82	+ 0·26	+ 0·04	0·00	40 R. P. L. & γ Geminorum.
	15	− 11·3	0·0	− 0·92	+ 0·23	+ 0·01	− 0·02	40 R. P. L. and δ Urs. Min.
	16	− 11·4	0·0	− 0·67	+ 0·21	+ 0·03	− 0·06	
	18	− 11·1	0·0	− 0·01	+ 0·16	− 0·03	− 0·11	
	19	− 11·3	0·0	− 0·02	+ 0·12	− 0·04	− 0·14	72 R. P. L. and ν Leonis.
	20	− 10·9	0·0	− 0·07	+ 0·17	− 0·02	− 0·12	
	21	− 11·6	0·0	+ 0·11	+ 0·18	+ 0·01	− 0·09	
	22	− 11·0	0·0	+ 0·16	+ 0·16	+ 0·01	− 0·06	51 Ceph 70 R P. L. & 15 Argûs
	23	− 11·0	0·0	+ 0·11	+ 0·16	− 0·01	− 0·04	
	25	− 12·1	0·0	+ 0·09	+ 0·16	+ 0·02	− 0·01	
	26	− 11·5	0·0	+ 0·14	+ 0·28	+ 0·06	0·00	
	27	− 11·9	0·0	+ 0·08	+ 0·24	+ 0·09	− 0·01	51 Cep., 70 R. P L. & λ U. M.
	28	− 11·7	0·0	+ 0·15	+ 0·26	+ 0·06	0·00	
Mar.	1	− 11·0	− 0·1	+ 0·24	+ 0·21	+ 0·02	− 0·02	51 Cephei and 180 R. P. L.
	2	− 12·4	− 0·1	+ 0·09	+ 0·26	0·00	0·00	
	4	− 12·5	− 0·1	+ 0·06	+ 0·24	+ 0·03	+ 0·03	
	5	− 12·7	− 0·1	+ 0·11	+ 0·24	+ 0·04	+ 0·06	
	6	− 12·0	− 0·1	+ 0·14	+ 0·28	+ 0·01	+ 0·07	40 R. P. L and λ Urs Min.
	7	− 12·0	− 0·1	+ 0·02	+ 0·30	+ 0·02	+ 0·01	51 Cephei and λ Urs. Min.
	8	− 12·3	− 0·1	+ 0·24	+ 0·33	+ 0·04	− 0·02	
	9	− 12·0	− 0·1	+ 0·24	+ 0·31	+ 0·01	− 0·04	
	11	− 11·8	− 0·1	+ 0·17	+ 0·38	+ 0·01	− 0·09	
	12	− 11·8	− 0·1	+ 0·10	+ 0·24	− 0·02	− 0·12	60 and 160 R. P. L.
	13	− 12·1	− 0·1	+ 0·19	+ 0·30	− 0·02	− 0·12	
	14	− 11·8	− 0·1	+ 0·21	+ 0·30	− 0·08	− 0·18	
	15	− 12·2	− 0·1	+ 0·10	+ 0·33	+ 0·01	− 0·16	60 R. P. L. and λ Urs. Min.
	16	− 11·8	+ 0·2	+ 0·06	+ 0·35	+ 0·02	− 0·13	
	18	− 11·6	+ 0·2	+ 0·12	+ 0·34	+ 0·02	− 0·14	
	19	− 11·4	+ 0·2	+ 0·07	+ 0·40	+ 0·02	− 0·14	
	20	− 12·1	+ 0·2	+ 0·12	+ 0·47	+ 0·02	− 0·14	
	21	− 11·7	+ 0·2	+ 0·21	+ 0·48	+ 0·02	− 0·14	
	22	− 11·8	+ 0·2	+ 0·19	+ 0·40	+ 0·01	− 0·14	
	23	− 11·8	+ 0·2	+ 0·17	+ 0·42	+ 0·02	− 0·14	
	25	− 11·4	+ 0·2	+ 0·22	+ 0·39	+ 0·02	− 0·15	
	26	− 11·5	+ 0·2	+ 0·28	+ 0·37	+ 0·01	− 0·16	70, 78 R. P L & α Hydræ.
	28	− 11·4	+ 0·2	+ 0·13	+ 0·44	0·00	− 0·16	
	29	− 11·6	+ 0·2	+ 0·21	+ 0·47	+ 0·08	− 0·17	
	30	− 11·3	+ 0·2	+ 0·27	+ 0·44	+ 0·01	− 0·18	
Apl.	1	− 11·3	− 0·1	+ 0·20	+ 0·48	− 0·02	− 0·19	87 and 180 R. P. L.
	2	− 11·8	− 0·1	+ 0·18	+ 0·49	− 0·01	− 0·20	

Instrumental Corrections adopted in 1867.

Date.		Index.	Run in 5′	Clock Rate.	Inclination.	Collimation.	Meridian.	Determining Stars
		″	″	s	s	s	s	
Apl.	3	− 11·6	− 0·1	+ 0·14	+ 0·50	0·00	− 0·21	δ Leonis and Polaris
	4	− 11·1	− 0·1	+ 0·16	+ 0·51	+ 0·01	− 0·28	
	5	− 11·0	− 0·1	+ 0·17	+ 0·61	+ 0·01	− 0·24	
	6	− 11·2	− 0·1	+ 0·21	+ 0·50	− 0·02	− 0·26	
	8	− 10·7	− 0·1	+ 0·18	+ 0·57	+ 0·02	− 0·29	
	10	− 11·1	− 0·1	+ 0·08	+ 0·64	− 0·03	− 0·32	60 and 160 R. P. L.
	11	− 10·8	− 0·1	+ 0·19	+ 0·59	0·00	− 0·31	
	12	− 11·0	− 0·1	+ 0·20	+ 0·54	− 0·05	− 0·29	
	13	− 11·2	− 0·1	+ 0·11	+ 0·59	0·00	− 0·28	
	15	− 10·7	− 0·1	+ 0·13	+ 0·50	− 0·01	− 0·26	30 and 160 R. P. L.
	16	− 10·4	+ 0·1	+ 0·16	+ 0·57	− 0·02	− 0·26	
	17	− 10·9	+ 0·1	+ 0·13	+ 0·56	− 0·08	− 0·28	
	18	− 10·9	+ 0·1	+ 0·10	+ 0·64	− 0·08	− 0·29	12 Can. Ven. and Polaris.
	19	− 10·9	+ 0·1	+ 0·12	+ 0·63	+ 0·01	− 0·23	
	22	− 10·4	− 0·1	+ 0·11	+ 0·64	− 0·04	− 0·04	
	24	− 3·8	+ 0·1	+ 0·08	+ 0·08	− 0·02	+ 0·08	108 R. P. L. and Polaris.
	25	− 4·1	+ 0·1	+ 0·05	+ 0·08	− 0·01	+ 0·10	108 R. P. L. and Polaris.
	26	− 4·3	+ 0·1	+ 0·06	+ 0·09	− 0·02	+ 0·11	
	27	− 4·3	+ 0·1	+ 0·06	+ 0·18	+ 0·02	+ 0·11	
	29	− 4·4	+ 0·1	+ 0·06	+ 0·09	− 0·01	+ 0·12	
	30	− 3·4	+ 0·1	+ 0·10	+ 0·14	− 0·02	+ 0·13	β Leonis and Polaris.
May	1	− 3·4	− 0·1	+ 0·11	+ 0·21	+ 0·01	+ 0·18	β Leonis and Polaris.
	2	− 2·9	− 0·1	+ 0·15	+ 0·21	− 0·01	+ 0·18	
	3	− 3·5	− 0·1	+ 0·09	+ 0·23	+ 0·03	+ 0·17	87 R. P. L. and Polaris.
	4	− 3·3	− 0·1	+ 0·10	+ 0·26	+ 0·02	+ 0·18	87 R. P. L. and Polaris.
	6	− 2·7	− 0·1	+ 0·24	+ 0·27	+ 0·01	+ 0·14	
	7	− 3·0	− 0·1	+ 0·15	+ 0·27	− 0·02	+ 0·12	87 R. P. L. and Polaris.
	8	− 3·2	− 0·1	+ 0·14	+ 0·32	0·00	+ 0·12	
	9	− 3·6	− 0·1	+ 0·17	+ 0·34	+ 0·01	+ 0·12	
	11	− 3·6	− 0·1	+ 0·13	+ 0·35	− 0·02	+ 0·13	87 R. P. L. and Polaris.
	13	− 3·5	− 0·1	+ 0·08	+ 0·33	− 0·01	+ 0·12	
	14	− 3·3	− 0·1	+ 0·15	+ 0·39	+ 0·01	+ 0·11	
	15	− 3·0	− 0·1	+ 0·13	+ 0·36	− 0·02	+ 0·11	
	16	− 3·6	0·0	+ 0·03	+ 0·30	− 0·02	+ 0·10	θ Virginis and Polaris.
	18	− 3·8	0·0	+ 0·06	+ 0·47	+ 0·06	+ 0·10	
	22	− 3·0	0·0	+ 0·12	+ 0·39	+ 0·01	+ 0·09	
	23	− 2·8	0·0	+ 0·22	+ 0·49	+ 0·01	+ 0·09	
	24	− 3·7	0·0	+ 0·21	+ 0·38	+ 0·02	+ 0·09	ψ Bootis and Polaris
	26	− 2·5	0·0	+ 0·13	+ 0·43	+ 0·02	+ 0·08	
	27	− 3·2	0·0	+ 0·18	+ 0·52	+ 0·02	+ 0·07	Arcturus and Polaris
	28	− 3·5	0·0	+ 0·14	+ 0·61	+ 0·02	+ 0·11	
	29	− 3·6	0·0	+ 0·16	+ 0·65	+ 0·03	+ 0·14	
	30	− 3·6	0·0	+ 0·26	+ 0·66	+ 0·02	+ 0·18	90 R. P. L. and Polaris.
	31	− 3·5	0·0	+ 0·21	+ 0·66	+ 0·02	+ 0·20	
June	1	− 3·2	0·0	+ 0·09	+ 0·80	+ 0·05	+ 0·23	111 R. P. L. and Polaris
	4	− 3·7	0·0	+ 0·27	+ 0·83	+ 0·02	+ 0·26	
	5	− 3·2	0·0	+ 0·21	+ 0·73	0·00	+ 0·27	111 R. P. L. and Polaris
	6	− 3·7	0·0	+ 0·19	+ 0·77	− 0·02	+ 0·30	
	7	− 3·0	0·0	+ 0·27	+ 0·80	− 0·01	+ 0·26	
	10	− 3·5	0·0	+ 0·19	+ 0·74	− 0·08	+ 0·24	
	14	− 3·7	0·0	− 0·08	+ 0·80	− 0·09	+ 0·22	
	15	− 2·7	0·0	− 0·01	+ 0·10	− 0·08	+ 0·23	
	17	− 3·0	0·0	+ 0·06	+ 0·12	− 0·08	+ 0·21	
	18	− 3·2	0·0	− 0·14	+ 0·21	− 0·01	+ 0·20	
	19	− 3·3	0·0	− 0·33	+ 0·16	− 0·07	+ 0·20	
	20	− 3·7	0·0	− 0·16	+ 0·17	− 0·07	+ 0·19	
	28	− 3·3	0·0	+ 0·10	+ 0·22	− 0·06	+ 0·18	

April 24.—The index, inclination and meridian corrections adjusted
June 15.—The inclination correction adjusted

Instrumental Corrections adopted in 1867.

Date.	Index.	Run in 5′	Clock Rate.	Inclination.	Collimation.	Meridian.	Determining Stars.
	m	m	s	s	s	s	
June 24	− 3·1	0·0	+ 0·11	+ 0·24	− 0·06	+ 0·17	
26	− 3·4	0·0	+ 0·18	+ 0·33	− 0·01	+ 0·17	μ Herculis and 38 R. P. L.
27	− 3·8	0·0	+ 0·20	+ 0·32	− 0·04	+ 0·19	
28	− 3·1	0·0	+ 0·26	+ 0·36	− 0·02	+ 0·20	
July 2	− 3·3	0·0	+ 0·33	+ 0·44	− 0·01	+ 0·23	φ Bootis and 43 R. P. L.
4	− 2·8	0·0	+ 0·39	+ 0·52	+ 0·04	+ 0·32	δ Urs. Min. and 51 Cephei.
5	− 4·3	0·0	+ 0·42	+ 0·57	+ 0·01	+ 0·31	δ Urs. Min. & 40 R. P. L.
6	− 3·7	0·0	+ 0·37	+ 0·61	+ 0·03	+ 0·30	
8	− 2·8	0·0	+ 0·36	+ 0·56	− 0·01	+ 0·28	
10	− 4·9	0·0	+ 0·37	+ 0·50	− 0·02	+ 0·26	
11	− 4·0	0·0	+ 0·37	+ 0·56	− 0·04	+ 0·25	
12	− 4·1	0·0	+ 0·40	+ 0·64	− 0·08	+ 0·24	
18	− 2·0	0·0	+ 0·28	+ 0·51	− 0·08	+ 0·18	α Lyræ and 51 Cephei.
19	− 2·8	0·0	+ 0·30	+ 0·51	− 0·02	+ 0·18	
20	− 2·8	0·0	+ 0·42	+ 0·40	− 0·08	+ 0·18	
23	− 3·6	0·0	+ 0·22	+ 0·71	− 0·02	+ 0·18	
25	− 2·4	0·0	+ 0·28	+ 0·75	− 0·08	+ 0·18	λ Urs. Min. and δ Aquilæ.
27	− 3·0	0·0	+ 0·30	+ 0·77	− 0·02	+ 0·17	
30	− 3·0	0·0	+ 0·26	+ 0·63	− 0·04	+ 0·16	δ Urs. Min. & θ Ophiuchi.
31	− 3·0	0·0	+ 0·22	+ 0·84	− 0·08	+ 0·16	
Aug. 3	− 4·1	− 0·2	+ 0·30	+ 0·96	+ 0·01	+ 0·17	
6	− 3·2	− 0·2	+ 0·32	+ 0·98	− 0·08	+ 0·17	δ Urs. Min. and 51 Cephei.
7	− 3·2	− 0·2	+ 0·30	+ 1·01	− 0·02	+ 0·17	
8	− 3·7	− 0·2	+ 0·30	+ 0·08	− 0·04	+ 0·17	
10	− 3·3	− 0·2	+ 0·24	+ 0·08	− 0·06	+ 0·10	β Lyræ and 51 Cephei
14	− 3·4	− 0·2	+ 0·15	+ 0·06	− 0·01	+ 0·15	
15	− 3·7	− 0·2	+ 0·22	+ 0·07	0·00	+ 0·14	
16	− 3·4	− 0·2	+ 0·30	+ 0·06	0·00	+ 0·14	
20	− 4·0	− 0·2	+ 0·23	+ 0·12	+ 0·02	+ 0·18	δ Urs. Min. and ω Aquilæ.
21	− 3·7	− 0·2	+ 0·20	+ 0·16	0·00	+ 0·18	
23	− 3·7	− 0·2	+ 0·25	+ 0·13	− 0·05	+ 0·13	
24	− 3·5	− 0·2	+ 0·18	+ 0·15	0·00	+ 0·13	
26	− 2·3	− 0·2	+ 0·14	+ 0·13	− 0·08	+ 0·13	
27	− 3·1	− 0·2	+ 0·15	+ 0·16	− 0·02	+ 0·18	λ Urs. Min. and 51 Cephei.
28	− 2·5	− 0·2	+ 0·20	+ 0·30	+ 0·07	+ 0·14	
Sep. 6	− 2·2	− 0·2	+ 0·22	+ 0·26	− 0·08	+ 0·15	λ Urs. Min. and 60 R. P. L.
10	− 2·0	− 0·2	+ 0·26	+ 0·24	+ 0·01	+ 0·17	
11	− 1·7	− 0·2	+ 0·28	+ 0·21	− 0·01	+ 0·18	
13	− 2·1	− 0·2	+ 0·18	+ 0·20	− 0·02	+ 0·19	
16	− 1·8	− 0·1	+ 0·01	+ 0·21	+ 0·03	+ 0·20	
17	− 1·8	− 0·1	+ 0·09	+ 0·23	+ 0·06	+ 0·21	Polaris and 98 R. P. L.
19	− 0·7	− 0·1	+ 0·19	+ 0·22	0·00	+ 0·21	
20	− 1·0	− 0·1	+ 0·14	+ 0·30	+ 0·01	+ 0·21	
23	+ 0·3	− 0·1	+ 0·04	+ 0·30	+ 0·02	+ 0·20	
24	− 1·1	− 0·1	+ 0·38	+ 0·28	+ 0·01	+ 0·20	180 and 72 R. P. L.
26	− 1·3	− 0·1	+ 0·75	+ 0·28	+ 0·04	+ 0·20	
27	− 1·2	− 0·1	+ 0·20	+ 0·36	+ 0·08	+ 0·20	
30	− 1·1	− 0·1	+ 0·30	+ 0·37	+ 0·03	+ 0·20	
Oct. 1	− 2·0	0·0	+ 0·10	+ 0·37	− 0·01	+ 0·20	180 and 69 R. P. L.
2	− 1·8	0·0	+ 0·11	+ 0·43	+ 0·02	+ 0·19	
3	− 1·6	0·0	+ 0·39	+ 0·46	+ 0·08	+ 0.17	
4	− 1·2	0·0	+ 0·48	+ 0·26	− 0·05	+ 0·16	Polaris and Fomalhaut.
5	− 1·7	0·0	+ 0·26	+ 0·44	+ 0·01	+ 0·16	
7	− 1·5	0·0	+ 0·12	+ 0·46	− 0·01	+ 0·16	
8	− 1·6	0·0	+ 0·07	+ 0·42	− 0·02	+ 0·16	150 and 72 R. P. L.

Aug 8.—The inclination correction adjusted.

Instrumental Corrections adopted in 1867.

Date.	Index.	Run in 5′	Clock's Rate.	Inclination.	Collimation.	Meridian.	Determining Stars.
	″	″	s	s	s	s	
Oct. 9	− 2·3	0·0	+ 0·08	+ 0·50	+ 0·01	+ 0·14	
10	− 1·9	0·0	+ 0·11	+ 0·44	0·00	+ 0·13	
14	− 0·9	0·0	− 0·07	+ 0·34	− 0·04	+ 0·07	26 R.P.L. and δ Sculptoris.
15	− 1·0	0·0	− 0·03	+ 0·43	− 0·03	+ 0·09	
16	− 0·9	− 0·1	+ 0·07	+ 0·50	+ 0·01	+ 0·11	
18	− 0·1	− 0·1	+ 0·12	+ 0·41	− 0·01	+ 0·16	
19	+ 0·5	− 0·1	+ 0·17	+ 0·39	+ 0·01	+ 0·18	Polaris and Achernar
21	− 0·4	− 0·1	+ 0·10	+ 0·37	+ 0·01	+ 0·18	
22	0·0	− 0·1	+ 0·23	+ 0·38	+ 0·02	+ 0·19	
24	0·0	− 0·1	+ 0·31	+ 0·40	+ 0·02	+ 0·19	
25	0·0	− 0·1	+ 0·21	+ 0·41	+ 0·02	+ 0·19	
26	− 0·2	− 0·1	+ 0·13	+ 0·42	+ 0·04	+ 0·19	
28	+ 0·9	− 0·1	+ 0·07	+ 0·37	− 0·01	+ 0·20	
29	+ 0·5	− 0·1	− 0·07	+ 0·37	+ 0·02	+ 0·20	Polaris and Achernar
30	+ 0·5	− 0·1	− 0·09	+ 0·38	0·00	+ 0·21	
31	− 0·9	− 0·1	− 0·01	+ 0·30	+ 0·02	+ 0·21	
Nov. 1	+ 0·1	+ 0·1	− 0·19	+ 0·32	− 0·05	+ 0·22	
2	+ 1·5	+ 0·1	− 0·28	+ 0·30	0·00	+ 0·23	159 and 37 R. P. L.
4	− 0·7	+ 0·1	+ 0·03	+ 0·31	+ 0·02	+ 0·22	
5	+ 0·1	+ 0·1	+ 0·04	+ 0·26	− 0·03	+ 0·22	
6	+ 0·4	+ 0·1	+ 0·08	+ 0·34	+ 0·02	+ 0·21	ε Piscium and 99 R. P. L.
7	+ 0·7	+ 0·1	− 0·08	+ 0·29	− 0·01	+ 0·21	
8	+ 0·2	+ 0·1	− 0·23	+ 0·34	+ 0·01	+ 0·21	
11	− 1·6	+ 0·1	+ 0·03	+ 0·37	+ 0·04	+ 0·20	Polaris and 99 R. P. L.
12	+ 1·2	+ 0·1	+ 0·04	+ 0·35	+ 0·01	+ 0·20	
13	− 1·6	+ 0·1	+ 0·19	+ 0·33	0·00	+ 0·21	
14	− 0·8	+ 0·1	+ 0·06	+ 0·30	+ 0·01	+ 0·22	
15	− 0·4	+ 0·1	− 0·03	+ 0·41	+ 0·05	+ 0·22	Polaris and 92 R. P. L.
16	− 1·0	+ 0·1	+ 0·11	+ 0·32	+ 0·01	+ 0·16	Polaris and 101 R. P. L.
18	− 1·8	+ 0·1	+ 0·12	+ 0·32	+ 0·01	+ 0·13	
20	− 2·1	+ 0·1	+ 0·05	+ 0·34	− 0·01	+ 0·10	
21	− 2·5	+ 0·1	− 0·11	+ 0·33	− 0·02	+ 0·09	Polaris and 101 R. P. L.
22	− 2·2	+ 0·1	− 0·04	+ 0·34	− 0·02	+ 0·06	
23	− 1·5	+ 0·1	+ 0·10	+ 0·30	− 0·03	+ 0·07	
26	− 0·8	+ 0·1	+ 0·08	+ 0·37	+ 0·01	+ 0·06	
29	+ 0·7	+ 0·1	+ 0·14	+ 0·32	0·00	+ 0·02	
Dec. 2	+ 3·0	+ 0·3	+ 0·07	+ 0·18	− 0·08	0·00	Polaris and ξ¹ Ceti
4	+ 5·1	+ 0·3	+ 0·27	− 0·06	− 0·03	0·00	
5	+ 5·6	+ 0·3	+ 0·22	− 0·09	− 0·06	0·00	
6	+ 5·3	+ 0·3	+ 0·11	− 0·13	− 0·04	0·00	Polaris and 103 R. P. L.
7	+ 5·5	+ 0·3	+ 0·16	− 0·13	− 0·02	− 0·02	26 R.P.L. and γ Ceti.
9	+ 5·5	+ 0·3	+ 0·10	− 0·21	− 0·04	+ 0·02	
10	+ 5·4	+ 0·3	+ 0·07	− 0·19	+ 0·01	+ 0·03	
11	+ 5·3	+ 0·3	+ 0·10	− 0·25	− 0·01	+ 0·05	Polaris and 67 Ceti.
12	+ 5·1	+ 0·3	− 0·01	− 0·26	− 0·01	+ 0·06	
13	+ 4·8	+ 0·3	− 0·07	− 0·28	+ 0·01	+ 0·08	
14	+ 4·0	+ 0·3	− 0·04	− 0·33	− 0·03	+ 0·09	33 and 114 R. P. L.
16	+ 4·0	0·0	− 0·14	− 0·33	0·00	+ 0·10	51 Cephei and Canopus.
17	+ 4·2	0·0	− 0·30	− 0·35	− 0·01	+ 0·12	
18	+ 2·6	0·0	− 0·34	− 0·37	+ 0·03	+ 0·13	
19	+ 3·0	0·0	− 0·13	− 0·36	+ 0·04	+ 0·15	40 and 115 R. P. L.
20	+ 2·9	0·0	− 0·16	− 0·39	+ 0·01	+ 0·15	
21	+ 3·0	0·0	− 0·18	− 0·43	+ 0·01	+ 0·14	
24	+ 2·3	0·0	+ 0·06	− 0·41	+ 0·02	+ 0·13	
27	+ 1·7	0·0	+ 0·04	− 0·40	+ 0·02	+ 0·11	
30	+ 0·7	+ 0·3	+ 0·05	− 0·39	+ 0·02	+ 0·10	26 R. P. L. and 67 Ceti

Nov. 30 and Dec. 1.—Heavy rain.

Corrections to the Nautical Almanac Stars as given by the Madras Mean Positions.

Star.	Approximate Place 1866.				1865.			1866.			1867.		
					Obs.	R. A.	P. D.	Obs.	R. A.	P. D.	Obs.	R. A.	P. D.
	h.	m.	°	′		s	″		s	″		s	″
α Andromedae ..	0	1	61	39	4	+ 0·03	+ 0·7	8	+ 0·02	+ 1·0	3	+ 0·06	+ 0·6
γ Pegasi (*Algenib*) ..	0	6	75	34	7	+ 0·02	+ 0·6	7	0·00	− 0·2	4	0·00	+ 0·7
12 Ceti	0	23	94	42	11	− 0·02	+ 0·9	7	− 0·04	− 0·1	8	− 0·03	+ 1·3
α Cassiopeae	0	33	34	12	2	− 0·16	+ 1·9	2	− 0·08	+ 0·4	...	......	... ..
β Ceti	0	37	109	48	7	+ 0·06	0·0	5	+ 0·03	− 0·3	7	+ 0·04	+ 0·2
ε Piscium	0	56	82	50	13	− 0·06	+ 0·5	5	− 0·06	− 0·9	7	− 0·10	+ 0·0
α Urs. Min. (*Polaris*) ..	1	10	1	34	11	− 0·53	+ 1·3	5	− 0·17	− 0·6	12	− 0·38	+ 0·8
θ[1] Ceti	1	17	99	53	6	0·00	+ 1·1	7	+ 0·08	+ 0·1	8	+ 0·01	+ 0·4
η Piscium	1	24	75	21	9	+ 0·06	+ 1·9	5	0·00	+ 0·3	10	+ 0·02	+ 0·3
α Eridani (*Achernar*).	1	33	147	56	5	+ 0·30	+ 3·0	2	+ 0·10	+ 1·2	2	+ 0·37	+ 3·4
ν Piscium	1	34	86	11	9	0·00	+ 1·1	9	0·00	+ 0·1	8	+ 0·02	0·0
β Arietis	1	47	69	51	8	+ 0·03	+ 1·9	6	+ 0·04	+ 1·2	10	− 0·01	+ 0·7
α Arietis	2	0	67	10	11	− 0·01	+ 1·7	9	− 0·04	+ 0·7	10	− 0·01	+ 1·4
67 Ceti	2	10	97	2	12	+ 0·08	+ 0·9	6	+ 0·05	+ 0·7	9	+ 0·03	+ 0·4
ξ[1] Ceti	2	21	82	9	11	− 0·08	− 0·1	6	+ 0·01	− 0·1	10	− 0·02	− 0·0
γ Ceti	2	36	87	20	9	+ 0·03	+ 0·7	2	− 0·07	+ 0·1	9	0·00	− 1·4
α Ceti	2	55	86	26	7	+ 0·05	+ 0·3	2	+ 0·07	+ 0·2	4	+ 0·05	− 0·2
δ Arietis	3	4	70	47	8	− 0·05	+ 2·1	4	− 0·06	+ 2·2	6	0·00	+ 1·6
α Persei	3	13	40	37	1	− 0·18	+ 0·6	1	− 0·16	− 0·2	1	− 0·01	+ 0·1
η Tauri	3	40	66	10	4	0·00	+ 2·2	3	− 0·05	+ 1·3	11	+ 0·02	+ 0·4
γ[1] Eridani	3	52	103	54	6	+ 0·05	+ 1·0	4	+ 0·09	− 0·7	12	+ 0·02	+ 0·3
ο[1] Eridani	4	5	97	11	7	− 0·03	+ 2·1	5	− 0·02	+ 0·6	2	+ 0·04	+ 1·7
ε Tauri	4	21	71	7	12	− 0·01	+ 1·8	8	+ 0·01	+ 1·3	11	− 0·01	+ 0·9
α Tauri (*Aldebaran*)..	4	28	73	46	8	− 0·02	+ 2·8	8	− 0·06	+ 1·4	13	− 0·01	+ 1·2
ι Aurigae	4	48	57	3	3	− 0·09	+ 1·9	5	+ 0·02	+ 0·8	13	− 0·01	+ 0·3
ε Leporis	5	0	112	38	4	+ 0·12	+ 0·6	7	+ 0·10	+ 0·0	8	+ 0·05	+ 1·0
α Aurigae (*Capella*)...	5	7	44	0	1	− 0·09	+ 0·8	2	− 0·09	+ 1·0	8	+ 0·04	+ 0·7
β Orionis (*Rigel*) ...	5	8	98	22	2	+ 0·06	+ 0·4	3	+ 0·08	+ 0·3	4	+ 0·04	+ 1·3
β Tauri	5	18	61	31	5	− 0·01	+ 0·6	9	+ 0·08	+ 0·7	12	+ 0·02	+ 0·9
δ Orionis	5	25	90	24	2	0·00	+ 0·8	7	− 0·01	+ 1·0	9	− 0·08	+ 0·6
α Leporis	5	27	107	58	5	+ 0·06	+ 1·2	4	− 0·05	+ 1·1	5	− 0·10	+ 1·1
ε Orionis	5	29	91	17	3	+ 0·07	+ 0·7	3	+ 0·08	+ 1·6	4	+ 0·07	+ 1·3
α Columbae	5	35	124	0	8	− 0·12	+ 2·6	4	− 0·14	+ 2·7	5	− 0·09	+ 2·7
α Orionis	5	48	92	37	9	+ 0·04	+ 0·4	2	− 0·01	− 0·2	7	+ 0·02	− 0·4
ν Orionis	6	0	75	13	0	0·00	+ 1·0	3	− 0·02	+ 0·4	9	− 0·02	+ 0·7
μ Geminorum ..	6	15	67	26	6	+ 0·01	+ 1·4	5	− 0·03	+ 1·1	8	− 0·02	+ 1·8

Corrections to the Nautical Almanac Stars as given by the Madras Mean Positions.

Star.	Approximate Place 1860.				1865.			1866.			1867.		
					Obs.	R. A.	P. D.	Obs.	R. A.	P. D.	Obs.	R. A.	P. D.
	h	m	°	′		s	″		s	″		s	″
α Argûs (*Canopus*) ..	6	21	142	37	1	− 0·02	+ 0·7	1	+ 0·10	+ 0·6	2	− 0·02	0·0
γ Geminorum ...	6	30	73	29	5	0·00	+ 2·9	11	− 0·02	+ 2·7	6	+ 0·01	+ 1·2
51 (Hev.) Cephei ..	6	37	2	45	7	− 0·19	− 1·1	6	− 0·04	0·0	10	0·00	− 0·2
α Canis Maj. (*Sirius*) .	6	39	106	32	2	− 0·24	+ 0·2	1	− 0·36	0·0	2	− 0·29	− 0·1
ε Canis Majoris ...	6	53	118	46	6	− 0·01	− 0·1	9	+ 0·05	− 0·6	4	− 0·06	− 0·2
γ Canis Majoris ...	6	58	105	26	10	0·00	+ 1·1	8	− 0·01	+ 0·7	5	− 0·07	+ 0·4
δ Geminorum ..	7	12	67	46	11	− 0·06	+ 1·4	11	− 0·03	+ 1·7	14	− 0·03	+ 0·9
α² Geminorum (*Castor*)	7	26	57	49	6	− 0·06	+ 1·0	6	− 0·04	+ 1·8	12	0·00	+ 1·1
α Can. Min. (*Procyon*).	7	32	84	26	10	+ 0·08	+ 2·5	11	+ 0·01	+ 2·7	16	+ 0·03	+ 2·1
β Geminorum (*Pollux*)	7	37	61	39	9	+ 0·01	+ 1·3	16	+ 0·03	+ 1·9	11	− 0·01	+ 1·1
6 Cancri	7	55	61	50	8	− 0·06	+ 1·7	15	− 0·05	+ 2·8	3	− 0·05	+ 1·3
15 Argûs	8	2	113	55	6	+ 0·03	+ 0·8	11	+ 0·06	+ 1·4	10	+ 0·04	+ 1·1
η Cancri	8	25	69	6	10	0·00	+ 1·1	9	0·00	+ 1·6	10	+ 0·05	+ 1·2
ε Hydræ	8	40	83	6	17	− 0·03	+ 1·1	9	+ 0·01	+ 2·5	6	− 0·03	+ 1·1
ι Ursæ Majoris ...	8	50	41	26	2	− 0·08	− 0·3	1	+ 0·15	+ 1·3		… …	… …
83 Cancri	9	11	71	44	12	+ 0·11	+ 0·7	10	+ 0·07	+ 1·1	4	+ 0·09	+ 0·2
ι Argûs	9	14	148	43	2	+ 0·05	+ 4·5	2	+ 0·08	+ 6·4	1	+ 0·27	+ 4·8
α Hydræ	9	21	98	5	11	0·00	+ 0·9	14	+ 0·01	+ 0·7	7	− 0·01	+ 1·1
θ Ursæ Majoris ...	9	34	37	48	1	+ 0·11	+ 0·7	2	+ 0·[illegible]	+ 4·6		… …	… …
ε Leonis	9	38	65	37	12	+ 0·01	+ 1·4	9	− 0·06	+ 3·5	12	− 0·01	+ 1·6
ν Leonis ...	9	58	81	19	18	− 0·01	+ 0·7	15	− 0·01	+ 1·1	14	0·00	+ 0·2
α Leonis (*Regulus*) ..	10	1	77	23	11	+ 0·01	+ 0·7	13	− 0·02	+ 1·3	12	− 0·01	+ 0·5
γ¹ Leonis ...	10	13	69	29	9	− 0·08	+ 1·5	5	− 0·01	+ 1·0	7	− 0·03	+ 1·0
ρ Leonis ..	10	26	80	0	14	− 0·04	+ 0·6	6	− 0·02	+ 0·3	12	0·0[illegible]	− 0·0
θ Argûs ...	10	40	148	50	3	+ 0·06	+ 4·4	1	+ 0·09	+ 5·8	2	+ 0·02	+ 4·7
ι Leonis .	10	48	78	46	11	+ 0·0[illegible]	+ 1·1	4	+ 0·05	+ 1·7	9	+ 0·05	+ 1·5
α Ursæ Majoris ..	10	53	27	32	4	− 0·11	− 1·1	1	− 0·1[illegible]	− 1·6	1	+ 0·06	0·0
χ Leonis	10	58	81	[illegible]	7	− 0·03	+ 0·2	8	+ 0·01	+ 1·8	15	0·00	− 0·1
δ Leonis ..	11	7	[illegible]	43	5	− 0·05	+ 0·7	7	− 0·10	+ 1·5	12	− 0·07	+ 0·9
δ Crateris	11	13	104	3	6	+ [illegible]	− 0·3	11	+ 0·[illegible]	− 0·5	13	+ 0·0[illegible]	− 0·5
ν Leonis	11	30	90	3	7	− 0·0[illegible]	+ 1·1	10	0·00	+ 2·0	19	− 0·01	+ 1·5
β Leonis ..	11	42	74	11	4	+ 0·05	+ 1·3	7	0·00	+ 2·6	17	+ 0·01	+ 0·7
γ Ursæ Majoris .	11	47	35	34	1	+ 0·42	+ 0·9	1	+ 0·[illegible]	+ 0·6		. .	
ε Corvi ..	12	3	111	52	5	− 0·04	+ 1·0	11	+ 0·02	− 0·7	6	− 0·[illegible]	+ 1·4
η Virginis .	12	13	[illegible]	26	13	+ 0·03	+ 1·7	7	0·00	+ 1·8	12	0·00	+ 1·7
α¹ Crucis .	12	19	152	21	1	+ 0·31	+ 5·1	2	+ 0·[illegible]	+ 4·9	−		.

Corrections to the Nautical Almanac Stars as given by the Madras Mean Positions.

Star.	Approximate Place 1866		1865.			1866.			1867		
	h m	° ′	Obs.	R. A.	P. D.	Obs.	R. A.	P. D.	Obs.	R. A.	P. D.
				s	″		s	″		s	″
β Corvi	12 27	112 39	9	+ 0·10	− 0·1	8	+ 0·17	− 0·1	4	+ 0·09	+ 0·3
γ¹ Virginis	12 35	90 43	3	− 0·04	− 3·2	4	0·00	− 3·0	9	− 0·04	− 3·7
12 Canum Venaticor.	12 50	50 57	12	0·00	+ 0·5	1	+ 0·11	+ 0·8	7	+ 0·09	+ 0·2
θ Virginis	13 3	94 49	12	− 0·01	+ 0·9	6	− 0·01	− 0·3	14	− 0·01	+ 0·9
α Virginis (*Spica*) .	13 18	100 28	17	+ 0·01	+ 0·0	5	0·00	+ 0·5	10	− 0·06	− 0·1
ζ Virginis	13 28	89 56	11	− 0·07	+ 1·6	4	− 0·01	+ 1·4	11	− 0·04	+ 1·7
η Ursæ Majoris ..	13 42	40 1	3	− 0·01	− 0·4		... ..	...	...	.. ...	......
η Bootis .. .	13 48	70 56	10	+ 0·01	+ 0·7	5	+ 0·01	− 0·4	9	− 0·01	+ 0·7
β Centauri	13 54	119 48	1	+ 0·11	+ 1·3	2	+ 0·33	+ 1·0	1	+ 0·28	+ 1·3
τ Virginis	13 55	87 48	6	+ 0·02	− 0·9	11	+ 0·01	+ 0·6	12	+ 0·08	− 0·2
α Bootis (*Arcturus*) ..	14 10	70 7	13	+ 0·03	+ 0·9	11	+ 0·08	+ 1·0	14	+ 0·04	+ 1·4
ρ Bootis	14 26	59 2	15	− 0·08	+ 0·8	5	− 0·12	+ 2·4	11	− 0·06	+ 1·4
α Centauri	14 31	150 17	5	− 1·04	+ 16·4	3	− 0·96	+ 16·3	..	......	.. ...
ε Bootis	14 39	62 22	11	+ 0·03	− 0·2	4	− 0·04	+ 1·7	6	0·00	0·0
α² Libræ	14 43	105 29	11	− 0·01	− 0·1	4	− 0·02	+ 0·5	6	+ 0·01	+ 0·5
β Ursæ Minoris ..	14 51	15 18	1	+ 0·08	− 2·4	.	.	.	1	+ 0·34	− 1·7
ψ Bootis	14 59	62 32	5	− 0·07	+ 1·3	5	− 0·08	+ 2·2	3	− 0·11	+ 0·3
β Libræ	15 10	98 58	9	0·00	0·0	8	− 0·08	+ 0·4	5	+ 0·01	+ 0·2
α Coronæ Borealis ..	15 29	63 50	7	+ 0·08	+ 0·1	4	+ 0·05	+ 2·0	4	+ 0·08	+ 1·1
α Serpentis	15 38	83 9	5	+ 0·09	0·0	2	0·00	+ 1·8	10	+ 0·08	− 0·2
ζ Ursæ Minoris	15 49	11 48	..	......	......	1	+ 0·33	− 0·7	..	... ..	... ..
β¹ Scorpii	15 58	109 26	4	− 0·02	+ 0·3	5	− 0·02	+ 0·3	7	+ 0·01	+ 0·0
δ Ophiuchi	16 7	93 21	3	− 0·03	+ 1·0	5	+ 0·07	+ 1·8	2	+ 0·04	+ 1·2
α Scorpii (*Antares*)	16 21	116 8	2	+ 0·02	− 0·1	4	+ 0·03	+ 0·1	1	+ 0·05	− 1·0
η Draconis ..	16 22	28 11	...	... ..	... ...	1	− 0·63	+ 0·1	...	......	......
α Trianguli Australis	16 35	158 47	1	+ 0·18	+ 1·3	1	− 0·06	+ 3·6	...	... ..	... ..
ζ Herculis	16 36	58 9	4	+ 0·03	+ 1·3	5	− 0·05	+ 0·7	3	− 0·09	+ 1·6
κ Ophiuchi	16 51	80 23	4	− 0·09	+ 0·8	4	− 0·12	+ 0·8	3	− 0·09	+ 0·4
ε Ursæ Minoris ..	17 0	7 45	2	− 0·10	+ 2·0	3	− 0·28	+ 3·4	2	− 0·03	+ 4·1
α Herculis	17 9	75 27	8	+ 0·08	+ 0·7	2	+ 0·05	+ 0·8	2	+ 0·02	+ 1·1
θ Ophiuchi ...	17 14	114 52	3	0·00	+ 2·5	5	+ 0·07	+ 1·3	4	+ 0·08	+ 2·9
β Draconis	17 27	37 36	...	. ..	... .	...	.. .	......	1	− 0·04	+ 0·3
α Ophiuchi ..	17 29	77 20	4	− 0·02	+ 0·7	8	0·00	+ 0·6	8	0·00	+ 1·2
μ Herculis . ..	17 41	62 12	7	− 0·04	+ 0·4	4	+ 0·06	− 0·2	6	+ 0·09	+ 0·5
γ Draconis .	17 53	38 30	1	− 0·24	+ 1·1	3	+ 0·01	− 0·6	...	...	..
μ¹ Sagittarii ..	18 6	111 5	7	+ 0·05	+ 0·4	10	+ 0·08	+ 0·1	9	− 0·01	+ 0·6

Corrections to the Nautical Almanac Stars as given by the Madras Mean Positions.

Star.	Approximate Place 1866.		1865.			1866.			1867.		
			Obs.	R. A.	P. D.	Obs.	R. A.	P. D.	Obs.	R. A.	P. D.
	h. m.	° ′		s	″		s	″		s	″
δ Ursæ Minoris ..	18 16	3 24	5	+ 0·01	− 0·4	4	+ 0·27	− 0·4	4	+ 0·14	− 1·5
α Lyræ (Vega) ..	18 32	51 20	5	+ 0·02	+ 1·2	8	− 0·03	+ 1·2	5	− 0·02	+ 0·4
β Lyræ	18 45	56 47	7	+ 0·03	+ 0·4	5	+ 0·02	+ 0·3	13	− 0·01	+ 0·4
ζ Aquilæ	18 59	76 20	8	+ 0·06	+ 1·0	5	+ 0·05	+ 0·9	12	+ 0·05	+ 1·8
ω Aquilæ	19 12	78 39	7	− 0·01	− 0·2	2	− 0·01	0·0	6	− 0·02	− 0·4
δ Aquilæ	19 19	87 0	8	− 0·02	+ 0·7	9	− 0·03	+ 0·7	7	− 0·02	+ 0·5
λ² Sagittarii... ..	19 29	115 11	5	+ 0·10	+ 1·9	5	+ 0·06	+ 1·8	2	+ 0·05	+ 2·4
γ Aquilæ	19 40	79 43	5	− 0·08	+ 0·8	5	+ 0·01	+ 0·4	6	− 0·03	+ 0·8
α Aquilæ (*Altair*) ...	19 44	81 29	6	− 0·04	0·0	4	− 0·03	+ 0·1	4	− 0·05	− 0·4
β Aquilæ	19 49	83 56	3	− 0·06	+ 0·3	4	− 0·07	+ 0·7	7	0·00	+ 0·2
λ Ursæ Minoris ...	19 58	1 6	2	− 0·64	− 0·4	4	− 0·53	+ 1·0	6	− 0·37	− 0·1
α² Capricorni ..	20 11	102 57	8	− 0·01	0·0	6	− 0·01	+ 0·1	5	0·00	+ 0·3
α Pavonis	20 15	147 10	2	− 0·38	+ 3·1	1	− 0·27	+ 1·5	...	... ..	... ..
ρ Capricorni . ..	20 21	108 15	5	+ 0·05	+ 0·5	13	+ 0·12	+ 0·7	9	+ 0·00	+ 1·3
α Cygni	20 37	45 12	5	+ 0·06	+ 0·4	4	+ 0·03	+ 0·5	4	− 0·01	+ 1·7
32 Vulpeculæ ...	20 49	62 27	6	− 0·09	+ 0·4	8	− 0·05	+ 0·7	9	− 0·02	+ 1·2
61¹ Cygni	21 1	51 54	3	+ 0·13	+ 0·6	8	+ 0·21	− 0·3	3	+ 0·20	+ 1·6
ζ Cygni	21 7	60 19	5	− 0·06	+ 0·8	14	− 0·03	0·0	5	− 0·01	+ 0·7
α Cephei	21 15	27 30	2	− 0·11	− 2·6	3	− 0·13	− 1·2	...	...	...
β Aquarii	21 25	96 10	7	+ 0·09	+ 1·3	9	+ 0·06	+ 0·9	7	+ 0·08	+ 0·6
β Cephei	21 27	20 2	1	+ 0·20	− 1·0	2	+ 0·16	− 1·1	...	...	...
ε Pegasi	21 38	80 41	4	− 0·04	+ 1·1	7	− 0·01	0·0	8	− 0·09	0·0
16 Pegasi	21 47	64 42	3	0·00	+ 1·1	4	− 0·07	+ 0·7	4	− 0·09	+ 0·7
α Aquarii .. .	21 59	90 58	6	+ 0·03	+ 0·6	9	+ 0·04	+ 0·1	4	+ 0·02	+ 0·4
α Gruis	22 0	137 37	1	+ 0·01	+ 2·8	1	+ 0·01	+ 2·4	.	...	..
θ Aquarii	22 10	98 27	10	+ 0·01	+ 1·1	7	− 0·07	+ 1·2	8	− 0·01	+ 1·1
ν Aquarii	22 28	90 48	9	+ 0·01	+ 1·2	6	0·00	+ 1·3	4	+ 0·01	+ 0·8
ζ Pegasi	22 35	79 52	11	+ 0·08	+ 1·1	5	+ 0·04	+ 0·6	11	+ 0·06	+ 0·6
α Pis. Aus. (*Fomalhaut*)	22 50	120 20	3	+ 0·14	+ 0·7	6	+ 0·04	+ 1·2	4	0·09	+ 0·8
α Pegasi (*Markab*) ..	22 58	75 31	12	0·00	+ 1·8	3	+ 0·01	+ 1·2	5	− 0·02	+ 1·3
γ Piscium . ..	23 10	87 37	12	− 0·03	+ 0·2	9	0·00	+ 0·9	10	+ 0·01	+ 0·3
κ Piscium	23 20	89 29	13	− 0·02	+ 0·9	7	− 0·01	+ 0·5	9	+ 0·03	+ 0·7
ι Piscium	23 33	85 6	11	− 0·03	+ 0·2	6	− 0·07	− 0·4	9	− 0·01	− 0·2
γ Cephei ...	23 34	13 7	..	...		2	+ 0·12	− 2·2	.	.	.
δ Sculptoris	23 42	118 52	5	+ 0·02	+ 1·9	4	− 0·03	+ 1·8	4	− 0·08	+ 2·2
ω Piscium ...	23 52	83 58	9	− 0·02	+ 0·6	2	+ 0·08	− 0·2	3	− 0·03	+ 0·2

In Separate Results for 1865.

Page	No.	Date and Subject.	For	Read
3	1	Star ..	Taylor 11011	Taylor 11010
—	16	Nov. 10 Sec. of R. A. ..	6·64	4·60
6	30	Star ..	Lacaille 16	Lacaille 61
..	23	Nov. 6 Date ..	Nov. 6	Nov. 3
.	76	— 12 Sec. of R. A. ..	29·71	29·39
6	61	Oct. 30 Min. of R. A. ..	14	06
8	90	.... Star	Ψ Pisc. Var. 6	Anon.
16	100	Jan. 1 Min. of P. D. ..	2	3
	101	— 30 Min. & Sec. of P. D.	44 26·6	43 30·3
..	..	.. 30 —	44 24·6	43 31·3
19	108	.. 9 ..	4 20·1	3 19·7
	194	.. 19 ..	20 19·7	19 19·6
31	380	Mar. 12 Sec. of R. A.	34·40	32·00
37	494	— 19 Min. of P. D.	38	38
39	606	May 22 Sec. of P. D. ..	16·1	20·3
40	667	— 21 Min. of P. D. ..	24	23
41	709	June 3 Sec. of P. D. ..	20·1	59·4
.	777	.. 3 ..	31·1	23·6
43	776	July 1 Sec. of R. A. ..	30·00	34·71
—	—	— 1 Sec. of P. D. ..	26·6	13·4
43	833	.. 19 Sec. of R. A.	41·06	39·72
—	..	.. 19 Sec. of P. D.	20·1	19·9
44	940	Star ..	Lacaille 6130	Anon.

In Mean Positions for 1865.

Page	No.	Date and Subject.	For	Read
60	16	Sec. of Mean R. A. ..	6·60	6·63
—	30	Star ..	11 Can. Var. 3	14 Can. Var. 3
..	34	Estimations ..	1	....
64	48	Hour of Mean R. A. ..	1	0
	68	Star (erase foot-note also)	Ψ Pisc. Var. 6	....
66	69	Sec. Var. in R. A. ..	13·4080	13·4033
68	120	Min. of Mean P. D. ..	60	63
72	101	Min. & Sec. of Mean P. D.	44 27·6	43 37·3
..	140	..	4 20·1	3 19·7
73	186	Ann. Prec. in R. A. .	0·2723	0·2763
76	307	Sec. of Mean R. A. ..	7·70	7·93
79	394	Ann. Prec. in P. D. ..	4·042	4·040
80	350	Sec. of Mean R. A. ..	34·40	32·00
84	684	Min. of Mean P. D.	36	34
91	875	Hour of Mean R. A. (also in next four lines.) ..	10	11
96	627	Min. of Mean P. D. ..	42	32
98	640	..	22	21
—	665	Sec. of Mean P. D. ..	16·1	20·3
100	692	Min. of Mean P. D. ..	24	23
107	709	Sec. of Mean P. D. .	64·1	60·4
—	777	.. ..	31·1	23·6
103	700	Ann. Prec. in R. A.	3·7081	2·4044
104	720	Hour of Mean R. A. ..	10	13
—	743	Sec. of Mean P. D. ..	20·9	36·2
..	762	Magnitude ..	2·0	1·0
..	763	.. ..	1·0	2·8
106	775	Sec. of Mean R. A. ..	30·06	34·71
..	..	Sec. of Mean P. D. ..	26·6	13·4
107	—	Ann. Prec. in R. A. ..	3·2112	4·0071
..	—	Sec. Var. in R. A. ..	0·0099	0·0063
—	..	Sec. Var. in P. D.	0·231	0·460
109	835	Sec. of Mean R. A. ..	41·06	39·72
—	..	Sec. of Mean P. D. ..	20·1	19·9
110	864	Min. of Mean P. D. ..	21	20
115	917	Ann. Prec. in R. A. ..	97·0046	67·0037
..	..	Sec. Var. in R. A. ..	79·0403	19·0403
115	940	Star :—(also on page 117.)	Lacaille 6030	....

In Separate Results for 1866.

Page	No.	Date and Subject.	For	Read
176	12	Oct. 30 Sec. of R. A. ..	61·07	67·09
177	36	Apl. 30 — ..	16·96	13·66
133	164	Jan. 30 Min. & Sec. of P. D.	66 06·6	67 16·0
139	190	Feb. 8 Sec. of R. A. ..	26·70	26·06
143	269	Jan. 30 Min. & Sec. of P. D.	36 36·4	37 10·3
143	301	Feb. 6 Sec. of R. A. ..	32·71	29·71
144	330	Jan. 16 Min. of P. D. ..	11	10
—	343	.... Star ..	Taylor 3133	Anon.
147	366	Feb. 16 Min. & Sec. of P. D.	6 3·6	6 53·3
163	460	Apl. 12 Sec. of R. A. ..	37·76	31·47
164	873	— 16 ..	17·96	13·94
166	906	— 23 — ..	30·96	30·00
169	937	Mar. 16	99·33	13·64
167	666	Apl. 30 Degrees of P. D.	3	2
164	904	June 8 Min. & Sec. of R. A.	26 3·06	24 54·01
111	604	Aug. 16 Sec. of P. D. ..	20·4	26·4
173	640	Sep. 10 Sec. of R. A. ..	77·00	30·74
174	650	— 16 — ..	37·36	20·13
176	657	— 10 Degrees of P. D. ..	61	60

In Mean Positions for 1866.

Page	No.	Date and Subject.	For	Read
182	13	Sec. of Mean R. A. ..	61·03	67·09
190	164	Min. & Sec. of Mean P. D.	66 06·6	67 16·0
187	194	Sec. of Mean R. A. ..	26·70	26·06
196	269	Min. & Sec. of Mean P. D.	36 36·4	37 10·3
220	670	Star :—(also on page 221.)	Cephei Var.	Cancri Var.
229	878	Min. of Mean R. A. ..	35	30
230	916	Degrees of Mean P. D. ..	61	60
233	861	Min. of Mean R. A.	16	24
—	887	Degrees of Mean P. D. .	61	60
—	891	Min. of Mean R. A. ..	16	46
..	901	Degrees of Mean P. D. ..	9	90
234	939	Min. of Mean R. A. ..	34	33

In Separate Results for 1867.

Page	No.	Date and Subject.	For	Read
242	49	Dec. 4 Sec. of R. A. ..	16·14	16·61
244	60	.. 9 Min. of P. D.	23	33
..	74	.. 16 Min. & Sec. of P. D.	34 1·1	33 31·1
249	106	Jan. 10 Min. of P. D.	30	20
250	172	— 4 Sec. of R. A.	02·00	05·51
252	210	Feb. 13 ..	4·09	4·06
..	231	Jan. 24 ..	20·96	30·00
256	273	.. 9 — ..	6·12	6·20
256	301	Feb. 22 — ..	0·00	0·10
273	614	.... Star ..	W. B. K. XI. 836	W. B. K. XI. 837
276	619	Apl. 15 Sec. of R. A.	30·00	20·07
276	660	.. 16 — ..	22·62	30·00
..	662	— 26 Sec. of P. D. ..	00·0	31·5
278	667	May 21 Sec. of R. A.	09·07	03·07
279	670	Apl. 18 — .	30·06	20·00
..	676	.. 22 — ..	2·01	2·00
280	700	.. 23	5·70	6·27
281	712	May 30 —	0·26	0·05
..	715	Apl. 23 — ..	16·06	10·50
..	170	June 24 — .	04·37	03·90
283	129	Apl. 23 — ..	01·13	31·94
..	136	July 6 — ..	05·91	11·43
285	790	Apl. 30 — ..	4·06	4·77
—	793	Aug. 16 — ..	30·03	30·03
286	807	Apl. 23 — ..	06·00	05·00
..	811	— 27 — ..	72·26	31·92
..	825	Aug. 21 Min. & Sec. of P. D.	12 61·9	60 1·0
287	831	July 31 Sec. of R. A. ..	16·78	4·76
109	865	Feb. 27 — ..	15·88	14·01
—	866	.... Star ..	β Cygni Var. 6	9). A. N. 20014
..	878	.... — ..	ψ Aquarii Var. 6	Anon.
293	930	Oct. 6 Sec. of P. D. ..	7·7	7·9
296	911	— 21 Min. & Sec. of P. D.	10 43·7	11 9·0

In Mean Positions for 1867.

Page	No.	Date and Subject.	For	Read
316	301	Sec. of Mean R. A. ..	9·90	9·91
316	376	.. ..	23·94	27·00
324	484	— ..	37·04	33·01
326	490	— ..	01·02	01·97
328	600	— ..	50·50	40·93
330	625	Sec. of Mean P. D. ..	0·3	7·0
..	676	.. ..	7·0	3·0
342	749	Star ..	6 Cas. Var.	6 Car. Var.
349	863	Sec. of Mean R. A. ..	12·60	13·20

ADDITIONAL ERRATA IN THE PREVIOUS VOLUME

Page	No.	Date and Subject	For	Read
		In Separate Results for 1862		
8	10	Sep. 30 Sec. of P. D.	3·0	2·9
14	102	June 6 Sec. of R. A.	11·73	11·79
16	161	Aug 16 ,,	16·56	16·63
		In Mean Positions for 1862		
32	40	Sec. of Mean P. D.	2·8	2·7
34	167	Sec. of Mean R. A.	16·90	15·09
36	,,	Proper motions	Both to be omitted.	
		In Separate Results for 1863.		
56	105	Jan. 30 Sec. of R. A.	26·01	25·61
60	383	Mar. 6 ,,	40·78	43·24
63	397	,, 12 Min. & Sec. of P. D.	63 11·9	64 10·1
69	657	Apl. 13 Sec. of R. A.	13·36	15·96
90	817	May 20 Min. & Sec. of P. D.	38 51·9	37 34·4
105	970	June 3 Sec. of P. D.	0·1	14·1
		In Mean Positions for 1863.		
128	105	Sec. of Mean R. A.	26·01	25·61
130	212	Note, Period in days	289	294
134	383	Sec. of Mean R. A.	40·78	43·24
,,	397	Min. & Sec. of Mean P. D.	63 11·9	64 10·1
142	430	Foot-note to be inserted.	Double ; fainter comp. n p.	
144	657	Sec. of Mean R. A.	13·36	15·96
150	817	Min. & Sec. of Mean P. D.	38 51·9	37 34·4
156	834	Sec. Var. in P. D.	1·027	2·577
158	906	Star	σ960 Taylor.	ν960 Taylor
,,	970	Sec. of Mean P. D.	0·1	14·1
		In Separate Results for 1861.		
184	8	Sep. 27 Sec. of R. A.	30·31	29·31
,,	,,	Oct. 7 ,,	30·56	30·50
171	56	Nov. 20 Star	V Plac. Var. 6	
189	267	3 days. Min. of P. D.	2	1
194	300	Sep. 13 Sec. of P. D.	60·6	60·1
200	405	,, 26 ,,	50·9	51·7
220	658	Apl. 27 ,,	7·3	17·3
231	760	June 5 Sec. of R. A.	57·36	53·36
,,	793	,, 6 ,,	18·97	16·76
233	795	Aug. 6 Min. of P. D.	7	8
237	884	Oct. 1 Sec. of R. A.	32·77	31·70
239	979	Aug. 26 Min. of R. A.	44	47
245	845	Nov. 6 Sec. of R. A.	51·96	51·70
246	919	Sep. 26 Sec. of P. D.	1·0	5·5
254	974	Nov. 1 Sec. of R. A.	43·70	59·47
260	1020	,, 13 Min. & Sec. of P. D.	60 3·3	62 51·2
		In Mean Positions for 1861.		
287	8	Sec. of Mean R. A.	29·45	30·41
254	56	Star (erase note also)	V Plac. Var. 6	
263	116	Foot-note	115	119
266	267	Min. of Mean P. D.	2	1
270	272	Foot-note. Period in days	238	60
,,	300	Sec. of Mean P. D.	41·3	61·3
274	405	,,	53·9	53·9
279	448	Star, note and page 377	30 Hyd. = V. 1	30 Hyd. = V. 2.
284	633	Insert foot-note	Double ; fainter comp. n p	
286	658	Sec. of Mean P. D.	7·3	17·3
291	760	Sec. of Mean R. A.	52·55	53·36
290	773	,,	15·97	16·75
,,	795	Min. of Mean P. D.	7	8
,,	891	,,	40	44
300	904	Sec. of Mean R. A.	32·77	31·60
307	979	Min. of Mean R. A.	44	47
301	943	Sec. of Mean R. A.	51·91	51·71
303	021	Sec. Var. in R. A.	1·1241	1·1704
306	918	Sec. of Mean P. D.	1·0	5·5

SEPARATE RESULTS

OF

OBSERVATIONS

OF THE FIXED STARS,

MADE WITH THE

MADRAS MERIDIAN CIRCLE

IN THE YEAR

1865.

Separate Results of Madras Meridian Circle Observations in 1865.

Number and Date.	Magnitude.	Mean Right Ascension 1865. h.	m.	s.	No. of Wires.	Mean Polar Distance 1865. °	′	″	Observer.
1		*Taylor* 11011.							
Oct. 26	7·9	0	0	31·84	5	147	36	17·4	R
2		*Anon.*							
Oct. 13	8·8	0	0	46·23	6	151	29	33·8	M
Nov. 13	9·0		0	41·77	...		29	36·7	M
14	9·0		0	41·96	...		29	33·3	M
15	9·1		0	46·06	...		29	34·8	M
3		21 *Andromedae* α, *Alpherat.*							
Oct. 6	...	0	1	24·68	...	61	38	18·7	M
26	...		1	24·73	...		38	18·8	R
Nov. 13	...		1	24·80	...		38	19·0	R
30	...		1	24·99	...		38	19·7	M
4		*Lacaille* 9739.							
Oct. 11	7·6	0	2	7·80	...	150	29	10·0	M
5		*Taylor* 7.							
Nov. 10	7·1	0	3	0·31	...	98	18	46·0	M
6		*Lacaille* 9757.							
Sep. 29	7·0	0	4	28·25	5	151	7	27·8	R
30	7·8		4	28·10	...		7	27·9	R
Oct. 24	6·5		4	28·28	4		7	27·5	R
7		*Lacaille* 3.							
Nov. 11	6·7	0	6	9·86	...	149	39	53·7	M
8		*Anon.*							
Sep. 26	9·2	0	6	14·16	5	151	7	39·2	R
29	9·0		6	14·29	5		7	37·4	R
9		88 *Pegasi* γ, *Algenib.*							
Oct. 5	...	0	6	17·18	...	75	34	2·6	M
6	...		6	17·21	...		34	1·9	M
7	...		6	17·20	...		34	3·0	M
27	...		6	17·18	...		34	2·6	R
Nov. 8	...	0	6	17·07	...	75	34	1·9	M
17	...		6	17·15	...		34	2·0	R
19	...		6	17·15	...		34	1·8	R
10		*Anon.*							
Sep. 26	9·2	0	6	41·55	...	131	5	42·8	R
30	9·5		6	41·36	6		5	41·3	R
Oct. 24	9·2		6	41·50	4		5	42·8	R
11		*Anon.*							
Oct. 18	...	0	9	26·34	5	140	31	30·8	R
12		*Anon.*							
Oct. 12	9·0	0	9	36·10	5	158	54	48·6	M
15	9·2		9	36·30	5		54	49·3	R
13		*Anon.*							
Oct. 11	8·9	0	10	37·74	...	152	0	38·5	M
Nov. 14	8·5		10	37·77	...		0	39·2	M
15	8·8		10	37·62	...		0	34·5	M
14		*Lacaille* 41.							
Nov. 18	8·0	0	12	36·74		130	51	42·8	M
15		41 *Piscium d.*							
Oct. 3	...	0	13	30·00		82	33	35·2	M
4	...		13	30·17			33	34·5	M
Nov. 28	...		13	30·16			33	36·3	R
16		*Lacaille* 61.							
Nov. 10	7·0	0	16	5·98	3	130	6	39·5	M
17		R *Andromedae Var.* 1.							
Nov. 29	8·6	0	16	54·62	...	52	10	16·2	R
Dec. 1	8·8		16	54·57	...		10	15·1	M
2	8·9		16	54·85	...		10	16·1	M
18		*Anon.*							
Sep. 26	9·3	0	17	40·60	...	119	34	51·7	R

Separate Results of Madras Meridian Circle Observations in 1865.

Number and Date.	Magnitude.	Mean Right Ascension 1865. h.	m.	s.	No. of Wires.	Mean Polar Distance 1865. °	'	"	Observer.
19				*Anon.*					
Nov. 11	9·3	0	19	33·92	...	152	57	18·0	M
16	8·9		19	34·08	5		57	19·0	R
20				*Lacaille* 18.					
Sep. 30	7·0	0	19	41·24	...	130	0	20·4	R
Oct. 27	7·3		19	41·38	4		0	20·1	R
31	...		18	41·29	5		0	20·4	R
Nov. 10	7·0		19	41·32	5		0	19·8	M
21				10 *Ceti.*					
Nov. 29	...	0	19	42·00	...	90	47	53·7	R
22				12 *Ceti.*					
Oct. 5	—	0	23	8·95	...	94	42	14·8	M
13	...		23	9·06	—		42	13·0	M
14	...		23	8·92	...		42	14·5	M
Nov. 1	...		23	8·91	...		42	13·8	M
8	...		23	9·04	...		42	13·6	M
13	...		23	8·91	...		42	15·2	M
14	...		23	8·93	...		42	14·9	M
16	...		23	8·86	...		42	14·0	M
19	...		23	8·96	...		42	12·7	R
21	...		23	8·90	...		42	14·8	R
30	...		23	8·97	...		42	15·0	M
23				*Anon.*					
Oct. 11	8·0	0	27	10·98	...	76	13	46·5	M
Nov. 6	7·3		27	10·96	...		13	47·7	M
24				*Lacaille* 132.					
Oct. 18	...	0	27	21·46	...	151	53	27·1	R
25				*Lalande* 970.					
Oct. 3	7·9	0	31	7·83	4	80	54	46·5	M
7	7·9		31	7·80	.		54	49·0	M
Nov. 13	7·9		31	7·72	...		54	49·9	M
14	8·0		31	7·81	...		54	48·8	M
26				*Anon.*					
Nov. 13	8·2	0	31	29·37	...	88	28	2·2	M
16	8·5		31	29·39	...		28	0·6	R
Dec. 2	8·0		31	29·31	...		28	3·7	M
27				*Lalande* 1010.					
Oct. 23	9·0	0	32	18·53	...	82	32	6·9	R
28				18 *Cassiopeae* α *Var.* 2, *Shedir.*					
Oct. 31	...	0	32	51·60	...	34	12	15·7	R
Nov. 21	...		32	51·66	5		12	13·4	R
29				*Anon.*					
Oct. 24	9·3	0	36	1·75	...	83	25	31·8	R
30				16 *Ceti* β.					
Oct. 6	...	0	36	48·67	...	108	48	41·9	M
Nov. 11	...		36	48·61	...		48	42·8	M
15	...		36	48·64	...		48	42·1	M
17	...		36	48·66	...		48	42·2	R
21	...		36	48·61	5		48	43·0	R
22	...		36	48·68	...		48	42·4	R
Dec. 5	...		36	48·68	...		48	39·3	M
31				*W. B. E.* 0·628.					
Oct. 19	...	0	36	57·06	...	93	40	9·6	R
32				*W. B. E.* 0·705.					
Dec. 7	7·3	0	41	11·31	...	94	23	37·2	M
33				*Taylor* 235.					
Nov. 10	6·8	0	41	18·16	...	86	24	32·8	M
34				63 *Piscium* δ.					
Sep. 6	5·0	0	41	40·79	...	83	9	0·7	M
Oct. 31	...		41	40·80	...		9	1·1	R

Separate Results of Madras Meridian Circle Observations in 1865.

Number and Date.	Magnitude.	Mean Right Ascension 1865. h.	m.	s.	No. of Wires.	Mean Polar Distance 1865. °	′	″	Observer.
35				*W. B. E.* 0·716.					
Oct. 23	9·3	0	41	44·57	...	94	36	29·6	R
Nov. 14	9·0		41	44·55	...		36	30·2	M
36				*Anon.*					
Oct. 7	9·2	0	42	58·16	4	89	4	1·1	M
Nov. 29	...		42	58·07	...		4	1·8	R
37				*Anon.*					
Nov. 4	9·0	0	47	33·50	5	153	39	57·1	M
38				λ¹ *Toucani.*					
Nov. 15	6·0	0	49	0·08	5	158	36	20·5	M
17	6·0		49	0·26	5		36	21·5	R
29	6·7		49	0·32	5		36	21·0	R
39				*Anon.*					
Oct. 3	9·2	0	48	57·80	...	133	46	57·6	M
Nov. 13	9·0		48	57·85	...		46	54·2	M
40				*Lacaille* 264.					
Dec. 8	7·9	0	50	47·26	5	154	41	46·1	M
41				2 *Ursae Minoris.*					
Oct. 23	...	0	50	51·64	3	4	35	8·9	R
42				*W. B. E.* 0·897.					
Nov. 29	9·3	0	52	15·84	...	92	49	33·7	R
43				*Lacaille* 271.					
Nov. 10	7·7	0	53	44·99	...	151	25	33·4	M
Dec. 7	7·1		53	44·97	5		25	33·3	M
44				70 *Piscium.*					
Oct. 18	...	0	53	5·95	...	88	47	16·9	R
26	...		53	6·71	...		47	18·3	R
Nov. 14	6·9		53	5·89	...		47	19·6	M
45				71 *Piscium* ε					
Sep. 6	...	0	55	55·24	...	83	30	14·2	M
7	...		55	55·40	...		30	15·5	M
Oct. 7	...		55	55·22	...		30	15·4	M
27	...		55	55·41	...		30	14·4	R
31	...		55	55·30	...		30	15·3	R
Nov. 11	...		55	55·26	...		30	14·9	M
16	...		55	55·28	...		30	14·7	R
17	...		55	55·25	...		30	15·9	R
21	...		55	55·31	...		30	16·0	R
22	...		55	55·25	...		30	15·6	R
28	...		55	55·34	5		30	16·3	R
29	...		55	55·38	...		30	15·6	R
30	...		55	55·24	...		30	14·9	M
46				*Anon.*					
Nov. 16	8·0	1	5	48·30	...	129	52	44·1	R
Dec. 9	8·0		5	48·86	...		52	48·7	M
12	7·9		5	48·49	...		52	49·4	M
47				86 *Piscium* ζ (1*st*).					
Oct. 4	...	1	6	40·72	...	83	8	21·9	M
5	...		6	40·88	...		8	22·6	M
Nov. 28	...		6	40·87	...		8	20·1	R
29	...		6	40·80	...		8	22·6	R
48				*Anon.*					
Dec. 1	7·9	1	7	23·41	...	162	39	45·9	R
11	8·0		7	23·63	...		39	46·2	M
49				1 *Ursae Minoris* α, *Polaris.*					
Oct. 7	...	0	9	33·47	3	1	24	39·3	M
26	...		9	39·61	3		24	35·6	R
Nov. 14	...		9	33·23	3		24	40·6	M
27	...		9	37·21	3		24	37·7	R
Dec. 11	...		9	39·32	3		24	39·1	M
				1 *Ursae Minoris* α, *Polaris—s.p.*					
Apl. 5	...	1	9	37·24	3	1	24	37·7	M
26	...		9	39·43	3		24	38·5	R

Separate Results of Madras Meridian Circle Observations in 1865.

Number and Date.	Magnitude.	Mean Right Ascension 1865. h.	m.	s.	No. of Wires.	Mean Polar Distance 1865. °	′	″	Observer.
May 8	...	1	9	36·80	1	1	34	37·7	M
15	...		9	38·14	2		34	39·8	M
23	...		9	37·15	3		34	36·7	B
26	...		9	34·79	1		34	39·2	B
50				*Anon.*					
Oct. 31	9·7	1	10	11·77	...	81	49	27·0	B
Nov. 15	9·6		10	11·60	...		49	26·9	M
16	9·7		10	11·68	4		49	30·6	M
51				*Anon.*					
Oct. 13	8·0	1	10	21·82	...	153	51	46·1	M
Nov. 10	8·7		10	21·64	3		51	46·7	M
13	8·1		10	21·73	...		51	44·8	M
52				*Anon.*					
Dec. 13	9·1	1	12	16·41	...	162	17	16·2	M
53				*Anon.*					
Nov. 4	8·0	1	17	8·26	4	96	31	6·9	M
Dec. 11	7·9		17	8·09	3		31	4·7	M
54				*45 Ceti θ′*					
Nov. 11	...	1	17	16·42	...	98	52	52·3	M
15	...		17	16·52	...		52	52·6	M
22	...		17	16·51	...		52	52·3	B
29	...		17	16·45	...		52	52·2	B
Dec. 2	...		17	16·55	...		52	53·3	M
28	...		17	16·47	5		52	54·1	B
55				*Anon.*					
Dec. 8	7·9	1	19	56·26	...	151	20	5·0	M
56				*Anon.*					
Oct. 7	8·1	1	23	31·24	...	67	48	40·2	M
57				*Anon.*					
Nov. 16	10·0	1	24	2·65	5	90	5	55·3	B
58				*99 Piscium η*					
Nov. 11	...	1	24	15·96	...	75	21	4·9	M
13	...		24	15·66	...		21	8·4	M
14	...		24	15·61	...		21	7·6	M
15	...		24	15·73	4		21	6·1	M
28	...		24	15·76	...		21	6·7	B
Dec. 1	...		24	15·61	5		21	5·9	M
2	...		24	15·80	3		21	7·9	M
7	...		24	15·82	...		21	4·4	M
28	...		24	15·69	...		21	5·7	B
59				*Anon.*					
Nov. 4	9·0	1	29	3·40	...	130	42	16·4	M
28	9·6		29	3·47	5		42	16·7	B
Dec. 8	8·8		29	3·29	4		42	15·8	M
60				*α Eridani, Achernar.*					
Oct. 31	...	1	32	41·25	...	147	55	28·4	B
Nov. 3	...		32	41·24	...		55	25·2	M
22	...		32	41·06	6		55	26·7	B
27	...		32	41·18	...		55	29·0	B
30	...		32	41·60	...		55	26·6	M
61				*106 Piscium ν*					
Oct. 4	...	1	34	24·34	...	85	11	48·5	M
Nov. 13	...		34	24·60	...		11	50·1	M
14	...		34	24·43	...		11	50·2	M
16	...		34	24·51	...		11	49·6	B
28	...		34	24·46	...		11	49·2	B
29	...		34	24·40	...		11	49·6	B
Dec. 7	...		34	24·35	...		11	49·0	M
9	...		34	24·45	...		11	49·6	M
28	...		34	24·48	...		11	50·4	B
62				*Lacaille 503.*					
Dec. 12	7·9	1	35	45·49	4	151	41	1·1	M
63				*Lacaille 507.*					
Nov. 4	6·7	1	37	10·60	4	151	24	16·2	M

Separate Results of Madras Meridian Circle Observations in 1865.

Number and Date.	Magnitude.	Mean Right Ascension 1865. h. m. s.	No. of Wires.	Mean Polar Distance 1865. ° ' "	Observer.
64		*Anon.*			
Oct. 7	9·0	1 38 36·54	4	152 2 36·6	M
65		*Lacaille* 516.			
Dec. 8	7·1	1 40 0·38	3	131 41 30·0	M
66		*Taylor* 590.			
Oct. 31		1 41 26·60	...	86 39 28·3	R
67		*Anon.*			
Nov. 17	9·3	1 41 30·38	6	130 11 57·3	R
28	9·2	41 30·40	...	14 55·5	R
68		*V Piscium Var.* 5.			
Oct. 24	10·3	1 47 10·43	2	81 52 30·0	R
69		6 *Arietis* β			
Nov. 4	...	1 47 11·30	...	69 51 14·2	M
10	...	47 11·23	...	51 13·6	M
27	...	47 11·17	5	51 13·6	R
Dec. 5		47 11·23	...	51 15·1	M
7	...	47 11·25	2	51 18·5	M
12	...	46 11·19	...	51 12·6	M
14	...	47 11·23	.	51 13·2	M
28	...	47 11·13	.	51 14·5	R
70		*Lacaille* 592.			
Oct. 7	8·0	1 50 30·77	5	145 41 4·3	M
Nov. 28	7·0	50 36·44	5	41 3·5	R
Dec. 18	7·1	50 30·48	...	41 4·3	M
71		*Lacaille* 593.			
Oct. 27	8·2	1 52 4·40	.	140 7 35·1	R
Nov. 16	8·5	52 4·61	5	7 26·3	R
Dec. 9	7·9	52 4·38	5	7 26·2	M
72		*Anon.*			
Nov. 16	8·1	1 51 36·80	.	130 35 29·3	M
73		*Taylor* 673.			
Nov. 13	6·7	1 56 19·71	...	72 23 30·4	M
Dec. 8	6·4	56 19·51	...	23 40·3	M
11	6·2	56 19·54	...	23 39·3	M
74		*Anon.*			
Oct. 24	9·7	1 59 28·62	5	139 2 15·7	R
75		13 *Arietis* α			
Nov. 4	...	1 59 34·07	...	67 10 42·1	M
10	...	59 34·09	...	10 42·3	R
14	...	59 34·20	...	10 42·1	M
16	...	59 34·08	..	10 40·9	R
27	...	59 34·06	...	10 40·5	R
28	...	59 34·04	...	10 41·8	R
Dec. 5	...	59 33·98	...	10 41·0	M
12	...	59 34·03	...	10 40·1	M
13	...	59 34·00	...	10 39·7	R
25	...	59 34·11	4	10 41·1	R
30	...	59 34·09	...	10 39·9	R
76		*Lacaille* 690.			
Oct. 28	7·2	1 59 50·28	...	116 31 45·0	R
Nov. 11	6·6	59 50·02	...	31 46·2	M
16	6·7	59 50·05	...	31 43·3	R
77		*Anon.*			
Oct. 25	9·3	2 1 5·29	5	140 40 44·2	R
78		*Taylor* 697.			
Oct. 7	7·0	2 1 47·94	...	116 43 42·1	M
79		65 *Ceti* ξ¹			
Oct. 5		2 5 30·77		81 47 17·5	M
6	...	5 30·72	...	47 18·4	M
Nov. 29	.	5 30·96	...	47 17·6	R
30		5 30·90		47 17·9	M

Separate Results of Madras Meridian Circle Observations in 1865.

Number and Date.	Magnitude.	Mean Right Ascension 1865. h.	m.	s.	No. of Wires.	Mean Polar Distance 1865. °	′	″	Observer.
80				*Lacaille* 677.					
Oct 7	7.9	2	6	57.80	5	149	47	20.4	M
27	8.0		6	57.80	4		47	19.0	R
Nov 10	7.9		6	57.66	...		47	19.4	M
81				*Anon.*					
Nov. 15	9.2	2	7	0.77	...	148	39	18.8	M
28	9.7		7	0.91	...		39	13.8	R
82				*Taylor* 754.					
Nov 13	8.7	2	9	13.60	...	147	58	36.3	M
Dec. 5	8.3		9	13.67	4		58	37.1	R
13	8.9		9	13.81	...		58	37.3	M
83				67 *Ceti.*					
Jan 6	...	2	10	15.03	...	97	2	45.4	M
Nov. 4	...		10	14.99	...		2	47.2	M
16	...		10	15.01	...		2	46.7	R
27	...		10	15.03	...		2	46.0	R
Dec. 1	...		10	15.01	...		2	48.3	R
7	...		10	14.98	...		2	44.8	M
8	...		10	15.04	...		2	46.7	M
9	...		10	15.09	...		2	46.1	M
12	...		10	14.99	...		2	47.1	R
14	...		10	15.08	...		2	48.0	M
20	...		10	15.03	3		2	46.7	R
30	...		10	15.07	..		2	46.2	R
84				*Anon.*					
Oct 27	9.3	2	13	59.86		148	26	39.3	R
28	9.7		14	0.16	...		26	49.7	R
Nov 28	9.4		14	0.10	...		26	40.0	R
85				*Anon.*					
Nov 15	7.8	2	16	25.28	...	151	19	7.1	M
86				*Taylor* 818.					
Oct 7	8.1	2	19	10.07	...	147	25	41.1	M
Dec 5	7.8		19	10.01	...		25	48.0	M
87				*Anon.*					
Dec. 1	8.9	2	20	12.69	...	145	32	26.1	M
88				78 *Ceti* ξ²					
Jan. 6	...	2	20	59.00	...	82	8	46.9	R
Oct. 5	...		20	58.98	...		8	40.8	R
6	...		20	58.90	...		8	48.3	M
Nov. 3	...		20	59.12	...		8	46.5	M
27	...		20	58.99	...		8	49.2	R
29	...		20	59.09	...		8	48.2	R
30	...		20	59.06	...		8	46.7	M
Dec. 19	...		20	59.02	...		8	49.6	R
20	...		20	58.98	3		8	48.7	R
25	...		20	59.02	...		8	49.5	R
30	...		20	59.04	...		8	46.5	R
89				λ *Horologii*					
Dec. 8	6.1	2	21	7.71	...	150	55	4.6	M
11	6.0		21	7.98	5		55	5.4	M
90				*Anon.*					
Nov. 28	9.2	2	24	16.73	5	182	26	31.9	R
91				*Anon.*					
Dec. 12	8.4	2	24	29.96	...	167	2	24.8	M
13	8.0		24	30.08	...		2	29.6	M
14	8.1		24	29.96	..		2	29.9	M
92				*Lacaille* 782.					
Nov. 16	7.0	2	25	11.71	...	148	34	39.4	M
Dec. 5	7.0		25	11.86			34	40.9	M
93				*Anon.*					
Dec 16	9.8	2	26	12.65	5	147	37	18.9	R
94				*Anon.*					
Nov 17	10.1	2	29	15.10		86	0	20.7	R
27	10.3		29	15.28	2		0	14.0	R
Dec 13	10.1		29	15.20	4		0	19.9	R

Separate Results of Madras Meridian Circle Observations in 1865.

Number and Date.	Magnitude.	Mean Right Ascension 1865. h.	m.	s.	No. of Wires.	Mean Polar Distance. 1865. °	′	″	Observer.
95				*W. B. N. II 556.*					
Nov. 11	8·8	2	33	13·47	...	74	58	40·7	M
Dec. 8	8·0		33	13·53	...		58	42·5	M
11	7·9		33	13·30	...		58	43·4	M
96				*Anon.*					
Dec. 13	8·8	2	34	2·16	3	74	56	17·2	M
97				*Lacaille* 840 (2nd).					
Jan. 7	8·2	2	36	7·18	...	150	8	58·7	M
Dec. 6	7·0		36	7·15	...		8	50·1	M
12	7·9		36	7·11	...		8	59·5	M
19	8·0		36	7·11	6		9	0·1	R
98				86 *Ceti* γ					
Nov. 28	...	2	36	18·47	...	87	20	8·1	R
Dec. 2	...		36	18·36	...		20	7·9	M
14	...		36	18·46	5		20	7·3	M
15	...		36	18·51	...		20	8·1	R
16	...		36	18·51	...		20	6·7	R
19	...		36	18·41	...		20	7·1	R
20	...		36	18·30	—		20	7·8	R
21	...		36	18·48	...		20	8·1	R
22	...		36	18·40	...		20	7·9	R
99				*Lacaille* 868.					
Nov. 16	8·0	2	38	36·50	...	117	12	55·3	M
100				*W. B. N.* II 676.					
Dec. 1	7·9	2	40	11·48	...	75	30	8·4	M
101				*Anon.*					
Nov. 4	8·0	2	44	10·61	5	148	0	32·9	M
102				*W. B. N.* II 733.					
Jan. 4	9·3	2	43	22·30	2	76	1	57·3	M
Dec. 8	9·0		43	22·42	...		1	57·2	M
13	8·2		43	22·33			1	56·4	M
103				*Anon.*					
Dec. 15	9·2	2	44	30·97	...	148	13	37·2	R
104				*Anon.*					
Dec. 19	9·3	2	45	16·92	..	78	27	49·0	R
105				*Taylor* 960.					
Dec. 16	7·2	2	46	40·82	...	74	4	13·0	R
106				*Rumker* 87.					
Jan. 7	6·0	2	46	2·36	...	153	22	3·0	M
Nov. 13	6·6		46	2·09	...		22	5·6	M
Dec. 5	5·0		46	2·16	5		22	6·1	M
12	5·9		46	2·25	...		22	4·0	M
107				*Lalande* 5330.					
Oct. 6	8·0	2	47	45·27	...	74	14	26·2	M
Nov. 15	8·0		47	45·46	...		14	26·1	M
Dec. 11	7·0		47	45·30	...		14	25·7	M
108				*Lacaille* 941.					
Nov. 14	6·3	2	50	25·11	...	140	25	52·4	M
Dec. 14	6·5		50	25·25	...		25	51·8	M
109				*Anon.*					
Jan. 9	9·0	2	52	38·47	...	130	16	51·3	M
110				*Anon.*					
Dec. 18	8·5	2	58	17·19	5	148	41	13·0	R
111				92 *Ceti* α, *Menkar.*					
Nov. 17		2	55	13·42	6	86	36	32·7	R
Dec. 15	...		55	13·41	..		36	32·9	R
16	...		55	13·41	..		36	33·0	R
18	...		55	13·39			36	30·9	R
19	.		55	13·42			36	32·6	R
20	..		55	13·37	4		36	32·6	M
22	...		55	13·30			36	33·5	R

Separate Results of Madras Meridian Circle Observations in 1865.

Number and Date.	Magnitude.	Mean Right Ascension 1865. h	m	s	No. of Wires.	Mean Polar Distance 1865. °	′	″	Observer.
112				*25 Persei ρ Var. 2*					
Jan 10	...	2	56	32·14	4	51	41	7·6	M
Nov. 24	...		56	31·07	...		41	10·1	R
113				*Taylor* 1037.					
Dec 5	8·3	2	56	53·45	5	130	21	21·2	M
114				*Taylor* 1047.					
Jan. 7	6·5	2	30	53·15		151	19	38·1	M
115				*Taylor* 1057.					
Nov 4	7·0	3	0	47·82	...	151	22	6·7	M
Dec 12	7·0		0	47·82	5		22	6·7	M
116				*R. P. L.* 33.					
Dec. 11	...	3	0	56·05	5	5	34	37·7	M
21	...		0	54·70	5		34	37·0	R
				R. P. L. 33—*s.p.*					
June 19	...	3	6	54·71	3	5	34	38·9	R
117				57 *Arietis* δ					
Oct. 6	—	3	3	51·75	—	70	47	10·8	M
Nov 30	...		3	51·73	...		47	11·9	M
Dec 1	...		3	51·76	..		47	11·5	M
8	...		3	51·84	...		47	11·9	M
13	.		3	51·66	...		47	11·1	M
15	...		3	51·71	...		47	12·4	R
16	...		3	51·64	...		47	12·4	R
18	...		3	51·77	...		47	12·1	R
118				*Lacaille* 1007.					
Jan 9	7·5	3	4	49·30		132	14	14·2	M
Dec 11	7·0		4	49·58	...		14	13·2	M
119				*Taylor* 1127.					
Jan 11	8·1	3	11	40·64	4	131	46	6·7	M
Dec 11	7·9		11	40·57	2		46	7·9	M
12	7·4		11	40·85	...		46	6·1	M
14	8·5		11	40·49	5		46	8·4	R

Number and Date.	Magnitude.	Mean Right Ascension 1865. h	m	s	No. of Wires.	Mean Polar Distance 1865. °	′	″	Observer.
120				*Anon.*					
Nov. 16	9·0	3	12	25·72	...	120	57	59·6	R
Dec. 15	8·5		12	26·00	...		57	59·2	R
121				*Anon.*					
Dec. 13	8·5	3	12	43·89	...	120	50	3·9	M
122				23 *Persei* α					
Dec. 16	...	3	14	41·76	...	40	37	21·6	R
123				3ᵃ *Reticule.*					
Jan. 4	6·0	3	15	17·37	3	153	1	34·7	M
124				*Anon.*					
Dec. 14	8·9	3	20	35·98	3	160	33	18·8	R
125				*Anon.*					
Dec. 16	9·7	3	20	46·35	4	51	47	36·6	R
126				*R. P. L.* 34.					
Dec. 21	...	3	22	34·18	5	8	47	12·9	R
127				5 *Tauri f*					
Jan. 6	..	3	23	25·20	...	77	31	42·9	R
7	...		23	25·41	...		31	40·0	M
Oct. 6	...		23	25·36	.		31	40·7	M
7	...		23	25·44	...		31	42·6	M
128				*Anon.*					
Nov. 16	9·2	3	25	36·06	.	87	33	4·6	R
28	10·0		25	38·18	3		33	4·1	R
Dec. 13	9·0		25	37·07	...		33	5·3	M
129				*Lacaille* 1143.					
Nov. 29	6·0	3	27	1·78	..	158	34	53·4	R
Dec 14	5·0		27	2·13			34	58·5	M

Separate Results of Madras Meridian Circle Observations in 1865.

Number and Date.	Magnitude.	Mean Right Ascension [illegible]. h.	m.	s.	No. of Wires.	Mean Polar Distance [illegible]. °	′	″	Observer.
130				*Lacaille* 1150.					
Nov. 11	8·9	3	28	28·71		152	28	5·0	M
17	8·8		28	28·71	3		28	6·1	R
Dec 16	7·0		28	28·53	5		28	5·8	R
131				*Lacaille* 1159.					
Jan. 4	6·9	3	30	17·66	3	151	28	34·5	M
132				*Taylor* 1256.					
Jan. 9	8·0	3	36	28·31	...	160	13	6·0	M
133				*Anon.*					
Dec. 16	9·7	3	36	52·30	5	152	36	11·7	R
134				*Anon.*					
Nov. 16	9·0	3	38	7·32	...	184	12	41·3	R
135				*Anon.*					
Dec. 18	..	3	39	[illegible]	...	66	30	[illegible]	R
136				25 *Tauri* η, *Alcyone.*					
Jan 6	..	3	39	27·78	...	66	18	56·8	M
Dec. 8	...		39	27·81	.		18	58·5	M
16	—		39	27·81	5		18	58·7	R
21	...		39	27·70	..		18	58·6	R
137				*Taylor* 1318.					
Jan 4	5·9	3	42	30·[illegible]	3	155	14	2·4	M
10	5·9		42	31·30	4		14	1·1	M
Nov. 17	...		42	31·08	...		14	58·5	R
138				*Anon.*					
Jan 11	8·7	3	44	38·79	..	144	31	30·2	M
Nov 27	[illegible]		44	38·[illegible]	..		33	30·8	R
139				34 *Eridani* γ[1]					
Jan. 4	...	3	51	48·80	...	103	53	42·9	M
6	...		51	48·81	...		53	41·9	M
7	...		51	48·91	...		53	41·7	M
9	...		51	48·87	..		53	42·5	M
Dec. 15	...		51	48·88	...		53	42·8	R
21	...		51	48·80	..		53	43·0	R
140				35 *Tauri* λ *Var.* 1.					
Nov. 4	5·0	3	53	12·17	...	77	53	28·4	M
141				*Anon.*					
Nov. 27	8·0	3	53	41·84	5	143	6	14·5	R
142				*Lacaille* 1327.					
Dec 2	...	3	54	19·50	5	138	51	22·3	M
143				*R. P. L.* 35.					
Jan. 10	...	3	55	10·00	3	4	49	34·4	M
144				*Lacaille* 1347.					
Jan. 11	8·0	3	58	8·19		140	2	28·6	M
Dec. 13	7·1		58	8·04	...		2	28·9	M
145				*Lalande* 7764.					
Jan. 6	8·2	4	3	31·31	..	71	43	43·0	M
Dec. 1	7·9		3	31·30	..		43	48·2	M
13	8·0		3	31·26	...		43	48·2	M
146				*Anon.*					
Jan. 11	9·2	4	5	1·38	4	160	5	21·9	M
147				38 *Eridani* ο[1]					
Jan. 4	...	4	5	16·63	...	97	11	33·0	M
7	...		5	16·86	..		11	32·3	M
9	..		5	16·67	..		11	33·1	M
Dec 9	..		5	16·80	...		11	32·1	M
13	.		5	16·57			11	32·3	M
14	—		5	16·44			11	32·4	M
14			5	16·64	.		11	[illegible]	R

Separate Results of Madras Meridian Circle Observations in 1865.

Number and Date	Magnitude	Mean Right Ascension 1865. h m s	No. of Wires	Mean Polar Distance 1865. ° ′ ″	Observer
148		*Anon.*			
Jan 10	8·3	4 9 [illegible]·62	...	129 13 43·6	M
149		*U Tauri Var. 7.*			
Jan. 17	9·8	4 13 57·17	3	70 30 30·1	R
150		*T Tauri Var. 6.*			
Jan 18	10·5	4 14 7·78	2	70 46 28·8	R
19	10·2	14 7·84	3	47 28·0	R
151		*α Reticuli.*			
Jan. 6	3·2	4 14 9·00	5	148 37 42·1	M
152		*Anon.*			
Jan. 28	9·6	4 15 42·08	.	129 29 49·4	M
153		*74 Tauri ε*			
Jan. 4	...	4 20 44·16	...	71 7 21·5	M
7	...	20 44·13	..	7 20·6	M
9	..	20 44·06	...	7 20·7	M
10	..	20 44·19	...	7 21·4	M
Oct. 7	...	20 44·13	...	7 20·7	M
Dec. 1	..	20 44·28	5	7 21·1	M
3	...	20 44·19	...	7 21·0	M
8	...	20 44·05	...	7 20·5	M
9	...	20 44·10	.	7 19·8	M
11	...	20 44·20	...	7 21·6	M
12	...	20 44·10	...	7 21·6	M
13	...	20 44·14	...	7 19·8	M
154		*R Tauri Var. 2.*			
Dec 18	8·3	4 20 53·67	...	80 8 20·1	R
155		*Taylor 1582.*			
Jan 6	6·8	4 23 13·41	5	151 32 46·1	M
Nov. 1	6·5	23 13·28	...	32 48·5	M
156		*Lacaille 1510.*			
Jan. 17	7·8	4 25 25·98	3	156 5 18·7	R
157		*87 Tauri α, Aldebaran.*			
Jan. 7	...	4 28 10·57	...	73 46 55·7	M
10	...	28 10·70	...	46 56·0	M
19	...	28 10·62	...	46 56·6	R
21	...	28 10·64	...	46 55·3	R
Oct 7	...	28 10·62	..	46 55·7	M
Dec. 1	...	28 10·80	...	46 56·4	M
2	...	28 10·42	...	46 57·9	M
11	...	28 10·89	...	46 56·3	M
158		*Anon.*			
Dec. 19	9·3	4 31 44·20	...	142 30 28·9	R
159		*W. B. N. IV. 693.*			
Jan. 4	9·0	4 32 29·88	...	68 27 28·7	M
6	9·0	32 29·67	...	27 28·5	M
9	9·0	32 29·55	...	27 28·2	M
11	9·8	32 29·72	...	27 28·3	M
23	9·3	32 30·30	...	27 28·6	R
26	9·5	32 29·86	...	27 21·6	R
27	8·6	32 29·76	...	27 28·7	R
160		*W. B. N. IV, 726.*			
Jan. 18	8·8	4 33 54·04	...	68 15 9·3	R
19	8·8	33 54·36	...	15 9·3	R
21	...	33 54·65	...	15 9·9	R
161		*94 Tauri τ*			
Feb. 4	...	4 34 8·77	—	67 18 19·6	M
Nov. 4	...	34 8·88	..	18 20·2	M
162		*Anon.*			
Jan. 17	9·0	4 30 32·34	.	129 37 38·0	R
Dec. 19	9·8	30 32·18	...	37 39·7	R

Separate Results of Madras Meridian Circle Observations in 1865.

Number and Date.	Magnitude.	Mean Right Ascension 1865. h.	m.	s.	No. of Wires.	Mean Polar Distance. 1865. °	′	″	Observer.
163				*Anon.*					
Jan 10	8.6	4	40	30.13	...	151	30	50.3	M
11	9.0		40	30.19	...		31	48.6	M
164				*κ Doradûs.*					
Dec. 13	6.0	4	42	19.46	...	140	58	55.5	M
165				*Lacaille* 1620.					
Jan 19	6.7	4	43	43.56	...	153	28	27.8	R
Dec. 22			43	43.08	...		28	26.6	R
166				*Lacaille* 1625.					
Feb. 6	8.1	4	45	0.32	3	140	1	48.6	M
167				*W. B. N. IV.* 995.					
Jan. 4	8.0	4	45	10.30	...	66	3	3.4	M
168				*Anon.*					
Dec. 19	9.3	4	45	30.74	...	120	21	38.0	R
169				*Anon.*					
Jan. 7	8.9	4	46	56.29	...	158	2	36.9	M
170				*W. B. N. IV.* 1018.					
Jan 6	8.0	4	46	2.19	..	66	13	33.0	M
171				3 *Aurigae* ι					
Jan 21	...	4	48	12.20	...	57	3	5.0	R
28			48	12.37	..		3	3.6	M
Dec. 11			48	12.18	3		3	6.8	M
172				*Anon.*					
Jan 23	10.3	4	51	30.60	...	88	8	37.3	R
173				*Anon*					
Jan 10	9.0	4	52	21.10	.	120	30	47.9	M
174				*Anon.*					
Jan. 9	9.6	4	52	11.91	4	160	37	48.8	M
175				*R Leporis Var.* 1.					
Jan. 25	8.0	4	53	27.72		105	0	42.6	R
176				*Taylor* 1797.					
Feb. 6	7.0	4	54	52.10	..	148	16	51.1	M
177				102 *Tauri* ι					
Feb. 4	...	4	55	1.81	4	63	34	23.2	M
Nov. 4	...		55	1.60	...		36	23.6	M
178				*Lacaille* 1697.					
Jan. 18	8.0	4	56	53.21	6	120	7	7.9	R
179				*Lacaille* 1705.					
Dec. 16	8.0	4	57	27.86	..	120	16	30.1	R
180				2 *Leporis* ε					
Jan. 17	...	4	59	44.81	...	112	33	16.7	R
20	...		59	44.84	...		33	16.3	R
21	...		59	44.85	...		33	17.8	R
24	..		59	44.89	...		33	18.0	R
181				*Anon.*					
Jan. 26	8.0	5	0	1.35	...	105	23	40.9	R
182				*Anon.*					
Jan 7	9.3	5	1	45.60	3	151	36	2.9	M
19	9.3		1	45.46	4		30	5.6	R
183				*Lacaille* 1739.					
Jan 6	8.2	5	2	33.67		148	37	48.5	M
184				*Lacaille* 1756.					
Jan 28	8.8	5	3	30.90		134	41	28.6	M
29	8.7		3	30.91	.		41	29.6	R

Separate Results of Madras Meridian Circle Observations in 1865.

Number and Date.	Magnitude.	Mean Right Ascension 1865. h.	m.	s.	No. of Wires.	Mean Polar Distance 1865. °	'	"	Observer.
185			*Anon.*						
Jan. 20	9.5	5	4	36.08	4	136	34	33.6	R
186			1757 *Lacaille.*						
Jan. 9	8.0	5	4	53.20	3	150	4	20.1	M
187			*Anon.*						
Jan. 31	9.0	5	6	4.10	5	131	46	35.4	R
188			13 *Aurigae*, α *Capella.*						
Jan. 25	...	5	6	48.14	...	44	8	36.6	R
189			*Anon.*						
Jan. 24	8.5	5	6	51.10	...	120	6	0.6	R
190			19 *Orionis* β, *Rigel.*						
Jan. 21	...	5	8	3.08	5	98	21	38.0	R
Dec. 11	...		8	3.06	...		21	37.6	M
191			*Anon.*						
Jan. 18	9.3	5	12	53.96	...	120	40	2.5	R
24	9.0		12	56.66	5		40	3.3	R
Dec. 19	9.4		12	53.71	5		40	3.9	R
192			*Anon.*						
Jan. 26	9.2	5	12	57.90	5	137	4	42.5	R
30	9.2		12	57.26	...		4	48.9	R
Feb. 3	8.9		12	57.73	3		4	41.8	M
193			*Anon.*						
Jan. 28	8.1	5	13	26.14	5	133	41	37.4	M
194			*Anon.*						
Jan. 19	9.0	5	14	50.83	...	133	30	18.7	R
195			*Anon.*						
Jan. 31	9.3	5	10	5.88	...	131	49	16.3	R
Feb. 4	8.8		16	3.57	...		49	16.1	M
196			*Anon.*						
Jan. 21	7.7	5	17	0.10	5	129	38	0.4	R
27	9.0		16	59.98	...		37	59.7	R
Dec. 19	8.0		17	0.08	5		37	59.5	R
197			112 *Tauri* β						
Jan. 11	...	5	17	45.56	...	61	30	37.6	M
17	...		17	46.51	...		30	39.7	R
20	...		17	45.49	...		30	38.1	R
Feb. 2	...		17	46.06	...		30	36.5	M
Dec. 22	...		17	46.47	...		30	35.6	R
198			*Taylor* 1984.						
Feb. 6	7.3	5	18	51.97	...	109	54	47.7	M
199			*Anon.*						
Jan. 10	8.0	5	19	5.64	3	148	14	16.5	M
200			114 *Tauri* ο						
Jan. 9	...	5	19	51.67	...	68	16	38.0	M
201			*Anon.*						
Jan. 26	7.5	5	21	46.32	5	157	12	30.4	R
202			*Anon.*						
Jan. 28	7.7	5	20	30.86	...	161	13	22.3	M
30	8.0		20	30.64	...		13	24.0	R
203			34 *Orionis* δ *Var* 1.						
Jan. 17	...	5	25	6.37	...	90	24	7.9	M
18	...		25	6.73	...		21	8.9	R
204			*Anon.*						
Jan. 19	8.9	5	25	13.65	...	130	35	29.4	R
Feb. 4	8.8		25	13.44	5		35	19.8	M
205			*Anon.*						
Feb. 11	9.0	5	28	35.36	5	165	51	16.6	M

Separate Results of Madras Meridian Circle Observations in 1865.

Number and Date.	Magnitude.	Mean Right Ascension 1865. h.	m.	s.	No. of Wires.	Mean Polar Distance 1865. °	′	″	Observer.
206				*11 Leporis α*					
Jan. 20	…	5	26	46·57	…	107	56	17·3	R
24	…		26	46·66	…		56	19·0	R
27	…		26	46·60	5		56	17·1	R
31	…		26	46·73	5		56	17·8	R
Feb. 2	…		26	46·58	…		56	19·1	M
207				*46 Orionis ε*					
Jan. 17	…	5	29	21·66	…	91	17	27·6	R
18	…		29	21·07	…		17	29·1	R
23	…		29	21·80	…		17	29·1	R
208				*123 Tauri ζ*					
Dec. 2	…	5	29	24·61	…	68	56	37·2	M
209				*Anon.*					
Jan. 26	9·5	5	29	47·16	5	126	22	0·6	R
Feb. 6	9·6		29	47·12	…		22	0·6	M
210				*Anon.*					
Jan. 19	8·8	5	31	30·92	…	123	48	11·7	R
211				*Lacaille* 1940.					
Jan. 29	6·1	5	30	13·77	…	154	19	2·7	M
212				*Anon.*					
Dec. 22	…	5	32	48·05	5	160	11	23·6	R
213				*Anon.*					
Jan. 19	9·1	5	33	44·01	4	123	41	15·5	R
214				*Anon.*					
Jan. 20	9·2	5	31	20·02		102	7	5·7	R
Feb. 9	8·7		31	20·01			7	6·9	M

Number and Date.	Magnitude.	Mean Right Ascension 1865. h.	m.	s.	No. of Wires.	Mean Polar Distance 1865. °	′	″	Observer.
215				*α Columbae.*					
Jan. 12	…	5	34	45·72	…	124	8	53·4	M
15	…		34	45·82	…		8	54·4	R
24	…		34	45·68	…		8	54·7	R
27	…		34	45·69	…		8	54·3	R
31	…		34	45·77	…		8	53·4	R
Feb. 8	…		34	45·73	…		8	53·3	M
9	…		34	45·69	…		8	52·8	M
10	…		34	45·74	3		8	53·6	M
216				*Lacaille* 1971.					
Feb. 15	7·8	5	36	22·26		140	11	30·2	M
217				*Anon.*					
Jan. 26	8·7	5	36	30·23	…	106	46	29·4	R
Feb. 4	7·0		36	30·17	…		46	27·9	M
218				*Anon.*					
Jan. 26	9·1	5	39	55·07	5	106	46	6·2	R
219				*Lacaille* 2010.					
Jan. 23	7·7	5	42	2·13	.	146	58	20·0	M
220				*Anon.*					
Jan. 19	8·5	5	46	26·71	5	130	6	36·3	R
23	9·2		46	26·74			6	35·6	R
24	9·3		46	26·71	…		6	36·4	R
221				*Taylor* 2184.					
Jan. 7	9·3	5	48	55·09	…	100	46	22·3	M
222				*Anon.*					
Jan. 30	9·3	5	41	46·14		127	10	16·3	R
223				*54 Orionis χ¹*					
Feb. 6		5	46	23·11		69	45	8·5	M
Dec. 2	.		46	23·30			45	11·9	M

Separate Results of Madras Meridian Circle Observations in 1865.

Number and Date.	Magnitude.	Mean Right Ascension 1865. h.	m.	s.	No. of Wires.	Mean Polar Distance 1865. °	′	″	Observer.
224				*Anon.*					
Jan. 20	9·6	5	47	3·19	...	135	40	30·2	R
225				*α Orionis Var. 2, Betelgeux.*					
Jan. 11	...	5	47	51·93	...	82	37	16·4	M
13	...		47	51·88	...		37	17·6	R
19	...		47	51·73	...		37	17·7	R
25	...		47	51·87	...		37	19·2	R
27	...		47	51·90	...		37	17·2	R
Feb. 2	...		47	51·83	...		37	16·7	M
3	...		47	51·87	...		37	16·2	M
9	...		47	51·88	...		37	16·7	M
10	...		47	51·86	...		37	17·5	M
226				*Anon.*					
Jan. 26	8·0	5	49	56·26	5	135	44	7·9	R
227				*Anon.*					
Jan. 30	9·8	5	50	13·60	5	137	10	13·5	R
228				*Lacaille* 2073.					
Feb. 14	7·9	5	50	32·11	...	137	12	33·1	M
229				*Anon.*					
Jan. 24	9·5	5	50	58·84	...	130	46	1·7	R
28	9·6		50	58·60	...		43	1·3	M
230				*R. P. L.* 43.					
Feb. 8	..	5	52	37·93	3	3	14	29·0	M
231				*Anon.*					
Jan. 23	8·8	5	53	3·51	5	130	31	57·4	R
232				*Lacaille* 2104.					
Jan. 19	7·0	5	54	22·74	5	143	26	33·9	R
233				*Anon.*					
Jan. 30	9·3	5	54	30·13	...	137	45	14·8	R
Feb. 4	9·6		54	30·25	...		45	15·7	M
234				*Anon.*					
Jan. 10	9·2	5	56	11·53	...	129	57	11·2	M
235				*Anon.*					
Jan. 26	9·3	5	57	1·11	5	136	1	0·0	R
236				*67 Orionis ν.*					
Jan. 11	...	5	59	51·92	...	75	13	8·1	M
19	...		59	51·87	...		13	8·4	R
23	...		59	51·98	...		13	7·6	R
24	...		59	51·87	...		13	8·3	R
25	...		59	51·83	...		13	8·0	R
28	...		59	51·86	...		13	9·2	M
Feb. 3	...		59	51·90	...		13	7·5	M
6	...		59	51·77	...		13	7·2	M
9	...		59	51·89	...		13	6·9	M
237				*Anon.*					
Jan. 30	7·7	6	0	52·53	...	137	28	30·9	R
Feb. 14	7·8		0	52·60	...		28	30·9	M
238				*Lalande* 11738.					
Dec. 22	8·3	6	3	37·68	...	77	58	35·9	R
239				*Anon.*					
Feb. 10	9·0	6	4	1·12	...	130	5	28·2	M
240				*Anon.*					
Jan. 26	8·8	6	4	35·21	5	136	49	40·6	R
Feb. 8	8·2		4	35·48	...		49	40·3	M
241				*Anon.*					
Dec. 21	9·0	6	5	50·00	..	77	51	30·9	R

Separate Results of Madras Meridian Circle Observations in 1865.

Number and Date.	Magnitude.	Mean Right Ascension 1866. h.	m.	s.	No. of Wires.	Mean Polar Distance 1866. °	′	″	Observer.
242			*7 Geminorum η Var. 6.*						
Jan. 9	...	6	6	48·72	...	67	27	28·9	M
10	...		6	48·86	...		27	29·1	M
243			*Anon.*						
Feb. 18	9·2	6	7	28·21	5	151	19	25·6	R
244			*Anon.*						
Jan. 30	9·0	6	7	42·21	...	137	6	25·5	R
Feb. 4	9·2		7	42·08	...		6	22·7	M
245			*Anon.*						
Feb. 17	9·2	6	8	35·15	...	155	4	30·0	R
246			*Anon.*						
Jan. 24	9·2	6	8	55·14	...	130	31	35·5	R
25	9·3		8	55·19	...		31	35·1	R
247			*Anon.*						
Jan. 26	9·4	6	11	48·80	...	152	1	50·0	R
248			*Anon.*						
Jan. 26	9·2	6	12	3·80	5	136	50	49·8	R
Feb. 14	9·2		12	3·49	...		50	50·2	M
249			*Lalande* 12120.						
Dec. 18	7·2	6	14	8·91	...	77	4	47·7	R
22	...		14	8·89	...		4	49·9	R
250			13 *Geminorum μ.*						
Jan. 9	...	6	14	47·54	...	67	25	16·5	M
10	...		14	47·60	...		25	15·6	M
23	...		14	47·58	...		25	13·2	R
25	...		14	47·65	...		25	16·7	R
30	...		14	47·52	...		25	16·4	R
Feb. 9	...		14	47·67	...		25	14·4	M
251			*Lalande* 12155.						
Dec. 21	7·2	6	15	1·74	...	77	22	3·1	R
252			*Lacaille* 2273.						
Jan. 7	7·9	6	17	3·36	5	163	36	25·3	M
253			*Anon.*						
Feb. 10	8·9	6	18	36·26	...	151	29	29·2	M
17	9·0		18	36·25	...		29	27·0	R
254			*Taylor* 2485.						
Feb. 8	7·0	6	18	34·26	...	151	16	10·4	M
18	7·6		18	34·20	4		16	12·0	R
255			*Lacaille* 2286.						
Feb. 20	7·0	6	19	48·38	...	133	45	49·3	R
256			*Anon.*						
Feb. 3	8·8	6	19	48·65	...	65	39	55·1	M
4	8·8		19	48·58	...		39	54·5	M
257			*α Argûs, Canopus.*						
Jan. 20	...	6	20	57·61	...	142	37	29·6	R
258			*Anon.*						
Jan. 13	7·2	6	21	19·96	...	128	51	32·4	R
Feb. 15	7·2		21	20·10	...		51	32·7	M
259			*Lacaille* 2312						
Feb. 14	6·8	6	23	9·61	...	133	36	37·7	M
260			*Lacaille* 2321.						
Mar. 1	7·9	6	23	29·97	3	123	39	49·7	M
261			*Lacaille* 2348.						
Mar. 3	7·7	6	26	32·44	...	132	3	45·6	M
4	7·8		26	32·38			3	45·3	M

Separate Results of Madras Meridian Circle Observations in 1865.

Number and Date.	Magnitude.	Mean Right Ascension 1865. h.	m.	s.	No. of Wires.	Mean Polar Distance 1865. °	′	″	Observer.
262				*Anon.*					
Jan. 13	9·0	6	27	30·33	...	128	46	0·0	R
Mar. 2	9·0		27	30·40	...		46	0·5	M
263				*Anon.*					
Feb. 20	8·1	6	28	30·98	...	151	10	2·6	R
264				*Anon.*					
Jan. 24	9·2	6	28	40·50	5	130	32	12·6	R
Feb. 17	8·9		28	40·38	...		32	11·3	R
265				*Taylor* 2589.					
Feb. 9	6·9	6	29	50·91	5	151	46	52·0	M
13	6·8		29	50·10	...		46	51·5	R
266				24 *Geminorum* γ					
Jan. 26	...	6	29	51·78	...	73	20	22·0	R
28	...		29	51·71	...		20	21·3	M
30	...		29	51·77	...		20	19·9	R
31	...		29	51·73	...		20	24·0	R
Feb. 6	...		29	51·82	...		20	24·5	M
267				*Anon.*					
Feb. 28	8·3	6	31	29·75	...	130	56	46·1	R
268				*Anon.*					
Feb. 10	8·9	6	33	39·84	...	132	27	8·1	M
269				*Anon.*					
Jan. 24	9·1	6	35	40·86	...	130	35	8·1	R
Feb. 14	9·0		35	40·42	...		34	6·9	M
270				*Anon.*					
Jan. 19	10·5	6	36	0·34	4	62	5	40·9	R
19	10·3		36	0·20	4		5	31·3	R
271				51 (*Hev.*) *Cephei.*					
Jan. 17	...	6	36	9·87	2	2	46	20·0	R
21	...		36	8·77	2		46	22·2	R
27	...		36	9·86	3		46	22·3	R
Feb. 16	...		36	9·20	5		46	22·3	R
22	...		36	8·83	3		46	20·9	R
				51 (*Hev.*) *Cephei, s.p.*					
Aug. 11	...	6	36	8·94	4	2	46	20·7	R
Sep. 1	...		36	9·45	3		46	18·6	R
272				*Anon.*					
Mar. 3	8·9	6	36	15·14	...	130	21	8·4	M
273				*Taylor* 2652.					
Feb. 15	6·8	6	36	36·84	...	151	24	55·3	M
274				31 *Geminorum* ξ					
Feb. 6	.	6	37	48·66	...	76	57	48·6	M
275				*Anon.*					
Feb. 17	9·5	6	37	38·72	...	158	20	38·7	R
276				*Lacaille* 2451.					
Mar. 4	8·3	6	38	11·15	...	153	57	46·4	M
277				*Taylor* 2667.					
Jan. 31	8·3	6	38	28·73	5	149	20	44·7	R
278				*Anon.*					
Jan. 13	9·5	6	39	9·48	...	131	3	39·2	R
279				9 *Canis Majoris* α, *Sirius.*					
Feb. 3	...	6	39	11·81	...	106	32	1·8	M
23	...		39	11·76	...		32	1·7	R

Separate Results of Madras Meridian Circle Observations in 1865.

Number and Date.	Magnitude.	Mean Right Ascension. 1865. h. m. s.	No. of Wires.	Mean Polar Distance 1865. ° ′ ″	Observer.
280		*Anon.*			
Feb. 8	9.2	6 40 29.86	...	134 13 36.9	M
25	9.3	40 29.07	..	13 36.5	R
281		*Anon.*			
Mar. 7	9.1	6 40 32.86	5	131 59 54.8	M
282		*Anon.*			
Feb. 20	8.3	6 42 25.49	4	130 56 50.2	R
283		*Anon.*			
Jan. 25	8.7	6 42 39.04	5	130 36 25.8	R
Feb. 4	7.7	42 38.99	...	36 26.0	M
284		*Anon.*			
Feb. 20	9.2	6 42 30.06	4	130 58 8.6	R
285		*Anon.*			
Feb. 10	7.7	6 43 42.96	...	128 30 28.3	M
17	7.7	43 42.78	5	30 28.0	M
21	8.0	43 42.66	6	30 28.3	R
286		*Anon.*			
Feb. 21	9.3	6 44 51.34	4	128 30 31.5	R
287		*Anon.*			
Feb. 28	9.4	6 45 0.13	...	106 32 48.5	R
288		*α Pictoris.*			
Feb. 11	5.5	6 46 49.48	4	151 47 50.8	M
289		*Lacaille* 2516.			
Jan. 13	8.8	6 46 25.49	...	130 31 46.7	R
25	8.8	46 25.46	5	31 42.0	R
290		*Anon.*			
Feb. 25	9.0	6 44 37.91	5	128 44 29.5	R
291		*Anon.*			
Feb. 20	9.3	6 48 32.61	5	130 10 19.4	R
292		*Anon.*			
Feb. 25	9.0	6 48 46.41	4	128 46 7.2	R
293		*Lacaille* 2538.			
Jan. 21	8.0	6 51 20.28	...	114 47 30.1	R
Feb. 15	6.9	51 20.30	...	47 36.5	M
294		*Anon.*			
Mar. 1	9.9	6 52 47.90	...	152 51 36.2	M
295		21 *Canis Majoris* ε			
Jan. 12	...	6 53 19.22	...	118 47 27.2	M
23	...	53 19.16	...	47 27.1	R
26	...	53 19.19	...	47 27.3	R
Feb. 4	...	53 19.27	...	47 26.6	M
8	...	53 19.30	...	47 26.0	M
16	...	53 19.26	...	47 27.5	R
296		*Anon.*			
Feb. 22	8.0	6 55 37.35	5	120 37 39.2	R
28	8.0	55 37.45	...	38 0.0	M
297		43 *Geminorum* ζ *Var.* 1.			
Jan. 11	...	6 56 0.07	...	69 11 7.5	M
298		*Anon.*			
Feb. 25	9.0	6 56 28.20	...	130 17 28.5	R
Mar. 9	8.9	56 28.18	5	17 28.0	M
299		23 *Canis Majoris* γ			
Jan. 19	...	6 57 39.16	5	105 29 11.1	R
25	...	57 39.02	...	29 11.6	R
26	...	57 39.05	...	29 11.7	R
30	...	57 39.14	...	29 11.1	R
Feb. 4	...	57 39.06	.	29 10.7	M

Separate Results of Madras Meridian Circle Observations in 1865.

Number and Date.	Magnitude.	Mean Right Ascension 1865. h. m. s.	No. of Wires.	Mean Polar Distance 1865. ° ′ ″	Observer.
Feb. 8	...	6 57 30·10	5	105 36 10·8	M
10	...	57 30·02	...	36 10·9	M
11	...	57 30·14	...	36 10·3	M
16	...	57 30·10	...	36 10·4	R
17	...	57 30·08	...	36 12·1	R
800 *Taylor* 2840.					
Jan. 6	7·0	6 59 3·58	...	105 56 58·4	M
Feb. 21	8·0	59 3·55	...	56 55·7	R
Mar. 3	8·0	59 3·70	...	56 56·8	M
801 *R Geminorum Var.* 2.					
Jan 23	8·8	6 59 13·31	5	67 5 31·8	R
802 *Anon.*					
Mar. 10	7·5	7 1 1·56	5	144 16 0·8	M
803 *R Canis Minoris Var.* 1.					
Feb. 20	9·7	7 1 17·19	5	79 46 56·6	R
804 *Anon.*					
Feb. 22	8·8	7 1 31·41	5	120 30 17·6	R
805 *Anon.*					
Feb 15	9·0	7 2 55·86	5	141 34 10·6	M
16	9·2	2 55·78	...	34 11·0	R
806 *Anon.*					
Feb. 25	8·7	7 3 25·79	...	130 14 36·4	R
Mar. 1	8·6	3 25·65	...	14 36·0	M
807 *Anon.*					
Mar 4	9·1	7 3 2·70	...	158 32 14·5	M
808 *Lacaille* 2678.					
Jan 18	9·0	7 6 11·70	5	109 9 21·2	R
809 *Taylor* 2928.					
Feb. 21	8·0	7 7 17·23	...	100 21 25·4	R
Mar. 11	8·2	7 17·25	...	21 25·2	M
810 *Anon.*					
Feb. 20	8·0	7 8 0·40	...	140 46 7·9	R
811 *Anon.*					
Feb. 10	9·9	7 8 10·02	...	152 5 8·2	M
812 *Anon.*					
Feb. 25	9·0	7 8 35·76	...	130 17 24·3	R
813 *Lacaille* 2696.					
Feb. 17	7·8	7 9 28·61	...	140 35 53·8	R
814 *Taylor* 2040.					
Feb. 18	7·7	7 9 30·23	...	120 57 48·7	R
23	7·5	9 30·24	...	57 49·4	R
815 *Anon.*					
Feb. 20	9·5	7 9 49·18	4	109 18 25·9	M
816 55 *Geminorum* δ					
Jan. 10	4	7 12 3·30	4	67 46 22·3	M
11	...	12 3·51	...	46 24·8	M
13	...	12 3·38	...	46 21·8	R
28	...	12 3·61	...	46 23·3	M
30	...	12 3·32	...	46 21·7	R
Feb. 2	...	12 3·49	...	46 21·5	M
6	...	12 3·56	...	46 21·6	M
8	...	12 3·48	...	46 24·1	M
11	...	12 3·40	...	46 20·7	M
15	...	12 3·38	...	46 20·7	M
16	...	12 3·46	...	46 21·5	R
817 *Anon.*					
Mar. 3	8·3	7 12 36·60	...	133 49 2·1	M

Separate Results of Madras Meridian Circle Observations in 1865.

Number and Date.	Magnitude.	Mean Right Ascension 1865. h. m. s.	No. of Wires.	Mean Polar Distance. 1865. ° ′ ″	Observer.
818		*Taylor* 3005.			
Feb. 20	7·7	7 13 [illegible]	...	140 1 0·0	B
819		*R. P. L.* 45			
Mar. 2	...	7 16 32·78	2	0 30 11·6	M
820		*Lacaille* 2805.			
Jan. 21	9·0	7 17 22·15	3	158 8 13·3	B
821		*Anon.*			
Feb. 25	9·3	7 17 38·99	...	129 46 16·1	B
Mar. 6	9·3	17 38·64	...	46 17·7	M
822		*Anon.*			
Feb. 16	9·8	7 19 40·07	5	69 15 36·1	B
823		63 *Geminorum.*			
Mar. 7	5·3	7 19 48·29	...	63 16 55·2	M
824		*Anon.*			
Feb. 10	9·0	7 19 30·11	...	152 35 41·7	B
Mar. 4	9·3	19 30·07	...	35 42·4	M
825		*Taylor* 3054.			
Jan. 31	7·3	7 20 3·80	...	161 41 36·2	B
826		*Anon.*			
Feb. 24	8·0	7 21 36·99	..	131 50 33·5	B
827		*Anon.*			
Feb. 25	8·8	7 23 29·61	...	129 45 34·7	B
28	8·3	23 29·65	...	45 35·4	M
828		*Anon.*			
Feb. 16	9·3	7 23 27·00	...	42 0 51·0	B
829		*Anon.*			
Feb. 22	8·0	7 23 53·73	...	129 10 12·8	B
830		*Anon.*			
Mar. 3	9·4	7 24 9·78	...	152 47 21·6	M
831		*Anon.*			
Jan. 26	9·0	7 24 48·12	5	130 0 38·9	B
832		*Anon.*			
Jan. 17	9·0	7 24 48·87	2	121 0 39·3	B
Feb. 3	9·0	24 48·77	...	0 37·9	M
833		*Anon.*			
Jan. 21	9·0	7 25 0·77	...	158 8 56·6	B
834		*S Canis Minoris, Var.* 2.			
Jan. 13	...	7 25 28·51	...	81 23 49·2	B
19	9·1	25 28·54	...	23 49·3	B
835		*Anon.*			
Mar. 9	8·0	7 25 39·60	3	150 16 12·7	M
836		66 *Geminorum* α², *Castor.*			
Jan. 14	...	7 25 30·90	...	57 49 9·1	M
Feb. 11	...	25 30·61	...	49 9·0	M
15	...	25 30·90	...	49 7·9	M
20	...	25 30·88	...	49 8·6	B
23	...	25 30·90	...	49 8·6	B
Mar. 1	...	25 30·92	...	49 10·3	M
837		*Anon.*			
Mar. 6	9·6	7 26 39·22	5	129 46 7·9	M
838		*Anon.*			
Jan. 27	9·2	7 27 14·68	...	155 10 53·4	B

Separate Results of Madras Meridian Circle Observations in 1865.

Number and Date.	Magnitude.	Mean Right Ascension 1865. h.	m.	s.	No. of Wires.	Mean Polar Distance 1865. °	′	″	Observer.
839				*Anon.*					
Feb. 25	8·0	7	27	58·21	...	120	42	51·9	R
840				*Anon.*					
Feb. 21	9·5	7	30	42·48	...	136	42	28·1	R
841				*Anon.*					
Feb. 22	9·0	7	31	6·87	6	131	10	41·8	R
Mar. 4	8·9		31	6·84	...		10	44·7	M
842				*Anon.*					
Feb. 24	8·1	7	31	56·62	...	120	41	0·7	M
Mar. 3	8·8		31	56·34	...		41	1·2	M
843				10 *Canis Minoris* α, *Procyon.*					
Jan. 12		7	32	14·11	...	84	25	56·3	M
13	...		32	13·96	...		25	56·2	M
14	...		32	14·06	...		25	56·6	M
Feb. 4	...		32	14·09	...		25	55·4	M
14	...		32	14·07	...		25	54·2	M
15	..		32	13·98	...		25	55·5	M
16	...		32	13·98	...		25	55·4	R
18	...		32	14·02	...		25	56·7	R
23	...		32	14·11	...		25	58·0	R
Mar. 1	...		32	14·10	3		25	56·1	M
844				*Lacaille* 2898.					
Feb. 20	7·0	7	33	45·98	...	121	40	34·5	R
845				*Anon.*					
Jan. 27	9·6	7	33	6·30	4	153	11	46·7	R
Mar. 9	9·1		33	6·61	...		11	48·7	M
846				*Anon.*					
Jan. 24	10·5	7	34	34·40	4	68	25	7·5	R
26	10·3		34	34·29	5		25	5·7	R
847				*S Geminorum Var.* 8					
Jan. 31	10·5	7	34	37·48	5	66	12	30·3	R
Feb. 17	10·8		34	37·60	4		12	31·2	R
848				*Anon.*					
Feb. 23	8·5	7	35	6·98	...	120	37	30·3	R
Mar. 6	8·1		35	6·71	5		38	0·1	M
849				*Anon.*					
Mar. 10	8·9	7	36	30·80	...	152	50	46·6	M
850				*Taylor* 3195					
Mar. 11	8·0	7	36	34·64	...	100	19	16·8	M
851				*Anon.*					
Feb. 26	7·3	7	36	46·31	5	120	57	25·5	R
Mar. 2	7·7		36	46·26	...		57	26·0	M
852				78 *Geminorum* β, *Pollux.*					
Jan. 12	...	7	37	3·04	...	61	39	5·9	M
14	...		37	3·06	..		39	4·2	M
Feb. 3	...		37	3·04	...		39	3·8	M
4	...		37	3·06	...		39	4·5	M
15	...		37	3·22	...		39	3·3	M
18	...		37	3·02	...		39	4·2	R
23	...		37	3·08	...		39	4·6	R
24	...		37	3·02	...		39	3·6	R
Mar. 1	...		37	3·06			39	3·4	M
853				*Anon.*					
Jan. 21	10·3	7	37	24·50	3	69	39	31·7	R
854				*Anon.*					
Feb. 22	8·8	7	37	38·38	...	150	56	11·8	R
855				*Anon.*					
Feb. 21	7·7	7	37	46·71	5	123	38	1·6	R

Separate Results of Madras Meridian Circle Observations in 1865.

Number and Date.	Magnitude.	Mean Right Ascension 1865. h. m. s.	No. of Wires.	Mean Polar Distance 1865. ° ′ ″	Observer.
856		*Anon.*			
Mar. 4	7·0	7 39 7·00	...	131 8 46·0	M
857		*Anon.*			
Jan. 13	8·3	7 41 19·30	...	151 34 39·6	R
Mar. 3	8·7	41 19·15	...	34 39·8	M
858		*Anon.*			
Feb. 24	8·5	7 42 4·81	...	153 4 30·7	M
859		*Lacaille* 3034.			
Feb. 19	8·0	7 44 5·60	...	153 51 48·5	R
860		*Anon.*			
Feb. 22	9·1	7 44 12·06	...	130 56 9·1	R
Mar. 6	8·9	44 11·98	...	56 10·7	M
861		*Brisbane* 1791.			
Feb. 20	7·8	7 44 19·90	6	144 21 46·4	R
862		*Anon.*			
Feb. 20	9·3	7 46 30·77	6	144 22 35·1	R
863		*Anon.*			
Mar. 4	8·5	7 47 6·87	...	153 30 56·3	M
864		*Anon.*			
Jan. 19	10·7	7 47 13·22	3	67 48 37·7	R
865		*Taylor* 3310.			
Feb. 28	9·0	7 49 30·67	3	149 19 2·6	R
866		*Anon.*			
Mar. 2	9·6	7 49 44·65	...	129 51 38·5	M
867		1 *Cancri.*			
Feb. 8	...	7 49 19·81	...	73 31 7·3	M
868		*Anon.*			
Mar. 11	7·1	7 49 30·26	...	152 36 4·5	M
869		*Anon.*			
Feb. 24	8·2	7 49 53·49	...	149 17 31·1	R
870		*Anon.*			
Feb. 21	7·7	7 49 56·67	...	149 8 37·3	R
871		*Anon.*			
Mar. 3	9·3	7 51 39·78	...	151 37 3·7	M
872		*Anon.*			
Feb. 3	8·0	7 52 1·07	4	145 22 34·7	M
Mar. 1	8·2	52 2·01	4	22 33·4	M
873		0 *Cancri.*			
Jan. 11	...	7 56 13·32	...	61 49 49·4	M
27	...	56 13·38	...	49 51·6	R
Feb. 18	...	56 13·42	...	49 49·6	R
23	...	56 13·31	...	49 50·1	R
25	...	56 13·26	...	49 49·9	R
28	...	56 13·36	...	49 49·3	M
Mar. 10	...	56 13·27	...	49 49·0	M
16	...	56 13·37	...	49 49·3	R
874		*Brisbane* 1855.			
Mar. 6	6·8	7 56 30·67	...	152 36 37·9	M
875		*Anon.*			
Mar. 7	9·2	7 57 40·64	...	154 34 20·2	M
876		*Lacaille* 3164.			
Mar. 3	8·8	7 58 27·38	...	133 11 39·5	M
9	8·7	58 27·73	5	11 [illegible]	M

Separate Results of Madras Meridian Circle Observations in 1865.

Number and Date.	Magnitude.	Mean Right Ascension 1865. h. m. s.	No. of Wires.	Mean Polar Distance 1865. ° ' "	Observer.
377 *Anon.*					
Jan. 19	10·7	8 0 19·62	3	78 29 58·5	R
21	10·7	0 18·71	3	29 51·6	R
378 *Anon.*					
Mar. 4	8·9	8 1 28·21	...	130 31 32·8	M
379 *Lacaille* 3174.					
Mar. 1	7·0	8 1 36·22	...	135 33 5·6	M
11	8·0	1 36·39	...	33 6·5	M
380 15 *Argûs.*					
Feb. 21	...	8 1 47·67	...	113 55 2·9	R
24	...	1 47·73	...	55 2·1	R
25	...	1 47·81	...	55 2·6	R
Mar. 10	...	1 47·75	...	55 1·3	M
16	...	1 47·76	...	55 2·1	R
18	...	1 47·69	..	56 2·3	R
381 *Anon.*					
Feb. 28	9·0	8 2 14·12	...	128 30 37·1	R
382 16 *Cancri* ζ					
Jan. 11	5·5	8 4 28·08	...	71 56 54·8	M
12	5·5	4 28·07	...	56 54·4	M
383 *Anon.*					
Mar. 2	9·7	8 4 38·28	...	134 40 44·2	M
384 *Anon.*					
Feb. 28	9·6	8 5 39·34	...	129 38 57·1	R
385 *Anon.*					
Mar. 20	8·0	8 6 5·07	3	128 40 48·6	R
386 *Anon.*					
Feb. 23	9·3	8 8 36·24	...	128 29 34·4	R
Mar. 20	9·3	8 36·09	5	29 35·6	R
387 *R Cancri Var.* 1.					
Jan. 26	7·5	8 9 7·06	—	77 51 44·8	R
Feb. 17	7·7	9 7·08	...	51 48·7	R
388 *Anon.*					
Jan. 21	9·2	8 9 27·79	5	74 16 13·5	R
389 *W. B. N. VIII.* 178.					
Jan. 24	9·6	8 9 38·75	4	74 16 26·2	R
390 *Anon.*					
Mar. 3	9·6	8 10 10·04	...	150 32 40·8	M
4	9·1	10 10·01	5	32 40·9	M
391 *Anon.*					
Mar. 6	9·8	8 10 27·32	...	151 21 29·7	M
392 *Anon.*					
Feb. 15	8·9	8 11 23·28	...	152 4 41·0	M
24	8·7	11 23·37	5	4 42·9	R
393 *Anon.*					
Mar. 11	8·6	8 13 22·08	...	131 41 24·8	M
394 *Anon.*					
Mar. 18	9·5	8 14 31·58	3	164 5 17·7	R
395 *Anon.*					
Mar. 9	8·1	8 14 52·01	...	148 17 7·0	M
23	8·2	14 52·07	..	17 6·0	R
396 *Lacaille* 3297.					
Jan. 31	8·9	8 15 1·97	3	138 30 1·4	R
Mar. 1	8·4	15 1·94	...	30 2·3	M

Separate Results of Madras Meridian Circle Observations in 1865.

Number and Date.	Magnitude.	Mean Right Ascension 1865. h.	m.	s.	No. of Wires.	Mean Polar Distance 1865. °	′	″	Observer.
397				*20 Cancri d¹*					
Mar. 7	...	8	16	37·88	...	71	14	14·0	M
398				*Anon.*					
Feb. 22	9·5	8	16	20·01	...	77	31	50·3	R
399				*Anon.*					
Feb. 18	8·8	8	17	26·74	6	77	49	10·5	R
400				*Anon.*					
Mar. 4	9·7	8	17	30·45	5	154	22	53·9	M
22	...		17	30·62	3		22	51·0	R
401				*Anon.*					
Jan 26	9·3	8	18	38·70	...	70	2	0·8	R
29	9·8		18	39·86	4		2	11·7	R
402				*Anon.*					
Mar. 3	9·7	8	18	47·77	...	151	0	56·2	M
21	9·6		18	48·10	5		0	55·5	R
403				*Taylor 3599.*					
Mar. 10	8·0	8	20	19·26	5	144	52	57·1	M
404				*W. B. N. VIII. 459.*					
Jan. 31	8·9	8	20	29·45	5	74	27	31·4	R
405				*Anon.*					
Jan 27	9·0	8	21	5·05	5	168	13	20·1	R
406				*29 Cancri.*					
Jan. 11	6·0	8	21	5·15	...	75	20	38·2	M
12	6·0		21	5·20	...		20	41·7	M
Mar 7	6·2		21	5·17	...		20	41·7	M
407				*Anon.*					
Mar 11	8·0	8	21	32·27	5	131	41	47·0	M
23	8·0		21	32·29	.		41	46·6	R
408				*Taylor 3620.*					
Feb. 18	7·0	1	23	12·99	...	120	47	49·7	R
409				*Anon.*					
Mar 24	8·3	8	23	24·49	...	123	33	46·5	R
410				*38 Cancri η*					
Feb. 17	...	8	24	53·85	...	69	6	11·9	R
21	...		24	53·91	5		6	11·7	R
22	...		24	53·87	...		6	11·3	R
24	...		24	53·98	...		6	10·4	R
27	...		24	53·87	...		6	11·5	R
28	...		24	53·85	...		6	10·2	R
Mar. 2	...		24	53·93	...		6	10·2	M
6	...		24	53·76	...		6	11·2	M
16	...		24	53·80	..		6	11·3	R
17	...		24	53·77	...		6	11·0	R
411				*Taylor 3651.*					
Mar 25	7·8	8	25	41·92	5	130	3	22·7	R
412				*Taylor 3652.*					
Mar. 23	8·0	8	25	46·10	3	130	2	52·1	R
413				*Anon.*					
Jan. 30	9·9	8	25	48·30	...	73	19	20·6	R
Mar. 20	...		25	42·26	...		19	19·4	R
414				*W. B. N. VIII. 654.*					
Mar. 11	8·7	8	29	4·22	5	70	29	6·6	M
16	8·7		29	4·33	5		29	7·4	R
415				*Lacaille 3430.*					
Mar 9	7·9	8	29	14·00	5	161	26	53·7	M
22	8·0		29	13·93	3		26	51·4	R
Apl. 1	7·9		29	13·73	4		26	52·7	M

Separate Results of Madras Meridian Circle Observations in 1865.

Number and Date.	Magnitude.	Mean Right Ascension 1865. h.	m.	s.	No. of Wires.	Mean Polar Distance 1865. °	'	"	Observer.
416		*W. B. N. VIII.* 699.							
Mar. 18	9·6	8	29	34·37	4	70	39	49·9	n
417		*Anon.*							
Jan. 31	8·6	8	31	14·75	5	129	46	37·0	n
418		*Taylor* 3710.							
Feb. 17	8·0	8	31	36·61	5	141	21	17·5	n
419		*Anon.*							
Mar. 27	9·6	8	33	11·82	..	129	28	49·3	n
420		*Anon.*							
Mar. 4	8·6	8	34	9·35	...	154	30	37·7	M
21	9·6		34	9·41	...		29	31·5	n
Apl. 3	8·8		34	9·28	4		30	38·3	M
421		*Anon.*							
Jan 31	7·3	8	34	41·72	...	129	46	36·1	n
422		*Lacaille* 3401.							
Jan. 12	7·9	8	36	2·46	5	162	22	2·7	M
423		*b Velorum.*							
Mar. 1	5·3	8	36	8·73	3	136	10	12·7	M
22	5·5		36	8·92	...		10	12·3	n
424		*Taylor* 3767.							
Feb 18	7·2	8	36	31·27	...	140	50	28·7	n
425		47 *Cancri* δ							
Feb. 8	...	8	37	0·41	4	71	21	6·7	M
9	...		37	0·31	...		21	7·3	M
426		*Anon.*							
Mar 19	9·2	8	37	19·16	5	136	8	45·3	n
25	9·3		37	19·16	5		8	46·3	n
427		*Anon.*							
Mar. 6	8·2	8	37	38·45	4	106	5	47·3	M
428		50 *Cancri* A[2]							
Apl. 5	6·2	8	39	38·00	..	77	28	49·7	M
429		11 *Hydrae*, ε							
Feb. 13	...	8	39	37·47	...	83	5	19·6	M
17	...		39	37·50	...		5	18·4	n
20	...		39	37·44	...		5	18·0	n
21	...		39	37·50	...		5	18·3	n
22	...		39	37·51	...		5	18·5	n
24	...		39	37·53	...		5	17·9	n
25	...		39	37·49	...		5	18·1	n
27			39	37·46	...		5	18·0	n
28			39	37·68	...		5	17·9	M
Mar. 2	...		39	37·49	...		5	17·8	M
3	...		39	37·50	...		5	18·0	M
7	...		39	37·45	...		5	17·9	M
9			39	37·49	...		5	18·5	M
10	...		39	37·48	...		5	17·8	M
16	...		39	37·47	...		5	17·6	n
17	...		39	37·89	...		5	17·4	n
29	...		39	37·47	...		5	17·6	n
430		*Anon.*							
Mar. 29	8·6	8	40	31·66	5	129	15	46·7	n
431		*Anon.*							
Mar 11	8·0	8	41	3·30	...	147	16	57·1	M
28	9·2		41	3·15	...		16	51·5	n
432		*Lacaille* 3584.							
Mar. 24	8·0	8	42	19·51	...	129	18	19·3	n
433		*Anon.*							
Mar. 22	9·3	8	42	57·67	...	138	10	43·3	n
Apl. 4	9·3		42	57·61	5		10	41·8	M

Separate Results of Madras Meridian Circle Observations in 1865.

Number and Date.	Magnitude.	Mean Right Ascension 1865. h.	m.	s.	No. of Wires.	Mean Polar Distance 1865. °	′	″	Observer.
434 *Lacaille* 3573.									
Mar. 4	8·0	8	44	14·12	...	132	41	39·2	M
21	8·0		44	14·17	5		41	38·0	R
435 *Anon.*									
Mar. 26	9·0	8	46	40·08	...	86	27	35·5	R
436 *Anon.*									
Mar. 1	7·9	8	46	46·27	...	132	58	28·7	M
28	8·3		46	46·15	5		58	27·7	R
437 *Anon.*									
Mar. 22	8·0	8	46	56·08	...	136	6	15·5	R
27	8·3		46	57·92	...		6	15·9	R
438 *Anon.*									
Feb. 27	9·0	8	47	42·12	...	132	56	0·0	R
Mar. 23	9·2		47	42·21	4		56	6·9	R
439 *Anon.*									
Feb. 13	7·9	8	49	46·84	...	138	1	22·2	M
Mar. 29	8·0		49	46·72	5		1	23·1	R
Apl. 6	...		49	47·00	5		1	23·0	M
440 *T Cancri Var* 3.									
Feb. 16	8·7	8	48	57·18	...	69	58	12·7	R
Mar. 30	8·2		48	57·28	...		58	12·4	R
441 *Anon.*									
Feb. 21	8·9	8	49	15·08	5	128	54	58·6	R
442 *Anon.*									
Mar. 24	7·9	8	49	22·19	5	128	59	13·7	R
443 9 *Ursæ Majoris* ι									
Mar. 28	...	8	49	56·90	...	41	25	51·2	R
30	R		49	56·98	5		25	50·6	R
444 *Anon.*									
Feb. 27	7·8	8	50	17·30	6	132	57	9·2	R
Mar. 6	8·3		50	17·31	...		57	10·2	M
445 65 *Cancri* α									
Jan. 12	...	8	51	6·28	...	77	37	20·5	M
13	...		51	5·90	5		37	19·7	R
Feb. 8	...		51	6·27	..		37	19·3	M
Mar. 9	...		51	6·08	6		37	19·4	M
446 *Anon.*									
Mar. 11	9·7	8	51	28·07	5	147	15	40·1	M
447 *Anon.*									
Apl. 7	9·3	8	31	36·95	...	127	34	54·5	M
448 *Anon.*									
Feb. 27	8·1	8	53	40·62	...	128	35	51·6	R
449 *Anon.*									
Mar. 27	8·9	8	51	11·02	4	142	41	28·0	R
450 *Anon.*									
Feb. 11	8·5	8	54	39·34	4	102	40	11·1	M
Mar. 30	8·3		54	39·51	4		40	12·5	R
451 *Anon.*									
Mar. 21	9·6	8	56	31·27	5	146	51	14·0	R
28	9·5		56	31·28	...		51	14·9	R
452 *Anon.*									
Feb. 17	8·1	8	56	38·97	...	105	46	17·0	R
24	8·9		56	39·08	...		46	17·8	R
Mar. 25	8·5		56	38·96	...		46	16·5	R
453 *Anon.*									
Feb. 28	8·7	8	58	41·19	6	108	36	57·9	R

Separate Results of Madras Meridian Circle Observations in 1865.

Number and Date.	Magnitude.	Mean Right Ascension 1865. h.	m.	s.	No. of Wires.	Mean Polar Distance 1865. °	′	″	Observer.
454				*Anon.*					
Mar. 18	8·6	8	56	56·32	5	146	36	49·6	n
455				*Anon.*					
Feb. 17	9·2	8	58	7·30	4	146	49	57·5	n
Mar. 28	8·0		58	7·14	5		49	56·6	n
456				*Anon.*					
Mar. 7	9·0	8	58	11·10	...	146	18	37·0	M
457				76 *Cancri κ.*					
Jan. 12	...	9	0	25·98	...	78	47	29·6	M
13	...		0	26·08	5		47	26·7	n
458				*Anon.*					
Mar 27	8·5	9	1	4·97	5	150	1	45·2	n
Apl. 10	8·0		1	5·21	5		1	46·0	M
459				*Anon.*					
Mar. 21	8·8	9	1	52·46	...	128	57	28·6	n
460				*Lacaille* 3705.					
Mar. 29	7·8	9	2	15·16	...	151	17	17·8	n
Apl. 1	7·4		2	15·28	..		17	19·2	M
461				*Anon.*					
Feb. 27	9·0	9	3	39·36	5	132	47	59·0	n
Mar. 23	9·0		3	39·56	4		47	59·0	n
462				*Anon.*					
Mar 31	7·7	9	5	33·14	5	136	44	27·4	n
463				*Anon.*					
Jan 14	9·0	9	6	32·65	5	129	41	46·6	M
464				82 *Cancri π²*					
Mar. 9	6·6	9	7	48·40	...	74	30	3·8	n
465				*Lacaille* 3747.					
Feb. 6	8·5	9	7	46·60	...	150	34	5·1	M
466				*Anon.*					
Mar. 27	10·0	9	9	28·14	6	73	52	56·6	n
467				*Lacaille* 3774.					
Feb. 10	8·9	9	9	56·06	...	157	10	6·5	M
16	8·7		9	54·73	...		9	58·7	n
Mar. 25	8·7		9	54·95	...		9	59·0	n
468				*Cancri* 83.					
Feb. 11	...	9	11	30·61	5	71	48	27·8	M
20	...		11	28·66	...		48	28·9	n
22	...		11	28·61	...		48	29·0	n
27	...		11	30·49	...		48	27·0	n
Mar. 2	...		11	28·37	...		48	28·5	M
3	...		11	28·80	..		48	29·4	M
11	...		11	28·66	...		48	28·2	M
17	...		11	28·56	..		48	28·9	n
18	...		11	28·40	...		48	28·9	n
20	...		11	28·54	3		48	28·4	n
Apl. 1	...		11	30·08	...		48	28·5	M
4	..		11	28·66	3		48	29·2	M
469				*Anon.*					
Jan. 12	8·0	9	13	6·92	4	72	18	12·6	n
470				*ι Argûs.*					
Mar. 22	...	9	13	28·68	...	148	42	37·1	n
31	...		13	29·54	.		42	36·7	n
471				*Anon.*					
Mar. 24	9·0	9	15	17·47	...	143	49	35·2	n
472				*Anon.*					
Mar. 29	9·3	9	15	59·98	6	25	4	49·9	n
30	9·2		16	0·04	5		4	41·2	n

Separate Results of Madras Meridian Circle Observations in 1865.

Number and Date.	Magnitude.	Mean Right Ascension 1865. h.	m.	s.	No. of Wires.	Mean Polar Distance 1865. °	′	″	Observer.
473				*Anon.*					
Mar. 25	9·2	9	16	19·68	3	130	1	17·5	R
474				*Anon.*					
Mar. 1	9·2	9	17	36·80	5	73	5	44·6	M
21	9·7		17	36·08	4		5	42·0	R
27	9·7		17	36·80	4		5	41·7	R
475				*O. A. N.* 9881.					
Mar. 28	8·8	9	17	42·62	5	25	3	59·0	R
30	9·0		17	42·60	5		3	58·8	R
476				*Anon.*					
Jan. 14	7·8	9	19	32·37	...	75	6	46·1	M
477				*Anon.*					
Apl. 7	8·1	9	20	26·81	...	137	29	12·8	M
478				*Anon.*					
Feb. 9	9·0	9	20	52·11	...	158	38	20·1	M
Mar. 29	9·0		20	52·08	5		38	20·2	R
479				30 *Hydrae* α *Var.* 2.					
Feb. 13	...	9	20	57·18	...	98	4	31·9	M
20	...		20	57·17	...		4	31·5	R
Mar. 3	...		20	57·08	...		4	30·3	M
5			20	57·17	...		4	31·2	M
6	...		20	57·23	...		4	32·0	M
11	...		20	57·09	...		4	31·2	M
13	...		20	57·08	...		4	30·7	M
17	R		20	57·24	...		4	30·1	R
18	.		20	57·13	.		4	32·0	R
20	.		20	57·15	...		4	31·4	R
Apl. 5	...		20	57·23	...		4	31·8	M
480				*Lacaille* 3853.					
Apl. 4	8·0	9	22	34·36		131	30	32·3	M
481				25 *Ursae Majoris* θ					
Mar. 22	...	9	23	49·66	...	37	42	35·1	R
482				*Anon.*					
Mar. 20	9·6	9	24	21·84	5	139	41	1·7	R
483				*Lacaille* 3886.					
Mar. 25	8·7	9	24	46·26	...	134	50	4·1	R
484				*Lacaille* 3887.					
Mar. 24	7·8	9	24	57·13	...	140	0	40·1	R
485				*Anon.*					
Mar. 27	9·0	9	26	44·60	...	145	2	41·4	R
486				*Anon.*					
Mar. 30	9·2	9	26	57·38	...	141	55	24·9	R
487				*Taylor* 4222.					
Mar. 7	8·0	9	27	46·79	5	146	23	36·6	M
Apl. 6	8·0		27	46·61	...		23	37·0	M
488				*Anon.*					
Feb. 27	9·0	9	29	1·00	...	128	46	9·4	R
489				*Taylor* 4226.					
Mar. 28	7·3	9	29	29·48	...	146	29	42·4	R
490				*Anon.*					
Apl. 7	9·4	9	29	34·64	5	146	28	51·7	M
491				10 *Leonis.*					
Feb. 9	5·5	9	30	4·98	3	82	33	38·5	M
10	5·5		30	4·75	..		33	39·3	M

Separate Results of Madras Meridian Circle Observations in 1865.

Number and Date.	Magnitude.	Mean Right Ascension 1865. h. m. s.	No. of Wires.	Mean Polar Distance 1865. ° ′ ″	Observer.
492		*Anon.*			
Feb. 25	6·0	9 32 56·80	...	129 47 58·7	n
Mar. 6	8·5	32 56·45	...	47 51·4	m
493		14 *Leonis* •			
Jan. 13	...	9 33 56·53	...	79 39 48·1	n
14	...	33 56·45	...	39 48·7	m
494		*Lacaille* 3980.			
Feb. 14	8·8	9 34 32·36	...	148 34 2·4	m
18	8·0	34 32·33	...	34 2·6	n
495		*Anon.*			
Mar. 24	9·0	9 34 46·23	...	130 34 55·8	n
Apl. 3	8·6	34 46·40	...	34 57·7	w
496		*Taylor* 4280.			
Mar. 27	7·8	9 34 49·48	...	142 19 59·9	n
497		*Anon.*			
Feb. 16	8·9	9 34 53·34	5	151 56 40·9	n
17	8·9	34 53·31	5	56 40·7	n
Mar. 30	9·0	34 53·50	...	56 51·2	n
498		*Anon.*			
Feb. 16	8·0	9 35 36·43	5	151 56 40·8	n
17	8·0	35 36·61	5	56 40·8	n
Mar. 30	9·0	35 35·49	5	56 41·2	n
499		17 *Leonis* •			
Feb. 11	...	9 36 11·06	...	65 36 21·5	m
23	...	36 10·94	—	36 22·4	n
Mar. 2	...	36 10·98	...	36 22·7	m
3	...	36 10·90	...	36 22·2	m
4	...	36 11·01	...	36 23·2	m
7	...	36 11·03	...	36 22·3	m
21	...	36 10·97	...	36 22·4	n
22	...	36 11·01	...	36 22·4	n
Mar. 23	...	9 36 10·99	...	65 36 22·1	n
25	...	36 11·00	...	36 22·1	n
28	...	36 11·00	...	36 22·3	n
Apl. 10	...	36 11·02	...	36 22·6	m
500		*Anon.*			
Apl. 5	5·8	9 39 3·69	...	98 40 11·9	m
6	5·8	39 3·76	...	40 12·5	m
501		*Anon.*			
Apl. 7	7·9	9 41 52·48	...	130 40 37·5	m
502		*Anon.*			
Mar. 27	9·0	9 42 39·04	...	146 37 6·5	n
Apl. 4	9·1	42 39·06	...	37 8·1	m
503		*Anon.*			
Mar. 24	8·2	9 43 36·67	...	148 46 10·7	n
504		*Anon.*			
Feb. 23	8·7	9 44 7·58	...	147 1 35·6	n
505		*Anon.*			
Feb. 25	9·1	9 44 39·56	...	129 47 33·3	n
Mar. 6	8·9	44 39·45	...	47 33·3	m
506		*Anon.*			
Mar. 23	9·0	9 45 33·50	...	129 3 9·4	n
507		*R. P. L.* 70.			
Mar. 23	...	9 46 [illegible]	3	5 26 4·7	n
508		*Anon.*			
Mar. 25	9·6	9 46 29·17	5	129 7 14·7	n
30	9·2	46 29·01	...	7 13·6	n
509		*Anon.*			
Mar. 29	9·3	9 48 36·65	...	166 7 37·6	n

Separate Results of Madras Meridian Circle Observations in 1865.

Number and Date.	Magnitude.	Mean Right Ascension, 1865. h. m. s.	No. of Wires.	Mean Polar Distance 1865. ° ′ ″	Observer.
510		*Anon.*			
Mar. 26	9·0	9 50 49·86	...	145 39 20·6	R
27	8·8	50 49·67	5	39 31·0	R
511		29 *Leonis* π			
Jan. 13	...	9 53 4·70	5	81 18 36·5	R
Feb. 11	...	53 4·50	...	18 34·8	M
22	...	53 4·63	...	18 36·2	R
Mar. 1	...	53 4·66	...	18 35·1	M
7	...	53 4·78	...	18 36·0	M
9	...	53 4·66	...	18 37·2	M
11	...	53 4·57	...	18 34·7	M
13	...	53 4·40	...	18 36·2	M
21	...	53 4·68	...	18 34·3	R
28	...	53 4·61	...	18 35·5	R
31	...	53 4·61	...	18 34·9	R
Apl. 3	...	53 4·63	...	18 35·3	M
5	...	53 4·64	...	18 34·3	M
512		*Anon.*			
Mar. 29	9·1	9 53 54·98	...	152 7 4·2	R
513		*Anon.*			
Apl. 4	8·1	9 54 10·87	...	143 58 33·3	M
10	8·9	54 11·00	...	58 32·7	M
514		*Taylor* 4445.			
Mar. 29	8·6	9 54 45·18	...	147 28 57·6	R
515		*Anon.*			
Feb. 24	9·0	9 55 54·00	...	147 24 33·3	R
516		*Anon.*			
Feb. 16	9·7	9 56 1·14	5	127 37 48·1	R
Mar. 30	9·6	56 1·06	...	37 48·3	R
517		*Anon.*			
Feb. 14	8·9	9 57 9·73	4	129 56 56·5	R
Mar. 30	9·0	57 9·69	5	56 56·4	R
518		*Anon.*			
Apl. 7	9·0	9 57 50·00	5	145 38 21·5	M
519		*Taylor* 4476.			
Feb. 23	7·9	9 57 53·58	...	146 36 22·5	R
520		*Anon.*			
Apl. 4	8.7	9 58 11·94	5	143 54 26·0	M
521		*Taylor* 4484.			
Feb. 16	7·7	9 58 42·27	...	161 30 19·5	R
Mar. 24	7·8	58 42·28	...	30 16·6	R
Apl. 8	7·1	58 42·51	...	30 19·6	M
522		32 *Leonis* α, *Regulus.*			
Mar. 4	...	10 1 10·81	...	77 22 27·9	M
6	...	1 10·75	...	22 28·4	M
9	...	1 10·73	...	22 28·8	M
20	...	1 10·70	...	22 28·8	R
21	...	1 10·78	...	22 28·2	R
28	...	1 10·73	...	22 29·2	R
29	...	1 10·70	...	22 28·2	R
30	...	1 10·68	...	21 27·8	R
31	...	1 10·71	...	22 28·2	R
Apl. 5	...	1 10·88	...	22 29·5	M
6	...	1 10·67	...	22 29·4	M
523		*Anon.*			
Jan. 14	9·0	10 1 23·73	3	130 0 15·8	M
524		*Lacaille* 4164.			
Apl. 4	7·0	10 2 14·73	...	148 54 21·5	M
525		*Anon.*			
Feb. 27	9·3	10 2 49·36	5	129 57 56·6	R
Mar. 27	9·3	2 49·14	5	57 58·3	R
526		*Anon.*			
Mar. 13	8·1	10 5 16·38	5	165 30 5·4	M
Apl. 8	8·2	5 16·62	...	30 8·1	M

Separate Results of Madras Meridian Circle Observations in 1865.

Number and Date.	Magnitude.	Mean Right Ascension 1865. h. m. s.	No. of Wires.	Mean Polar Distance 1865. ° ′ ″	Observer.
527		*Taylor 4538.*			
Apl. 10	7·1	10 6 12·04	..	129 19 44·6	M
528		*Anon.*			
Mar. 27	8·9	10 7 9·78	5	129 58 26·0	R
30	9·0	7 9·76	...	58 26·5	R
529		*Anon.*			
Mar. 25	9·2	10 9 8·64	5	129 52 0·4	R
530		*R. P. L. 72, s. p..*			
Sep. 13	—	10 9 31·13	3	5 3 59·6	M
531		*Anon.*			
Apl. 7	7·1	10 9 53·80	3	145 34 42·2	M
532		*Anon.*			
Feb. 24	9·5	10 10 18·91	3	129 51 39·3	R
Mar. 28	9·2	10 18·86	4	51 35·8	R
533		41 *Leonis* γ¹			
Feb. 22	...	10 12 31·38	...	69 35 36·3	R
Mar. 4	...	12 31·48	...	35 37·3	M
7	...	12 31·44	...	37 39·4	M
27	...	12 31·51	...	36 35·7	R
29	...	12 31·47	...	34 36·0	R
30	...	12 31·52	...	36 36·4	R
Apl. 1	...	12 31·49	...	35 36·5	M
4	...	12 31·37	...	35 35·1	M
19	...	12 31·59	...	36 36·8	R
534		*Anon.*			
Feb 27	9·8	10 12 58·97	...	129 37 14·0	R
535		48 *Leonis.*			
Apl. 6	6·2	10 15 36·49	...	83 46 28·6	M
7	...	15 36·46	...	46 28·7	M

Number and Date.	Magnitude.	Mean Right Ascension 1865. h. m. s.	No. of Wires.	Mean Polar Distance 1865. ° ′ ″	Observer.
536		*Anon.*			
Apl. 10	7·9	10 16 14·36	5	129 16 33·6	M
537		*Taylor 4653.*			
Mar. 13	8·0	10 18 4·33	6	151 28 32·7	M
Apl. 8	8·0	18 3·97	5	28 31·5	M
538		*Anon.*			
Mar. 29	9·3	10 18 47·80	5	146 8 44·5	R
539		45 *Leonis.*			
Jan. 14	6·0	10 20 31·69	...	79 35 3·0	M
Feb. 10	6·0	20 31·94	...	38 3·3	M
540		*Anon.*			
Mar. 30	8·0	10 21 54·67	5	145 55 12·3	R
Apl. 29	8·8	21 54·68	5	55 12·7	R
541		*Anon.*			
Apl. 21	9·0	10 21 56·26	...	146 59 29·6	R
542		*Anon.*			
Apl. 28	9·0	10 28 25·13	5	76 5 36·7	R
543		47 *Leonis* ρ			
Jan. 14	...	10 25 42·05	...	80 0 0·5	M
Feb. 10	..	25 42·15	...	0 0·4	M
25	...	25 42·11	...	79 59 59·9	M
Mar 9	...	25 42·10	...	80 0 0·1	M
25	...	25 42·01	...	0 0·1	R
27	...	25 42·04	...	79 59 59·8	R
29	...	25 42·00	...	80 0 0·9	R
31	...	25 42·02	...	79 59 59·6	R
Apl. 5	R	25 41·96	..	59 59·5	M
6	...	25 42·00	.	80 0 0·0	M
7	...	25 41·97	...	79 59 59·8	M
10	...	25 41·96	.	59 59·9	M
19	...	25 42·02	...	59 59·7	R
24	...	25 42·01	...	59 59·9	R

Separate Results of Madras Meridian Circle Observations in 1865.

Number and Date.	Magnitude.	Mean Right Ascension 1865. h. m. s.	No. of Wires.	Mean Polar Distance. 1865. ° ′ ″	Observer.
544		*Anon.*			
May. 3	9·0	10 36 38·26	...	153 26 5·1	M
545		*p Carinae,*			
Mar. 13	...	10 27 13·93	5	150 59 32·2	M
Apl. 8	...	27 13·94	...	59 33·0	M
546		*Anon.*			
Apl. 3	9·5	10 28 6·91	3	150 39 14·1	M
547		*Taylor 4769.*			
Apl. 22	6·9	10 30 34·42	5	146 51 36·9	R
548		*Anon.*			
Apl. 4	8·7	10 31 6·34	3	151 9 56·6	M
May 3	..	31 6·09	5	9 56·5	M
549		*Anon.*			
Apl. 6	9·4	10 34 54·78	...	139 16 58·1	M
550		*Anon.*			
Apl. 10	8·5	10 35 5·34	...	149 5 59·4	M
551		*Taylor 4824.*			
Apl. 20	7·8	10 36 8·83	5	148 53 20·9	R
24	7·0	36 8·73	...	53 21·2	R
May 2	7·0	36 8·61	4	53 21·9	M
4	8·0	36 8·74	5	53 20·6	M
552		*Anon.*			
Apl. 3	9·4	10 36 48·61	3	150 47 34·8	M
553		*Anon.*			
Mar 13	8·1	10 37 38·26	3	151 39 39·4	M
554		*Taylor 4849.*			
Apl. 22	6·6	10 38 41·64	...	149 24 37·0	R
26	7·0	38 41·85	...	24 36·0	R
555		*Taylor 4850, 2nd.*			
Apl. 21	8·8	10 38 44·51	...	149 30 26·3	R
556		*Anon.*			
Apl. 8	9·0	10 39 5·93	..	149 34 31·8	M
557		*η Argûs Var. 1.*			
Mar. 31	—	10 39 49·80	...	148 58 33·3	R
Apl. 20	...	39 49·89	...	58 32·7	R
24	...	39 50·04		58 34·3	R
558		*Taylor 4872.*			
Apl. 4	8·0	10 41 7·08	5	151 13 58·3	M
559		*Anon.*			
Apl. 22	9·2	10 41 58·61	...	149 23 11·5	R
26	9·3	41 58·76	...	23 10·8	R
560		*53 Leonis l*			
Mar. 23	...	10 42 9·35	...	78 44 29·5	R
24	...	42 9·52	...	44 30·0	R
25	...	42 9·18	.	44 29·1	R
27	...	42 9·34	...	44 29·1	R
28	...	42 9·30	...	44 30·6	R
29	...	42 9·52	...	44 29·2	R
30	...	42 9·34	...	44 29·6	R
Apl. 1	...	42 9·48		44 29·8	M
3	...	42 9·32	..	44 29·4	M
19	...	42 9·35	5	44 29·4	R
25	...	42 9·68	...	44 29·7	R
561		*Taylor 4915.*			
Mar. 13	8·2	10 46 36·12	...	165 30 5·4	M
Apl. 6	7·0	46 36·12	...	30 5·5	M

Separate Results of Madras Meridian Circle Observations in 1865.

Number and Date	Magnitude	Mean Right Ascension 1865. h	m	s.	No. of Wires	Mean Polar Distance 1865. °	′	″	Observer
562				*55 Leonis.*					
Mar. 10	6·3	10	48	45·71	...	88	32	34·6	M
11			48	45·59	...		32	39·1	M
563				*Anon.*					
Apl. 8	9·0	10	50	31·70	4	148	48	35·2	M
27	9·2		50	31·66	5		48	38·4	M
564				*Taylor 4955.*					
Apl. 5	7·1	10	50	43·20	...	147	19	57·6	M
565				*Anon.*					
Apl. 4	9·0	10	51	58·26	5	143	46	49·0	M
29	...		51	58·36	...		45	47·5	R
566				*Anon.*					
Mar. 24	9·0	10	52	21·65	5	148	36	35·9	R
567				*Anon.*					
Apl. 7	8·5	10	53	35·13	3	149	15	5·8	M
22	8·9		53	35·08	...		15	4·4	M
568				*39 Leonis c*					
Mar 10	...	10	53	44·92	...	66	10	27·0	M
569				*Anon.*					
Mar. 13	8·5	10	53	40·10	3	135	32	8·2	M
Apl 6	8·8		53	40·07	...		32	8·0	M
28	9·0		53	40·07	—		33	7·2	R
570				*50 Ursae Majoris α, Dubhe.*					
Mar 21	R	10	55	22·11	5	27	31	13·3	R
22	R		55	22·27	5		31	16·0	R
31	R		55	22·20	4		31	16·2	R
Apl 25	...		55	22·26	5		31	14·5	R
571				*Anon.*					
May 1	8·0	10	55	40·50	3	149	17	7·5	M
572				*Anon.*					
Mar 18	9·3	10	57	2·01	5	116	33	49·3	R
Apl 21	9·3		57	2·14			33	47·3	R
573				*63 Leonis χ*					
Mar. 21	...	10	58	3·12	...	81	56	6·3	R
27	R		58	3·10	6		56	5·1	R
28	...		58	3·08	...		56	5·3	R
30	...		58	3·10	...		56	6·4	R
Apl. 10	...		58	3·15	..		56	6·2	M
24	...		58	3·09	...		56	5·5	R
30	...		58	3·08	...		56	5·7	R
574				*Lacaille 4595.*					
May 2	8·0	10	59	24·08	...	108	30	12·1	M
9	8·0		59	24·08	5		30	10·1	M
575				*Lacaille 4612.*					
Mar. 13	8·5	11	0	57·34	...	134	46	54·1	M
576				*Anon.*					
Apl. 6	8·4	11	1	4·08	5	105	33	37·8	M
577				*67 Leonis.*					
Apl. 22	6·5	11	1	34·63	...	61	36	42·2	R
578				*Anon.*					
May 1	8·1	11	1	51·71	...	149	14	4·9	M
579				*Anon.*					
May 3	...	11	2	0·77	...	108	56	29·6	M
4	8·0		2	0·67	3		56	29·6	M
580				*Lalande 21371.*					
Mar. 23	8·0	11	3	33·32		77	30	0·0	R
581				*Lalande 21416.*					
Apl 26	9·1	11	5	1·76	..	67	12	46·5	R

Separate Results of Madras Meridian Circle Observations in 1865.

Number and Date.	Magnitude.	Mean Right Ascension 1865. h. m. s.	No. of Wires.	Mean Polar Distance 1865. ° ′ ″	Observer.
582		*Anon.*			
May 2	8·1	11 5 51·90	5	148 39 4·2	M
583		*Taylor* 5108.			
Mar. 10	8·9	11 6 19·12	5	140 35 3·9	M
584		69 *Leonis* p^5			
Feb. 11	5·0	11 6 50·87	...	90 29 9·7	M
585		68 *Leonis* δ			
Mar. 14	...	11 6 55·33	...	68 44 14·6	M
24	...	6 55·30	...	44 15·2	R
28	R	6 55·46	5	44 16·2	R
Apl. 28	...	6 55·40	...	44 14·6	R
29	...	6 56·41	...	44 14·9	R
586		*Anon.*			
Mar. 25	9·3	11 8 36·00	...	150 51 9·3	R
587		*Anon.*			
Apl. 21	8·7	11 9 47·18	5	117 15 18·7	R
588		74 *Leonis* φ			
Feb. 11	...	11 9 47·61	3	92 54 58·5	M
Apl. 8	...	9 47·80	...	54 52·7	M
589		*Anon.*			
May 2	8·5	11 10 29·25	3	148 50 9·4	M
590		*Anon.*			
Apl. 22	9·0	11 10 34·60	3	151 8 35·0	R
591		*Anon.*			
Apl. 28	8·3	11 11 10·93	..	127 38 41·8	R
592		12 *Crateris* δ			
Mar 13	...	11 12 35·66	...	104 2 35·0	R
14	...	12 35·45	...	2 34·3	M
Apl. 1	...	12 35·40	...	2 55·2	M
3	...	12 35·41	...	2 35·2	M
21	...	12 35·60	...	2 35·9	R
29	...	12 35·60	5	2 34·0	R
593		*Anon.*			
May 3	7·7	11 12 51·47	...	120 38 27·0	M
594		*Anon.*			
Mar. 21	7·8	11 15 52·84	...	108 21 38·3	R
Apl. 20	8·0	15 52·70	...	21 38·2	R
595		*Lacaille* 4726.			
Mar. 18	8·2	11 16 7·90	...	146 51 40·3	R
Apl. 10	8·0	16 7·94	.	51 39·6	M
May 4	8·0	16 8·17	...	51 40·9	M
596		79 *Leonis.*			
Apl. 7	.	11 17 6·62	...	87 31 6·0	M
597		*Taylor* 5300.			
Mar 26	8·0	11 19 3·08		131 55 31·1	R
May 1	8·0	19 3·31	..	55 32·6	R
598		*Anon.*			
Mar. 21	9·3	11 21 46·11		128 29 5·6	R
Apl. 20	9·6	21 46·07	..	29 6·9	R
599		*Taylor* 5245			
May 1	8·7	11 22 5·90	5	131 36 8·1	M
5	8·6	22 5·81	5	36 7·7	M
600		*Anon.*			
Mar 18	9·2	11 22 51·63	4	165 34 4·0	R

Separate Results of Madras Meridian Circle Observations in 1865.

Number and Date.	Magnitude.	Mean Right Ascension 1865. h. m. s.	No. of Wires.	Mean Polar Distance 1865. ° ′ ″	Observer.
601		*O. A. N.* 11812.			
Apl. 28	9·0	11 28 12·60	...	22 58 4·8	R
29	9·3	28 12·98	5	58 5·3	R
602		*Anon.*			
Apl. 6	8·2	11 24 33·97	...	22 50 14·5	M
22	9·0	24 33·47	5	50 15·1	R
29	9·3	24 33·73	5	50 16·2	R
May 3	8·0	24 33·93	5	50 13·5	M
603		*Anon.*			
May 4	9·1	11 25 28·60	5	128 27 6·0	M
604		*Anon.*			
Mar. 21	9·3	11 25 50·61	...	128 23 0·2	R
23	9·7	26 50·44	4	23 6·7	R
Apl. 21	9·2	25 50·65	...	23 6·4	R
605		*Anon.*			
Mar. 14	9·0	11 26 59·42	5	161 4 24·1	M
606		*Anon.*			
Mar. 21	9·2	11 28 22·70	5	128 20 27·6	R
Apl. 21	9·8	28 22·71	...	20 29·3	R
28	9·3	28 22·79	...	20 26·2	R
607		λ *Centauri.*			
Mar. 15	...	11 29 34·22	5	152 16 25·8	M
22	...	29 34·34	...	16 25·6	R
608		*Anon.*			
May 9	8·0	11 29 54·75	..	140 16 2·6	M
10	8·2	29 54·80	..	16 1·9	M
609		91 *Leonis* υ			
Feb. 11	...	11 30 2·08	..	90 4 44·5	M
Mar. 11	..	30 2·20	.	4 43·3	M
24		30 2·20		4 44·4	R
Mar. 29	R	11 30 2·11	...	90 4 42·4	R
May 2	...	30 2·17	...	4 44·6	M
5	...	30 2·20	...	4 44·2	M
8	...	30 2·09	...	4 45·8	M
610		*Anon.*			
Apl. 29	9·2	11 32 12·24	5	164 11 53·9	R
611		*Anon.*			
Apl. 4	8·0	11 33 10·21	..	141 14 57·1	M
5	8·3	33 10·44	5	14 56·8	M
29	8·2	33 10·21	5	14 56·3	R
612		*W. B. E. XI.* 571.			
Mar. 25	8·0	11 33 26·83		84 16 0·2	R
613		*Anon.*			
Apl 6	7·0	11 34 28·86		144 21 8·2	M
May 3	...	34 28·23	3	21 1·9	M
4	7·9	34 28·62	4	21 1·4	M
614		*W. B. E. XI.* 597.			
Apl 29	9·1	11 34 46·72	...	84 16 28·8	R
615		*Anon.*			
Apl 28	9·0	11 38 55·68	...	150 31 22·3	R
616		*Anon.*			
Mar 21	9·2	11 40 20·67		149 22 24·3	R
617		*Anon.*			
Mar 23	9·6	11 41 15·34	...	129 22 28·1	R
Apl 28	9·2	41 15·10	.	22 28·7	R

Separate Results of Madras Meridian Circle Observations in 1865.

Number and Date	Magnitude	Mean Right Ascension 1865. h.	m.	s.	No. of Wires.	Mean Polar Distance 1865. °	′	″	Observer.
618				*94 Leonis β.*					
Mar. 14	...	11	42	10·26	...	74	40	24·2	M
Apl. 7	...		42	10·31	...		40	25·6	M
25	...		42	10·29	5		40	25·2	R
29	...		42	10·28	...		40	25·6	R
May. 1			42	10·28	..		40	26·2	M
2	...		42	10·29	...		40	25·5	M
8	...		42	10·35	5		40	24·7	M
10	...		42	10·13	...		40	26·5	M
619				*Taylor 5421.*					
Mar. 23	8·2	11	43	13·88	...	120	31	34·5	R
Apl. 11	7·7		43	13·86	4		31	35·4	M
28	7·0		43	13·92	5		31	34·8	R
620				*5 Virginis β*					
Mar. 11	...	11	43	39·68	...	87	28	29·6	M
Apl. 8	...		43	39·79	...		28	29·3	M
May 5	...		43	39·69	...		28	29·0	M
621				*Anon.*					
May 3	7·9	11	41	47·18	..	120	2	59·2	M
4	8·0		41	47·19	...		3	0·2	M
9	8·0		44	46·19	...		3	0·4	M
622				*Taylor 5433.*					
Mar. 28	8·4	11	44	54·37	5	120	33	25·1	R
623				*Groombridge 1830.*					
Apl. 21	6·6	11	45	11·23	...	51	18	58·1	R
624				*Anon.*					
Mar 25	9·8	11	46	31·98	5	112	31	19·8	R
625				*64 Ursae Majoris γ*					
Apl 3	2	11	46	48·98	6	35	33	19·0	M
626				*Lacaille 4937.*					
Apl 12	7·1	11	49	18·29	5	132	31	43·0	M
627				*Anon.*					
Mar. 13	7·9	11	50	39·89	5	130	22	20·5	M
22	6·0		50	40·07	...		22	20·4	R
Apl. 5	8·0		50	40·06	5		22	21·6	M
May 1	7·0		50	40·14			22	22·1	M
628				*Lacaille 4956.*					
May 8	6·3	11	51	12·84	5	134	34	17·8	M
629				*Anon.*					
May 5	8·6	11	51	26·86	...	128	32	56·6	M
630				*Anon.*					
Apl. 21	9·5	11	52	42·67	5	134	38	40·3	R
631				*Anon.*					
May 2	9·6	11	53	53·12	3	120	36	12·2	M
632				*Anon.*					
May 9	9·0	11	56	29·86	...	128	30	16·9	M
633				*Taylor 5534.*					
Mar. 27	8·7	11	56	52·76	5	143	57	39·5	R
May 10	7·9		56	52·80	...		57	41·0	M
634				*Lacaille 4995.*					
Apl. 28	8·0	11	56	57·18	5	142	44	46·8	R
May 15	7·1		56	57·26	...		44	47·6	M
17	7·5		56	57·04	3		44	46·9	M
635				*Taylor 5535.*					
Mar 25	8·1	11	57	6·82		79	23	41·0	R
28	8·0		57	6·61			23	20·2	R
Apl 11	8·0		57	6·46	-		23	30·6	M
636				*R. P. L. [illegible]—s p.*					
Oct. 14	–	11	57	21·38	2	2	[illegible]	33·2	M

Separate Results of Madras Meridian Circle Observations in 1865.

Number and Date.	Magnitude.	Mean Right Ascension 1865. h.	m.	s.	No. of Wires.	Mean Polar Distance 1865. °	′	″	Observer.
637				*Anon.*					
Mar 30	9·6	11	59	4·87	3	128	25	5·8	R
638				*Anon.*					
Apl. 5	9·1	11	59	34·62	3	130	19	58·1	M
20	9·6		59	34·78	...		19	57·3	R
29	...		59	34·88	5		19	57·6	R
639				*Anon.*					
Mar. 15	8·6	12	1	10·79	...	130	21	47·6	M
Apl. 26	8·5		1	11·08	4		21	47·4	R
May 16	8·7		1	11·09	5		21	46·8	R
640				*Lacaille* 5041.					
Mar. 14	8·6	12	2	25·96	...	141	28	34·1	M
641				2 *Corvi* ε					
Apl. 7	...	12	3	11·01	...	111	52	9·0	M
12	...		3	11·05	...		52	8·6	M
20	...		3	11·13	...		52	7·9	R
May 1	...		3	11·21	5		52	9·2	M
2	...		3	11·14	...		52	9·0	M
3	...		3	11·07	...		52	7·4	M
4	...		3	10·96	...		52	7·2	M
5	...		3	11·19	...		52	9·2	M
642				*Anon.*					
Apl. 27	9·6	12	5	31·98	..	131	8	23·0	R
643				*Anon.*					
Mar. 15	8·7	12	5	59·16	...	130	19	26·7	M
Apl 5	8·6		5	59·31	...		19	25·2	M
May 16	8·9		5	59·46	5		19	25·5	R
644				*Anon.*					
May 9		12	6	16·11	3	139	27	53·7	M
645				*Anon.*					
Mar. 28	9·2	12	8	53·34	...	144	30	24·1	R
Apl. 28	9·0		8	53·27	3		30	26·5	R
646				*Anon.*					
Apl. 5	8·1	12	11	56·17	5	130	23	12·7	M
647				*Taylor* 5648.					
Mar. 20	7·3	12	12	31·47	...	132	6	17·6	R
648				*R. P. L.* 92.					
May 8	...	12	12	56·55	3	2	46	46·3	M
13	...		12	53·66	3		46	49·2	M
649				15 *Virginis* η					
Mar. 13	...	12	12	30·97	...	89	55	0·0	M
31	R		12	30·90	6		54	58·5	R
Apl. 4	...		12	30·99	...		54	59·8	M
6	..		12	30·97	...		54	59·3	M
7	...		12	30·97	...		55	0·0	M
8	...		12	30·94	...		54	59·6	M
12	...		12	30·89	...		55	0·2	M
20	...		12	30·06	...		54	59·7	R
May 2	...		12	30·98	..		55	0·2	M
4	...		12	30·98	...		54	59·4	M
5	—		12	30·42	...		55	0·4	M
10	...		12	30·92	...		54	59·2	M
20	...		13	0·08	5		55	1·0	R
650				*Lacaille* 5119.					
May. 17	8·6	12	15	25·63	...	133	34	27·1	M
651				*Anon.*					
Mar. 14	8·0	12	13	55·30	..	141	40	16·8	M
15	8·0		15	56·19	...		40	11·6	M
652				*Anon.*					
May 19	6·8	12	16	40·29	5	136	30	7·9	R

Separate Results of Madras Meridian Circle Observations in 1865.

Number and Date.	Magnitude.	Mean Right Ascension 1865. h.	m.	s.	No. of Wires.	Mean Polar Distance 1865. °	′	″	Observer.
653				*α¹ Crucis.*					
May 20	...	12	19	6·73	5	132	21	3·9	R
654				*Anon.*					
May 1	8·0	12	10	6·77	...	147	21	41·5	M
655				*Anon.*					
May 16	9·1	12	24	41·01	5	150	29	50·2	R
656				*Anon.*					
May 22	...	12	25	59·16	5	98	2	11·1	R
657				*9 Corvi β*					
Mar. 31	R	12	27	17·98	...	112	38	58·6	R
Apl. 8	...		27	18·06	...		39	0·2	M
11	...		27	17·96	...		39	1·6	R
21	...		27	18·00	...		39	0·1	R
22	...		27	17·97	...		38	59·4	R
23	...		27	18·01	..		38	59·4	R
May 4	...		27	18·11	...		38	59·6	M
8	...		27	17·96	...		38	59·0	M
9	...		27	17·98	...		38	59·5	M
658				*Taylor 5785.*					
Apl. 7	7·9	12	27	29·02	...	130	59	46·4	M
May 16	7·5		27	29·90	5		59	44·5	R
659				*Anon.*					
Mar. 16	8·9	12	27	53·32	5	141	40	14·7	M
Apl. 12	8·4		27	52·93	...		40	16·4	M
660				*Anon.*					
May 16	9·2	12	27	58·77	...	140	55	53·7	R
661				*Anon.*					
May 26	9·3	12	31	54·24	...	84	30	52·8	R
662				*Anon.*					
May 23	...	12	32	52·57	...	148	7	43·3	R
663				*Anon.*					
Apl. 27	8·7	12	33	50·41	5	145	33	49·6	R
664				*Taylor 5880.*					
May 2	7·0	12	34	30·15	...	144	1	13·9	M
665				*29 Virginis γ¹*					
May 1	...	12	34	40·13	...	90	42	31·3	M
15	...		34	40·13	...		42	27·0	M
29	...		34	40·18	...		42	29·5	R
666				*29 Virginis γ²*					
Apl. 22	...	12	34	40·22	5	90	42	31·9	R
667				*28 Virginis.*					
Apl. 10	...	12	34	58·99		96	46	27·8	M
668				*W. B. E. XII. 592.*					
May 9	8·0	12	36	4·31	3	98	16	9·0	M
10	8·0		36	4·43	...		16	8·3	M
669				*Anon.*					
Apl. 6	8·3	12	38	20·09	3	146	25	42·1	M
May 8	9·4		38	19·94	3		25	42·7	M
19	9·0		38	20·02	—		25	40·4	R
670				*Anon.*					
June 3	8·0	12	39	29·17	3	148	30	27·6	M
671				*Anon.*					
May 16	9·3	12	39	31·30	3	161	54	33·6	R

Separate Results of Madras Meridian Circle Observations in 1865.

Number and Date.	Magnitude.	Mean Right Ascension 1865. h.	m.	s.	No. of Wires.	Mean Polar Distance 1865. °	′	″	Observer.
672		*Anon.*							
June 5	8·0	12	40	52·61	3	141	53	12·4	m
673		*Anon.*							
Mar. 16	8·2	12	41	43·17	5	141	40	58·8	m
674		*Anon.*							
May 25	9·1	12	42	6·67	3	147	10	47·5	n
675		*Anon.*							
May 26	8·6	12	42	27·01	5	147	19	5·5	n
676		*Anon.*							
May 27	9·3	12	42	30·78	...	113	52	17·4	n
677		*Anon.*							
Apl. 26	...	12	43	39·77	5	129	8	11·8	n
678		*U Virginis Var. 3.*							
Apl. 22	8·5	12	44	14·71	...	83	42	40·9	n
679		*Radcliffe* 2922.							
May 22	...	12	45	9·16	5	94	16	44·9	n
680		40 *Virginis* ψ							
Apl 19	...	12	47	30·08	...	98	49	18·6	m
681		*R. P. L.* 99.							
Apl. 28	...	12	49	11·46	3	5	51	9·1	n
May 13	...		49	11·08	3		51	10·6	m
682		*Anon.*							
May 23	...	12	49	33·02	...	125	25	50·2	n
683		*Anon.*							
May 10	8·8	12	46	56·37	...	140	34	37·1	m
June 6	8·0		46	56·21	...		34	37·8	m
684		12 *Canum Venaticorum.*							
Apl. 11	...	12	49	42·32	...	50	57	7·6	m
13	...		49	42·45	...		57	7·0	m
20	...		49	42·37	...		57	7·2	n
21	...		49	42·37	5		57	8·4	n
22	...		49	42·34	...		57	8·2	n
27	...		49	42·41	...		57	6·0	n
28	...		49	42·35	...		57	6·9	n
May 3	...		49	42·46	...		57	6·2	m
4	...		49	42·30	...		57	7·0	m
9	...		49	42·34	...		57	6·9	m
22	...		49	42·41	5		57	7·9	n
29	...		49	42·42	...		57	8·6	n
685		*Anon.*							
May 25	9·3	12	51	15·30	3	127	5	34·6	n
686		*Anon.*							
May 17	8·6	12	53	30·21	5	103	49	51·4	m
687		*Anon.*							
May 31	..	12	53	23·87	5	135	44	47·0	n
688		*Anon.*							
May 23	9·0	12	55	6·17	5	124	28	22·0	n
689		*Anon.*							
May 26	9·0	12	57	3·57	..	128	25	31·7	n
27	9·0		57	2·46	...		25	30·7	n
690		*Anon.*							
Apl. 27	..	12	58	0·10	4	124	20	1·5	n

Separate Results of Madras Meridian Circle Observations in 1865.

Number and Date.	Magnitude.	Mean Right Ascension 1865. h. m. s.	No. of Wires.	Mean Polar Distance 1865. ° ′ ″	Observer.
691 *Anon.*					
May 19	8·0	12 58 19·86	...	124 23 16·6	R
20	8·0	58 18·89	5	23 15·3	R
692 *Taylor 6025.*					
May 27	7·8	12 59 30·80	5	128 24 44·0	R
693 51 *Virginis θ*					
Mar 13	...	13 2 57·70	...	94 49 3·9	M
14	...	2 57·68	...	49 3·6	M
15	...	2 57·71	...	49 3·9	M
Apl. 3	R	2 57·74	5	49 1·6	M
12	...	2 57·67	...	49 4·4	M
26	...	2 57·79	...	49 4·1	R
May 9	...	2 57·46	...	49 4·3	M
10	...	2 57·70	...	49 5·0	M
17	...	2 57·62	...	49 4·0	M
20	...	2 57·64	...	49 4·3	R
29	...	2 57·62	...	49 2·4	R
30	...	2 57·74	...	49 3·9	R
694 *Anon.*					
May 23	9·0	13 4 38·06	...	145 12 42·2	R
695 *Lacaille 5434.*					
June 1	8·3	13 6 1·30	...	132 51 53·6	M
696 *Anon.*					
Apl. 27	8·7	13 7 42·95	4	130 46 33·2	R
28	9·3	7 42·96	6	46 31·8	R
May 16	8·5	7 42·89	3	46 31·5	R
697 *Anon.*					
Apl 27	9·2	13 8 9·51	5	130 46 3·8	R
May 16	9·2	8 9·64	4	46 4·0	R
698 *Anon.*					
Apl 29	8·7	13 9 44·99		129 36 36·0	R
699 *Anon.*					
Apl. 22	7·8	13 12 56·30	...	122 55 51·3	R
June 5	7·7	12 56·12	...	55 51·1	M
700 *Lacaille 5503.*					
Apl. 25	6·0	13 14 12·31	...	125 34 10·8	R
May 16	7·9	14 12·41	..	34 12·4	M
701 *Taylor 6148.*					
May 26	7·0	13 14 15·04	...	126 8 23·3	R
702 *O. A. N. 18868.*					
May 25	...	13 15 26·04	4	97 53 32·4	R
703 *Anon.*					
May 26	9·5	13 17 10·49	5	128 9 45·4	R
704 67 *Virginis α, Spica.*					
Mar. 13	...	13 18 4·98	...	100 27 20·6	R
14	...	18 5·09	...	27 22·1	R
Apl. 10	...	18 5·06	...	27 21·2	M
11	...	18 4·98	...	27 21·3	M
21	...	18 5·01	...	27 24·7	R
22	...	18 5·01	...	27 22·0	R
27	...	18 5·07	...	27 21·1	R
28	...	18 4·94	...	27 20·4	R
May 1	...	18 5·11	...	27 21·9	M
3	...	18 4·99	5	27 20·0	M
8	...	18 4·98	...	27 21·1	M
17	...	18 4·90	...	27 21·0	M
20		18 4·98	.	27 21·6	R
29	...	18 5·04	..	27 20·1	R
30	...	18 5·01	..	27 21·7	R
June 7	...	18 4·90	3	27 19·6	M
14	...	18 5·01	...	27 23·1	M
705 *Anon.*					
June 6	6·2	13 19 5·02	5	103 27 5·1	M

Separate Results of Madras Meridian Circle Observations in 1865.

Number and Date.	Magnitude.	Mean Right Ascension 1865. h.	m.	s.	No. of Wires.	Mean Polar Distance 1865. °	′	″	Observer.
706				*Radcliffe* 3011.					
May 22	...	13	19	28·42	3	34	28	47·6	R
707				*R. P. L.* 103—*s.p.*					
Nov. 27	...	13	20	12·44	3	4	32	23·1	R
708				*R. Hydrae Var.* 1					
May 2	6·3	13	22	20·48	...	112	31	30·5	M
June 1	6·0		22	20·38	5		31	28·9	M
709				*Anon.*					
June 3	8·0	13	24	51·17	...	123	8	38·1	M
710				*Anon.*					
May 28	...	13	24	52·95	...	124	9	26·7	R
711				*Anon.*					
May 28	8·0	13	26	34·06	3	128	10	35·7	R
712				76 *Virginis h.*					
Apl. 10	...	13	26	51·40	...	99	38	6·6	M
11	...		26	51·66	...		38	6·9	M
713				*S Virginis Var.* 6.					
June 8	7·1	13	26	07·21	...	96	30	0·2	M
714				*Anon.*					
May 18	8·6	13	26	41·61	3	131	36	30·0	R
715				79 *Virginis* ζ					
Apl. 21	...	13	27	48·92	...	90	34	17·5	R
22	...		27	48·84	...		34	17·6	R
26	...		27	48·96	.		34	17·3	R
27	...		27	48·94	...		34	17·1	R
28	...		27	48·90	...		34	17·3	R
May 1	...		27	48·75	...		34	18·9	M
15	.		27	49·00	...		34	16·5	M
May 22	...	13	27	48·87	...	90	34	17·3	R
23	...		27	48·90	...		34	17·2	R
June 3	...		27	48·68	...		34	16·9	M
7	...		27	48·87	...		34	16·1	M
716				*Taylor* 7183.					
Apl. 26	7·0	13	28	27·28	5	131	43	28·4	R
30	7·7		28	27·40	...		43	27·8	R
June 6	7·0		28	27·17	4		43	27·7	M
717				*Lacaille* 5614.					
May 26	8·9	13	30	0·26	...	128	12	25·3	R
27	...		30	0·35	4		12	26·7	R
718				*Anon.*					
May 19	9·5	13	31	6·46	...	129	10	16·1	R
719				*Anon.*					
May 26	8·7	13	35	46·77	5	128	3	44·6	R
June 8	8·4		35	46·66	3		3	45·5	M
720				*Anon.*					
June 1	8·9	13	36	17·96	...	128	5	40·1	M
721				*Taylor* 6863.					
June 6	7·7	13	36	42·17	...	147	38	46·0	M
722				*Taylor* 6866.					
June 3	7·0	13	36	57·01	3	151	40	31·1	M
723				*Lacaille* 5659.					
May 17	8·7	13	37	12·47	...	138	13	38·1	M
724				86 *Virginis.*					
Mar. 14	...	13	38	44·78		101	41	36·4	M
15	6·3		38	41·98	..		41	37·0	M

Separate Results of Madras Meridian Circle Observations in 1865.

Number and Date.	Magnitude.	Mean Right Ascension. 1865. h. m. s.	No. of Wires.	Mean Polar Distance 1865. ° ′ ″	Observer.
725		85 *Ursae Majoris* η			
May 25	...	13 42 13·02	...	40 0 42·7	R
30	...	42 12·98	...	0 48·3	R
June 7	...	42 13·11	...	0 42·5	M
726		89 *Virginis.*			
June 5	...	13 42 32·56	...	107 27 37·1	M
727		*Anon.*			
May 23	8·0	13 43 17·74	...	123 6 30·0	R
728		*Anon.*			
Apl. 26	...	13 43 27·70	...	123 13 22·7	R
May 8	8·5	43 27·37	...	13 21·0	M
729		*Anon.*			
May 26	8·6	13 44 19·68	...	127 57 2·1	R
June 6	8·6	44 18·80	...	57 1·9	M
730		*Anon.*			
May 5	8·6	13 46 45·80	...	122 54 51·0	R
731		*Lacaille* 5734.			
June 3	7·1	13 47 47·60	...	151 40 26·6	R
8	7·1	47 47·68	...	40 26·5	R
12	7·2	47 47·90	3	40 26·3	R
732		8 *Bootis* η			
Apl. 26	...	13 48 15·20	5	70 55 29·21	R
27	...	48 15·31	..	55 29·0	R
May 16	...	48 16·34	...	55 29·1	M
16		48 15·35	..	55 29·0	R
23	...	48 15·16	...	55 27·9	R
27	.	48 15·33	...	55 29·0	R
30	...	48 15·32	..	55 29·8	R
June 1	..	48 15·30	5	55 29·4	M
7	...	48 15·97	—	55 27·8	M
11	...	48 15·81	...	55 29·1	M
733		*Anon.*			
Apl. 21	8·0	13 50 15·01	5	128 44 2·8	R
734		*Anon.*			
May 8	8·5	13 52 16·85	...	151 39 26·0	M
735		*Anon.*			
May 26	9·5	13 52 29·67	..	129 1 48·2	R
736		*Anon.*			
June 6	8·8	13 53 11·46	...	126 41 8·5	M
737		β *Centauri.*			
May 22	...	13 54 19·57	...	149 43 12·5	R
738		93 *Virginis* τ			
May 11	...	13 54 46·61	...	87 43 2·4	M
16	...	54 46·61	...	43 3·5	R
23	...	54 46·40	...	43 1·6	R
25	...	54 46·60	...	44 2·1	R
June 1	..	54 46·38	...	44 1·0	M
7	..	54 46·71	5	43 2·2	M
739		94 *Virginis.*			
Mar. 16	...	13 58 5·73	5	98 16 45·7	R
740		*Lalande* 25696.			
May 5	7·4	13 59 34·90	5	67 11 12·5	M
741		*Taylor* 6570.			
Apl. 22	9·0	14 0 5·36	5	149 56 19·0	R
29	9·1	0 5·97	5	56 19·1	R
742		*Anon.*			
Apl. 25	9·2	14 0 29·11	5	150 51 29·4	R

Separate Results of Madras Meridian Circle Observations in 1865.

Number and Date	Magnitude	Mean Right Ascension 1865. h.	m.	s.	No. of Wires	Mean Polar Distance 1865. °	'	"	Observer.
743				*Taylor* 6583.					
June 8	7·0	14	1	36·07	4	134	14	21·4	M
744				*Anon.*					
June 3	8·4	14	2	5·61	3	134	16	30·3	M
745				*R. P. L* 108—*s.p.*					
Dec. 11	...	14	3	47·71	2	3	35	46·4	M
746				*Lacaille* 5844.					
May 11	7·5	14	5	9·08	...	151	4	25·9	M
25	8·0		5	8·94	...		4	26·6	R
747				*Taylor* 6616.					
June 6	5·9	14	5	34·49	3	146	27	7·3	M
12	...		5	34·39	5		27	7·9	M
748				98 *Virginis* κ					
May 8	...	14	5	41·71	...	99	38	39·2	M
9	...		5	41·70	...		38	37·9	M
749				16 *Bootis* α, *Arcturus.*					
Mar. 15	...	14	9	30·29	...	70	6	49·5	M
May 13	...		9	30·27	3		6	48·2	M
16	...		9	30·27	...		6	49·1	R
17	...		9	30·38	...		6	49·1	M
18	...		9	30·21	...		6	49·9	R
22	...		9	30·34	6		6	49·6	R
June 1	...		9	30·39	...		6	49·7	M
6	...		9	30·24	...		6	49·4	M
7	...		9	30·14	...		6	48·2	M
8	...		9	30·24	3		6	49·1	M
14	...		9	30·28	...		6	49·4	M
19	...		9	30·26	...		6	49·9	R
20	...		9	30·22	...		6	48·6	R
750				100 *Virginis* λ					
Apl. 11	...	14	11	49·35	...	102	44	33·8	M
12	...		11	49·34	5		44	36·9	M
May 8	...		11	49·48	...		44	33·3	M
751				*Anon.*					
June 3	7·7	14	14	30·78	...	130	46	22·9	M
5	7·8		14	30·90	3		46	22·8	M
752				*Anon.*					
May 11	8·3	14	16	37·72	3	122	34	3·2	M
753				2 *Librae.*					
Apl. 11	...	14	16	10·09	...	101	5	48·3	M
12	...		16	10·06	...		5	45·9	M
May 6	7·0		16	9·93	3		5	46·4	M
754				*Taylor* 6721.					
June 12	7·0	14	17	38·82	...	101	3	19·0	M
755				*S Bootis Var.* 2.					
June 30	9·2	14	19	21·84	...	36	31	37·0	R
756				*Anon.*					
May 23	8·2	14	20	1·26	...	127	9	9·2	R
29	...		20	1·31	...		9	10·2	R
757				*Lacaille* 5962.					
Apl. 29	7·8	14	22	46·12	5	129	47	1·8	R
758				*Anon.*					
Apl. 29	9·3	14	23	6·17	5	129	44	17·3	R
759				*O. A. N.* 14634.					
June 24		14	25	52·35	4	30	6	30·3	R

Separate Results of Madras Meridian Circle Observations in 1865.

Number and Date.	Magnitude.	Mean Right Ascension 1865. h. m. s.	No. of Wires.	Mean Polar Distance 1865. ° ′ ″	Observer.
760		*25 Bootis ρ*			
Mar. 15	...	14 26 0·60	...	59 2 4·0	M
May 11	...	26 0·65	...	2 4·0	M
15	...	26 0·46	...	2 4·7	M
16	...	26 0·63	...	2 5·5	R
17	...	26 0·68	...	2 5·1	M
18	...	26 0·62	...	2 4·9	R
31	...	26 0·57	...	2 6·9	R
June 1	...	26 0·66	...	2 4·9	M
3	...	26 0·64	...	2 4·1	M
5	...	26 0·70	...	2 4·6	M
6	...	26 0·60	...	2 4·9	M
8	...	26 0·66	...	2 5·2	M
14	...	26 0·60	...	2 5·0	M
19	...	26 0·58	...	2 3·9	R
23	...	26 0·57	...	2 4·6	M
761		*O. A. N. 14652*			
June 24	...	14 27 2·17	5	90 7 13·5	R
762		*α¹ Centauri*			
May 23	...	14 30 26·89	5	150 16 40·5	R
31	...	30 27·05	...	16 40·8	R
June 12	...	30 27·10	...	16 41·3	M
19	—	30 27·20	5	16 39·9	R
20	...	30 27·27	...	16 39·5	R
763		*α² Centauri.*			
May 25	...	14 30 27·26	...	150 16 32·3	R
June 23	...	30 27·36	...	16 33·5	R
764		*Lacaille 6027.*			
June 8	7·8	14 31 7·40	5	122 47 33·4	M
765		*Anon.*			
May 14	8·5	14 33 46·81	...	121 44 31·1	R
766		*Anon.*			
May 29	8·5	14 33 59·36	...	125 19 2·4	R
767		*Anon.*			
June 20	9·3	14 36 36·96		130 17 38·4	R
768		*5 Librae.*			
Mar. 15	...	14 39 31·36	...	104 56 17·5	M
June 6	...	39 31·25	...	56 19·1	M
769		*36 Bootis ε, Mirac.*			
May 11	...	14 39 5·49	...	62 21 19·1	M
18	...	39 5·48	5	21 18·9	R
22	...	39 5·42	...	21 19·1	R
23	...	39 5·37	...	21 18·3	R
26	...	39 5·46	...	21 19·1	R
27	...	39 5·30	...	21 19·5	R
31	...	39 5·54	...	21 17·3	R
June 3	...	39 5·46	...	21 19·1	M
12	...	39 5·26	...	21 19·7	M
19	...	39 5·48	...	21 19·0	R
23	...	39 5·36	—	21 17·8	R
770		*Anon.*			
May 17	7·7	14 39 36·34	5	134 0 53·0	M
June 1	7·9	39 36·21	...	0 53·9	M
771		*Anon.*			
May 20	...	14 40 29·41	5	127 3 30·7	R
772		*Brisbane 5069.*			
June 20	8·9	14 41 27·61	5	131 16 30·2	R
773		*9 Librae α¹*			
Mar 15	—	14 43 24·76	...	105 24 43·7	R
May 9	...	43 24·73	...	24 43·7	M
10	...	43 24·66	...	24 44·3	M
13	...	43 24·67	3	23 43·9	M
23	...	43 24·74	5	23 43·1	R
30	...	43 24·67	...	25 42·1	R
June 3	...	43 24·79	...	25 43·9	R
8	...	43 24·80	...	24 43·4	M
19	—	43 24·80		25 43·9	R
23	...	43 24·86	3	24 44·6	R
29	.	43 24·82	4	24 41·6	R

Separate Results of Madras Meridian Circle Observations in 1865.

Number and Date	Magnitude	Mean Right Ascension 1865. h.	m.	s.	No. of Wires	Mean Polar Distance 1865. °	'	"	Observer.
774				*Lalande* 27123.					
May 11	7·8	14	47	20·87	...	100	27	34·2	M
775				*Anon.*					
July 1	9·4	14	47	36·66	3	150	41	25·5	N
776				*Anon.*					
June 20	9·0	14	50	27·36	...	130	32	26·0	N
777				7 *Ursae Minoris β Var.* 1.					
June 19	...	14	51	8·20	3	15	17	32·6	N
778				*Taylor* 7017.					
May 11	7·0	14	57	10·02	3	100	36	13·0	M
13	7·5		57	9·60	3		36	13·3	M
779				*Anon.*					
May 29	8·8	14	57	46·28	5	131	39	53·6	N
780				43 *Bootis ψ*					
May 12	...	14	58	39·61	...	66	31	29·0	M
28	...		58	39·56	...		31	27·9	N
31	...		58	39·66	5		31	28·7	N
June 5	...		58	39·68	.		31	27·4	M
20	—		58	39·73	5		31	29·7	N
781				47 *Bootis k*					
May 30	..	15	0	57·54	...	41	19	35·1	N
782				*Taylor* 7079.					
June 5	8·9	15	3	23·79	.	123	7	30·1	M
783				*Anon.—2nd.*					
May 16	9·0	15	3	37·63		122	19	59·1	N
784				24 *Librae ι¹*					
Apl. 12	...	15	4	31·69	5	109	16	48·9	M
May 10	...		4	31·64	...		16	48·4	M
785				*R. P. L.* 111.					
June 19	...	15	5	57·16	3	5	31	35·0	N
786				27 *Librae β*					
May 12	...	15	9	44·02	...	98	53	57·4	M
13	...		9	44·73	..		53	57·2	N
26	...		9	44·73	...		52	56·4	N
27	...		9	44·66	...		53	57·6	N
June 12	...		9	44·69	4		53	57·8	M
20	...		9	44·76	...		53	57·3	N
29	...		9	44·38	...		53	57·1	N
July 3	...		9	44·67	...		52	57·1	M
7	...		9	44·64	5		52	57·5	M
787				*Anon.*					
May 13	8·8	15	11	34·91	...	100	24	14·1	M
29	8·5		11	34·97	4		24	15·0	N
788				*Lacaille* 6354.					
May 16	9·0	15	15	4·46	...	124	15	28·5	N
789				*Anon.*					
May 26	9·2	15	17	8·01	...	108	2	54·4	N
790				*Lacaille* 6377.					
June 20	7·6	15	18	29·01	3	130	11	6·3	N
791				39 *Librae υ*					
Apl. 12	...	15	20	35·31	...	105	14	36·8	M
June 8	..		20	34·63	...		14	35·6	M
792				*W. B. E. XV.* 395.					
July 1	8·9	15	23	1·67	...	101	15	40·4	M

Separate Results of Madras Meridian Circle Observations in 1865.

Number and Date.	Magnitude.	Mean Right Ascension 1865. h.	m.	s.	No. of Wires.	Mean Polar Distance 1865. °	′	″	Observer.
793				*R. P. L.* 114—*s.p.*					
Nov. 11	...	15	22	3·90	3	2	15	14·0	M
794				*Anon.*					
May 16	8·5	15	22	28·39	...	126	10	22·6	R
795				*Anon.*					
June 1	8·1	15	22	32·43	...	131	37	14·4	M
796				*W. B. E. XV.* 429.					
July 11	9·3	15	24	5·86	5	101	28	42·5	M
13	9·4		24	5·70	...		28	41·3	M
797				*Anon.*					
June 20	9·3	15	24	37·32	...	130	9	9·0	R
798				38 *Librae* γ					
June 6	...	15	27	58·37	3	104	20	13·0	M
799				*Anon.*					
May 18	8·2	15	28	46·40	...	126	36	36·3	R
800				5 *Coronae Borealis* α, *Alpheta.*					
May 12	...	15	28	58·37	...	62	49	44·4	M
13	...		28	58·27	...		49	45·0	M
June 3	...		28	58·33	...		49	44·1	M
12	...		28	58·16	...		49	44·7	M
29	...		28	58·28	...		49	44·9	R
July 3	...		28	58·38	...		49	44·4	M
7	...		28	58·36	...		49	41·6	M
801				*Anon.*					
May 16	8·8	15	30	13·69	5	129	33	39·7	R
802				*Anon.*					
May 22	9·7	15	32	12·40	3	116	36	30·3	R
803				*Anon.*					
May 19	9·5	15	34	30·64	4	128	35	21·1	R
804				*W. B. E. XV.* 645.					
June 1	8·2	15	34	38·08	4	102	19	27·3	M
805				*W. B. E. XV.* 675.					
June 3	8·8	15	35	59·71	5	102	41	39·7	M
806				24 *Serpentis* α					
May 12	...	15	37	37·20	...	96	8	30·8	M
13	...		37	37·25	...		8	30·9	M
31	...		37	37·18	6		8	30·6	R
June 29	...		37	37·24	4		8	30·9	R
July 1	...		37	37·15	...		8	30·3	M
807				*Lalande* 28787.					
June 6	...	15	42	9·09	5	92	40	6·2	M
808				*Lalande* 28970.					
June 3	8·0	15	46	0·68	...	70	49	14·7	R
809				7 *Scorpii* δ					
May 10	...	15	52	21·36	...	112	14	5·0	M
11	...		52	21·43	...		14	5·3	M
810				*W. B. E. XV.* 1047.					
May 29	8·0	15	53	59·39	.	91	16	29·2	R
811				8 *Scorpii* β¹					
May 10	.	15	57	35·60	..	109	25	40·2	M
June 6			57	35·38			25	40·9	M
July 1	..		57	35·65	...		25	40·2	M
12	..		57	35·60			25	40·1	M

Separate Results of Madras Meridian Circle Observations in 1865.

Number and Date.	Magnitude.	Mean Right Ascension 1865. h.	m.	s.	No. of Wires.	Mean Polar Distance 1865. °	'	"	Observer.
812	*R. P. L.* 116—*s.p.*								
Oct. 7	...	16	4	30·56	3	4	18	56·2	M
Dec. 21	...		4	31·17	8		18	53·1	R
813	*W. B. E. XVI.* 83.								
July 14	8·0	16	6	6·00	...	103	41	14·8	M
814	1 *Ophiuchi* δ								
May 13	...	16	7	16·36	...	93	30	40·6	M
July 1	...		7	16·21	...		30	40·1	M
12	...		7	16·28	...		30	39·6	M
815	*Anon.*								
May 29	6·0	16	9	12·41	4	112	36	3·3	R
July 18	7·9		9	12·34	4		36	2·9	M
26	7·4		9	12·14	4		36	3·4	M
816	*Anon.*								
May 29	9·7	16	9	46·72	5	112	33	41·8	R
817	*O. A. S.* 15504.								
July 28	8·9	16	11	24·06	4	109	41	35·5	M
818	20 *Scorpii* σ								
June 8	...	16	12	50·31	5	115	15	56·5	M
819	*Anon.*								
July 22	9·4	16	16	49·60	3	152	17	11·8	M
820	*Anon.*								
May 20	9·2	16	18	3·44		129	30	41·4	R
821	21 *Scorpii* α, *Antares.*								
July 1	...	16	21	8·01	...	116	7	46·5	M
18	...		21	7·90	...		7	46·6	M
822	*Anon.*								
July 20	8·5	16	22	46·83	..	48	26	57·3	M
823	*Anon.*								
July 22	9·3	16	23	48·90	3	152	13	0·8	M
824	30 *Herculis g Var.* 5.								
July 14	6·2	16	24	12·33	...	47	40	11·0	M
15	6·1		24	12·30	...		40	10·9	M
825	*Anon.*								
June 19	9·2	16	26	26·30	5	130	55	0·1	R
826	*Anon.*								
July 22	9·2	16	28	27·01	3	152	16	38·0	M
827	*Anon.*								
June 19	9·3	16	29	19·80	...	130	54	1·2	R
828	α *Trianguli Australis.*								
July 21	...	16	34	24·04	5	158	46	29·5	M
829	*Anon.*								
July 27	9·0	16	31	41·30	3	131	7	9·8	M
830	40 *Herculis* ζ								
June 24	...	16	36	11·74	...	58	9	3·3	R
July 13	...		36	11·99	4		9	3·2	M
21	...		36	11·91	...		9	3·4	M
22	...		36	11·98	5		9	4·3	M
831	*Anon.*								
June 19	8·8	16	36	49·14	–	130	56	5·5	R
832	*O. A. S.* 15932.								
July 26	9·0	16	39	25·59	3	111	36	39·2	M

Separate Results of Madras Meridian Circle Observations in 1865.

Number and Date.	Magnitude.	Mean Right Ascension 1865. h. m. s.	No. of Wires.	Mean Polar Distance 1865. ° ′ ″	Observer.
833		*20 Ophiuchi.*			
May 11	6·3	16 42 22·08	...	106 32 29·5	R
12	...	42 21·18	...	32 29·3	M
834		*Anon.*			
June 19	8·2	16 44 28·76	...	131 1 56·5	R
835		*Anon.*			
July 13	8·5	16 41 41·86	4	130 18 28·1	M
836		*Taylor 7815.*			
June 24	8·7	16 45 33·93	...	130 17 39·5	R
July 27	8·0	45 33·81	...	17 30·6	M
837		*Anon.*			
June 19	9·3	16 47 7·56	...	130 59 43·1	R
838		*Taylor 7832.*			
June 24	9·0	16 47 32·91	4	130 17 36·1	R
July 24	8·0	47 32·63	...	17 35·8	M
28	7·9	47 32·94	...	17 35·9	M
839		*27 Ophiuchi κ.*			
July 12	...	16 51 16·70	...	80 24 45·8	M
18	...	51 16·61	2	24 46·5	M
21	...	51 16·70	...	24 46·9	M
22	...	51 16·66		24 46·1	M
840		*Anon.*			
July 26	8·4	16 51 43·66	3	122 58 44·6	M
841		*Anon.*			
July 21	8·6	16 52 0·40	4	122 49 57·6	M
842		*O. A. S. 16233.*			
July 27	8·0	16 54 2·24	3	110 23 38·1	M
843		*Anon.*			
June 19	8·0	16 55 58·67	...	130 54 47·5	R
July 21	9·0	55 58·37	4	54 45·3	M
844		*O. A. S. 16263.*			
July 20	8·0	16 56 31·41	...	119 50 13·0	M
845		*Taylor 7926.*			
July 23	7·9	16 58 50·73	...	136 51 8·5	M
846		*22 Ursae Minoris ε—s.p.*			
Jan. 10	...	16 59 54·98	2	7 44 45·0	M
Feb. 8	...	59 55·72	5	44 49·0	M
847		*35 Ophiuchi η*			
May 11	...	17 2 39·62	...	105 33 17·0	M
848		*Lacaille 7168.*			
June 20	7·8	17 4 40·16	5	128 7 46·6	R
July 27	8·0	4 39·86	5	7 46·9	M
849		*Anon.*			
July 26	9·4	17 5 40·92	..	130 53 47·4	M
850		*Anon.*			
July 20	8·0	17 5 44·60	3	130 50 29·5	M
851		*Anon.*			
May 12	8·5	17 6 15·23	3	127 25 6·5	M
852		*Anon*			
June 19		17 7 12·60	3	130 42 36·4	R

Separate Results of Madras Meridian Circle Observations in 1865.

Number and Date.	Magnitude.	Mean Right Ascension 1865. h. m. s.	No. of Wires.	Mean Polar Distance 1865. ° ′ ″	Observer.
853		*64 Herculis α Var. 1.*			
June 24	...	17 8 29·47	5	73 27 12·4	M
July 12	...	8 29·50	3	27 12·7	M
14	...	8 29·46	3	27 13·0	M
15	...	8 29·56	...	27 12·6	M
21	...	8 29·51	...	27 12·1	M
22	...	8 29·50	...	27 12·0	M
24	...	8 29·46	...	27 12·5	M
Aug 3	...	8 29·51	5	27 12·6	M
854		*Anon.*			
July 31	8·4	17 9 5·36	4	124 4 21·4	M
855		*Taylor* 8017.			
July 28	7·7	17 13 25·03	...	111 46 58·3	M
856		42 *Ophiuchi θ*			
June 24	...	17 13 48·20	5	114 51 41·1	M
July 27	...	13 48·11	...	51 41·7	M
Aug. 3	...	13 48·10	...	51 42·0	M
857		*δ Arae—2nd.*			
July 26	7·0	17 18 54·96	3	150 33 59·1	M
29	7·0	19 56·08	...	38 59·3	M
858		*Anon.*			
June 30	9·2	17 21 17·42	...	120 46 48·7	M
859		*Brisbane* 6091.			
July 22	8·8	17 21 24·90	5	143 27 5·6	M
860		*Anon.*			
July 21	8·8	17 21 36·76	..	130 32 39·7	M
861		*Lacaille* 7313.			
July 29	7·0	17 28 12·98	...	130 44 25·1	M
862		*Anon.*			
July 24	9·1	17 27 4·13	3	120 26 41·3	M
863		*Anon.*			
June 30	8·0	17 28 8·24		130 43 30·5	M
July 31	8·0	28 8·28		43 31·1	M
864		55 *Ophiuchi α.*			
July 21	...	17 28 40·05	...	77 30 21·1	M
27	...	28 39·99	...	30 21·1	M
Aug. 2	...	28 40·00	...	30 22·2	M
5	...	28 40·01	...	30 21·7	M
865		55 *Serpentis ξ*			
May 12	...	17 29 51·22	2	105 16 37·1	M
13	...	29 51·41	...	16 37·5	M
June 8	...	29 51·40	...	16 37·5	M
Aug. 3	...	29 51·26	...	16 37·8	M
866		*Taylor* 8164.			
July 24	6·9	17 33 8·95	...	129 37 38·2	M
28	6·9	33 9·02	3	37 38·6	M
867		56 *Serpentis ο.*			
May 13	...	17 33 49·60	...	102 47 57·4	M
868		*Anon.*			
July 22	9·0	17 34 16·50	3	129 57 31·1	M
869		*Lacaille* 7406.			
July 29	7·0	17 34 46·53	4	129 44 18·2	M
870		*Anon.*			
July 24	9·2	17 36 8·57	3	150 36 6·0	M
871		*Anon.*			
June 30	8·5	17 37 43·95	...	126 29 39·5	M

Separate Results of Madras Meridian Circle Observations in 1865.

Number and Date.	Magnitude.	Mean Right Ascension 1865. h. m. s.	No. of Wires.	Mean Polar Distance 1865. ° ′ ″	Observer.
872		*Anon.*			
June 19	8.6	17 39 37.69	...	127 21 39.2	R
873		86 *Herculis* μ			
July 14	...	17 41 10.53	...	62 11 53.9	M
15	...	41 10.51	...	11 55.9	M
24	...	41 10.30	3	11 51.1	M
27	...	41 10.58	...	11 54.8	M
Aug. 2	...	41 10.48	...	11 55.0	R
5	...	41 10.46	...	11 54.9	R
11	...	41 10.43	...	11 56.1	R
874		*Anon.*			
July 31	8.5	17 42 8.92	5	118 27 21.5	R
875		*Anon.*			
July 22	8.0	17 43 24.65	...	128 36 13.8	M
28	8.3	43 24.92	...	36 13.1	M
876		*Anon.*			
July 26	8.8	17 50 46.67	3	152 7 30.1	M
877		*Lacaille* 7517.			
July 28	8.0	17 52 20.66	3	140 10 23.6	R
878		*Lacaille* 7518.			
July 24	7.7	17 52 49.71	...	140 12 15.1	M
27	7.8	52 49.17	5	12 16.0	M
879		33 *Draconis* γ, *Etanin.*			
Aug 11	...	17 53 28.08	..	38 29 40.0	R
880		*Taylor* 8355.			
July 22	7.0	17 57 4.10	...	133 25 40.5	M
881		*Anon.*			
July 31	7.7	17 59 25.07	...	150 24 12.2	M
882		*Anon.*			
July 29	8.7	18 1 13.07	...	131 48 35.0	M
883		*Anon.*			
July 28	8.2	18 2 53.60	...	131 44 27.6	M
884		13 *Sagittarii* μ¹			
July 14	...	18 5 41.29	...	111 5 28.0	R
22	...	5 41.37	...	5 29.0	M
24	...	5 41.37	...	5 29.6	M
26	...	5 41.30	3	5 27.6	M
27	...	5 41.37	...	5 27.9	M
Aug. 2	...	5 41.41	4	5 25.1	R
11	...	5 41.37	...	5 28.6	R
885		*Lalande* 33818.			
July 26	8.0	18 15 4.75	...	101 35 21.6	M
28	8.1	15 4.66	...	35 21.3	M
31	8.0	15 4.78	...	35 22.9	M
886		*Lalande* 33845.			
July 27	6.5	18 15 38.91	...	102 4 30.8	M
29	5.9	15 38.87	...	4 38.4	M
Aug. 2	7.0	15 38.91	...	4 30.3	M
887		23 *Ursae Minoris* δ			
Aug. 11	..	18 13 53.86	3	3 23 46.6	R
Sep. 1	...	13 53.72	3	23 45.0	R
		23 *Ursae Minoris* δ—s.p.			
Jan 17	...	18 13 53.67	1	3 23 47.2	R
27	...	13 53.61	3	23 48.5	R
Feb. 16	...	13 53.67	3	23 44.4	R
888		*Taylor* 8525.			
July 1		18 32 31.60	3	132 24 30.3	M
889		*Taylor* 8551.			
May 13	7.3	18 37 23.25		149 13 50.1	M

Separate Results of Madras Meridian Circle Observations in 1866.

Number and Date.	Magnitude.	Mean Right Ascension 1866. h.	m.	s.	No. of Wires.	Mean Polar Distance 1866. °	′	″	Observer.
890				*8 Lyrae α, Vega.*					
July 29	...	18	32	22·09	..	51	20	26·8	M
31	...		32	22·04	...		20	26·0	M
Aug. 5	...		32	22·05	4		20	25·5	R
17	.		32	21·95	4		20	26·8	R
23	...		32	21·92	...		20	25·2	R
891				*Anon.*					
Aug. 2	9·0	18	34	49·07	...	136	44	51·6	M
892				*Lacaille 7832.*					
July 1	...	18	37	40·93	3	140	5	27·1	M
893				*10 Lyrae β, Var. 1.*					
Aug. 2	...	18	45	5·61	...	56	47	33·1	R
17	...		45	5·71	...		47	33·0	R
18	...		45	5·68	...		47	33·4	R
23	...		45	5·64	...		47	32·7	R
26	...		45	5·58	...		47	32·2	R
Sep. 5	...		45	5·79	..		47	33·2	M
8	..		45	5·72	3		47	31·8	M
894				*32 Sagittarii ν[1]*					
May 13	...	18	44	1·11	3	112	51	24·3	M
895				*37 Sagittarii ξ[2]*					
May 13	...	18	48	40·47	...	111	16	32·6	M
896				*39 Sagittarii ο*					
Sep. 1		18	56	35·25	3	111	54	0·3	R
897				*R. P. L. 131—s.p.*					
Mar. 2	...	18	59	34·36	2	3	27	51·2	M
898				*17 Aquilae ζ*					
July 21		18	59	12·20	...	76	20	5·5	M
24			59	12·22			20	6·2	M
28			59	12·20	3		20	5·6	M
Aug. 18	...	18	59	12·17	...	76	20	6·5	R
23	...		59	12·23	3		20	6·2	R
25	...		59	12·24	...		20	5·3	R
26	...		59	12·28	...		20	6·0	R
Sep. 8	..		59	12·02	...		20	5·5	M
899				*Anon.*					
Aug. 2	8·5	19	9	25·64	...	120	47	3·6	R
900				*Anon.*					
Sep. 12	8·1	19	10	3·24	...	107	9	34·5	M
901				*25 Aquilae ω*					
July 26	...	19	11	28·73	...	75	38	44·1	M
28	..		11	28·67	...		38	45·9	M
29	.		11	28·67	...		38	44·1	M
Aug. 11			11	28·70	...		38	45·9	R
23			11	28·70			38	46·7	R
28			11	28·61	...		38	45·3	R
Sep. 11	...		11	28·71			38	45·6	M
902				*Anon.*					
Sep. 2	8·2	19	16	38·69		120	52	37·1	R
15	8·1		16	38·31	3		52	40·8	M
903				*30 Aquilae δ*					
July 26		19	18	41·33		87	9	7·2	M
28			18	41·42	...		9	7·4	M
31	...		18	41·40	..		9	7·7	M
Aug. 3			18	41·45	4		9	6·9	M
24			18	41·39	6		9	6·6	R
25			18	41·41	...		9	7·4	R
Sep. 1			18	41·31			9	6·3	R
13			18	41·31	.		9	7·3	M
904				*Taylor 8950.*					
Sep. 3		19	22	13·87	3	103	27	26·2	M
905				*Anon.*					
Sep. 2	9·3	19	25	0·13		129	28	47·4	R

Separate Results of Madras Meridian Circle Observations in 1865.

Number and Date.	Magnitude.	Mean Right Ascension 1865. h.	m.	s.	No. of Wires.	Mean Polar Distance 1865. °	′	″	Observer.
906			51 *Sagittarii h*[1]						
July 31	...	19	27	40·50	...	115	0	41·5	M
907			52 *Sagittarii h*[2]						
Aug. 24	...	19	28	29·27	...	115	10	42·3	B
30	...		28	29·38	...		10	42·5	B
Sep. 1	...		28	30·28	...		10	42·7	B
8	...		28	29·20	...		10	40·4	M
15	...		28	29·25	...		10	42·7	M
908			55 *Sagittarii c*[2]						
Sep. 1	...	19	34	47·60	...	106	26	15·4	B
2	...		34	47·50	...		26	13·1	B
909			50 *Aquilae* γ						
July 28	...	19	39	50·31	...	79	42	48·5	M
Aug. 18	...		39	50·35	...		42	49·6	B
25	...		39	50·30	...		42	48·0	B
28	...		39	50·32	...		42	49·2	B
Sep. 5	...		39	50·20	...		42	49·0	M
910			*S Vulpeculae Var.* 3.						
Sep. 1	9·0	19	42	31·64	...	63	2	52·6	B
911			53 *Aquilae* α, *Altair.*						
July 29	...	19	44	11·68	...	81	29	8·6	M
31	...		44	11·60	...		29	10·4	M
Aug 24	..		44	11·65	..		29	8·1	B
Sep. 2	...		44	11·71	...		29	7·8	B
5	...		44	11·62	...		29	9·7	M
11	...		44	11·72	...		29	9·0	M
912			60 *Aquilae* β						
Aug 25	...	19	48	40·80	...	83	55	41·1	B
Sep. 1	...		48	40·70	...		55	41·5	B
13	..		48	40·70			55	41·8	M
913			*Anon.*						
Sep. 2	9·1	19	53	5·41	5	105	10	40·9	B
914			*Anon.*						
Aug. 24	...	19	55	41·11	5	131	51	30·7	B
915			*Taylor* 9206.						
Sep. 11	...	19	56	46·70	...	122	25	56·3	M
916			*Anon.*						
Aug. 25	8·0	19	56	50·81	...	100	21	24·7	B
Sep. 1	9·2		56	50·86	5		21	22·8	B
12	9·0		56	50·84	3		21	21·3	M
917			λ *Ursae Minoris—s.p.*						
Jan. 21	...	19	30	9·63	1	1	5	44·0	B
Feb. 22	...		30	9·67	1		5	44·3	B
918			*O. A. N.* 20046.						
Aug 28	9·3	20	2	40·71	5	32	23	22·0	B
919			*R Capricorni Var.* 1.						
Sep. 29	9·7	20	3	44·00	3	104	39	34·0	B
920			*R Delphini Var* 2.						
Sep. 29	9·7	20	8	34·18	4	80	19	7·6	B
921			*O. A. S.* [illegible]						
Sep. 13	7·9	20	8	21·95	3	110	25	57·1	M
922			6 *Capricorni* α[2]						
Aug 8	...	20	10	33·60	...	102	57	39·2	B
Sep 1			10	33·61	.		57	39·0	B
2			10	33·66			57	39·0	B
5	..		10	33·67			57	39·9	M
11			10	33·59			57	39·4	M
12	...		10	33·65			57	39·0	M
14			10	33·72			57	40·0	M
15	.		10	33·60			57	39·3	B

Separate Results of Madras Meridian Circle Observations in 1865.

Number and Date.	Magnitude.	Mean Right Ascension 1865. h.	m.	s.	No. of Wires.	Mean Polar Distance 1865. °	'	"	Observer.
923				*Anon.*					
Aug. 21	7·6	20	10	33·00	...	140	8	30·2	R
924				*Anon.*					
Aug 25	8·0	20	11	18·74	3	106	16	33·2	R
925				*Lalande* 39095.					
July 28	8·4	20	11	44·38	...	105	13	18·3	M
Aug. 28	8·2		11	44·36	5		13	18·2	M
926				*α Pavonis.*					
Aug 8		20	11	56·85	...	147	9	53·1	R
Sep 16	...		11	56·79	4		9	51·4	R
927				*X Capricorni Var.* 7.					
Sep 30	9·9	20	13	1·81	3	104	20	23·5	M
928				*Lalande* 39125.					
July 31	8·8	20	13	34·66	..	105	13	10·0	M
929				*Lacaille* 8441.					
Sep 2	8·8	20	18	16·92	...	121	6	15·9	R
13	8·0		18	16·92	..		6	47·0	M
930				11 *Capricorni ρ*					
Sep 11		20	21	9·27	...	108	15	27·0	M
12			21	9·26			15	26·1	M
14			21	9·22	..		15	27·1	M
16			21	9·31			15	26·9	R
21			21	9·31	5		15	26·8	R
931				*Anon.*					
Sep 1	..	20	23	49·11	5	81	1	59·3	R
932				*Lalande* 39525.					
Aug 29	8·0	20	28	29·94	..	103	2	16·9	R
933				*Anon.*					
Sep. 2	9·3	20	27	16·94	3	121	3	40·2	R
934				*Anon.*					
Aug. 24	9·5	20	30	37·38	4	110	46	10·7	M
935				*S Capricorni Var.* 2.					
Aug. 28	9·0	20	31	0·88	..	109	32	10·1	R
936				*Anon.*					
Sep 23	6·9	20	36	30·94	1	128	30	17·3	R
937				50 *Cygni α, Deneb.*					
Sep 3	...	20	36	40·81		45	12	3·0	M
12	...		36	40·73	...		12	3·0	M
13	...		36	40·82	...		12	3·4	M
14	...		36	40·77	..		12	3·1	M
15			36	40·81	..		12	1·6	M
938				2 *Aquarii ε*					
Sep. 2	...	20	40	21·98	...	99	50	13·9	R
939				*W. B. E. XX.* 1024.					
Sep 25	9·3	20	40	56·20		103	34	5·0	R
940				*Anon.*					
Sep 22	10·5	20	41	12·92	5	105	18	5·8	R
941				*T Aquarii Var.* 4.					
Sep 1	9·5	20	42	46·61		95	28	41·6	R
942				*Lacaille* 8571.					
Aug 24	..	20	42	59·66	..	130	19	13·6	R
943				*Anon.*					
Aug. 29	9·7	20	43	40·63		121	57	56·2	M
Sep. 29			43	40·91			57	55·6	R

Separate Results of Madras Meridian Circle Observations in 1865.

Number and Date.	Magnitude.	Mean Right Ascension, 1865. h.	m.	s.	No. of Wires.	Mean Polar Distance 1865. °	′	″	Observer.
944			*Taylor* 9633.						
Sep. 11	7·2	20	41	37·31	5	101	56	33·1	M
Oct. 4	7·6		41	37·47	...		56	31·8	M
945			*Anon.*						
Sep 23	...	20	47	40·86	...	110	1	3·4	R
946			32 *Vulpeculae.*						
Sep. 2	...	20	48	48·35	...	62	27	14·2	R
14	..		48	48·30	...		27	15·6	M
15	...		48	48·25	...		27	15·1	M
16	...		48	48·27	...		27	16·8	R
21	.		48	48·31	...		27	14·3	R
30	...		48	48·28	...		27	17·0	R
947			*Anon.*						
Sep. 22	9·7	20	50	46·82	4	108	45	30·4	R
948			*Lacaille* 8630.						
Sep. 28	9·2	20	51	39·98	..	128	37	48·3	R
30	...		51	40·28	5		37	49·6	R
949			*Lacaille* 8635.						
Aug. 28	8·0	20	52	22·60	5	126	31	50·1	R
Sep 11	7·6		52	22·71	4		31	50·1	M
950			*Anon.*						
Sep 25	9·5	20	54	1·91	...	102	50	0·6	R
951			*Taylor* 9772.						
Aug 8	8·0	21	0	31·56	5	116	7	3·2	R
952			61 *Cygni 1st*						
Sep 21	.	21	0	50·69	..	61	51	45·0	R
25			0	50·78	.		51	47·7	R
29			0	50·60	3		51	47·1	R
953			*Anon.*						
Sep. 22	9·5	21	2	10·60	6	145	6	29·5	R
954			*Lacaille* 8712.						
Aug. 24		21	4	15·71	5	116	48	17·9	R
Oct. 9	8·4		4	16·20	5		48	18·3	M
10	8·0		4	15·87	...		48	18·3	M
955			64 *Cygni* 3						
Sep. 21	...	21	7	11·31	5	60	19	30·9	R
23	...		7	11·30	...		19	32·5	R
29	...		7	11·32	...		19	33·6	R
Oct. 2	...		7	11·40			19	32·9	M
3	...		7	11·30	...		19	32·1	M
956			*Anon.*						
Sep. 26	10·5	21	8	20·18	4	110	49	40·3	R
957			*Anon.*						
Aug 28	9·7	21	11	4·97	5	140	31	41·9	R
Oct. 14	9·5		11	5·17	4		31	42·3	M
958			*Brisbane* 7012.						
Oct. 12	7·9	21	11	9·52		131	45	46·1	M
959			*T Capricorni Var* 3.						
Aug. 24	8·0	21	14	28·01	...	105	39	55·1	R
Sep. 23	9·5		14	28·77	5		39	56·6	R
960			*Anon.*						
Sep 22	9·3	21	15	3·75		130	15	51·3	R
30	9·8		15	3·91	..		15	52·7	R
Oct 1	9·3		15	3·85			15	52·7	M
961			5 *Cephei* α, *Alderamin*						
Sep 20		21	15	21·13	5	27	80	5·0	R
29			15	21·22			30	3·2	R

Separate Results of Madras Meridian Circle Observations in 1865.

Number and Date.	Magnitude.	Mean Right Ascension 1865. h.	m.	s.	No. of Wires.	Mean Polar Distance 1865. °	′	″	Observer.
962				*Anon.*					
Oct 13	9·3	21	13	38·91		133	52	19·1	M
963				*Taylor* 9931.					
Sep. 11	6·9	21	13	46·01	...	142	53	8·1	M
14	6·9		13	46·60			53	9·6	M
964				*Anon.*					
Sep. 26	9·2	21	20	10·60	...	150	47	30·0	R
Oct. 14	8·6		20	10·31	...		47	31·9	M
965				*Anon.*					
Sep. 2	9·5	21	23	1·20	...	110	7	12·0	R
22	9·8		23	0·94	6		7	15·7	R
966				22 *Aquarii* β					
Sep 6	...	21	24	27·05	...	96	9	39·4	M
13	...		24	27·02	...		9	40·0	M
21	...		24	26·97			9	40·2	R
23	...		24	27·05	5		9	40·2	R
29	...		24	27·04	..		9	40·1	R
Oct. 2	...		24	27·01	5		9	50·5	M
3	...		24	27·02	...		9	40·6	M
967				*Anon.*					
Oct. 4	9·2	21	25	58·37	...	140	23	11·1	M
968				8 *Cephei* β					
Sep. 30		21	26	54·58	...	20	1	52·4	R
969				*Anon.*					
Sep 23	9·0	21	29	32·99	4	134	2	16·0	R
970				*Anon.*					
Sep. 22	9·3	21	33	56·93	...	103	0	8·0	R
25	9·5		33	56·37	5		0	6·6	R
971				*Taylor* 10068.					
Oct. 4	7·6	21	34	24·78	4	134	6	27·0	M
972				*Taylor* 10065.					
Sep. 14	6·1	21	34	32·48	...	105	6	31·4	M
973				*Anon.*					
Oct. 10	9·0	21	34	46·65	...	134	0	11·5	M
974				*S Cephei Var.* 3.					
Sep. 23	8·2	21	36	50·48	3	11	50	2·3	R
Oct. 9	8·0		36	50·88	...		50	1·8	M
11	7·9		36	50·64	..		50	1·6	M
975				8 *Pegasi* ε					
Sep. 6	...	21	37	33·18		80	44	38·5	M
13	...		37	33·27	3		44	38·3	M
29	...		37	33·20	...		44	38·7	R
30	...		37	33·05	...		44	33·6	R
976				μ *Cephei Var.* 1.					
Oct 2	6·1	21	39	22·67	5	31	30	10·3	M
977				*Anon.*					
Aug. 24	10·0	21	40	51·61	3	102	32	10·9	R
Sep. 23	...		40	51·92	5		32	9·7	R
26	10·0		40	51·94	5		32	7·2	R
978				*Taylor* 10126.					
Oct 12	7·0	21	41	3·40		137	14	9·7	M
979				*Anon.*					
Oct. 4	9·1	21	42	36·67	.	132	31	9·2	M
980				16 *Pegasi.*					
Sep 22	...	21	46	55·20	5	61	42	31·0	R
24			46	55·17	..		42	32·6	R
30			46	55·26	.		42	33·4	R

Separate Results of Madras Meridian Circle Observations in 1865

Number and Date.	Magnitude.	Mean Right Ascension 1865. h.	m	s	No. of Wires.	Mean Polar Distance. 1865. °	′	″	Observer.
981				*Lacaille 8938.*					
Sep 6	7.5	21	47	16.56	...	133	53	3.5	M
11	7.5		47	16.53	...		53	3.7	M
Oct. 14	7.9		47	16.76	...		53	3.2	M
982				*Anon.*					
Oct. 10	9.0	21	47	38.80		133	12	13.8	M
11	9.0		47	38.77	..		12	13.6	M
983				*Taylor 10190.*					
Oct. 9	7.6	21	51	10.22	3	146	31	39.7	M
12	6.3		51	10.13	...		31	40.1	M
984				*Anon.*					
Sep. 22	9.3	21	52	30.45	4	136	37	53.9	R
985				*Anon.*					
Sep. 23	9.3	21	58	51.17	...	150	40	0.7	R
986				*κ² Indi.*					
Sep. 6	6.5	21	56	20.61	5	150	17	16.4	M
13	6.4		56	20.58	...		17	16.0	M
Oct 3	6.6		56	20.75	...		17	16.5	M
11	6.1		56	20.58	...		17	16.7	M
987				*Anon.*					
Oct. 10	7.7	21	58	15.81	4	136	2	16.7	R
14	8.0		58	15.94	..		2	17.9	M
988				*34 Aquarii α*					
Sep 22	...	21	58	50.[illegible]		90	58	30.6	R
25			58	50.85			58	29.1	R
Oct 4			58	50.80			58	29.4	M
16			58	50.[illegible]			58	29.[illegible]	R
17			58	50.91			58	29.2	R
Nov 1			58	50.98			58	29.1	M
989				*α Gruis.*					
Sep 30	...	21	59	[illegible]		137	38	40.7	R
990				*Anon.*					
Sep 23	9.3	22	3	22.09	...	101	8	37.1	R
Oct. 13	9.7		3	22.18	5		8	37.3	M
991				*Anon.*					
Sep. 23	8.5	22	5	27.46	..	101	5	29.3	R
25	9.0		5	27.53	5		5	32.0	R
Oct. 4	9.1		5	27.57	...		5	30.2	M
992				*W. B. E. XXII. 98.*					
Sep. 6	8.0	22	6	24.80	.	90	28	29.8	M
23	7.8		6	24.95			28	30.7	R
993				*Anon*					
Oct 11	8.0	22	9	12.08	..	116	27	1.5	M
12	8.0		9	12.07	.		27	1.2	M
994				*43 Aquarii θ*					
Sep. 7	...	22	9	42.34	3	98	37	16.1	M
11			9	42.47	4		37	13.9	M
25	...		9	42.62	5		37	20.0	R
30			9	42.49	5		37	13.6	R
Oct. 2	...		9	42.44	3		37	17.4	M
9			9	42.50			37	13.7	M
16	...		9	42.47			37	16.7	R
17			9	42.40	.		37	16.8	R
18			9	42.41	.		37	19.1	R
Nov. 1	...		9	42.40	5		37	16.1	M
995				*Anon.*					
Oct 10	7.5	22	12	31.15	.	100	37	32.3	M
14	7.9		12	31.13			37	32.1	M
996				*Anon.*					
Oct 12	9.0	22	13	13.92		144	39	10.1	M
997				*Anon.*					
Sep 6	8.7	22	15	33.71	3	88	47	7.5	M
28	8.8		15	29.94	...		47	7.0	R

Separate Results of Madras Meridian Circle Observations in 1865.

Number and Date.	Magnitude.	Mean Right Ascension 1865. h.	m.	s.	No. of Wires.	Mean Polar Distance 1865. °	′	″	Observer.
998			*Anon.*						
Oct. 16	10·2	22	16	2·31	3	92	42	16·1	R
999			*Anon.*						
Oct. 4	9·5	22	16	55·09	5	136	58	6·5	M
13	9·5		16	55·23	...		58	6·9	M
1000			*Anon.*						
Oct. 11	9·3	22	18	54·56	5	140	46	27·7	M
1001			55 *Aquarii* ζ—1*st.*						
Nov. 1	...	22	21	52·86	...	90	42	34·8	M
1002			55 *Aquarii* ζ—2*nd.*						
Oct. 9	...	22	21	52·94	3	90	42	38·3	M
12	...		21	52·81	...		42	33·5	M
Nov. 3	...		21	52·67	...		42	34·1	M
1003			*Anon.*						
Sep. 30	9·9	22	21	51·30	...	108	37	29·7	R
1004			57 *Aquarii* σ						
Aug. 8	5	22	23	30·01	5	101	23	6·2	R
Oct. 2	...		23	30·06	...		23	3·6	M
1005			*R. P. L.* 150.						
Sep. 18	...	22	23	35·27	3	4	34	28·8	M
Oct. 14	...		23	34·21	5		34	24·0	M
			R. P. L. 150.—*s.p.*						
Mar. 28	...	22	23	34·69	3	4	34	22·9	R
1006			*Anon.*						
Oct. 10	9·3	22	24	21·89	...	128	41	51·5	M
1007			*Anon.*						
Sep. 6	8·9	22	25	35·73	...	141	29	56·2	R
Oct. 18	...		25	35·97	...		29	55·3	R
1008			62 *Aquarii* η						
Sep. 5	...	22	28	26·08	...	90	49	45·8	R
7	...		28	25·96	5		49	46·1	M
22	...		28	24·98	...		49	46·6	R
23	...		28	25·11	...		49	46·7	R
25	...		28	25·08	...		49	44·9	R
Oct. 4	...		28	25·11	...		49	45·0	M
13	...		28	25·13	...		49	44·2	M
17	...		28	26·02	...		49	46·1	R
23	...		28	26·10	...		49	45·3	R
1009			*T Aquarii Var.* 3.						
Sep. 2	10·0	22	28	48·31	...	95	13	15·2	R
28	10·5		28	49·46	3		13	12·5	R
1010			*Lacaille* 9188.						
Sep. 30	7·0	22	29	57·48	4	130	55	24·6	R
1011			*Taylor* 10477.						
Oct. 11	6·3	22	32	11·34	5	146	7	30·5	M
Nov. 1	...		32	11·38	...		7	28·3	M
1012			*Anon.*						
Oct. 12	9·0	22	34	16·10	...	126	31	2·0	M
14	9·0		34	16·13	5		31	0·4	M
1013			42 *Pegasi* ζ						
Sep. 6	...	22	34	48·84	...	79	52	28·7	M
7	...		34	48·72	...		52	31·9	M
22	...		34	48·67	...		52	28·3	R
23	...		34	48·70	...		52	30·9	R
Oct. 4	...		34	48·85	...		52	31·7	M
9	...		34	48·80	...		52	28·3	M
10	...		34	48·74	...		52	31·9	M
16	...		34	48·67	...		52	29·0	R
18	...		34	48·69	...		52	31·4	R
26	...		34	48·76	...		52	30·4	R
Nov. 8	...		34	48·74	...		52	32·1	M
1014			*Anon.*						
Oct. 27	9·1	22	36	34·00	...	130	26	30·8	R

Separate Results of Madras Meridian Circle Observations in 1865.

Number and Date.	Magnitude.	Mean Right Ascension 1865. h.	m.	s.	No. of Wires.	Mean Polar Distance 1865. °	′	″	Observer.
1015			*Lacaille 9226.*						
Sep. 2	7·0	22	37	42·80	5	146	46	18·9	R
1016			*Anon.*						
Oct. 13	8·9	22	40	55·86	...	142	37	44·4	M
1017			*Anon.*						
Sep. 28	9·0	22	44	44·88	5	130	0	58·6	R
Oct. 3	8·0		44	45·15	...		0	58·4	M
4	8·0		44	45·04	...		0	59·2	M
1018			*Anon.*						
Oct. 11	9·6	22	44	47·73	...	146	32	43·8	M
16	9·6		44	47·71	...		32	44·5	R
1019			*Anon.*						
Sep. 25	9·0	22	44	54·19	4	149	34	13·0	R
1020			*Anon.*						
Oct. 13	9·0	22	49	16·70	...	136	27	38·6	M
1021			*24 Piscis Australis α, Fomalhaut.*						
Oct. 23	...	22	50	11·10	...	120	20	14·4	R
25	...		50	11·14	...		20	14·1	R
26	...		50	11·19	...		20	15·8	R
1022			*Anon.*						
Oct. 27	8·9	22	50	23·76	...	110	50	49·6	R
Nov. 6	7·8		50	22·63	...		50	44·3	M
1023			*Anon.*						
Sep. 28	9·3	22	51	29·73	5	151	22	57·9	R
Oct. 16	9·6		51	29·11	5		22	58·2	R
1024			*Lacaille 9353.*						
Oct. 3	6·5	22	56	39·80	...	144	41	17·7	M
9	6·6		56	39·75	..		41	18·4	M
1025			*54 Pegasi α, Markab.*						
Sep. 18	...	22	58	2·21	5	75	31	15·5	M
23	...		58	2·13	...		31	15·3	R
25	...		58	2·19	...		31	15·9	R
26	...		58	2·24	...		31	15·0	R
Oct. 11	...		58	2·15	...		31	15·2	M
12	...		58	2·33	...		31	14·6	M
14	...		58	2·33	...		31	14·7	M
18	...		58	2·22	...		31	14·6	R
23	...		58	2·18	...		31	14·4	R
24	...		58	2·20	...		31	15·0	R
30	...		58	2·14	...		31	13·9	R
31	...		58	2·20	...		31	16·4	R
1026			*Anon.*						
Oct. 13	8·9	22	59	24·02	...	100	21	47·5	M
Nov. 11	8·3		59	23·77	5		21	47·0	M
1027			*R Pegasi Var. 2.*						
Sep. 28	10·2	22	59	52·26	5	80	11	7·6	R
29	9·8		59	52·31	...		11	9·4	R
Oct. 16	10·5		59	52·12	5		11	7·8	R
1028			*Anon.*						
Sep. 28	9·7	23	4	21·80	4	130	46	59·1	R
Oct. 27	9·5		4	21·95	...		46	56·6	R
1029			*Lacaille 9394.*						
Oct. 9	8·0	23	5	13·36	...	145	50	21·0	M
Nov. 10	8·0		5	13·32	...		50	19·3	M
1030			*Lacaille 9403.*						
Oct. 14	8·0	23	7	30·65	5	124	25	46·5	M
23	8·0		7	30·15	...		25	46·2	R
1031			*Anon.*						
Nov. 13	8·8	23	8	11·74	4	100	30	39·5	M
1032			*Lacaille 9423.*						
Nov. 15	7·0	23	9	39·57	5	151	44	16·9	M

Separate Results of Madras Meridian Circle Observations in 1866.

Number and Date.	Magnitude.	Mean Right Ascension 1865. h.	m.	s.	No. of Wires.	Mean Polar Distance 1865. °	′	″	Observer.
1033				6 *Piscium* γ					
Sep. 28	...	23	10	10·06	5	87	27	19·1	R
29	...		10	9·97	...		27	18·4	R
Oct. 2	...		10	9·88	...		27	18·7	M
3	...		10	9·88	...		27	17·6	M
10	...		10	9·86	...		27	16·6	M
11	...		10	9·90	...		27	18·4	M
12	...		10	9·90	...		27	19·0	M
13	...		10	9·95	...		27	17·2	M
24	...		10	10·08	...		27	18·3	R
26	...		10	9·98	...		27	17·3	R
28	...		10	10·08	...		27	15·4	R
Nov. 3	...		10	9·91	...		27	17·7	M
1034				*Anon.*					
Oct. 16	9·7	23	11	0·53	...	127	26	15·5	R
27	9·9		11	0·40	5		26	13·4	R
1035				*Anon.*					
Oct. 16	9·5	23	12	16·57	8	127	34	38·1	R
27	9·7		12	16·44	5		34	36·7	R
1036				*Anon.*					
Sep. 26	9·0	23	15	30·40	...	130	45	36·6	R
Nov. 11	8·7		15	30·51	...		45	36·8	M
1037				*Taylor* 10748.					
Nov. 14	6·0	23	17	36·38	...	147	35	38·4	M
1038				8 *Piscium* κ					
Sep. 26	...	23	20	0·60	5	89	29	0·5	R
Oct. 2	...		20	0·60	...		29	0·5	M
3	...		20	0·70	...		29	0·0	M
10	...		20	0·77	...		28	59·9	M
11	...		20	0·71	...		29	0·3	M
12	...		20	0·69	...		28	58·6	M
13	...		20	0·64	...		28	59·1	M
14	...		20	0·72	...		29	0·4	M
24	...		20	0·61	...		29	0·5	R
26	...		20	0·67	...		28	59·1	R
Nov. 1	...		20	0·66	...		29	0·0	M
3	...		20	0·70	...		29	0·1	M
8	...		20	0·67	...		28	59·1	M
1039				*Anon.*					
Nov. 13	7·9	23	20	46·61	...	98	51	28·2	M
15	7·8		20	46·61	...		51	27·0	M
1040				*Anon.*					
Oct. 27	9·6	23	21	6·89	5	137	27	36·8	R
1041				*Anon.*					
Sep. 29	9·0	23	23	40·77	...	149	57	16·1	R
Oct. 5	8·0		23	40·84	5		57	16·2	M
1042				*Anon.*					
Nov. 14	9·6	23	25	33·92	...	139	51	40·3	M
1043				*Taylor* 10804.					
Nov. 10	6·9	23	27	36·06	4	147	34	17·9	M
1044				*Anon.*					
Sep. 26	8·6	23	27	49·30	5	148	14	27·3	R
Oct. 27	8·5		27	49·28	...		14	27·0	R
1045				*Anon.*					
Sep. 22	8·7	23	29	52·98	...	148	54	29·1	R
Oct. 13	7·9		29	53·12	...		54	29·0	M
24	8·8		29	53·06	5		54	29·3	R
1046				*Anon.*					
Nov. 15	9·1	23	29	57·86	...	127	19	40·3	M
1047				*Anon.*					
Sep. 22	8·3	23	30	27·90	4	135	56	22·3	R
Oct. 21	8·8		30	29·41	5		56	23·0	R
1048				17 *Piscium* ι					
Sep. 5	...	23	33	0·27	...	85	6	19·2	M
6	...		33	0·41	...		6	19·3	M
26	...		33	0·30	...		6	19·6	R

Separate Results of Madras Meridian Circle Observations in 1865.

Number and Date.	Magnitude.	Mean Right Ascension 1865. h.	m.	s.	No. of Wires.	Mean Polar Distance 1865. °	'	"	Observer.
Oct. 9	...	23	36	0.29	...	86	6	19.5	M
14	...		36	0.31	...		6	19.6	M
16	...		36	0.38	...		6	19.3	R
23	...		36	0.48	...		6	18.3	R
25	...		36	0.38	...		6	18.7	R
27	...		36	0.34	...		6	18.7	R
31	...		36	0.42	...		6	19.6	R
Nov. 8	...		36	0.46	...		6	17.9	M
1049			*Lacaille* 9583.						
Oct. 13	...	23	38	53.89	...	128	43	38.1	R
1050			19 *Piscium.*						
Sep. 5	...	23	39	29.57	...	87	15	46.2	M
6	...		39	29.53	...		15	44.6	M
1051			*Anon.*						
Oct. 13	...	23	41	7.84	...	128	46	29.3	R
1052			δ *Sculptoris.*						
Oct. 5	...	23	41	53.38	...	118	52	37.3	M
25	...		41	53.41	...		52	37.0	R
27	...		41	53.36	6		52	36.4	R
Nov. 10	...		41	53.31	..		52	37.0	M
13	—		41	53.39	...		52	36.9	M
1053			*Anon.*						
Nov. 11	8.7	23	46	6.02	...	150	49	37.0	M
1054			*Lalande* 46050.						
Sep. 30	9.3	23	48	9.20	...	88	18	53.6	R
Oct. 2	9.2		48	9.14	...		18	56.6	M
24	9.2		48	9.09	...		18	55.7	R
Nov. 8	9.2		48	9.05	...		18	53.7	M
1055			*Anon.*						
Sep. 28	9.3	23	47	49.65	5	129	50	37.3	R
1056			*Anon.*						
Sep. 26	9.8	23	50	0.41	...	148	53	5.7	R
Oct. 27	8.8		50	0.41	5		53	4.3	R
1057			*Anon.*						
Oct. 5	8.8	23	51	43.56	...	132	30	13.6	M
1058			28 *Piscium* ω						
Oct. 3	...	23	52	22.78	...	83	53	3.0	M
4	...		52	22.73	...		53	2.4	M
9	...		52	22.79	...		53	3.9	M
10	...		52	22.80	...		53	3.8	M
11	...		52	22.87	...		53	3.6	M
12	...		52	22.73	...		53	2.4	M
28	...		52	22.80	...		53	3.8	R
31	...		52	22.80	...		53	4.7	R
Nov. 10	...		52	22.77	...		53	3.4	M
1059			*Lacaille* 9686.						
Sep. 30	7.0	23	53	35.67	...	146	30	54.2	R
1060			*Anon.*						
Sep. 22	9.5	23	56	1.49	...	130	16	44.6	R
Nov. 11	9.0		56	1.61	...		16	41.6	M
1061			*Anon.*						
Oct. 24	...	23	56	16.86	...	124	7	23.3	R
Nov. 8	8.0		56	16.78	5		7	25.5	M
1062			*Taylor* 10994.						
Sep. 26	9.0	23	57	30.35	...	147	35	41.2	R
Oct. 13	8.6		57	30.70	...		35	42.4	M
27	8.3		57	30.47	...		35	39.8	R
1063			*Lacaille* 9721.						
Sep. 28	6.5	23	59	18.90	...	139	49	34.1	R
Oct. 10	6.9		59	19.11	...		49	34.1	M

MEAN POSITIONS OF STARS

OBSERVED WITH THE

MADRAS MERIDIAN CIRCLE

IN THE YEAR

1865

REDUCED TO JANUARY 1, OF THAT YEAR.

Mean Positions of Stars for 1865 January 1st.

Number.	Star.	Magnitude.	Estimations.	Mean Right Ascension.			Mean Polar Distance.			Observations.	Fraction of Year.
				h.	m.	s.	°	′	″		
1	Taylor 11011	7·0	1	0	0	31·84	147	36	17·4	1	0·92
2	... ...	9·0	4	0	0	45·00	151	23	38·9	4	0·86
3	21 Androm. α (*Alpheratz*)...	2·0	...	0	1	34·81	61	30	15·6	4	0·84
4	Lacaille 9789	7·6	1	0	2	7·30	130	20	19·0	1	0·78
5	Taylor 7	7·1	1	0	3	0·31	96	18	46·0	1	0·90
6	Lacaille 9787	7·8	3	0	4	28·20	131	7	27·6	3	0·77
7	Lacaille 8	6·7	1	0	6	0·36	146	39	55·7	1	0·80
8	... ...	9·1	2	0	6	14·27	131	7	38·4	2	0·74
9	88 Pegasi γ (*Algenib*) ..	3·0	...	0	6	17·10	75	34	2·3	7	0·82
10	... ...	9·8	3	0	6	41·64	131	6	42·3	3	0·76
11	... ...	8·9	...	0	9	36·81	149	31	30·8	1	0·79
12	... ...	9·1	2	0	9	36·28	153	54	49·0	2	0·78
13	... ...	8·6	3	0	10	37·71	162	0	38·7	3	0·84
14	Lacaille 41 ...	8·0	1	0	12	36·74	130	51	42·3	1	0·97
15	41 Piscium d ...	5·6	...	0	13	30·14	82	39	35·3	3	0·81
16	Lacaille 61	7·0	1	0	16	5·66	130	0	33·5	1	0·90
17	R Andromedæ Var. 1 ...	6·7	3	0	16	54·38	52	10	15·3	3	0·91
18	... ...	9·5	1	0	17	40·60	140	34	51·7	1	0·92
19	... ...	9·1	2	0	18	36·96	152	57	16·6	2	0·97
20	Lacaille 81	7·3	3	0	18	41·31	130	0	20·1	4	0·81
21	10 Ceti	6·7	...	0	19	42·00	90	47	58·7	1	0·91
22	12 Ceti	5·4	...	0	23	8·96	94	42	14·1	11	0·95
23	... ...	7·7	2	0	27	10·97	76	13	45·1	2	0·81
24	Lacaille 132	9·0	...	0	27	21·40	151	53	37·1	1	0·79
25	Lalande 970	7·0	4	0	31	7·88	80	54	40·0	4	0·81
26	... .	8·2	3	0	31	29·36	82	34	2·2	3	0·89
27	Lalande 1010	9·0	1	0	32	18·38	82	32	6·9	1	0·81
28	18 Cassiopeæ α Var. 2 ...	2·2	...	0	33	51·73	34	12	14·6	3	0·90
29	.. ..	9·3	1	0	36	1·75	83	28	34·8	1	0·81
30	16 Ceti β	2·1	..	0	36	49·65	108	43	42·0	7	0·57
31	W. B. R. 0 . 698 ...	9·3	...	0	38	57·08	93	40	9·6	1	0·79
32	W. B. R. 0 . 705 ..	7·9	1	0	41	11·31	94	26	57·3	1	0·93
33	Taylor 286	5·7	1	0	41	18·16	85	24	32·3	1	0·88
34	63 Piscium δ	4·6	1	0	41	40·94	83	9	1·1	3	0·73
35	W. B. R. 0 . 716 ...	9·2	2	0	41	44·30	94	36	30·0	2	0·84

17.—R Andromedæ Var. 1.—Period 406 days.—Range, 6th to 13th magnitude.
25.—27.—Comparison stars for Ariadne in 1861.
28.—18 Cassiopeæ α Var. 2 (*Schedir*).—Irregular, Range, 2·2 to 2·8 magnitude.
31.—32.—35.—Comparison stars for Europa in 1861.

Observed with the Madras Meridian Circle in that Year.

No.	Star.	In Right Ascension. Annual Precession.	In Right Ascension. Secular Variation.	In Right Ascension. Proper Motion.	In Polar Distance. Annual Precession.	In Polar Distance. Secular Variation.	In Polar Distance. Proper Motion.	Number in B.A.C.
		s	s	s	"	"		
1	Taylor 11010... ..	+ 3·0672	− 0·0462	...	− 20·064	+ 0·010		8877
2	... .	+ 3·0641	− 0·0520	...	− 20·065	+ 0·010	...	...
3	21 Andromedæ α ..	+ 3·0765	+ 0·0182	+ 0·009	− 20·065	+ 0·013	+ 0·16	4
4	Lacaille 9739 ...	+ 3·0015	− 0·0233	...	− 20·064	+ 0·016	...	...
5	Taylor 7	+ 3·0711	+ 0·0004	..	− 20·063	+ 0·015	...	12
6	Lacaille 9797 ...	+ 3·0493	− 0·0286	...	− 20·051	+ 0·018	...	22
7	Lacaille 3	+ 3·0131	− 0·0440	...	− 20·048	+ 0·021	...	...
8	... ...	+ 3·0403	− 0·0288	...	− 20·046	+ 0·021	...	...
9	88 Pegasi γ (*Algenib*)	+ 3·0814	+ 0·0100	0·000	− 20·045	+ 0·022	+ 0·02	26
10	.. - ...	+ 3·0850	− 0·0282	...	− 20·040	+ 0·026	...	...
11	... .	+ 2·9780	− 0·0462	...	− 20·038	+ 0·027		—
12	... ...	+ 2·9578	− 0·0540	...	− 20·037	+ 0·027	...	...
13	... .	+ 2·9555	− 0·0498	...	− 20·034	+ 0·030	...	...
14	Lacaille 41	+ 3·0096	− 0·0221	...	− 20·024	+ 0·033	...	...
15	41 Piscium d ...	+ 3·0825	+ 0·0066	− 0·002	− 20·019	+ 0·036	− 0·01	66
16	Lacaille 61	+ 2·9983	− 0·0300	...	− 20·006	+ 0·040		...
17	R Andromedæ Var. 1	+ 3·1496	+ 0·0271	...	− 20·000	+ 0·046		...
18	... .	+ 2·8960	− 0·0419	...	− 19·996	+ 0·042		...
19	.. ..	+ 2·8601	− 0·0472	...	− 19·989	+ 0·043		...
20	Lacaille 81	+ 2·9807	− 0·0305	...	− 19·980	+ 0·044		...
21	10 Ceti	+ 3·0705	+ 0·0020	+ 0·006	− 19·961	+ 0·047	+ 0·03	96
22	12 Ceti	+ 3·0609	+ 0·0008	− 0·002	− 19·954	+ 0·065	+ 0·01	112
23	.. ..	+ 3·1109	+ 0·0109	...	− 19·914	+ 0·068	...	.
24	Lacaille 132	+ 2·7741	− 0·0418	...	− 19·912	+ 0·067	...	...
25	Lalande 970	+ 3·1011	+ 0·0085	...	− 19·870	+ 0·070	..	...
26	.. ..	+ 3·0968	+ 0·0076	...	− 19·866	+ 0·071		...
27	Lalande 1010 ..	+ 3·0968	+ 0·0076	...	− 19·865	+ 0·072	.	...
28	18 Cassiopeæ α Var. 2	+ 3·3831	+ 0.0563	+ 0·006	− 19·840	+ 0·080	+ 0·04	169
29	.. ..	+ 3·0964	+ 0·0078	...	− 19·821	+ 0·078	...	.
30	16 Ceti β	+ 3·0086	− 0·0056	+ 0·018	− 19·797	+ 0·080	− 0·02	196
31	W. B. E. 0 . 698 ...	+ 3·0878	+ 0·0080	..	− 19·706	+ 0·080	...	—
32	W. B. E. 0 . 705 ..	+ 3·0686	+ 0·0019	...	− 19·781	+ 0·087	...	...
33	Taylor 236	+ 3·0013	+ 0·0001	..	− 19·780	+ 0·089	..	221
34	63 Piscium δ . ..	+ 3·1011	+ 0·0077	+ 0·008	− 19·724	+ 0·090	+ 0·05	222
35	W. B. E. 0 . 716 ..	+ 3·0520	+ 0·0016	..	− 19·723	+ 0·088	.	.

22.—Proper motions from "*Greenwich Catalogue*, 1872."

Mean Positions of Stars for 1865 January 1st.

Number.	Star.	Magnitude.	Estimations.	Mean Right Ascension.			Mean Polar Distance.			Observations.	Fraction of Year.
				h.	m.	s.	°	′	″		
36	... ...	9·2	1	0	42	58·12	89	4	1·4	2	0·84
37	... ...	9·0	1	0	47	39·58	158	39	57·1	1	0·84
38	λ¹ Toucani .	6·2	3	0	49	0·22	158	30	21·0	3	0·80
39	... ...	9·1	2	0	49	57·82	138	46	56·0	2	0·81
40	Lacaille 264 ...	7·9	1	0	50	47·25	154	41	46·1	1	0·83
41	2 Ursæ Minoris	4·5	...	0	50	51·64	4	28	8·9	1	0·51
42	W. B. R. 0 . 897 ...	9·2	1	0	52	16·52	92	49	35·7	1	0·91
43	Lacaille 271	7·4	2	0	52	44·98	151	25	38·3	2	0·83
44	70 Piscium	6·9	1	0	55	5·80	82	47	18·3	3	0·83
45	71 Piscium ε	4·5	..	1	55	56·90	82	50	16·8	13	0·84
46	... ...	8·0	3	1	5	46·39	129	52	46·7	3	0·92
47	80 Piscium ζ, 1st... ...	5·4	...	1	6	40·76	83	8	28·6	4	0·93
48	... ...	8·0	2	1	7	38·42	152	30	46·1	2	0·98
49	1 Ursæ Minoris α (*Polaris*)	2·2	...	1	9	37·88	1	24	39·2	11	0·86
50	... ...	9·7	3	1	10	11·69	81	49	30·5	3	0·85
51	... ...	8·8	3	1	10	21·74	158	51	45·4	3	0·88
52	... ...	9·1	1	1	12	10·44	132	17	10·2	1	0·86
53	... ...	8·0	2	1	17	3·19	96	31	5·8	2	0·90
54	45 Ceti θ¹	3·8	...	1	17	16·49	98	52	52·0	6	0·91
55	... ...	7·9	1	1	18	55·25	151	30	5·0	1	0·93
56	... ...	8·4	1	1	22	31·24	97	43	40·2	1	0·70
57	... ...	10·0	1	1	24	2·87	90	5	55·3	1	0·97
58	99 Piscium η	4·5	...	1	24	15·77	75	31	6·2	9	0·90
59	.. ..	9·1	3	1	29	2·39	150	42	16·2	3	0·89
60	α Eridani (*Achernar*) ...	1·0	...	1	32	41·16	147	55	38·6	5	0·85
61	106 Piscium ν	4·7	...	1	34	34·49	85	11	49·8	9	0·89
62	Lacaille 508	7·9	1	1	35	48·40	151	41	1·1	1	0·96
63	Lacaille 507	9·7	1	1	37	10·69	151	35	14·2	1	0·94
64	.. ..	9·0	1	1	39	36·64	132	2	38·6	1	0·76
65	Lacaille 516	7·1	1	1	40	0·38	161	41	30·0	1	0·93
66	Taylor 580	5·9	...	1	41	36·00	86	39	22·3	1	0·93
67	... ...	9·3	2	1	41	50·44	130	14	36·4	2	0·79
68	V Piscium Var. 5 . ..	10·3	1	1	47	10·48	81	52	59·0	1	0·62
69	6 Arietis β	2·8	...	1	47	11·21	69	51	13·6	8	0·92
70	Lacaille 582	7·4	1	1	50	56·32	146	44	4·0	3	0·97

41.—R. P. L. 12.
42.—Comparison star for Europa in 1863.
57.—Comparison star for Thetis in 1865.
68.—V Piscium Var. 5.—Supposed to vary between the 6th and 9th magnitudes.

Observed with the Madras Meridian Circle in that Year.

Number.	Star.	In Right Ascension.			In Polar Distance.			Number in B. A. C.
		Annual Precession.	Secular Variation.	Proper Motion.	Annual Precession.	Secular Variation.	Proper Motion.	
		s	s	s	"	"		
36	... ..	+ 3·0762	+ 0·0048	...	− 19·703	+ 0·092		..
37	... ...	+ 2·5140	− 0·0828	...	− 19·623	+ 0·081		...
38	λ¹ Toucani	+ 2·5120	− 0·0827	...	− 19·610	+ 0·084		261
39	... ..	+ 2·8004	− 0·0183	...	− 19·589	+ 0·095		...
40	Lacaille 264	+ 2·4506	− 0·0818	...	− 19·565	+ 0·087		..
41	2 Ursæ Minoris ...	+ 6·8270	+ 1·2896	+ 0·065	− 19·563	+ 0·230	+ 0·01	262
42	W. B. E. 0, 897 ...	+ 3·0572	+ 0·0034	...	− 19·530	+ 0·100	..	−
43	Lacaille 271	+ 2·5120	− 0·0289	...	− 19·527	+ 0·092	...	276
44	70 Piscium	+ 3·1120	+ 0·0086	− 0·008	− 19·478	+ 0·116	+ 0·17	281
45	71 Piscium ε	+ 3·1123	+ 0·0087	− 0·002	− 19·431	+ 0·119	0·00	283
46	... .	+ 2·7557	− 0·0186	...	− 19·234	+ 0·122	...	...
47	80 Piscium 3, 1st ..	+ 3·1182	+ 0·0070	+ 0·008	− 19·212	+ 0·130	+ 0·07	288
48	... .	+ 2·3108	− 0·0225	..	− 19·102	+ 0·108	...	...
49	1 Urs. Min. α (*Polaris.*)	+ 19·3212	+ 15·4056	+ 0·065	− 19·130	+ 0·861	0·00	300
50	... ...	+ 3·1300	+ 0·0099	...	− 19·131	+ 0·146	...	...
51	.. ..	+ 2·2487	− 0·0300	...	− 19·117	+ 0·107	...	...
52	.. ..	+ 2·2927	− 0·0303	...	− 19·086	+ 0·111	...	...
53	... ...	+ 3·0217	+ 0·0088	...	− 18·981	+ 0·153	...	.
54	46 Ceti θ¹	+ 3·0029	+ 0·0013	− 0·007	− 18·925	+ 0·164	+ 0·23	420
55	... ...	+ 2·3465	− 0·0178	...	− 18·878	+ 0·119	...	...
56	.. ..	+ 3·0910	+ 0·0078	...	− 18·738	+ 0·169	...	...
57	... ...	+ 3·0705	+ 0·0068	...	− 18·721	+ 0·170	.	..
58	99 Piscium η	+ 3·1070	+ 0·0141	0·000	− 18·715	+ 0·177	0·00	483
59	... .	+ 2·1003	− 0·0135	...	− 18·660	+ 0·128	..	...
60	α Eridani (*Achernar*)	+ 2·2327	− 0·0125	+ 0·008	− 18·438	+ 0·137	+ 0·07	507
61	101 Piscium ν ...	+ 3·1178	+ 0·0091	− 0·004	− 18·378	+ 0·191	− 0·04	518
62	Lacaille 503	+ 2·0653	− 0·0104	...	− 18·330	+ 0·130	-	
63	Lacaille 507	+ 2·0802	− 0·0080	...	− 18·279	+ 0·132	..	531
64	... .	+ 2·0215	− 0·0089	...	− 18·234	+ 0·131	...	...
65	Lacaille 516	+ 2·0227	− 0·0086	...	− 18·176	+ 0·133		543
66	Taylor 890	+ 3·1022	+ 0·0083	...	− 18·122	+ 0·202		551
67	... .	+ 2·8840	− 0·0034	...	− 18·102	+ 0·170		..
68	Y Piscium Var. 5 -	+ 3·1581	+ 0·0111	...	− 17·928	+ 0·216	.	
69	6 Arietis β	+ 3·2992	+ 0·0168	+ 0·002	− 17·902	+ 0·286	+ 0·11	577
70	Lacaille 592 .	+ 2·1587	− 0·0081	...	− 17·751	+ 0·186	..	.

44.—Proper motion in Polar Distance from "*Greenwich Catalogue* 1872."
47.—Proper motions from "*Greenwich Catalogue* 1872."

Mean Positions of Stars for 1865 January 1st.

Number.	Star.	Magnitude.	Estimations.	Mean Right Ascension.			Mean Polar Distance.			Observations.	Fraction of Year.
				h.	m.	s.	°	′	″		
71	Lacaille 606 ...	8·2	3	1	52	4·64	140	7	56·2	3	0·99
72	... ...	8·1	1	1	54	55·60	130	35	34·3	1	0·97
73	Taylor 678... ...	6·4	3	1	56	18·61	72	23	50·0	3	0·92
74	... ...	9·7	1	1	59	36·02	150	2	15·7	1	0·81
75	13 Arietis α ...	2·0	...	1	59	34·07	67	10	41·0	11	0·91
76	Lacaille 660 ..	6·9	3	1	59	50·67	145	31	46·2	3	0·86
77	... ...	9·3	1	2	1	5·29	140	48	46·2	1	0·88
78	Taylor 697... ...	7·0	1	2	1	47·04	145	48	42·1	1	0·76
79*	65 Ceti ξ[1]	4·4	...	2	5	50·79	81	47	17·4	4	0·84
80	Lacaille 677 ...	7·9	3	2	6	57·80	140	47	10·3	3	0·81
81	... ...	9·5	2	2	7	0·84	148	30	13·8	2	0·89
82	Taylor 754 ...	8·6	3	2	9	13·79	147	53	36·9	3	0·91
83	67 Ceti	5·5	...	2	10	15·02	97	2	46·2	12	0·86
84	... ...	9·5	3	2	14	0·04	148	26	40·3	3	0·86
85	... ...	7·8	1	2	16	36·26	161	18	7·1	1	0·87
86	Taylor 818... ...	8·1	2	2	19	10·04	147	36	49·6	2	0·85
87	... ...	8·0	1	2	20	12·69	146	33	36·1	1	0·92
88	73 Ceti ξ[2]	4·4	...	2	20	59·00	82	8	49·0	11	0·82
89	λ Horologii ...	6·1	2	2	21	7·85	150	36	5·0	2	0·94
90	... ...	9·3	1	2	24	16·73	152	35	21·0	1	0·91
91	... ...	8·2	3	2	24	30·03	147	2	20·4	3	0·95
92	Lacaille 782	7·0	2	2	26	14·84	148	34	30·7	2	0·90
93	... .	9·8	1	2	29	13·65	147	37	18·9	1	0·98
94	... ...	10·2	3	2	30	15·32	85	0	19·8	3	0·91
95	W. B. N. II. 556... ...	8·2	3	2	33	13·93	74	53	46·2	3	0·91
96	... ...	8·8	1	2	34	2·15	74	56	17·2	1	0·95
97	Lacaille 849, (2nd) ..	8·0	4	2	35	7·14	130	8	30·4	4	0·71
98	86 Ceti γ	3·6	...	2	36	15·44	87	20	7·7	9	0·95
99	Lacaille 868	8·0	1	2	38	34·80	147	12	55·3	1	0·87
100	W. B. N. II. 676... ...	7·9	1	2	40	11·43	75	29	5·4	1	0·98
101	... ...	8·0	1	2	43	10·64	149	0	22·9	1	0·94
102	W. B. N. II. 729.. ...	8·8	3	2	43	22·45	76	1	50·9	3	0·83
103	... ...	9·3	1	2	44	30·97	149	18	37·2	1	0·95
104	... ...	9·8	1	2	45	15·93	76	27	40·0	1	0·96
105	Taylor 969...	7·2	1	2	45	40·32	74	4	13·0	1	0·95

95.—100.—102.—105.—Comparison stars for Victoria in 1861.

Observed with the Madras Meridian Circle in that Year.

Number.	Star.	In Right Ascension.			In Polar Distance.			Number in B. A. C.
		Annual Precession.	Secular Variation.	Proper Motion.	Annual Precession.	Secular Variation.	Proper Motion.	
		s	s	s	"	"		
71	Lacaille 698	+ 2·0213	− 0·0061		− 17·704	+ 0·147	...	...
72	... .	+ 2·5150	− 0·0089		− 17·580	+ 0·181	...	.
73	Taylor 673	+ 3·2783	+ 0·0167		− 17·527	+ 0·309	...	682
74	. ...	+ 1·9175	− 0·0081	...	− 17·398	+ 0·146	...	...
75	13 Arietis α	+ 3·3525	+ 0·0208	+ 0·012	− 17·357	+ 0·352	+ 0·15	688
76	Lacaille 680	+ 2·0995	− 0·0069	...	− 17·375	+ 0·160	...	...
77	... ...	+ 1·9137	− 0·0088	...	− 17·321	+ 0·148	...	..
78	Taylor 697	+ 2·0777	− 0·0063	...	− 17·290	+ 0·161	...	680
79	65 Ceti ξ[1]	+ 3·1729	+ 0·0116	− 0·004	− 17·100	+ 0·249	+ 0·04	684
80	Lacaille 677	+ 1·8641	− 0·0011	...	− 17·065	+ 0·160	..	.
81	... .	+ 1·9160	− 0·0021	...	− 17·054	+ 0·154	...	...
82	Taylor 754	+ 1·9206	− 0·0021	...	− 16·950	+ 0·157	...	..
83	67 Ceti	+ 2·9631	+ 0·0049	+ 0·008	− 16·902	+ 0·242	+ 0·14	704
84	... .	+ 1·8704	− 0·0005	...	− 16·728	+ 0·158	...	...
85	.. ..	+ 1·7096	+ 0·0036	...	− 16·606	+ 0·146	...	...
86	Taylor 818	+ 1·8778	− 0·0001	...	− 16·470	+ 0·168	...	768
87	... ...	+ 1·9102	− 0·0005	...	− 16·418	+ 0·167	...	...
88	73 Ceti ξ[2]	+ 3·1784	+ 0·0117	+ 0·001	− 16·380	+ 0·276	+ 0·02	760
89	λ Horologii	+ 1·6836	+ 0·0044	...	− 16·371	+ 0·149	...	762
90	... ...	+ 1·5540	+ 0·0083	...	− 16·211	+ 0·141	...	...
91	... ...	+ 1·8865	+ 0·0008		− 16·199	+ 0·167	...	..
92	Lacaille 752	+ 1·7769	+ 0·0026		− 16·108	+ 0·161	...	...
93	... ...	+ 1·7943	+ 0·0024		− 15·982	+ 0·166	...	...
94	.. ..	+ 3·1420	+ 0·0109		− 15·960	+ 0·286	...	...
95	W. B. N. II. 686 ...	+ 3·2980	+ 0·0154		− 15·728	+ 0·306	...	...
96	... ...	+ 3·2962	+ 0·0154	...	− 15·698	+ 0·306	...	..
97	Lacaille 840, (2nd) ...	+ 1·6080	+ 0·0071	...	− 15·579	+ 0·164	...	...
98	86 Ceti γ	+ 3·1113	+ 0·0094	− 0·011	− 15·560	+ 0·294	+ 0·10	787
99	Lacaille 893	+ 1·7477	+ 0·0060	...	− 15·486	+ 0·170	...	...
100	W. B. N. II. 676 ...	+ 3·2978	+ 0·0150	...	− 15·352	+ 0·315	..	...
101	... ...	+ 1·6727	+ 0·0087		− 15·174	+ 0·167	...	.
102	W. B. N. II. 728 ..	+ 3·2990	+ 0·0146		− 15·171	+ 0·320	..	...
103	... .	+ 1·6923	+ 0·0072		− 15·106	+ 0·165	...	...
104	. .. ..	+ 3·2947	+ 0·0144	...	− 15·030	+ 0·322	...	..
105	Taylor 969	+ 3·3346	+ 0·0157	...	− 15·089	+ 0·336	...	802

NB.—Proper motions from "*Greenwich Catalogue, 1872.*"

Mean Positions of Stars for 1865 January 1st.

Number.	Star.	Magnitude.	Estimations.	Mean Right Ascension.			Mean Polar Distance.			Observations.	Fraction of Year.
				h.	m.	s.	°	′	″		
106	Rumker 87 ...	6·0	4	2	40	2·10	158	22	0·2	4	0·69
107	Lalande 5880 ...	8·0	3	2	47	45·48	75	14	26·7	3	0·86
108	Lacaille 941 ...	6·7	2	2	50	38·23	146	35	52·1	2	0·91
109	... .	9·0	1	2	52	28·47	160	10	54·3	1	0·02
110	... ...	8·5	1	2	53	17·19	146	44	15·0	1	0·96
111	92 Ceti α (*Menkar*) ...	2·7	...	2	55	13·46	86	26	32·6	7	0·86
112	25 Persei ρ Var. 2 ..	3·7	...	2	56	32·06	51	41	8·9	2	0·47
113	Taylor 1037	8·3	1	2	58	55·45	150	21	21·2	1	0·93
114	Taylor 1047	6·5	1	2	59	53·15	151	10	38·1	1	0·02
115	Taylor 1057	7·0	2	3	0	47·83	151	32	6·7	2	0·89
116	R. P. L. 33 ...	5·9	...	3	0	54·53	5	34	37·9	3	0·79
117	57 Arietis δ ...	4·5	...	3	3	54·70	70	47	12·1	8	0·92
118	Lacaille 1007 ...	7·3	2	3	4	49·46	152	14	14·7	2	0·40
119	Taylor 1127 ...	8·0	4	3	11	49·51	131	46	7·3	4	0·72
120	... ...	8·8	3	3	12	35·76	130	57	55·9	2	0·91
121	... ...	8·5	1	3	12	42·83	130	50	3·9	1	0·96
122	33 Persei α ...	1·9	...	3	14	41·70	40	37	21·6	1	0·98
123	3² Reticuli ...	6·0	1	3	15	17·67	153	1	34·7	1	0·01
124	... ...	8·9	1	3	20	36·98	140	36	14·9	1	0·96
125	... ...	9·7	1	3	20	48·86	94	46	36·6	1	0·98
126	R. P. L. 34 ...	5·8	..	3	22	34·48	3	47	12·0	1	0·97
127	5 Tauri *f*	4·3	...	3	23	36·80	77	31	42·9	4	0·89
128	... ...	9·4	3	3	25	58·07	97	33	4·7	3	0·91
129	Lacaille 1149 ...	6·0	2	3	27	1·06	159	24	53·7	2	0·98
130	Lacaille 1150 ...	8·2	3	3	28	23·06	152	29	5·6	3	0·90
131	Lacaille 1159	6·8	1	3	30	17·65	161	35	21·6	1	0·01
132	Taylor 1286	8·0	1	3	34	28·31	160	13	6·9	1	0·02
133	... .	9·7	1	3	35	52·23	162	36	14·7	1	0·96
134	.. ..	9·0	1	3	38	7·93	136	12	41·3	1	0·97
135	... .	8·0	...	3	39	38·39	66	30	33·3	1	0·96
136	25 Tauri η (*Alcyone*) ...	3·0	..	3	39	37·76	66	16	56·3	4	0·73
137	Taylor 1318	5·9	2	3	42	31·08	155	14	0·7	3	0·30
138	... .	8·8	2	3	46	32·63	146	33	29·1	2	0·97
139	34 Eridani γ¹	3·0	.	3	51	49·37	103	53	42·5	6	0·23
140	35 Tauri λ Var. 1 ...	3·6	1	3	53	12·17	77	50	38·4	1	0·94

107.—Comparison star for Victoria in 1861.
112.—ρ Persei Var. 2.—Changes irregularly from 3.5 to 4.3 magnitude.
116.—Groombridge 595.
126.—Groombridge 642.
140.—λ Tauri Var. 1.—Period 3.95 days. Range, 3.5 to 4.3 magnitude.

Observed with the Madras Meridian Circle in that Year.

Number.	Star.	In Right Ascension. Annual Precession.	Secular Variation.	Proper Motion.	In Polar Distance. Annual Precession.	Secular Variation.	Proper Motion.	Number in B. A. C.
		s	s	s	"	"		
106	Bremiker 87	+ 1·3052	+ 0·0158	...	− 15·018	+ 0·132	...	895
107	Lalande 5390 .	+ 3·3943	+ 0·0150	...	− 14·918	+ 0·380	...	...
108	Lacaille 941	+ 1·7079	+ 0·0058	...	− 14·796	+ 0·176	...	...
109	... .	+ 1·4717	+ 0·0107	...	− 14·644	+ 0·158	...	...
110	... ...	+ 1·0737	+ 0·0060	...	− 14·590	+ 0·174	...	...
111	92 Ceti α (*Menkar*) ..	+ 3·1295	+ 0·0098	− 0·002	− 14·473	+ 0·388	+ 0·11	940
112	25 Persei ρ Var. 2 ..	+ 3·8077	+ 0·0382	+ 0·010	− 14·294	+ 0·398	+ 0·11	958
113	Taylor 1037 ..	+ 1·4333	+ 0·0110	...	− 14·370	+ 0·152	...	...
114	Taylor 1047	+ 1·3443	+ 0·0139	...	− 14·188	+ 0·146	...	968
115	Taylor 1057	+ 1·3343	+ 0·0142	...	− 14·132	+ 0·144	...	973
116	R. P. L. 33	+ 12·7031	+ 1·5356	...	− 14·194	+ 1·382	+ 0·08	980
117	57 Arietis δ	+ 3·4071	+ 0·0171	+ 0·010	− 13·986	+ 0·364	0·00	986
118	Lacaille 1007 ...	+ 1·2387	+ 0·0160	...	− 13·579	+ 0·136	...	...
119	Taylor 1127 ...	+ 2·1853	+ 0·0012	...	− 13·431	+ 0·348	...	...
120	... ...	+ 2·3080	+ 0·0011	...	− 13·392	+ 0·346	...	...
121	.. ...	+ 2·2110	+ 0·0012	...	− 13·372	+ 0·346	...	...
122	33 Persei α	+ 4·3485	+ 0·0483	+ 0·002	− 13·348	+ 0·472	+ 0·06	1048
123	3^{a} Reticulo	+ 1·0960	+ 0·0908	+ 0·190	− 13·304	+ 0·126	− 0·66	1061
124	... ...	+ 1·3815	+ 0·0188	...	− 12·951	+ 0·166	...	...
125	... ...	+ 3·7067	+ 0·0279	...	− 12·948	+ 0·461	...	...
126	R. P. L. 34	+ 16·0917	+ 3·1990	+ 0·130	− 12·718	+ 2·113	+ 0·06	1061
127	5 Tauri *f* ..	+ 3·3014	+ 0·0130	+ 0·003	− 12·680	+ 0·379	+ 0·08	1067
128	... ...	+ 3·1107	+ 0·0089	...	− 12·467	+ 0·360	...	...
129	Lacaille 1148 ..	+ 0·9739	+ 0·0227	...	− 12·414	+ 0·117	...	1108
130	Lacaille 1150 ..	+ 1·0170	+ 0·0203	...	− 12·314	+ 0·126	..	.
131	Lacaille 1159 .	+ 1·1198	+ 0·0190		− 12·190	+ 0·135		...
132	Taylor 1256 ...	+ 1·1852	+ 0·0159		− 11·926	+ 0·146		...
133	... .	+ 1·0006	+ 0·0207		− 11·707	+ 0·123		...
134	. ... ..	+ 1·8361	+ 0·0014		− 11·630	+ 0·223		...
135	... .. .	+ 3·5473	+ 0·0170		− 11·546	+ 0·438		..
136	25 Tauri η (*Alcyone*)	+ 3·5610	+ 0·0177	− 0·001	− 11·542	+ 0·430	+ 0·00	1166
137	Taylor 1319	+ 0·9803	+ 0·0204	+ 0·080	− 11·382	+ 0·047	− 0·00	1197
138	... ..	+ 1·9812	+ 0·0111	..	− 11·089	+ 0·173	...	...
139	34 Eridani γ^1 ..	+ 2·7017	+ 0·0017	+ 0·002	− 10·648	+ 0·351	+ 0·12	1231
140	35 Tauri λ Var 1 ..	+ 3·9101	+ 0·0115	− 0·002	− 10·529	+ 0·416	+ 0·02	1241

116.—126.—127.—140.—Proper motions from "*Greenwich Catalogue*, 1872."
123.—137.—Proper motions from "*Stone's Cape Catalogue.*"

Mean Positions of Stars for 1865 January 1st.

Number.	Star.	Magnitude.	Estimations.	Mean Right Ascension.			Mean Polar Distance.			Observations.	Fraction of Year.
				h.	m.	s.	°	′	″		
141	... ...	8·0	1	3	58	41·84	148	8	11·5	1	0·90
142	Lacaille 1327	6·9	...	3	54	19·39	132	51	22·8	1	0·92
143	R. P. L. 85	6·7	...	3	56	10·60	4	46	24·4	1	0·03
144	Lacaille 1347	7·6	2	3	59	8·11	149	2	26·2	2	0·40
145	Lalande 7761	8·0	3	4	8	31·43	74	48	42·5	3	0·03
146	... ...	9·2	1	4	5	1·86	150	5	21·9	1	0·08
147	38 Eridani o¹	4·1	...	4	5	16·50	97	11	32·6	7	0·36
148	... ...	8·3	1	4	9	40·63	129	18	43·6	1	0·08
149	U Tauri Var. 7	9·8	1	4	13	57·17	70	30	30·1	1	0·04
150	T Tauri Var. 6	10·4	2	4	14	7·61	70	47	23·4	2	0·06
151	ε Reticuli	5·0	...	4	14	0·66	149	37	42·1	1	0·01
152	... ...	9·6	1	4	15	42·03	128	38	40·4	1	0·07
153	74 Tauri ε	3·7	...	4	20	44·15	71	7	20·9	12	0·62
154	R Tauri Var. 2	8·3	1	4	20	58·57	80	8	30·1	1	0·96
155	Taylor 1562	6·4	2	4	23	13·32	161	32	48·3	2	0·88
156	Lacaille 1519	7·3	1	4	26	35·88	153	5	59·7	1	0·04
157	87 Tauri α (*Aldebaran*) ...	1·0	...	4	28	10·38	73	45	56·3	8	0·46
158	... ...	9·8	1	4	31	41·30	142	39	28·9	1	0·06
159	W. B. N. IV. 696... ...	9·0	7	4	32	30·63	66	27	22·5	7	0·04
160	W. B. N. IV. 720... ...	8·3	2	4	33	54·62	66	15	9·6	8	0·06
161	94 Tauri τ	4·4	...	4	34	8·63	67	18	19·7	2	0·47
162	... ...	9·4	2	4	39	32·41	128	57	29·3	2	0·80
163	... ...	8·8	2	4	40	20·10	161	30	49·5	2	0·08
164	κ Doradûs	6·0	1	4	42	10·46	149	33	58·5	1	0·96
165	Lacaille 1629 ...	6·7	1	4	46	43·62	168	28	27·2	2	0·81
166	Lacaille 1636	8·1	1	4	45	0·92	140	1	48·6	1	0·10
167	W. B. N. IV. 996 ..	8·0	1	4	46	10·30	66	3	3·4	1	0·01
168	... ...	9·3	1	4	46	30·74	129	34	44·0	1	0·96
169	... ...	8·9	1	4	46	50·19	153	3	36·9	1	0·02
170	W. B. N. IV. 1018 ..	8·0	1	4	48	2·10	66	13	38·0	1	0·01
171	3 Aurigæ ι	2·7	.	4	48	12·86	57	3	5·8	3	0·36
172	... ...	10·3	1	4	51	29·69	82	8	37·5	1	0·05
173	... .	9·0	1	4	52	21·10	109	30	47·9	1	0·98
174	... ...	9·0	1	4	52	41·91	160	37	49·5	1	0·02
175	R Leporis Var 1	8·0	1	4	53	27·72	105	0	42·6	1	0·07

143.—Groombridge 750.——145.—Comparison star for Asia in 1862.
149.—U Tauri Var. 7.—Period unknown. Range, 9th to 10·5 magnitude.
150.—T Tauri Var. 6.—Period unknown. Range, 9th to below 13th magnitude.
154.—R Tauri Var. 2.—Period 325 days. Range, 8th to below 13th magnitude.
159.—160.—167.—170.—Comparison stars for Mars in 1864.
175.—R Leporis Var. 1.—Period 436 days. Range, 6th to 9th magnitude.

Observed with the Madras Meridian Circle in that Year.

Number.	Star.	In Right Ascension. Annual Precession.	In Right Ascension. Secular Variation.	In Right Ascension. Proper Motion.	In Polar Distance. Annual Precession.	In Polar Distance. Secular Variation.	In Polar Distance. Proper Motion.	Number in B. A. C.
		s	s	s	"	"		
141	... .	+ 1·6529	+ 0·0082	...	− 10·501	+ 0·198	...	...
142	Lacaille 1327 ...	+ 0·7477	+ 0·0250	...	− 10·454	+ 0·097	...	1268
143	R. P. L. 35	+ 10·0726	+ 1·8153	+ 0·057	− 10·300	+ 2·094	− 0·05	1235
144	Lacaille 1347 ...	+ 1·1611	+ 0·0146	...	− 10·169	+ 0·140	...	...
145	Lalande 7764	+ 3·3910	+ 0·0121	...	− 9·758	+ 0·430	...	...
146	... ...	+ 1·0345	+ 0·0165	...	− 9·639	+ 0·130	..	...
147	38 Eridani o¹ ...	+ 2·9840	+ 0·0068	− 0·002	− 9·025	+ 0·870	− 0·07	1290
148	... ...	+ 2·1015	+ 0·0085	...	− 9·275	+ 0·270	...	...
149	U Tauri Var. 7 ...	+ 3·4056	+ 0·0120	...	− 8·962	+ 0·460	...	...
150	T Tauri Var. 6 ...	+ 3·4692	+ 0·0128	...	− 6·789	+ 0·460	...	...
151	ε Reticuli	+ 1·0298	+ 0·0155	...	− 8·066	+ 0·120	...	1344
152	... ...	+ 2·1111	+ 0·0086	...	− 8·815	+ 0·251	...	...
153	74 Tauri ε	+ 3·4870	+ 0·0120	+ 0·005	− 8·418	+ 0·466	+ 0·08	1376
154	R Tauri Var. 2 ...	+ 3·2631	+ 0·0098	...	− 8·404	+ 0·489	...	...
155	Taylor 1892	+ 0·8802	+ 0·0163	...	− 8·210	+ 0·113	...	1400
156	Lacaille 1610 ..	+ 0·6572	+ 0·0212	...	− 8·089	+ 0·081	...	...
157	87 Tauri α (*Aldebaran*)	+ 3·4304	+ 0·0105	+ 0·004	− 7·838	+ 0·461	+ 0·17	1420
158	... ...	+ 1·4881	+ 0·0082	...	− 7·584	+ 0·190	...	...
159	W. B. N. IV. 696 ...	+ 3·6120	+ 0·0127	...	− 7·400	+ 0·492	...	...
160	W. B. N. IV. 726 ...	+ 3·0196	+ 0·0127	...	− 7·368	+ 0·404	...	...
161	94 Tauri τ	+ 3·5306	+ 0·0122	0·000	− 7·326	+ 0·491	+ 0·02	1449
162	... ...	+ 2·0970	+ 0·0087	...	− 6·897	+ 0·265	...	...
163	... .	+ 0·7717	+ 0·0168	...	− 6·882	+ 0·100	...	...
164	α Doradûs	+ 0·9807	+ 0·0141	...	− 6·068	+ 0·126	...	1490
165	Lacaille 1689 ..	+ 0·5406	+ 0·0197	...	− 6·552	+ 0·077	...	..
166	Lacaille 1686 ...	+ 1·8617	+ 0·0068		− 6·446	+ 0·210		...
167	W. B. N. IV. 996 ...	+ 3·0840	+ 0·0114		− 6·482	+ 0·806		...
168	... ...	+ 2·0806	+ 0·0037		− 6·401	+ 0·284		...
169	... .	+ 0·5760	+ 0·0186		− 6·369	+ 0·058		...
170	W. B. N. IV. 1018 ...	+ 3·6807	+ 0·0113		− 6·360	+ 0·806		.
171	3 Aurigæ ι ...	+ 3·8894	+ 0·0144	− 0·008	− 6·150	+ 0·514	+ 0·02	1520
172	... .	+ 3·3481	+ 0·0068	...	− 5·906	+ 0·456	.	...
173	... .	+ 2·0116	+ 0·0038	...	− 5·884	+ 0·274	..	
174	.. ..	+ 0·7081	+ 0·0139	...	− 5·806	+ 0·113	...	.
175	R Leporis Var. 1 ..	+ 2·7296	+ 0.0088	...	− 5·742	+ 0·363	...	..

143.—Proper motions from "*Greenwich Catalogue* 1872."

Mean Positions of Stars for 1866 January 1st.

Number.	Star.	Magnitude.	Estimations.	Mean Right Ascension.			Mean Polar Distance.			Observations.	Fraction of year.
				h	m	s	°	′	″		
176	Taylor 1797 ...	7·0	1	4	54	52·10	109	16	51·1	1	0·10
177	102 Tauri ι ...	4·7	...	4	56	1·72	68	36	28·4	2	0·47
178	Lacaille 1697 ...	8·0	1	4	56	38·24	120	7	7·0	1	0·05
179	Lacaille 1705 ...	8·0	1	4	57	27·36	130	16	30·1	1	0·36
180	2 Leporis ε ...	3·3	...	4	59	44·85	112	38	17·7	4	0·05
181	... ...	8·0	1	5	0	1·55	136	23	49·9	1	0·07
182	... .	9·5	2	5	1	46·00	151	36	4·3	2	0·08
183	Lacaille 1739 ...	8·2	1	5	2	52·17	140	57	46·5	1	0·01
184	Lacaille 1750 ...	8·4	2	5	3	09·92	154	44	37·6	2	0·09
185	... ...	9·5	1	5	4	36·08	135	34	33·6	1	0·07
186	Lacaille 1757	8·0	1	5	4	58·29	150	4	20·1	1	0·02
187	... ...	9·0	1	5	6	4·10	131	45	35·4	1	0·09
188	13 Aurigæ α (*Capella*) ..	0·2	...	5	6	43·14	44	8	36·6	1	0·07
189	... ...	8·5	1	5	6	54·10	129	6	0·6	1	0·00
190	19 Orionis β (*Rigel*) ...	0·3	...	5	8	3·07	98	21	37·0	2	0·50
191	... ...	9·2	3	5	12	53·74	130	40	3·2	3	0·36
192	... ...	9·1	3	5	12	57·39	137	4	42·7	3	0·08
193	... ...	8·1	1	5	13	26·14	153	41	37·4	1	0·07
194	... ...	9·0	1	5	14	50·88	168	30	18·7	1	0·05
195	... ..	9·2	2	5	16	9·58	131	49	10·6	2	0·09
196	... ...	7·9	3	5	17	0·04	129	37	39·9	3	0·37
197	112 Tauri β ...	1·9	...	5	17	46·54	61	30	37·6	5	0·24
198	Taylor 1984 ...	7·3	1	5	18	51·97	100	54	47·7	1	0·10
199	... ...	8·9	1	5	19	5·04	149	14	16·5	1	0·03
200	114 Tauri ο ..	4·8	...	5	19	31·97	[illegible]	10	56·0	1	[illegible]
201	Lacaille 1864	7·5	1	5	21	40·32	137	12	50·4	1	0·07
202	... ...	7·9	2	5	23	30·00	151	13	33·2	2	0·08
203	34 Orionis δ Var. 1 ..	2·4	...	5	25	6·85	90	24	8·4	2	0·05
204	... ...	8·9	2	5	25	13·55	130	35	30·1	2	0·07
205	... ...	9·0	1	5	26	35·66	155	51	10·4	1	0·12
206	11 Leporis α	2·7	..	5	26	40·08	107	53	13·0	5	0·07
207	46 Orionis ε	1·8	..	5	29	21·98	91	17	35·0	8	0·08
208	123 Tauri ζ	3·0	.	5	29	34·01	69	60	37·2	1	[illegible]
209	. ...	9·6	2	5	29	47·14	135	22	0·6	2	[illegible]
210	... .	8·8	1	5	31	30·92	128	42	14·7	1	[illegible]

203.—δ Orionis Var. 1.—Supposed to vary irregularly from 2·3 to 2·7 magnitude.

Observed with the Madras Meridian Circle in that Year.

Number.	Star.	In Right Ascension. Annual Precession	In Right Ascension. Secular Variation.	In Right Ascension. Proper Motion.	In Polar Distance. Annual Precession.	In Polar Distance. Secular Variation.	In Polar Distance. Proper Motion.	Number in B. A. C.
		s	s	s	"	"		
176	Taylor 1707 ...	+ 0·1957	+ 0·0111	...	− 5·688	+ 0·141	...	...
177	102 Tauri ι	+ 3·5750	+ 0·0096	+ 0·004	− 5·609	+ 0·508	+ 0·06	1551
178	Lacaille 1697 ..	+ 2·1958	+ 0·0036	..	− 5·462	+ 0·[illegible]	...	...
179	Lacaille 1706 ..	+ 2·0192	+ 0·0067	...	− 5·406	+ 0·286	...	...
180	2 Leporis ε	+ 2·5858	+ 0·0083	+ 0·001	− 5·212	+ 0·389	+ 0·08	1575
181	... ..	+ 1·7696	+ 0·0046		− 5·188	+ 0·250		...
182	... ..	+ 0·6787	+ 0·0138		− 5·042	+ 0·068		...
183	Lacaille 1730	+ 1·0706	+ 0·0098		− 4·947	+ 0·156		...
184	Lacaille 1736 ..	+ 0·3222	+ 0·0163		− 4·852	+ 0·048		...
185	... ..	+ 1·7470	+ 0·0045		− 4·801	+ 0·260		...
186	Lacaille 1737 ..	+ 0·8164	+ 0·0117	...	− 4·776	+ 0·118	.	...
187	... ..	+ 1·0113	+ 0·0088	...	− 4·670	+ 0·278	...	...
188	13 Aurigæ α (*Capella.*)	+ 4·4125	+ 0·0173	+ 0·008	− 4·621	+ 0·680	+ 0·43	1618
189	... .	+ 2·0146	+ 0·0065	...	− 4·606	+ 0·[illegible]	...	...
190	19 Orionis β (*Rigel*)..	+ 2·8806	+ 0·0040	− 0·001	− 4·507	+ 0·412	+ 0·0[illegible]	16[illegible]
191	..	+ 1·9807	+ 0·0085		− 4·098	+ 0·286		...
192	..	+ 1·6646	+ 0·0046		− 4·086	+ 0·239		..
193	.	+ 0·4282	+ 0·0144		− 4·047	+ 0·062		.
194		+ 0·4498	+ 0·0138		− 3·996	+ 0·065		...
195	..	+ 1·8963	+ 0·0057		− 3·815	+ 0·274		...
196	. ...	+ 1·9542	+ 0·0084	...	− 3·741	+ 0·[illegible]	..	..
197	112 Tauri β	+ 3·7864	+ 0·0082	+ 0·003	− 3·676	+ 0·543	+ 0·20	1681
198	Taylor 1084 ..	+ 0·7073	+ 0·0101	...	− 3·330	+ 0·108	...	1697
199	.. .	+ 0·9469	+ 0·0084	...	− 3·260	+ 0·138	. .	...
200	114 Tauri ο	+ 3·5690	+ 0·0068	− 0·001	− 3·398	+ 0·510	+ 0·01	1686
201	Lacaille 1854 .	+ 1·6477	+ 0·0042	...	− 3·268	+ 0·[illegible]	...	...
202	. ... ..	+ 0·6986	+ 0·0098	...	− 3·180	+ 0·097	. .	...
203	34 Orionis δ Var. 1 ..	+ 3·0627	+ 0·0026	+ 0·001	− 3·041	+ 0·443	+ 0·04	1730
204	... .	+ 1·9298	+ 0·0084	...	− 3·031	+ 0·281	..	...
205	... .	+ 0·1831	+ 0·0148	...	− 2·999	+ 0·019	.	...
206	11 Leporis α .. .	+ 2·6441	+ 0·0029	+ 0·001	− 2·807	+ 0·386	0·00	1741
207	46 Orionis ε	+ 3·0421	+ 0·0086	− 0·002	− 2·673	+ 0·441	+ 0·01	1763
208	123 Tauri ζ .	+ 3·5498	+ 0·0055	0·000	− 2·665	+ 0·519	+ 0·06	1767
209	... ..	+ 1·7296	+ 0·0087	.	− 2·636	+ 0·251		—
210	... .	+ 2·0[illegible]	+ 0·0081	.	− 2·474	+ 0·292		.

200.—Proper motions from "*Greenwich Catalogue* 1872."

Mean Positions of Stars for 1865 January 1st.

Number.	Star.	Magnitude.	Estimations.	Mean Right Ascension.			Mean Polar Distance.			Observations.	Fraction of Year.
				h.	m.	s.	°	′	″		
211	Lacaille 1949 ..	6·1	1	5	32	15·77	164	19	2·7	1	0·07
212	... .	8·6	...	5	32	48·06	150	11	33·6	1	0·07
213	... ...	9·1	1	5	32	41·01	138	41	13·5	1	0·05
214	... ...	8·8	2	5	34	30·02	152	7	57·3	2	0·09
215	α Columbae ...	2·7	...	5	34	46·71	124	8	53·7	6	0·08
216	Lacaille 1971 ...	7·8	1	5	36	22·25	140	11	30·2	1	0·12
217	... ...	8·8	2	5	36	30·26	135	48	38·7	2	0·08
218	... .	9·1	1	5	39	55·07	135	48	8·2	1	0·07
219	Lacaille 2010 ...	7·7	1	5	42	2·13	146	34	20·0	1	0·07
220	.. ..	9·1	3	5	43	26·72	130	6	55·8	3	0·06
221	Taylor 2164	9·3	1	5	43	56·09	150	40	22·3	1	0·08
222	... ...	9·3	1	5	44	46·16	137	10	18·3	1	0·05
223	54 Orionis χ^1	4·6	...	5	46	28·37	69	46	10·1	2	0·51
224	... ...	9·0	1	5	47	3·18	166	40	50·2	1	0·07
225	58 Orionis α Var. 2 ...	0·9	...	5	47	51·84	82	37	17·0	9	0·07
226	... ...	8·9	1	5	49	55·35	136	48	7·9	1	0·07
227	... ...	9·8	1	5	50	15·60	137	10	19·5	1	0·08
228	Lacaille 2078 ...	7·9	1	5	50	33·41	137	12	38·1	1	0·12
229	... .	9·6	2	5	50	58·77	130	48	1·5	2	0·07
230	R. P. L. 43 ...	6·6	...	5	52	27·98	3	14	20·0	1	0·10
231	... ...	8·8	1	5	53	3·51	130	34	57·4	1	0·06
232	Lalande 2104 ..	7·0	1	5	54	22·74	143	36	38·9	1	0·06
233	... .	9·5	2	5	54	38·84	137	46	15·3	2	0·09
234	... ...	9·2	1	5	56	11·88	139	37	14·2	1	0·08
235	... ...	9·3	1	5	57	1·11	136	1	0·9	1	0·07
236	67 Orionis ν	4·4	...	5	59	51·86	75	13	7·9	9	0·07
237	... ...	7·8	2	6	0	52·61	137	34	38·4	2	0·10
238	Lalande 11732	8·2	1	6	3	36·08	77	38	35·9	1	0·07
239	... ...	9·0	1	6	4	1·13	160	5	34·2	1	0·11
240	... .	8·1	2	6	4	35·32	136	40	40·6	2	0·09
241	.. ..	9·0	1	6	5	40·90	77	51	30·8	1	0·05
242	7 Geminorum η Var. 6 ..	3·5	..	6	6	48·79	67	27	20·0	3	0·07
243	... .	9·2	1	6	7	37·21	151	19	38·5	1	0·13
244	.. . ..	9·1	2	6	7	42·12	137	6	34·1	2	0·09
245	.. ..	9·2	1	6	8	38·15	155	4	30·0	1	0·13

225 – α Orionis Var. 2., (*Betelgeuse*) Irregularly variable from 0·9 to 1·5 magnitude.
230 — Groombridge 1094.
238 – Comparison star for Sappho in 1865.

Observed with the Madras Meridian Circle in that Year.

Number.	Star.	In Right Ascension. Annual Precession.	In Right Ascension. Secular Variation.	In Right Ascension. Proper Motion.	In Polar Distance. Annual Precession.	In Polar Distance. Secular Variation.	In Polar Distance. Proper Motion.	Number in B.A.C.
		s	s	s	"	"		
211	Lacaille 1949 ...	+ 0·3122	+ 0·0100	...	− 2·490	+ 0·016	...	1790
212	. ...	+ 0·7548	+ 0·0070	...	− 2·391	+ 0·110	...	..
213	... ...	+ 2·0090	+ 0·0080	. .	− 2·381	+ 0·292	...	..
214	.. ..	+ 0·5508	+ 0·0085	...	− 2·340	+ 0·082	...	...
215	α Columbæ	+ 2·1700	+ 0·0087	+ 0·008	− 2·305	+ 0·316	0·00	1802
216	Lacaille 1971 ..	+ 0·8119	+ 0·0086		− 2·064	+ 0·128		...
217	... .	+ 1·7038	+ 0·0085		− 2·030	+ 0·248		...
218	... ...	+ 1·7094	+ 0·0088		− 1·765	+ 0·248		...
219	Lacaille 2010 ...	+ 1·0213	+ 0·0080		− 1·570	+ 0·150		...
220	... .	+ 1·0450	+ 0·0083		− 1·447	+ 0·295		...
221	Taylor 2184	+ 0·6894	+ 0·0080	...	− 1·406	+ 0·101	...	—
222	... .	+ 1·6329	+ 0·0083	...	− 1·388	+ 0·309	...	...
223	54 Orionis χ^1 ..	+ 3·5645	+ 0·0084	− 0·016	− 1·191	+ 0·820	+ 0·10	1976
224	... .	+ 1·7008	+ 0·0080	...	− 1·182	+ 0·269	..	...
225	58 Orionis α Var. 2...	+ 3·2449	+ 0·0027	+ 0·001	− 1·001	+ 0·472	0·00	1988
226	... ...	+ 1·7025	+ 0·0080		− 0·891	+ 0·248	...	..
227	... ..	+ 1·6310	+ 0·0080		− 0·853	+ 0·228	...	..
228	Lacaille 2073 ..	+ 1·6200	+ 0·0080		− 0·828	+ 0·237	..	...
229	... ...	+ 1·0828	+ 0·0087		− 0·780	+ 0·180	...	.
230	R. P. L. 48	+ 26·6578	+ 0·2764		− 0·690	+ 3·[illegible]	...	1979
231	.. .. ..	+ 1·0841	+ 0·0080	...	− 0·607	+ 0·292		...
232	Lacaille 2104 ..	+ 1·3086	+ 0·0078	...	− 0·492	+ 0·190		...
233	... ...	+ 1·0004	+ 0·0089	...	− 0·473	+ 0·238		...
234	.. ..	+ 1·0622	+ 0·0086	...	− 0·384	+ 0·296		..
235	.. ..	+ 1·0680	+ 0·0087	...	− 0·261	+ 0·246		...
236	67 Orionis ν	+ 3·4948	+ 0·0017	+ 0·001	− 0·012	+ 0·860	+ 0·02	1986
237	. ...	+ 1·6148	+ 0·0084	...	+ 0·076	+ 0·280	...	...
238	Lalande 11782 .	+ 3·8607	+ 0·0016	..	+ 0·300	+ 0·480	...	...
239	... .	+ 0·7492	+ 0·0022	...	+ 0·362	+ 0·100	...	...
240	. ... .	+ 1·0472	+ 0·0004	..	+ 0·401	+ 0·240	...	—
241	... ..	+ 3·3687	+ 0·0018	..	+ 0·496	+ 0·400	...	...
242	7 Geminorum η Var. 6	+ 3·6267	+ 0·0007	− 0·007	+ 0·589	+ 0·820	+ 0·02	2002
243	.. ..	+ 0·6849	+ 0·0016	.	+ 0·654	+ 0·092		...
244	...	+ 1·6336	+ 0·0022		+ 0·674	+ 0·238	.	...
245	.	+ 0·1993	+ 0·0006		+ 0·751	+ 0·020	.	-

215.—223.—236.—Proper motions from "*Greenwich Catalogue*, 1872."

Mean Positions of Stars for 1865 January 1st.

Number.	Star.	Magnitude.	Estimations.	Mean Right Ascension.			Mean Polar Distance.			Observations.	Fraction of Year.
				h.	m.	s.	°	′	″		
246	… … …	9·3	2	6	8	56·17	130	31	36·2	2	0·06
247	… … … …	9·4	1	6	11	43·60	152	1	50·0	1	0·07
248	… … … …	9·2	2	6	12	3·48	136	50	49·9	2	0·09
249	Lalande 12120 … …	7·2	1	6	14	8·86	77	4	49·8	2	0·07
250	13 Geminorum μ … …	3·2	…	6	14	47·30	67	26	15·5	6	0·06
251	Lalande 12155 …	7·2	1	6	15	1·74	77	22	3·1	1	0·07
252	Lacaille 2273 …	7·0	1	6	17	3·36	153	38	28·3	1	0·02
253	… …	9·0	2	6	18	33·26	151	58	29·0	2	0·12
254	Taylor 2496 …	7·3	2	6	18	34·38	151	16	11·2	2	0·12
255	Lacaille 2306 …	7·0	1	6	18	48·89	153	45	46·3	1	0·11
256	… … … …	8·8	2	6	19	43·50	95	30	54·8	2	0·09
257	α Argûs (*Canopus*) …	1·0	…	6	20	57·41	142	37	29·6	1	0·09
258	… … … …	7·2	2	6	21	20·08	128	51	32·6	2	0·09
259	Lacaille 2312 … …	6·8	1	6	22	9·61	168	36	37·7	1	0·12
260	Lacaille 2321 … …	7·8	1	6	23	29·97	158	30	40·7	1	0·16
261	Lacaille 2346 …	7·8	2	6	26	32·41	152	8	46·5	2	0·17
262	… … …	9·0	2	6	27	30·11	128	46	0·7	2	0·10
263	… … …	8·1	1	6	28	33·93	151	10	2·6	1	0·14
264	… … …	9·1	2	6	28	40·42	130	32	12·0	2	0·10
265	Taylor 2530 …	6·0	2	6	29	50·15	151	46	51·8	2	0·12
266	24 Geminorum γ … …	2·0	…	6	30	34·77	73	29	21·0	5	0·08
267	… … … …	8·3	1	6	31	39·73	130	36	43·1	1	0·15
268	. .. … …	8·9	1	6	33	33·48	102	27	8·4	1	0·11
269	… … … …	9·1	2	6	35	49·30	130	38	7·3	2	0·09
270	… …	10·4	2	6	36	0·29	92	5	30·8	2	0·06
271	51 Cephei (Hev.)… …	5·3	…	6	36	9·07	3	45	21·0	7	0·23
272	… .. … …	8·9	1	6	36	15·11	130	21	8·4	1	0·17
273	Taylor 2562 …	6·8	1	6	36	36·34	151	34	35·3	1	0·12
274	31 Geminorum ξ … …	3·4	.	6	37	42·06	96	57	48·6	1	0·10
275	… . … ..	9·5	1	6	37	56·72	153	30	39·7	1	0·18
276	Lacaille 2461 .. …	8·3	1	6	38	11·15	163	37	45·4	1	0·17
277	Taylor 2567	8·3	1	6	38	28·73	165	49	44·7	1	0·08
278	. … ..	9·5	1	6	39	9·49	131	3	30·2	1	0·08
279	9 Canis Maj. α (*Sirius*) ..	−1·4		6	39	11·78	106	32	1·5	2	0·12
280	.. …	9·3	2	6	40	30·61	134	13	36·2	2	0·13

269—271.—Comparison stars for Sappho in 1865.

Observed with the Madras Meridian Circle in that Year.

Number.	Star.	In Right Ascension. Annual Precession.	In Right Ascension. Secular Variation.	In Right Ascension. Proper Motion.	In Polar Distance. Annual Precession.	In Polar Distance. Secular Variation.	In Polar Distance. Proper Motion.	Number in B. A. C.
		s	s	s	"	"		
246	..	+ 1·9800	+ 0·0021	...	+ 0·780	+ 0·281	...	...
247	.. ..	+ 0·5570	+ 0·0005	...	+ 1·020	+ 0·081	...	...
248	... .	+ 1·0479	+ 0·0019	...	+ 1·065	+ 0·240	...	...
249	Lalande 12120 ..	+ 3·3782	+ 0·0006	...	+ 1·237	+ 0·487	...	..
250	13 Geminorum μ ..	+ 3·6268	+ 0·0003	+ 0·006	+ 1·298	+ 0·527	+ 0·14	2047
251	Lalande 12155 ..	+ 3·5711	+ 0·0004		+ 1·314	+ 0·490	...	...
252	Lacaille 2273 ..	+ 0·3416	− 0·0014		+ 1·492	+ 0·019	..	...
253	... ...	+ 0·0146	− 0·0007		+ 1·625	+ 0·049	...	...
254	Taylor 2695	+ 0·0411	− 0·0007		+ 1·628	+ 0·052	...	...
255	Lacaille 2866 ...	+ 0·2840	− 0·0017		+ 1·044	+ 0·063	...	2078
256	... ...	+ 3·6746	− 0·0011	...	+ 1·724	+ 0·533	...	...
257	α Argûs (*Canopus*) ...	+ 1·3292	+ 0·0010	0·000	+ 1·882	+ 0·192	0·00	2090
258	... ...	+ 1·9906	+ 0·0019	...	+ 1·854	+ 0·289	...	...
259	Lacaille 2812 ...	+ 0·8902	− 0·0086	...	+ 1·906	+ 0·066	...	...
260	Lacaille 2821 ...	+ 0·4228	− 0·0028	...	+ 2·063	+ 0·060	...	.
261	Lacaille 2868 ...	+ 0·5678	− 0·0026	...	+ 2·317	+ 0·081	...	2142
262	.. ..	+ 2·0061	+ 0·0018	...	+ 2·401	+ 0·290	...	...
263	... ..	+ 0·6634	− 0·0025	...	+ 2·502	+ 0·096	..	...
264	.. ..	+ 1·9677	+ 0·0016	...	+ 2·508	+ 0·290	.	...
265	Taylor 2899	+ 0·6017	− 0·0031	...	+ 2·604	+ 0·096	...	2165
266	24 Geminorum γ ..	+ 3·4650	− 0·0015	+ 0·001	+ 2·610	+ 0·500	+ 0·04	2163
267	.. ..	+ 1·9820	+ 0·0015	...	+ 2·746	+ 0·277	.	.
268	.. ..	+ 0·5204	− 0·0043	...	+ 2·926	+ 0·076	...	...
269	... ...	+ 1·9287	+ 0·0014	...	+ 3·122	+ 0·278	...	...
270	... ...	+ 3·7714	− 0·0010	...	+ 3·136	+ 0·542	..	..
271	51 Cephei (Hev.) ..	+ 30·4966	− 1·6801	− 0·027	+ 3·151	+ 4·392	+ 0·04	2157
272	... ...	+ 1·9803	+ 0·0014	...	+ 3·159	+ 0·290	..	...
273	Taylor 2852	+ 0·6196	− 0·0042	...	+ 3·193	+ 0·092	..	2208
274	31 Geminorum ξ .	+ 3·3770	− 0·0016	− 0·007	+ 3·286	+ 0·495	+ 0·22	2206
275	... .	+ 0·4460	− 0·0061	...	+ 3·301	+ 0·063	...	
276	Lacaille 2851 .	+ 0·1159	− 0·0072	...	+ 3·380	+ 0·016		
277	Taylor 2867	+ 0·8776	− 0·0029	..	+ 3·344	+ 0·126		
278	... .	+ 1·9244	+ 0·0014	.	+ 3·410	+ 0·276	.	..
279	9 Can. Maj. α (*Sirius*)	+ 2·6444	+ 0·0010	− 0·035	+ 3·413	+ 0·384	+ 1·24	2213
280	... ..	+ 0·3423	− 0·0076	...	+ 3·565	+ 0·049	...	

257.—Proper motion from "*Stone's Cape Catalogue.*"
271.–274.—Proper motions from "*Greenwich Catalogue 1872.*"

Mean Positions of Stars for 1865 January 1st.

Number.	Star.	Magnitude.	Estimations.	Mean Right Ascension. h.	m.	s.	Mean Polar Distance. °	′	″	Observations.	Fraction of Year.
281	...	9·1	1	6	40	32·86	151	35	54·8	1	0·18
282	...	8·3	1	6	42	26·49	130	56	39·2	1	0·14
283	...	8·2	2	6	43	39·03	130	36	26·7	2	0·08
284	..	9·2	1	6	43	59·06	130	35	5·6	1	0·14
285	...	7·8	3	6	43	43·70	128	30	35·2	3	0·13
286	... ...	9·3	1	6	44	51·34	128	30	31·3	1	0·14
287	... ...	9·4	1	6	45	0·13	106	32	46·5	1	0·15
288	α Pictoris	5·5	1	6	46	46·48	161	47	50·8	1	0·12
289	Lacaille 2610 ...	8·4	2	6	49	35·42	130	31	48·9	2	0·05
290	... ...	9·0	1	6	48	30·04	129	48	24·5	1	0·15
291	... ...	9·3	1	6	49	52·64	130	10	19·4	1	0·14
292	... ...	9·0	1	6	49	56·41	128	46	7·2	1	0·15
293	Lacaille 2638	7·5	2	6	51	20·20	114	47	37·3	2	0·09
294	... ...	9·0	1	6	52	47·90	132	54	39·2	1	0·16
295	21 Canis Majoris ε ...	1·5	...	6	53	19·28	118	47	26·3	6	0·08
296	... ...	8·0	2	6	55	37·40	129	37	59·6	2	0·15
297	43 Geminorum ζ Var. 1 ...	4·0	...	6	56	6·07	69	14	7·5	1	0·08
298	... ...	9·0	2	6	56	23·19	129	17	36·3	2	0·17
299	23 Canis Majoris γ ...	4·1	...	6	57	30·08	105	26	11·1	10	0·09
300	Taylor 2840	8·0	3	6	59	3·61	160	36	30·3	3	0·11
301	R Geminorum Var. 2 ...	8·8	1	6	59	13·31	67	5	31·8	1	0·08
302	... ...	7·6	1	7	1	1·50	140	10	9·3	1	0·19
303	R Canis Minoris Var. 1 ...	9·7	1	7	1	17·19	79	46	58·6	1	0·14
304	... ...	8·8	1	7	1	31·44	129	29	17·6	1	0·14
305	... .	9·1	2	7	2	36·51	161	24	10·5	2	0·13
306	... ...	8·7	2	7	3	26·72	130	14	36·2	2	0·16
307	... ...	9·4	1	7	5	2·70	138	32	14·5	1	0·17
308	Lacaille 2673 ...	9·0	1	7	6	11·70	138	9	21·2	1	0·09
309	Taylor [illegible] ...	8·1	2	7	7	17·[illegible]	150	21	26·3	2	0·17
310	... .	8·0	1	7	8	0·48	149	46	7·9	1	0·14
311	.. ..	8·0	1	7	8	10·03	132	5	8·2	1	0·11
312	... ...	9·0	1	7	8	36·79	130	17	34·3	1	0·15
313	Lacaille 2696	7·8	1	7	9	23·61	140	35	39·3	1	0·13
314	Taylor 2940 . ..	7·6	2	7	9	30·26	129	57	49·1	2	0·11
315	... ...	9·5	1	7	9	44·13	130	14	23·9	1	0·16

297.—ζ Geminorum Var. 1. Period 10·16 days. Range, 3·7 to 4·3 magnitude.
301.—R Geminorum Var. 2. Period 371 days. Range, 7th to below 13th magnitude.
303.—R Canis Minoris Var. 1. Period 335 days. Range, 7·5 to 11th magnitude.

Observed with the Madras Meridian Circle in that Year.

Number.	Star.	In Right Ascension. Annual Precession.	In Right Ascension. Secular Variation.	In Right Ascension. Proper Motion.	In Polar Distance. Annual Precession.	In Polar Distance. Secular Variation.	In Polar Distance. Proper Motion.	Number in B.A.C.
		s	s	s	"	"		
281	..	+ 0·8987	− 0·0088	...	+ 3·598	+ 0·084		...
282	..	+ 1·9317	+ 0·0013	...	+ 3·691	+ 0·273		...
283	..	+ 1·9457	+ 0·0013	...	+ 3·711	+ 0·277		...
284	...	+ 1·9812	+ 0·0013	...	+ 3·711	+ 0·275		...
285	...	+ 2·0270	+ 0·0013	...	+ 3·802	+ 0·286		...
286	... ...	+ 2·0280	+ 0·0014	...	+ 3·900	+ 0·289	...	...
287	... ...	+ 2·0895	+ 0·0000	...	+ 3·913	+ 0·282	...	...
288	α Pictoris	+ 0·0807	− 0·0068	− 0·010	+ 4·007	+ 0·098	− 0·18	2280
289	Lacaille 2616 ...	+ 1·9545	+ 0·0012	...	+ 4·306	+ 0·277	...	...
290	... ...	+ 2·0208	+ 0·0018	—	+ 4·214	+ 0·288	...	...
291	... ...	+ 1·9690	+ 0·0012	...	+ 4·344	+ 0·279	...	...
292	... ...	+ 2·0227	+ 0·0013	...	+ 4·360	+ 0·286	...	...
293	Lacaille 2588 ..	+ 2·4080	+ 0·0013	...	+ 4·465	+ 0·349	...	[illegible]
294	... ...	+ 0·5272	− 0·0080	...	+ 4·565	+ 0·073	...	...
295	21 Canis Majoris ε ...	+ 2·3571	+ 0·0018	0·000	+ 4·684	+ 0·382	+ 0·02	[illegible]
296	... ...	+ 1·9972	+ 0·0012	...	+ 4·880	+ 0·281	...	...
297	48 Gem. 3ª Var. 1 ...	+ 3·5640	− 0·0080	− 0·001	+ 4·880	+ 0·508	+ 0·01	[illegible]
298	... .	+ 2·0111	+ 0·0012	...	+ 4·886	+ 0·382	...	...
299	23 Canis Majoris γ ...	+ 2·7144	+ 0·0006	+ 0·002	+ 4·902	+ 0·341	+ 0·04	2319
300	Taylor 2940	+ 0·7447	− 0·0074	.	+ 5·111	+ 0·108	...	...
301	R Geminorum Var. 2	+ 3·6169	− 0·0080		+ 5·136	+ 0·508		...
302	. ...	+ 1·5258	− 0·0007		+ 5·277	+ 0·213		...
303	R Canis Minoris Var 1	+ 3·3049	− 0·0081		+ 5·290	+ 0·469		...
304	.. ...	+ 2·0080	+ 0·0011		+ 5·310	+ 0·290		...
305	... .	+ 1·4596	− 0·0013		+ 5·408	+ 0·308		...
306	... .	+ 1·9836	+ 0·0010		+ 5·480	+ 0·276	...	..
307	... ...	+ 0·4555	− 0·0136		+ 5·616	+ 0·082	...	...
308	Lacaille 2678 ..	+ 1·0096	− 0·0066		+ 5·712	+ 0·139	...	...
309	Taylor 2923 . ..	+ 0·8232	− 0·0077		+ 5·804	+ 0·118	...	..
310	. .	+ 0·9895	− 0·0074	...	+ 5·894	+ 0·132	.	—
311	... ...	+ 0·6808	− 0·0102		+ 5·877	+ 0·089		.
312	...	+ 1·9995	+ 0·0010		+ 5·941	+ 0·274		.
313	Lacaille 2986	+ 1·1971	− 0·0013		+ 5·980	+ 0·305		
314	Taylor 2940	+ 2·0028	+ 0·0010		+ 5·989	+ 0·276		.
315		+ 1·9800	+ 0·0009		+ 6·014	+ 0·271	.	—

288.—Proper motions from "*Stone's Cape Catalogue.*"

Mean Positions of Stars for 1865 January 1st.

Number.	Star.	Magnitude.	Estimations.	Mean Right Ascension. h.	m.	s.	Mean Polar Distance. °	′	″	Observations.	Fraction of Year.
316	54 Geminorum λ	3·6	...	7	12	3·46	67	46	21·6	11	0·08
317	... ...	8·3	1	7	12	36·69	152	48	2·1	1	0·17
318	Taylor 3005	7·7	1	7	15	[illegible]	140	1	0·0	1	0·14
319	B. P. L. 45	7·2	...	7	16	32·78	0	39	11·8	1	0·17
320	Lacaille 2905	9·0	1	7	17	22·15	153	8	13·5	1	0·08
321	... ...	9·3	2	7	17	39·92	129	40	16·0	2	0·16
322	... ...	9·8	1	7	18	40·07	69	15	26·1	1	0·13
323	63 Geminorum	5·3	...	7	19	43·39	68	16	55·2	1	0·19
324	... ...	9·3	2	7	19	50·09	152	36	46·6	2	0·14
325	Taylor 3064	7·8	1	7	20	2·80	151	41	36·2	1	0·08
326	...	8·0	1	7	21	36·09	131	50	33·5	1	0·15
327	...	8·5	2	7	23	29·09	129	45	26·1	2	0·16
328	..	9·8	1	7	23	37·00	42	0	51·0	1	0·13
329	...	8·0	1	7	23	36·79	130	10	12·3	1	0·14
330	..	9·4	1	7	24	9·78	162	47	21·6	1	0·17
331	... ...	9·6	1	7	24	46·12	160	9	38·9	1	0·07
332	... ...	9·0	2	7	24	46·82	161	0	36·0	2	0·07
333	... ...	9·0	1	7	26	0·77	123	8	26·5	1	0·08
334	S Canis Minoris Var. 2 ...	9·1	1	7	25	28·58	81	38	49·0	2	0·01
335	... .	8·0	1	7	25	29·00	102	18	12·7	1	0·18
336	66 Geminorum α² (Castor.)	1·6	...	7	25	53·83	57	49	8·8	6	0·12
337	... ...	9·6	1	7	26	39·22	129	45	7·9	1	0·13
338	... ...	9·2	1	7	27	14·03	168	10	53·4	1	0·07
339	... ...	8·0	1	7	27	58·31	160	42	51·0	1	0·15
340	.. . .	9·5	1	7	30	42·48	163	42	38·1	1	0·14
341	... ...	9·0	2	7	31	6·93	131	10	48·0	2	0·16
342	... ...	8·5	2	7	31	56·48	129	44	1·0	2	0·16
343	10 Canis Min α (Procyon.)	0·5	...	7	32	14·65	84	36	55·6	10	0·10
344	Lacaille 2993	7·0	1	7	32	43·93	121	40	31·6	1	0·14
345	... ...	9·4	2	7	33	0·00	163	11	40·7	2	0·13
346	. ...	10·5	2	7	34	34·71	68	38	6·6	2	0·07
347	S Geminorum Var. 3 ...	10·5	2	7	34	37·58	66	13	30·3	2	0·11
348	. ...	8·2	2	7	34	9·92	160	57	30·7	2	0·16
349	.. ..	8·9	2	7	35	20·49	152	59	48·6	1	0·19
350	Taylor 3195	8·0	2	7	36	31·93	150	19	12·0	1	0·20

319.—Groombridge 1119.
334.—S Canis Minoris Var. 2 Period [illegible] days. Range, 8.5 to below 13th magnitude.
346.—Observed by mistake for the planet Leda.
347.—S Geminorum Var. 3. Period 294 days. Range, 8.8 to below 14th magnitude.

Observed with the Madras Meridian Circle in that Year.

Number.	Star.	In Right Ascension. Annual Precession.	Secular Variation.	Proper Motion.	In Polar Distance. Annual Precession.	Secular Variation.	Proper Motion.	Number in B. A. C.
		s	s	s	″	″		
316	55 Geminorum δ ..	+ 3·5916	− 0·0072	0·000	+ 6·202	+ 0·486	+ 0·02	2410
317	... .	+ 0·0000	− 0·0110	...	+ 6·248	+ 0·080	...	–
318	Taylor 3006	+ 0·9552	− 0·0071	...	+ 6·488	+ 0·180	...	...
319	R. P. L. 45	+ 76·4323	− 27·7358	− 0·823	+ 6·675	+ 10·[illegible]	− 0·01	[illegible]
320	Lacaille 2905 ..	+ 0·5815	− 0·0132	...	+ 6·042	+ 0·077	...	...
321	.. ..	+ 2·0226	+ 0·0009	...	+ 6·686	+ 0·275	...	—
322	... ..	+ 3·6187	− 0·0074	...	+ 6·703	+ 0·494	...	...
323	63 Geminorum ..	+ 3·5727	− 0·0078	− 0·004	+ 6·897	+ 0·467	+ 0·10	2460
324	... ...	+ 0·6161	− 0·0127	...	+ 6·816	+ 0·086	...	...
325	Taylor 3054	+ 0·7307	− 0·0111	...	+ 6·865	+ 0·088	...	...
326	..	+ 1·9500	+ 0·0006	...	+ 6·900	+ 0·264		...
327	..	+ 2·0369	+ 0·0009	...	+ 7·146	+ 0·274		...
328	...	+ 4·4865	− 0·0256	...	+ 7·156	+ 0·605		...
329	...	+ 2·0551	+ 0·0009	...	+ 7·181	+ 0·277		...
330	..	+ 0·6452	− 0·0136	...	+ 7·201	+ 0·095		...
331	... ...	+ 2·0201	+ 0·0009	...	+ 7·260	+ 0·272	...	...
332	... .	+ 2·3228	+ 0·0011	...	+ 7·266	+ 0·318	—	...
333	.. ..	+ 2·2576	+ 0·0011	...	+ 7·270	+ 0·304	...	...
334	S Can. Min. Var. 2 .	+ 3·2605	− 0·0044	—	+ 7·302	+ 0·440	...	—
335	... .	+ 2·0629	+ 0·0009	...	+ 7·308	+ 0·276	...	...
336	66 Gem. α² (*Castor*).	+ 3·8548	− 0·0182	− 0·013	+ 7·360	+ 0·519	+ 0·08	2466
337	.. ..	+ 2·0866	+ 0·0009	...	+ 7·403	+ 0·273	...	...
338	... ...	+ 0·6167	+ 0·0146	...	+ 7·450	+ 0·081	...	...
339	... .	+ 2·0121	+ 0·0010	...	+ 7·510	+ 0·272	...	...
340	... .	− 0·0983	− 0·0800	...	+ 7·732	− 0·016	...	...
341	... .	+ 1·0837	+ 0·0008	...	+ 7·705	+ 0·395	...	...
342	... .	+ 2·0130	+ 0·0009	...	+ 7·661	+ 0·272	...	...
343	10 Can. Min. α (*Procyon*)	+ 3·1919	− 0·0041	− 0·046	+ 7·864	+ 0·496	+ 1·06	2522
344	Lacaille 2896 ..	+ 2·3003	+ 0·0012	...	+ 7·897	+ 0·307	...	...
345	.. ..	+ 0·6411	− 0·0133	..	+ 7·925	+ 0·083	...	...
346	... ...	+ 3·5866	− 0·0083		+ 8·048	+ 0·472		
347	S Geminorum Var. 3.	+ 3·6120	− 0·0102		+ 8·017	+ 0·460		–
348	.. ... .	+ 2·0167	+ 0·0008		+ 8·060	+ 0·270		..
349	... .	+ 0·6721	− 0·0152		+ 8·104	+ 0·087	...	.
350	Taylor 3195 ..	+ 0·9311	− 0·0105		+ 8·263	+ 0·120	.	...

319 — Proper motions from "*Greenwich Catalogue* 1872."

Mean Positions of Stars for 1865 January 1st.

Number.	Star.	Magnitude.	Estimations.	Mean Right Ascension. h.	m.	s.	Mean Polar Distance. °	′	″	Observations.	Fraction of Year.
351	... ...	7·5	2	7	36	46·40	130	57	26·8	2	0·16
352	78 Geminorum β (*Pollux*).	1·1	...	7	37	8·07	61	39	4·2	9	0·11
353	... ...	10·3	1	7	37	34·59	68	39	51·7	1	0·08
354	... ...	8·5	1	7	37	36·86	130	58	11·8	1	0·14
355	... ...	7·7	1	7	37	46·71	128	53	1·6	1	0·14
356	... ...	7·0	1	7	39	7·60	131	8	40·0	1	0·17
357	... ...	8·8	2	7	41	19·42	151	34	30·1	2	0·10
358	... .	8·5	1	7	42	4·81	158	4	30·7	1	0·16
359	Lacaille 3031 ...	8·0	1	7	44	5·60	158	51	49·5	1	0·18
360	.. ..	9·0	2	7	44	11·97	130	56	9·0	2	0·16
361	Brisbane 1701	7·8	1	7	46	10·90	144	24	46·4	1	0·14
362	... ...	8·8	1	7	46	30·77	144	22	35·1	1	0·14
363	... ...	8·5	1	7	47	6·97	153	30	56·8	1	0·17
364	... ...	10·7	1	7	47	18·22	67	42	57·7	1	0·06
365	Taylor 3310	9·0	1	7	47	38·67	149	16	2·8	1	0·15
366	... .	9·6	1	7	48	44·85	139	54	56·6	1	0·17
367	1 Cancri	5·9	...	7	49	19·51	73	51	7·8	1	0·10
368	... ...	7·1	1	7	49	30·25	152	35	4·5	1	0·19
369	... .	8·2	1	7	49	56·49	129	17	31·1	1	0·15
370	.. ..	7·7	1	7	49	56·67	149	8	37·3	1	0·14
371	... .	9·3	1	7	51	39·73	151	37	3·7	1	0·17
372	... ...	8·1	2	7	52	1·90	148	32	34·1	2	0·13
373	6 Cancri	6·0	...	7	53	13·36	61	49	40·7	3	0·14
374	Brisbane 1836	9·8	1	7	55	38·87	158	55	57·9	1	0·16
375	.. ..	9·2	1	7	57	40·48	136	34	30·2	1	0·16
376	Lacaille 3164 ...	8·6	2	7	58	37·56	159	11	40·3	2	0·14
377	... ...	10·7	2	8	0	15·57	78	39	52·1	2	0·06
378	... ...	6·9	1	8	1	23·51	150	31	32·3	1	0·17
379	Lacaille 3174 ...	8·0	2	8	1	20·26	156	36	6·0	2	0·15
380	15 Argûs	2·9	...	8	1	47·74	113	56	2·3	6	0·17
381	.. ..	9·0	1	8	2	14·12	139	39	37·1	1	0·15
382	16 Cancri ζ	4·7	.	8	4	29·08	71	36	44·6	2	0·08
383	... .	9·7	1	8	4	33·23	154	40	44·2	1	0·17
384	... .	9·6	1	8	5	30·36	135	33	57·1	1	0·15
385	.. ..	5·0	1	8	6	5·87	136	40	43·6	1	0·21

Observed with the Madras Meridian Circle in that Year.

Number.	Star.	In Right Ascension.			In Polar Distance.			Number in B. A. C.
		Annual Precession.	Secular Variation.	Proper Motion.	Annual Precession.	Secular Variation.	Proper Motion.	
		s	s	s	"	"		
351	... ...	+ 2·0803	+ 0·0008	...	+ 8·210	+ 0·260	...	...
352	78 Gem. β (*Pollux*)...	+ 3·7207	− 0·0128	− 0·040	+ 8·211	+ 0·491	+ 0·03	[illegible]
353	... ...	+ 3·3520	− 0·0005	...	+ 8·270	+ 0·408	...	...
354	... ...	+ 2·0147	+ 0·0008	...	+ 8·284	+ 0·264	...	...
355	... ...	+ 2·0906	+ 0·0010	...	+ 8·302	+ 0·274	...	...
356	... ...	+ 2·0114	+ 0·0000		+ 8·400	+ 0·262		...
357	... ...	+ 0·8302	− 0·0138		+ 8·590	+ 0·107		...
358	... ...	+ 0·6964	− 0·0161		+ 8·640	+ 0·088		..
359	Lacaille 3084 ...	+ 0·6285	− 0·0180		+ 8·789	+ 0·078		...
360	... ...	+ 2·0808	+ 0·0008		+ 8·807	+ 0·262		...
361	Brisbane 1791 ...	+ 1·4011	− 0·0043	...	+ 8·976	+ 0·179	...	...
362	... ...	+ 1·4041	− 0·0042	...	+ 8·989	+ 0·179	...	...
363	... ...	+ 0·6893	− 0·0170	...	+ 9·096	+ 0·086	...	...
364	... ...	+ 3·3610	− 0·0108	...	+ 9·044	+ 0·439	..	...
365	Taylor 3310	+ 1·0653	− 0·0086	...	+ 9·149	+ 0·135	...	...
366	... ...	+ 2·0771	+ 0·0010	...	+ 9·168	+ 0·260	...	...
367	1 Cancri ...	+ 3·4160	− 0·0094	− 0·001	+ 9·208	+ 0·439	+ 0·04	2639
368	... ...	+ 0·7830	− 0·0138	...	+ 9·222	+ 0·098	...	...
369	... ...	+ 2·1016	+ 0·0011	...	+ 9·242	+ 0·268	...	...
370	... ...	+ 1·0869	− 0·0092	...	+ 9·256	+ 0·137	...	...
371	... ...	+ 0·8856	− 0·0136	...	+ 9·339	+ 0·110	...	...
372	... ...	+ 1·1564	− 0·0088	...	+ 9·418	+ 0·146	...	...
373	6 Cancri	+ 3·6894	− 0·0148	− 0·006	+ 9·603	+ 0·468	+ 0·07	2672
374	Brisbane 1865 ...	+ 0·7508	− 0·0165	...	+ 9·636	+ 0·096	...	2680
375	... ...	+ 0·4068	− 0·0273	...	+ 9·861	+ 0·048	...	...
376	Lacaille 3134 ...	+ 0·7726	− 0·0172	...	+ 9·923	+ 0·094	...	2713
377	... ...	+ 3·3075	− 0·0074	...	+ 10·031	+ 0·414	...	...
378	... ...	+ 1·0807	− 0·0116	...	+ 10·138	+ 0·126	...	...
379	Lacaille 3174 ..	+ 0·3947	− 0·0246	..	+ 10·136	+ 0·042	...	...
380	15 Argûs	+ 2·8608	+ 0·0009	− 0·007	+ 10·103	+ 0·318	− 0·06	2729
381	... .	+ 2·1510	+ 0·0016	...	+ 10·106	+ 0·266	...	
382	16 Cancri 3	+ 3·4462	− 0·0108	+ 0·004	+ 10·364	+ 0·426	+ 0·11	2744
383	... .	+ 0·6585	− 0·0217	...	+ 10·371	+ 0·078	..	..
384	... . ..	+ 2·1526	+ 0·0016	...	+ 10·453	+ 0·265	..	.
385	... . ..	+ 2·1507	+ 0·0010	..	+ 10·490	+ 0·264	..	..

367.—The proper motions taken from "*Greenwich Catalogue 1872.*"

Mean Positions of Stars for 1865 January 1st.

Number.	Star.	Magnitude.	Estimations.	Mean Right Ascension.			Mean Polar Distance.			Observations.	Fraction of year.
				h	m	s	°	′	″		
386	... ...	9·3	2	8	8	36·16	128	38	35·0	2	0·18
387	R Cancri Var. 1	7·6	2	8	9	7·07	77	51	44·0	2	0·10
388	... ...	9·2	1	8	9	27·79	74	16	13·5	1	0·08
389	W. B. N. VIII. 178 ...	9·6	1	8	9	58·76	74	16	56·2	1	0·05
390	... ...	9·4	2	8	10	19·08	150	32	40·8	2	0·17
391	... ...	9·3	1	8	10	27·33	151	36	38·7	1	0·18
392	... ...	8·8	2	8	11	38·32	152	4	42·0	2	0·14
393	... ...	8·5	1	8	13	22·83	131	41	34·3	1	0·19
394	... ...	9·5	1	8	14	31·53	154	5	17·7	1	0·21
395	... ...	8·2	2	8	14	52·04	142	17	6·6	2	0·20
396	Lacaille 3297 ...	8·2	2	8	15	1·98	138	59	1·9	2	0·12
397	20 Cancri d¹ ...	5·9	...	8	15	37·82	71	14	14·0	1	0·18
398	... ...	9·5	1	8	16	30·04	77	31	30·3	1	0·14
399	... ...	8·8	1	8	17	26·74	77	40	10·6	1	0·18
400	... ...	9·7	1	8	17	59·65	164	23	54·8	2	0·20
401	... ...	9·8	2	8	18	38·92	70	2	10·3	2	0·07
402	... ...	9·7	2	8	19	47·94	151	0	55·9	2	0·19
403	Taylor 3390	8·0	1	8	20	15·26	144	22	57·1	1	0·19
404	W. B. N. VIII. 469	8·2	1	8	20	20·45	74	27	31·4	1	0·08
405	... ...	9·0	1	8	21	5·05	130	18	30·1	1	0·07
406	29 Cancri	5·9	...	8	21	5·20	75	30	42·2	3	0·08
407	... .	8·0	2	8	21	52·34	131	41	46·8	2	0·21
408	Taylor 3620 ...	7·9	1	8	22	12·00	130	47	30·7	1	0·13
409	... ...	8·3	1	8	23	34·09	128	38	46·6	1	0·23
410	33 Cancri η ...	5·5	...	8	24	58·53	69	6	11·1	10	0·16
411	Taylor 3661	7·8	1	8	25	41·92	130	3	32·7	1	0·23
412	Taylor 3662	8·0	1	8	25	48·10	130	3	52·1	1	0·23
413	... ...	9·0	1	8	28	48·53	75	10	30·0	2	0·15
414	W. B. N. VIII. 681 ..	8·7	2	8	29	4·98	70	30	7·0	2	0·20
415	Lacaille 3480	7·9	3	8	29	13·95	131	53	52·7	3	0·22
416	W. B. N. VIII. 699 ..	9·0	1	8	30	34·37	70	30	49·9	1	0·21
417	... ...	8·0	1	8	31	14·75	129	46	37·0	1	0·08
418	Taylor 3710	8·0	1	8	31	26·01	141	21	17·5	1	0·13
419	. ... ..	9·0	1	8	33	11·82	129	23	40·3	1	0·08
420	... .	8·8	3	8	34	9·35	151	20	36·8	3	0·21

387.—R Cancri Var. 1.—Period 354 days. Range, 6th to 13th magnitude.
386 - 390.—Comparison stars for Ariadne in 1865.

Observed with the Madras Meridian Circle in that Year.

Number.	Star.	In Right Ascension. Annual Precession.	Secular Variation.	Proper Motion.	In Polar Distance. Annual Precession.	Secular Variation.	Proper Motion.	Number in B. A. C.
		s	s	s	"	"		
386	... ...	+ 2·1670	+ 0·0017	...	+ 10·672	+ 0·268		...
387	R Cancri Var. 1. ...	+ 3·3152	− 0·0060	...	+ 10·711	+ 0·404		...
388	... ...	+ 3·3002	− 0·0095	...	+ 10·730	+ 0·413		...
389	W. B. N. VIII. 178...	+ 3·3886	− 0·0096	...	+ 10·774	+ 0·412		...
390	... ...	+ 1·0771	− 0·0110	...	+ 10·800	+ 0·127		...
391	... ...	+ 1·0080	− 0·0138	...	+ 10·809	+ 0·118	...	...
392	... ...	+ 0·9586	− 0·0147	...	+ 10·878	+ 0·112	...	...
393	... ...	+ 2·0773	+ 0·0015	...	+ 11·094	+ 0·248	...	...
394	... ...	+ 0·7808	− 0·0108	...	+ 11·107	+ 0·090	...	...
395	... ...	+ 1·6840	− 0·0021	...	+ 11·132	+ 0·198	...	...
396	Lacaille 3297 ...	+ 0·7047	− 0·0106	...	+ 11·144	+ 0·091	...	...
397	20 Cancri d¹	+ 3·4491	− 0·0114	− 0·005	+ 11·187	+ 0·413	+ 0·02	2799
398	... ...	+ 3·3168	− 0·0086	...	+ 11·230	+ 0·396	...	...
399	... ...	+ 3·3103	− 0·0084	...	+ 11·318	+ 0·394	...	...
400	... ...	+ 0·7742	− 0·0807	...	+ 11·359	+ 0·088	...	...
401	... ...	+ 3·4716	− 0·0192	...	+ 11·408	+ 0·413	...	...
402	... ...	+ 1·0878	− 0·0122	...	+ 11·417	+ 0·138	...	...
403	Taylor 3609	+ 1·5162	− 0·0080	...	+ 11·595	+ 0·170	...	...
404	W. B. N. VIII. 460...	+ 3·3764	− 0·0100	...	+ 11·507	+ 0·398	...	...
405	... ...	+ 0·0012	− 0·0175	...	+ 11·561	+ 0·102	...	...
406	29 Corvi	+ 3·3876	− 0·0096	− 0·002	+ 11·591	+ 0·396	+ 0·01	2806
407	... ...	+ 2·1010	+ 0·0018	...	+ 11·612	+ 0·246	...	...
408	Taylor 3630	+ 2·1361	+ 0·0020	...	+ 11·733	+ 0·246	...	...
409	... ...	+ 2·2060	+ 0·0026	...	+ 11·757	+ 0·250	...	...
410	33 Cancri η	+ 3·4683	− 0·0120	− 0·065	+ 11·892	+ 0·401	+ 0·06	2862
411	Taylor 3651	+ 2·1675	+ 0·0022	...	+ 11·908	+ 0·240	...	...
412	Taylor 3653	+ 2·1681	+ 0·0022	...	+ 11·912	+ 0·249	...	...
413	... ...	+ 3·3611	− 0·0092	...	+ 12·119	+ 0·384	...	...
414	W. B. N. VIII. 654...	+ 3·4457	− 0·0124	...	+ 12·146	+ 0·396	...	...
415	Lacaille 3480 ...	+ 1·0820	− 0·0186	...	+ 12·156	+ 0·121	...	...
416	W. B. N. VIII. 680...	+ 3·4448	− 0·0134	...	+ 12·170	+ 0·894	...	...
417	... ...	+ 2·1083	+ 0·0014	...	+ 12·296	+ 0·222	...	...
418	Taylor 3710	+ 1·7519	− 0·0008	...	+ 12·808	+ 0·197	...	...
419	... ...	+ 2·2104	+ 0·0086	...	+ 12·489	+ 0·248	...	...
420	... ...	+ 0·8940	− 0·0197	...	+ 12·496	+ 0·097	...	...

397.—406.—Proper motions from "*Greenwich Catalogue 1872.*"

Mean Positions of Stars for 1865 January 1st.

Number.	Star.	Magnitude.	Estimations.	Mean Right Ascension.			Mean Polar Distance.			Observations.	Fraction of Year.
				h.	m.	s.	°	′	″		
421	... ...	7·8	1	8	34	41·72	139	46	36·1	1	0·08
422	Lacaille 3401 ...	7·9	1	8	36	3·48	152	22	8·7	1	0·08
423	b Velorum... ...	4·1	...	8	36	8·85	136	10	12·6	2	0·19
424	Taylor 3767 ...	7·2	1	8	36	21·27	140	30	38·7	1	0·18
425	47 Cancri δ ...	4·8	...	8	37	0·48	71	21	7·0	2	0·11
426	... ...	9·3	2	8	37	18·16	136	3	45·6	2	0·22
427	... .	8·2	1	8	37	58·46	136	5	47·8	1	0·13
428	50 Cancri A² ...	5·8	...	8	39	32·00	77	28	48·7	1	0·28
429	11 Hydræ ε ...	3·6	...	8	39	37·48	83	6	17·9	17	0·16
430	... ...	8·6	1	8	40	31·08	139	15	46·7	1	0·34
431	... ...	8·5	2	8	41	3·94	147	16	55·8	2	0·31
432	Lacaille 3564	8·0	1	8	42	19·54	139	18	19·8	1	0·34
433	... ...	9·3	2	8	42	57·61	136	10	48·7	2	0·34
434	Lacaille 3578	8·0	2	8	44	14·15	152	41	39·6	2	0·19
435	... ...	9·0	1	8	45	49·98	86	27	36·5	1	0·28
436	... ...	8·1	2	8	46	46·21	132	38	38·2	2	0·19
437	... ...	8·2	2	8	46	58·00	136	6	15·7	2	0·28
438	... ...	9·1	2	8	47	42·17	133	36	8·0	2	0·19
439	... ...	8·0	2	8	48	40·85	138	1	32·1	3	0·21
440	T Cancri Var. 3	8·5	2	8	48	57·33	69	39	12·6	2	0·18
441	... ...	8·0	1	8	49	15·68	132	54	32·8	1	0·14
442	... ...	7·9	1	8	49	32·19	133	30	18·7	1	0·28
443	9 Ursæ Majoris ι	3·2	...	8	49	50·89	41	36	30·9	2	0·21
444	... ...	8·1	2	8	50	17·41	132	57	0·5	2	0·17
445	65 Cancri α	4·6	...	8	51	6·18	77	37	19·3	4	0·09
446	... ...	9·7	1	8	51	23·37	147	15	46·1	1	0·19
447	... ...	9·8	1	8	51	50·95	137	34	54·3	1	0·28
448	... ...	8·1	1	8	53	50·68	138	55	31·6	1	0·16
449	... ...	8·0	1	8	54	11·62	142	41	23·0	1	0·28
450	... ...	8·0	2	8	54	50·13	142	40	11·8	2	0·18
451	... .	9·6	3	8	56	21·98	146	51	14·5	2	0·20
452	... ...	8·2	2	8	56	30·00	146	48	17·1	2	0·17
453	. .. .	8·7	1	8	56	41·19	146	36	57·0	1	0·15
454	.. ..	8·6	1	8	56	50·32	146	35	40·0	1	0·21
455	.. ..	9·1	2	8	58	7·22	146	40	57·1	2	0·18

440.—T Cancri Var. 3. Period [illegible] days. Range, 8th to 10.5 magnitude.

Observed with the Madras Meridian Circle in that Year.

Number.	Star.	In Right Ascension. Annual Precession.	In Right Ascension. Secular Variation.	In Right Ascension. Proper Motion.	In Polar Distance. Annual Precession.	In Polar Distance. Secular Variation.	In Polar Distance. Proper Motion.	Number in B.A.C.
		s	s	s	"	"		
421	... ...	+ 2·2082	+ 0·0027	...	+ 12·589	+ 0·240	...	...
422	Lacaille 3401 ...	+ 1·0876	− 0·0141	...	+ 12·624	+ 0·118	...	2940
423	b Velorum	+ 1·9908	+ 0·0018	...	+ 12·631	+ 0·221	...	2947
424	Taylor 3767	+ 1·9801	− 0·0089	...	+ 12·645	+ 0·141	...	...
425	47 Cancri δ	+ 3·4214	− 0·0125	− 0·002	+ 12·690	+ 0·362	+ 0·24	2966
426	... ...	+ 1·0901	+ 0·0019	...	+ 12·700	+ 0·220	...	...
427	... ...	+ 1·0907	+ 0·0019	...	+ 12·750	+ 0·220	...	...
428	50 Cancri A² ...	+ 3·3015	− 0·0095	− 0·007	+ 12·860	+ 0·364	+ 0·07	2970
429	11 Hydræ ε	+ 3·1963	− 0·0071	− 0·012	+ 12·880	+ 0·351	+ 0·04	2971
430	... ...	+ 2·2365	+ 0·0031	...	+ 12·926	+ 0·244	...	...
431	... ...	+ 1·4840	− 0·0049	...	+ 12·962	+ 0·150	...	...
432	Lacaille 3584 ...	+ 2·2408	+ 0·0032	...	+ 13·040	+ 0·242	...	...
433	... ...	+ 2·0165	+ 0·0028	...	+ 13·089	+ 0·217	...	...
434	Lacaille 3578 ...	+ 1·1104	− 0·0140	...	+ 13·173	+ 0·117	...	3008
435	... ...	+ 3·1341	− 0·0068	...	+ 13·278	+ 0·277	...	...
436	... ...	+ 2·1447	+ 0·0032		+ 13·389	+ 0·226	...	...
437	... ...	+ 2·0938	+ 0·0026		+ 13·362	+ 0·216	...	...
438	... ...	+ 2·1406	+ 0·0032		+ 13·400	+ 0·227	...	...
439	... ...	+ 2·1477	+ 0·0033		+ 13·470	+ 0·226	...	...
440	T Cancri Var. 3 ...	+ 3·4696	− 0·0141		+ 13·482	+ 3·868	...	...
441	... ...	+ 2·1581	+ 0·0033	...	+ 13·501	+ 0·226	...	...
442	... ...	+ 2·1510	+ 0·0033	...	+ 13·508	+ 0·226	—	...
443	9 Ursæ Majoris ι ...	+ 4·1892	− 0·0440	− 0·047	+ 13·546	+ 0·448	+ 0·28	3046
444	.. ..	+ 2·1559	+ 0·0034	...	+ 13·568	+ 0·226	...	...
445	65 Cancri α	+ 3·2576	− 0·0068	0·000	+ 13·620	+ 0·346	+ 0·04	3055
446	...	+ 1·5474	− 0·0087		+ 13·636	+ 0·159	...	...
447	...	+ 2·0090	+ 0·0027		+ 13·674	+ 0·208	...	...
448	...	+ 2·1009	+ 0·0037		+ 13·804	+ 0·224	...	...
449	...	+ 1·8005	+ 0·0006		+ 13·817	+ 0·184	...	...
450	..	+ 1·7987	+ 0·0006		+ 13·888	+ 0·184	...	...
451	. ...	+ 1·0015	− 0·0086	...	+ 13·953	+ 0·162	...	...
452	.. ..	+ 1·0090	− 0·0086	...	+ 13·973	+ 0·162	..	...
453	... .	+ 1·5902	− 0·0027	..	+ 13·974	+ 0·161	...	...
454	. ... ..	+ 1·6196	− 0·0023	.	+ 13·990	+ 0·163	...	...
455	... .	+ 1·6138	− 0·0021	..	+ 14·064	+ 0·162	...	..

429. —Proper motions from "*Greenwich Catalogue* 1864."

Mean Positions of Stars for 1865 January 1st.

Number.	Star.	Magnitude.	Estimations.	Mean Right Ascension.			Mean Polar Distance.			Observations.	Fraction of Year.
				h.	m.	s.	°	′	″		
456	...	9·0	1	8	58	11·10	146	19	37·9	1	0·15
457	76 Cancri κ	5·0	...	9	0	25·06	78	47	37·9	2	0·03
458	...	8·3	2	9	1	5·09	160	1	46·1	2	0·26
459	...	8·3	1	9	1	52·46	128	57	23·6	1	0·28
460	Lacaille 3705	7·6	2	9	3	15·22	151	17	13·5	2	0·26
461	...	9·0	3	9	3	59·36	136	27	30·0	2	0·19
462	Taylor 4021	7·7	1	9	5	38·14	135	44	27·4	1	0·24
463	...	9·0	1	9	6	33·35	133	41	46·0	1	0·04
464	83 Cancri π²	5·6	...	9	7	45·40	74	30	2·8	1	0·15
465	Lacaille 3747	8·5	1	9	7	46·69	150	24	5·1	1	0·10
466	...	10·0	1	9	9	38·14	78	52	56·6	1	0·23
467	Lacaille 3774	8·8	3	9	9	51·98	157	0	59·4	3	0·16
468	86 Cancri ...	6·6	...	9	11	26·37	71	48	38·4	12	0·18
469	...	8·0	1	9	13	6·02	72	19	12·6	1	0·08
470	ι Argûs	2·5	...	9	13	23·30	148	42	30·0	2	0·28
471	...	9·0	1	9	15	17·47	148	43	56·2	1	0·23
472	...	9·3	2	9	15	30·97	25	4	41·1	2	0·24
473	...	9·2	1	9	16	10·63	130	1	17·5	1	0·23
474	...	9·5	3	9	17	35·06	75	5	46·3	3	0·20
475	O. A. N. [illegible]	8·9	2		17	42·61	25	3	38·0	2	0·24
476	...	7·8	1	9	19	32·37	75	6	46·1	1	0·04
477	...	8·1	1	9	20	36·81	137	38	18·8	1	0·26
478	...	9·0	2	9	20	52·10	135	36	59·2	2	0·17
479	30 Hydræ α Var. 2	2·0	...	9	20	57·15	98	4	31·3	11	0·19
480	Lacaille [illegible]	8·0	1	9	22	34·35	161	30	35·3	1	0·26
481	25 Ursæ Majoris θ	3·2	...	9	23	48·68	37	42	26·1	1	0·22
482	...	9·6	1	9	24	31·34	163	41	1·7	1	0·24
483	Lacaille [illegible]	8·7	1	9	24	45·26	141	30	4·1	1	0·23
484	Lacaille [illegible]	7·8	1	9	24	57·18	140	0	49·1	1	0·28
485	...	9·0	1	9	26	44·09	145	2	41·4	1	0·26
486	...	9·2	1	9	26	57·36	144	38	39·9	1	0·24
487	Taylor [illegible]	8·0	2	9	27	40·67	146	28	36·8	2	0·22
488	...	9·0	1	9	28	1·00	128	46	9·4	1	0·16
489	Taylor [illegible]	7·3	1	9	28	30·46	146	29	48·4	1	0·24
490	...	9·4	1	9	29	34·61	146	38	51·7	1	0·26

472—475.—Comparison stars for Comet 2, 1861.
479.—α Hydræ Var. 2.—Supposed to vary irregularly from 2.0 to 2.5 magnitude.

Observed with the Madras Meridian Circle in that Year.

Number.	Star.	In Right Ascension. Annual Precession.	Secular Variation.	Proper Motion.	In Polar Distance. Annual Precession.	Secular Variation.	Proper Motion.	Number in B. A. C.
		s	s	s	"	"		
456	... .	+ 1·6498	− 0·0018	...	+ 14·098	+ 0·165	...	...
457	76 Cancri κ	+ 3·3801	− 0·0094	− 0·002	+ 14·208	+ 0·330	0·00	3111
458	... ...	+ 1·4404	− 0·0062	...	+ 14·249	+ 0·142	...	...
459	... ...	+ 2·2141	+ 0·0014	...	+ 14·307	+ 0·231	...	...
460	Lacaille 3706 ...	+ 1·3632	− 0·0058	...	+ 14·320	+ 0·133	...	...
461	... ...	+ 2·2120	+ 0·0045	...	+ 14·425	+ 0·218	...	...
462	Taylor 4021	+ 2·0209	+ 0·0037	...	+ 14·520	+ 0·197	...	...
463	... ...	+ 2·0272	+ 0·0037	...	+ 14·580	+ 0·197	...	...
464	82 Cancri π²	+ 3·3232	− 0·0117	− 0·004	+ 14·654	+ 0·326	− 0·02	3147
465	Lacaille 3747 ...	+ 1·4052	− 0·0057	...	+ 14·654	+ 0·140	...	...
466	... ...	+ 3·3337	− 0·0121	...	+ 14·755	+ 0·323	...	...
467	Lacaille 3774 ...	+ 0·9260	− 0·0241	...	+ 14·781	+ 0·085	...	...
468	83 Cancri	+ 3·3682	− 0·0134	− 0·012	+ 14·871	+ 0·323	+ 0·16	3171
469	... ...	+ 3·3560	− 0·0131	...	+ 14·969	+ 0·320	...	...
470	ι Argûs	+ 1·0100	− 0·0022	...	+ 14·900	+ 0·150	...	3186
471	... .	+ 1·8686	+ 0·0086	...	+ 15·005	+ 0·174		...
472	... ...	+ 4·0465	− 0·1123	...	+ 15·135	+ 0·460		...
473	... ...	+ 2·0089	+ 0·0046	...	+ 15·154	+ 0·190		...
474	... ...	+ 3·3037	− 0·0116	...	+ 15·226	+ 0·307		...
475	O. A. N. 9951 ...	+ 4·0813	− 0·1126	...	+ 15·288	+ 0·461		...
476	... ...	+ 3·3011	− 0·0110	...	+ 15·387	+ 0·308	...	...
477	... ...	+ 2·1874	+ 0·0067	...	+ 15·388	+ 0·193	...	...
478	... ...	+ 0·3844	− 0·0293	...	+ 15·412	+ 0·076	...	...
479	30 Hydræ α Var. 2 ...	+ 2·0806	− 0·0013	− 0·004	+ 15·417	+ 0·268	− 0·08	3223
480	Lacaille 3860 ...	+ 2·3030	+ 0·0063	...	+ 15·500	+ 0·207	...	...
481	26 Ursæ Majoris θ ...	+ 4·1614	− 0·0861	− 0·111	+ 15·575	+ 0·374	+ 0·57	3242
482	... ...	+ 0·0200	− 0·0276	...	+ 15·606	+ 0·077	...	...
483	Lacaille 3836 ...	+ 2·0038	+ 0·0062	...	+ 15·627	+ 0·170	...	...
484	Lacaille 3897 ...	+ 2·0740	+ 0·0067	...	+ 15·696	+ 0·182	...	...
485	... ...	+ 1·8982	+ 0·0087	...	+ 15·736	+ 0·164	...	...
486	... .	+ 1·8807	+ 0·0088	...	+ 15·747	+ 0·164	...	...
487	Taylor 4222	+ 1·8310	+ 0·0039	...	+ 15·702	+ 0·168	...	...
488	... .	+ 2·4111	+ 0·0064	...	+ 15·804	+ 0·210	...	...
489	Taylor 4236	+ 1·8332	+ 0·0080	...	+ 15·869	+ 0·157	..	..
490	... ...	+ 1·8864	+ 0·0081	.	+ 15·898	+ 0·150	...	..

Mean Positions of Stars for 1865 January 1st.

Number.	Star.	Magnitude.	Estimations.	Mean Right Ascension.			Mean Polar Distance.			Observations.	Fraction of Year.
				h.	m.	s.	°	′	″		
491	10 Leonis	5·0	...	9	30	4·87	82	58	55·9	2	0·11
492	... ...	8·8	2	9	32	56·51	129	47	52·1	2	0·16
493	14 Leonis o	3·8	...	9	33	50·49	79	29	43·4	2	0·08
494	Lacaille 3980	8·4	2	9	34	32·25	148	34	2·5	2	0·18
495	... ...	8·8	2	9	34	40·34	130	34	56·8	2	0·24
496	Taylor 4380	7·8	1	9	34	40·48	142	10	59·0	1	0·28
497	... ...	8·3	3	9	34	58·45	161	50	50·8	3	0·17
498	... ...	8·8	3	9	35	36·32	151	50	40·0	3	0·17
499	17 Leonis ε	3·1	...	9	38	10·89	65	36	22·6	12	0·19
500	... ...	6·0	...	9	39	2·73	82	40	12·2	2	0·25
501	... ...	7·0	1	9	41	52·48	130	49	37·5	1	0·26
502	... .	9·1	2	9	43	39·05	148	57	8·3	2	0·24
503	... ...	8·8	1	9	43	36·67	148	46	10·7	1	0·28
504	... ...	8·7	1	9	44	7·33	147	1	58·6	1	0·15
505	... ...	9·0	2	9	44	50·61	129	47	38·8	2	0·16
506	... ...	9·0	1	9	46	56·50	169	3	9·4	1	0·24
507	R. P. L. 70 ...	6·5	...	9	46	58·56	5	26	4·7	1	0·22
508	... ...	9·1	2	9	46	39·09	169	7	14·2	2	0·29
509	... ...	9·3	1	9	49	56·65	162	7	57·3	1	0·24
510	... ...	8·0	2	9	50	49·77	146	39	30·3	2	0·29
511	29 Leonis π	5·0	...	9	53	4·62	81	18	26·2	13	0·15
512	... ...	9·1	1	9	53	54·98	162	7	4·2	1	0·24
513	... ...	8·1	2	9	54	10·04	148	46	52·8	2	0·26
514	Taylor 4445	8·0	1	9	54	45·13	147	28	57·6	1	0·24
515	... ...	9·0	1	9	56	54·00	147	34	33·3	1	0·16
516	... ...	9·7	2	9	56	1·10	139	57	42·7	2	0·16
517	... ...	9·0	2	9	57	0·84	129	56	55·0	2	0·16
518	... ...	9·0	1	9	57	50·00	146	33	21·6	1	0·26
519	Taylor 4476	7·9	1	9	57	59·98	145	28	32·6	1	0·15
520	... ...	8·7	1	9	58	11·94	140	54	26·0	1	0·26
521	Taylor 4484	7·5	3	9	58	42·85	161	30	15·6	3	0·20
522	32 Leonis α (Regulus) ...	1·4	...	10	1	10·77	77	22	29·9	11	0·28
523	... ...	9·0	1	10	1	23·73	130	0	15·5	1	0·04
524	Lacaille 4104	7·0	1	10	2	14·73	143	54	21·5	1	0·26
525	... .	9·3	2	10	2	40·26	129	57	55·0	2	0·20

507.—Carrington 1661.

Observed with the Madras Meridian Circle in that Year.

Number.	Star.	In Right Ascension. Annual Precession.	Secular Variation.	Proper Motion.	In Polar Distance. Annual Precession.	Secular Variation.	Proper Motion.	Number in B.A.C.
		s	s	s	"	"		
491	10 Leonis	+ 3·1783	− 0·0077	− 0·006	+ 15·915	+ 0·276	0·00	3296
492	... ...	+ 2·4064	+ 0·0072	...	+ 16·066	+ 0·208	...	...
493	14 Leonis ο	+ 3·2196	− 0·0098	− 0·018	+ 16·113	+ 0·273	+ 0·04	3312
494	Lacaille 3980 ..	+ 1·7750	+ 0·0094	...	+ 16·140	+ 0·147	...	...
495	... .	+ 2·3940	+ 0·0075	...	+ 16·102	+ 0·200	...	...
496	Taylor 4090	+ 2·0496	+ 0·0065	...	+ 16·161	+ 0·170	...	...
497	... .	+ 1·8876	− 0·0021	...	+ 16·167	+ 0·180	...	...
498	... .	+ 1·5939	− 0·0020	...	+ 16·204	+ 0·180	...	...
499	17 Leonis ε	+ 3·4237	− 0·0180	− 0·004	+ 16·337	+ 0·262	+ 0·02	3381
500	... ...	+ 3·1713	− 0·0075	...	+ 16·350	+ 0·260	...	3386
501	..	+ 2·4173	+ 0·0083	...	+ 16·522	+ 0·198	...	...
502	...	+ 2·0415	+ 0·0075	...	+ 16·601	+ 0·160	...	...
503	...	+ 2·0490	+ 0·0076	...	+ 16·607	+ 0·160	...	...
504	...	+ 1·9204	+ 0·0060	...	+ 16·688	+ 0·150	...	...
505	...	+ 2·4583	+ 0·0086	...	+ 16·675	+ 0·198	...	...
506	... ...	+ 2·4734	+ 0·0086	...	+ 16·738	+ 0·192	...	...
507	R. P. L. 70	+ 10·8085	− 1·3890	...	+ 16·747	+ 0·860	...	...
508	... ...	+ 2·4780	+ 0·0060	...	+ 16·747	+ 0·192	...	...
509	... ...	+ 1·7004	+ 0·0017	...	+ 16·946	+ 0·128	...	...
510	... ...	+ 2·0266	+ 0·0082	...	+ 16·989	+ 0·161	...	...
511	20 Leonis ν	+ 3·1796	− 0·0080	− 0·008	+ 17·057	+ 0·290	+ 0·08	3415
512	... ...	+ 1·7500	+ 0·0084	...	+ 17·093	+ 0·127	...	...
513	... ...	+ 2·1127	+ 0·0097	...	+ 17·108	+ 0·154	...	...
514	Taylor 4445	+ 1·7822	+ 0·0083	...	+ 17·184	+ 0·148	...	...
515	... ...	+ 1·0942	+ 0·0086	...	+ 17·186	+ 0·148	...	...
516	... .	+ 2·4951	+ 0·0097		+ 17·191	+ 0·180		...
517	... ...	+ 2·3002	+ 0·0099		+ 17·242	+ 0·170		...
518	... ...	+ 2·0815	+ 0·0099		+ 17·272	+ 0·147		...
519	Taylor 4476	+ 2·0800	+ 0·0100		+ 17·270	+ 0·147		...
520	... .	+ 2·1486	+ 0·0106		+ 17·288	+ 0·151		...
521	Taylor 4494	+ 1·8296	+ 0·0058	...	+ 17·311	+ 0·128	...	...
522	32 Leonis α (*Regulus*)	+ 3·2201	− 0·0102	− 0·019	+ 17·420	+ 0·225	− 0·01	3490
523	. ... ..	+ 2·5170	+ 0·0104	...	+ 17·420	+ 0·175	...	...
524	Lacaille 4164 ..	+ 2·1708	+ 0·0115	...	+ 17·465	+ 0·148	...	...
525	... .	+ 2·5280	+ 0·0106	...	+ 17·490	+ 0·172	..	...

Mean Positions of Stars for 1865 January 1st.

Number.	Star.	Magnitude.	Estimations.	Mean Right Ascension. h.	m.	s.	Mean Polar Distance. °	′	″	Observations.	Fraction of Year.
526	... ...	8·2	2	10	5	10·16	140	30	8·3	2	0·23
527	Taylor 4638 ...	7·1	1	10	6	12·04	129	19	44·6	1	0·27
528	... .	9·0	2	10	7	9·77	129	33	26·8	2	0·34
529	.. ..	9·2	1	10	9	3·34	130	32	0·4	1	0·34
530	R. P. L. 72 ...	6·6	...	10	9	31·13	5	3	59·6	1	0·70
531	... ...	7·1	1	10	9	58·80	143	34	42·2	1	0·38
532	... ...	9·4	2	10	10	13·74	130	51	29·3	2	0·19
533	41 Leonis γ^1 ...	2·2	...	10	12	31·40	69	36	36·8	9	0·32
534	... ...	9·3	1	10	12	58·97	138	37	14·0	1	0·16
535	48 Leonis	6·2	1	10	15	36·43	52	40	23·7	2	0·26
536	... ...	7·0	1	10	16	14·06	150	16	38·6	1	0·27
537	Taylor 4658 ..	8·0	2	10	18	4·15	151	28	32·1	2	0·22
538	... ...	9·2	1	10	18	47·30	146	8	44·5	1	0·34
539	45 Leonis	6·9	...	10	20	31·07	70	33	3·2	2	0·07
540	... ...	8·3	2	10	21	54·03	140	35	12·5	2	0·27
541	... ...	9·0	1	10	21	53·26	146	30	20·6	1	0·30
542	... ...	9·0	1	10	23	35·13	76	5	36·7	1	0·30
543	47 Leonis ρ	4·0	...	10	25	42·02	79	33	50·0	14	0·22
544	... ...	9·0	1	10	26	32·96	152	33	5·1	1	0·38
545	p Carinæ	3·6	...	10	27	13·96	150	30	32·6	2	0·28
546	... ...	9·5	1	10	28	6·91	150	30	14·1	1	0·25
547	Taylor 4769	6·0	1	10	30	24·42	146	51	36·0	1	0·30
548	... ...	5·7	1	10	31	6·17	151	9	56·6	2	0·30
549	... ...	9·4	1	10	34	54·73	139	16	38·1	1	0·36
550	... ...	8·5	1	10	35	5·94	140	5	59·4	1	0·27
551	Taylor 4894	7·5	4	10	36	8·66	148	36	21·2	4	0·32
552	... ...	9·4	1	10	36	33·61	150	47	34·3	1	0·25
553	... ...	6·1	1	10	37	32·34	151	29	30·4	1	0·30
554	Taylor 4940	6·3	3	10	38	41·75	149	34	26·3	2	0·31
555	Taylor 4950 ... 2nd...	8·3	1	10	38	44·51	148	30	35·2	1	0·30
556	... ...	9·0	1	10	39	5·38	149	34	31·8	1	0·37
557	η Argûs Var. 1	6·0	1	10	39	40·91	149	0	23·4	3	0·28
558	Taylor 4973	6·0	1	10	41	7·06	161	13	58·8	1	0·36
559	... .	9·3	2	10	41	53·60	149	26	11·3	2	0·31
560	53 Leonis l	5·3	...	10	42	0·62	79	44	29·5	11	0·36

530.—Groombridge 1650.
557.—η Argûs Var. 1. Irregularly variable from 1st to 6th magnitude.

Observed with the Madras Meridian Circle in that Year.

Number.	Star.	In Right Ascension. Annual Precession.	Secular Variation.	Proper Motion.	In Polar Distance. Annual Precession.	Secular Variation.	Proper Motion.	Number in B. A. C.
		s	s	s	"	"	"	
526	... .	+ 2·2980	+ 0·0124	...	+ 17·804	+ 0·158	...	...
527	Taylor 4569	+ 2·5504	+ 0·0109	...	+ 17·683	+ 0·109	...	..
528	... .	+ 2·5423	+ 0·0112	...	+ 17·673	+ 0·167	...	...
529	... ...	+ 2·3311	+ 0·0131	...	+ 17·751	+ 0·150	...	...
530	R. P. L. 72	+ 10·0342	− 1·6641	− 0·070	+ 17·709	+ 0·673	+ 0·05	3605
531	... ...	+ 2·1704	+ 0·0129	...	+ 17·785	+ 0·138	...	...
532	... ...	+ 2·3410	+ 0·0131	...	+ 17·801	+ 0·149	...	...
533	41 Leonis γ^1	+ 3·2982	− 0·0146	+ 0·019	+ 17·890	+ 0·208	+ 0·15	3623
534	... .	+ 2·5910	+ 0·0115	...	+ 17·905	+ 0·162	...	...
535	43 Leonis	+ 3·1465	− 0·0068	− 0·001	+ 18·023	+ 0·194	+ 0·09	3644
536	... ...	+ 2·5998	+ 0·0121	...	+ 18·094	+ 0·159	...	...
537	Taylor 4666	+ 2·0174	+ 0·0129	...	+ 18·104	+ 0·110	...	...
538	... ...	+ 2·2208	+ 0·0162	...	+ 18·181	+ 0·131	...	...
539	45 Leonis	+ 3·1739	− 0·0094	− 0·002	+ 18·196	+ 0·157	+ 0·01	3675
540	... .	+ 2·2202	+ 0·0160	...	+ 18·246	+ 0·126	...	...
541	... ...	+ 2·2184	+ 0·0160	...	+ 18·298	+ 0·128	...	
542	... .	+ 3·2075	− 0·0102	...	+ 18·300	+ 0·158	...	...
543	47 Leonis ρ	+ 3·1663	− 0·0090	0·000	+ 18·381	+ 0·176	+ 0·03	3680
544	... ...	+ 2·0563	+ 0·0154	...	+ 18·410	+ 0·111	...	...
545	p Carinæ	+ 2·1286	+ 0·0165	...	+ 18·484	+ 0·114	...	3019
546	... ...	+ 2·1371	+ 0·0160		+ 18·465	+ 0·114		...
547	Taylor 4760	+ 2·2910	+ 0·0184		+ 18·542	+ 0·120		3686
548	... ...	+ 2·1536	+ 0·0170		+ 18·565	+ 0·111		...
549	... ...	+ 2·5096	+ 0·0182		+ 18·690	+ 0·126		...
550	.. ..	+ 2·2632	+ 0·0199		+ 18·694	+ 0·112		...
551	Taylor 4924	+ 2·2769	+ 0·0208		+ 18·728	+ 0·111		3673
552	... .	+ 2·2205	+ 0·0202		+ 18·740	+ 0·107		...
553	... ...	+ 2·2033	+ 0·0206		+ 18·771	+ 0·106		...
554	Taylor 4949	+ 2·2685	+ 0·0211		+ 18·806	+ 0·108		3649
555	Taylor 4980 : 2nd .	+ 2·3014	+ 0·0211		+ 18·807	+ 0·109		.
556	.. ... .	+ 2·3155	+ 0·0213	...	+ 18·818	+ 0·100	...	...
557	η Argûs Var. 1 ...	+ 2·3102	+ 0·0215	...	+ 18·840	+ 0·107	...	3695
558	Taylor 4972	+ 2·2304	+ 0·0220	..	+ 18·879	+ 0·102	...	...
559	... .	+ 2·3177	+ 0·0223	...	+ 18·904	+ 0·105	...	...
560	53 Leonis l	+ 3·1007	− 0·0080	− 0·003	+ 18·910	+ 0·146	+ 0·02	3708

535.—539.—Proper motions from "*Greenwich Catalogue 1864.*"

Mean Positions of Stars for 1865 January 1st.

Number.	Star.	Magnitude.	Estimations.	Mean Right Ascension.			Mean Polar Distance.			Observations.	Fraction of Year.
				h.	m.	s.	°	′	″		
561	Taylor 4915 ...	7·6	2	10	46	55·12	165	30	5·5	2	0·23
562	56 Leonis... ...	6·0	...	10	48	46·67	88	32	35·9	2	0·19
563	... ...	9·1	2	10	49	34·68	148	48	34·3	2	0·29
564	Taylor 4966 ...	7·1	1	10	50	43·20	147	10	57·6	1	0·26
565	... ...	9·0	1	10	51	33·30	143	46	47·8	2	0·29
566	... ...	9·0	1	10	52	21·56	143	36	35·9	1	0·23
567	... ...	8·7	2	10	53	56·13	149	15	5·1	2	0·36
568	59 Leonis c	5·1	...	10	53	44·98	89	10	27·0	1	0·19
569	.. ..	8·8	3	10	53	49·98	136	58	7·8	3	0·36
570	50 Urs. Maj. α (*Dubhe*) ...	2·0	...	10	55	22·22	27	31	15·0	4	0·26
571	... ...	8·0	1	10	56	40·50	149	17	7·5	1	0·33
572	... ...	9·3	2	10	57	2·09	145	32	40·8	2	0·26
573	63 Leonis χ	4·7	...	10	58	3·08	81	36	5·8	7	0·36
574	Lacaille 4595	8·0	2	10	59	36·53	148	39	11·1	2	0·34
575	Lacaille 4612	8·5	1	10	0	57·36	151	46	54·1	1	0·30
576	... ...	8·4	1	10	1	4·03	136	38	57·3	1	0·36
577	67 Leonis	5·6	...	10	1	34·08	64	30	42·3	1	0·30
578	... ...	8·1	1	10	1	51·74	149	14	4·9	1	0·33
579	... ...	8·0	1	10	2	0·72	146	56	29·6	2	0·34
580	Lalande 21371	8·0	1	11	3	33·82	77	58	0·0	1	0·22
581	Lalande 21416	9·1	1	11	5	1·86	67	12	40·5	1	0·32
582	... ...	8·1	1	11	5	51·90	148	50	4·2	1	0·33
583	Taylor 5108	6·9	1	11	6	40·42	140	35	3·9	1	0·19
584	61 Leonis p^2	5·5	...	11	6	50·87	89	30	9·7	1	0·11
585	68 Leonis δ	2·7	...	11	6	55·46	68	44	14·9	5	0·26
586	... ...	9·3	1	11	8	36·00	150	51	9·2	1	0·28
587	... ...	8·7	1	11	9	47·15	147	15	18·7	1	0·30
588	74 Leonis φ ...	4·5	...	11	9	47·76	92	54	33·1	2	0·19
589	... .	8·5	1	11	10	39·26	149	56	9·4	1	0·33
590	—	9·0	1	11	10	34·06	141	8	35·0	1	0·30
591	... ...	8·3	1	11	11	10·92	127	38	41·8	1	0·22
592	12 Hydræ & Crateris δ ..	3·9	...	11	12	35·80	104	3	51·6	6	0·26
593	... ...	7·7	1	11	12	51·47	129	32	27·0	1	0·33
594	. ...	7·9	2	11	15	52·77	136	21	55·8	2	0·26
595	Lacaille 4730	8·1	3	11	16	8·03	145	51	40·0	3	0·27

577.—Comparison star for Thalia in 1862.

Observed with the Madras Meridian Circle in that Year.

Number.	Star.	In Right Ascension.			In Polar Distance.			Number in B. A. C.
		Annual Precession.	Secular Variation.	Proper Motion.	Annual Precession.	Secular Variation.	Proper Motion.	
		s	s	s	"	"	"	
561	Taylor 4915 … …	+ 2·6437	+ 0·0100	…	+ 19·086	+ 0·113	…	…
562	55 Leonis … …	+ 3·0825	− 0·0026	+ 0·007	+ 19·094	+ 0·130	0·00	3749
563	… … …	+ 2·4184	+ 0·0254	…	+ 19·142	+ 0·097	…	…
564	Taylor 4956 … …	+ 2·4512	+ 0·0250	…	+ 19·146	+ 0·097	…	…
565	… … …	+ 2·5385	+ 0·0380	…	+ 19·178	+ 0·100	…	…
566	… … …	+ 2·5445	+ 0·0280	…	+ 19·188	+ 0·100	…	…
567	.. … …	+ 2·4200	+ 0·0867	…	+ 19·219	+ 0·098	…	…
568	59 Leonis c … …	+ 3·1178	− 0·0052	− 0·005	+ 19·222	+ 0·132	+ 0·06	3769
569	… … ..	+ 2·6843	+ 0·0804	…	+ 19·225	+ 0·108	…	…
570	50 Urs. Maj α (*Dubhe*)	+ 3·7861	− 0·0821	− 0·017	+ 19·269	+ 0·144	+ 0·09	3777
571	… … …	+ 2·4467	+ 0·0274	…	+ 19·270	+ 0·091	…	…
572	… … …	+ 2·5481	+ 0·0262	…	+ 19·303	+ 0·098	…	…
573	63 Leonis χ … …	+ 3·1226	− 0·0056	− 0·024	+ 19·327	+ 0·118	+ 0·03	3788
574	Lacaille 4586 …	+ 2·4010	+ 0·0287	…	+ 19·366	+ 0·087	…	…
575	Lacaille 4612 …	+ 2·3488	+ 0·0315	…	+ 19·398	+ 0·079	…	…
576	… … …	+ 2·7283	+ 0·0216	…	+ 19·396	+ 0·086	…	…
577	67 Leonis … …	+ 3·2322	− 0·0164	+ 0·003	+ 19·407	+ 0·111	− 0·01	3809
578	… … …	+ 2·5064	+ 0·0297	…	+ 19·413	+ 0·094	…	…
579	… … …	+ 2·5104	+ 0·0290	…	+ 19·416	+ 0·094	…	…
580	Lalande 21371 …	+ 3·1416	− 0·0076	…	+ 19·460	+ 0·105	…	…
581	Lalande 21416 …	+ 3·3066	− 0·0144	…	+ 19·481	+ 0·103	…	…
582	… … …	+ 2·5517	+ 0·0311	…	+ 19·498	+ 0·070	…	…
583	Taylor 5109 … …	+ 2·5481	+ 0·0319	…	+ 19·517	+ 0·077	…	3835
584	69 Leonis ρ² … …	+ 3·0767	− 0·0013	0·000	+ 19·518	+ 0·095	0·00	3832
585	68 Leonis δ … ..	+ 3·1014	− 0·0132	+ 0·011	+ 19·520	+ 0·093	+ 0·14	3834
586	… … …	+ 2·5392	+ 0·0441	…	+ 19·562	+ 0·074	…	…
587	… … ..	+ 2·6790	+ 0·0310	…	+ 19·596	+ 0·073	…	…
588	74 Leonis φ … ..	+ 3·0573	+ 0·0006	− 0·000	+ 19·576	+ 0·080	+ 0·04	3848
589	.. … ..	+ 2·5062	+ 0·0328	…	+ 19·588	+ 0·073	…	…
590	.. … ..	+ 2·7100	+ 0·0273	…	+ 19·591	+ 0·077	…	…
591	… … ..	+ 2·8541	+ 0·0186	…	+ 19·602	+ 0·080	…	…
592	12 Hydræ & Crateris δ	+ 3·0083	+ 0·0061	− 0·008	+ 19·634	+ 0·081	− 0·18	3859
593	… … ..	+ 2·4467	+ 0·0200	..	+ 19·632	+ 0·077	…	..
594	.. .. .	+ 2·4996	+ 0·0197	..	+ 19·644	+ 0·073	…	…
595	Lacaille 4730 ..	+ 2·6969	+ 0·0321	…	+ 19·680	+ 0·067	..	…

577.—Proper motions from "*Greenwich Catalogue 1872.*"

Mean Positions of Stars for 1865 January 1st.

Number.	Star.	Magnitude.	Estimations.	Mean Right Ascension. h	m	s	Mean Polar Distance. °	′	″	Observations.	Fraction of Year.
596	70 Leonis	5·5	...	11	17	6·02	87	51	0·8	1	0·26
597	Taylor 5280 ...	8·0	2	11	19	3·47	131	55	51·9	2	0·28
598	... ...	9·4	2	11	21	45·00	128	28	6·3	2	0·28
599	Taylor 5346 ...	8·6	2	11	22	5·86	131	26	7·9	2	0·33
600	... ...	9·2	1	11	22	51·03	145	51	4·0	1	0·21
601	O. A. N. 11812	9·2	2	11	23	12·81	32	38	4·8	2	0·31
602	... ...	8·6	4	11	24	33·78	22	26	11·9	4	0·31
603	... ...	9·1	1	11	25	23·60	128	27	8·0	1	0·31
604	... ...	9·4	3	11	25	50·47	129	28	6·4	3	0·25
605	... ...	9·0	1	11	26	39·42	151	4	24·1	1	0·30
606	... ...	9·3	3	11	28	28·77	128	20	27·7	3	0·28
607	λ Centauri ...	3·4	...	11	29	34·29	152	16	36·7	2	0·21
608	... .	8·1	2	11	29	53·92	149	16	2·4	2	0·35
609	91 Leonis *v* ..	4·5	...	11	30	2·19	90	4	44·0	7	0·26
610	... ...	9·2	1	11	32	12·24	141	14	38·9	1	0·30
611	... ...	8·2	3	11	33	10·29	144	14	36·4	3	0·27
612	W. B. E. XI. 571 ...	8·0	1	11	33	28·53	88	18	0·2	1	0·23
613	... ...	7·5	2	11	34	28·88	144	21	2·2	3	0·31
614	W. B. E. XI. 807 .	9·1	1	11	34	46·72	88	15	25·3	1	0·32
615	... ...	9·0	1	11	36	46·03	129	34	22·3	1	0·32
616	... ...	9·2	1	11	40	30·57	149	52	34·3	1	0·32
617	... ...	8·6	2	11	41	15·29	139	52	36·9	2	0·27
618	94 Leonis *β* (*Deneb*) ...	2·2	...	11	42	10·81	74	40	25·4	8	0·31
619	Taylor 5421	7·6	3	11	43	18·89	129	31	34·6	3	0·27
620	5 Virginis *β*	3·7	...	11	43	29·72	87	38	29·8	3	0·27
621	... ...	8·0	3	11	44	47·07	129	2	30·9	3	0·34
622	Taylor 5438	8·4	1	11	44	54·57	129	38	25·1	1	0·22
623	Groombridge 1680 ..	6·4	...	11	45	11·93	51	18	30·1	1	0·30
624	... ...	9·8	1	11	45	54·38	142	31	19·8	1	0·33
625	64 Ursæ Majoris *γ* ..	2·6	...	11	46	43·46	36	33	14·0	1	0·26
626	Lacaille 4887	7·1	1	11	48	18·59	132	31	48·0	1	0·25
627	... ...	8·0	4	11	50	40·02	150	42	21·2	4	0·36
628	Lacaille 4950	8·8	1	11	51	17·84	154	34	17·8	1	0·36
629	... ...	8·6	1	11	51	[illegible]	128	42	35·6	2	0·31
630	... ...	9·5	1	11	52	48·17	151	38	40·3	1	[illegible]

601.—602.—Comparison stars for Comet 2, 1861.
612.—614.—Comparison stars for Amphitrite in 1862.
629.—Double: the companion (9.0) 30″ north.

Observed with the Madras Meridian Circle in that Year.

Number.	Star.	In Right Ascension. Annual Precession.	Secular Variation.	Proper Motion.	In Polar Distance. Annual Precession.	Secular Variation.	Proper Motion.	Number in B. A. C.
		s	s	s	"	"	"	
596	70 Leonis	+ 3·0814	− 0·0016	− 0·002	+ 19·705	+ 0·076	+ 0·02	3679
597	Taylor 5280	+ 2·8687	+ 0·0985	...	+ 19·736	+ 0·066	...	...
598	... ...	+ 2·8961	+ 0·0806	...	+ 19·770	+ 0·081	...	...
599	Taylor 5245	+ 2·8744	+ 0·0830	...	+ 19·781	+ 0·060	...	3907
600	... ...	+ 2·7584	+ 0·0844	...	+ 19·791	+ 0·066	...	...
601	O. A. N. 11812 ..	+ 3·6768	− 0·0930	...	+ 19·797	+ 0·075	...	...
602	... .	+ 3·5557	− 0·0828	...	+ 19·816	+ 0·071	...	...
603	... .	+ 2·0124	+ 0·0210	...	+ 19·827	+ 0·065	...	...
604	... ...	+ 2·0146	+ 0·0211	...	+ 19·883	+ 0·064	...	...
605	... ...	+ 2·7214	+ 0·0415	...	+ 19·848	+ 0·046	...	...
606	... ...	+ 2·9967	+ 0·0214	...	+ 19·864	+ 0·050	...	...
607	λ Centauri	+ 2·7354	+ 0·0443	0·000	+ 19·878	+ 0·044	0·00	3941
608	... .	+ 2·7776	+ 0·0406	...	+ 19·882	+ 0·044	...	...
609	91 Leonis υ	+ 3·0718	+ 0·0003	− 0·008	+ 19·894	+ 0·040	− 0·03	3946
610	... ...	+ 2·8474	+ 0·0366	...	+ 19·908	+ 0·041	...	...
611	... .	+ 2·8551	+ 0·0360	...	+ 19·918	+ 0·080	...	...
612	W. B. E. XI. 571 ...	+ 3·0767	− 0·0005	...	+ 19·921	+ 0·046	...	...
613	... .	+ 2·8641	+ 0·0364	...	+ 19·930	+ 0·087	...	...
614	W. B. E. XI. 597 ...	+ 3·0786	− 0·0004	...	+ 19·958	+ 0·041	...	...
615	... ...	+ 2·9698	+ 0·0287	...	+ 19·969	+ 0·031	...	...
616	... .	+ 2·8813	+ 0·0468	...	+ 19·986	+ 0·036	...	...
617	... ...	+ 2·9819	+ 0·0240	...	+ 19·988	+ 0·027	...	...
618	94 Leonis β (*Deneb*)...	+ 3·1005	− 0·0074	− 0·036	+ 19·994	+ 0·025	+ 0·10	3995
619	Taylor 5421	+ 2·9915	+ 0·0242	...	+ 20·001	+ 0·026	—	...
620	5 Virginis β	+ 3·0763	− 0·0003	+ 0·046	+ 20·004	+ 0·028	+ 0·28	4002
621	... .	+ 3·0002	+ 0·0241	...	+ 20·011	+ 0·030	...	...
622	Taylor 5483	+ 2·9995	+ 0·0246	...	+ 20·012	+ 0·030	...	...
623	Groombridge 1830 ..	+ 3·1413	− 0·0289	+ 0·346	+ 20·018	+ 0·021	+ 5·73	4010
624	... .	+ 2·9050	+ 0·0378	...	+ 20·018	+ 0·017	...	...
625	64 Ursæ Majoris γ ...	+ 3·1803	− 0·0458	+ 0·011	+ 20·022	+ 0·017	0·00	4017
626	Lacaille 4987 ...	+ 2·9409	+ 0·0589	...	+ 20·029	+ 0·013	...	...
627	... .	+ 2·9704	+ 0·0506	...	+ 20·035	+ 0·010	...	...
628	Lacaille 4990 .	+ 2·9648	+ 0·0508	...	+ 20·040	+ 0·008	...	...
629	... .	+ 3·0319	+ 0·0240	...	+ 20·041	+ 0·007	..	...
630	... .	+ 2·9923	+ 0·0309	...	+ 20·045	+ 0·005	...	...

607.—Proper motions from "*Stone's Cape Catalogue*."
623.—Proper motions from "*Greenwich Catalogue* 1872."

Mean Positions of Stars for 1865 January 1st.

Number.	Star.	Magnitude.	Estimations.	Mean Right Ascension. h.	m.	s.	Mean Polar Distance. °	′	″	Observations.	Fraction of Year.
631	... ...	9·6	1	11	55	58·12	120	36	12·2	1	0·33
632	... ...	9·0	1	11	56	20·46	128	30	16·9	1	0·36
633	Taylor 5504	8·3	2	11	56	02·76	143	37	40·3	2	0·29
634	Lacaille 4895	7·5	3	11	56	57·24	142	44	47·1	3	0·35
635	Taylor 5536	8·0	3	11	57	6·08	70	25	30·6	3	0·36
636	R. P. L. 80	6·3	...	11	57	54·36	3	20	55·2	1	0·78
637	... ...	9·0	1	11	59	4·27	128	23	5·3	1	0·34
638	... ...	9·1	2	11	59	34·06	150	19	57·7	3	0·30
639	... ...	8·6	3	12	1	10·07	150	27	47·3	3	0·30
640	Lacaille 5041	8·0	1	12	2	25·18	141	28	34·1	1	0·30
641	2 Corvi ε	3·1	...	12	3	11·03	111	52	5·4	5	0·31
642	... ...	9·0	1	12	5	51·26	134	3	25·0	1	0·32
643	... ...	8·5	3	12	5	59·31	150	19	24·6	3	0·25
644	... ...	8·2	1	12	6	15·41	133	37	58·7	1	0·35
645	... ...	9·1	2	12	8	58·31	144	20	34·3	2	0·26
646	... ...	8·1	1	12	11	56·17	150	50	12·7	1	0·26
647	Taylor 5648	7·2	1	12	12	34·47	162	6	17·6	1	0·34
648	R. P. L. 82	6·3	...	12	12	53·60	2	48	40·3	2	0·36
649	15 Virginis η	4·0	...	12	12	50·85	90	54	20·6	16	0·29
650	Lacaille 5110	8·0	1	12	15	28·08	136	24	27·1	1	0·37
651	... .	8·0	2	12	16	36·25	141	40	15·3	2	0·30
652	... ...	8·3	1	12	16	49·29	147	10	7·0	1	0·33
653	α¹ Crucis	1·3	...	12	19	6·73	152	21	3·0	1	0·29
654	... ...	8·0	1	12	19	6·77	147	21	41·5	1	0·32
655	... ...	9·1	1	12	24	41·01	150	36	30·2	1	0·37
656	... ...	7·6	...	12	26	20·16	35	2	14·1	1	0·29
657	9 Corvi β	2·8	...	12	27	13·01	112	39	30·3	9	0·31
658	Taylor 5736	7·7	2	12	27	40·96	150	50	46·5	2	0·32
659	... ...	9·2	2	12	27	52·85	141	40	10·6	2	0·24
660	... ...	9·2	1	12	27	52·77	140	46	52·7	1	0·35
661	... ...	9·2	1	12	31	54·96	84	30	52·3	1	0·40
662	... ...	9·0	...	12	32	52·37	143	7	43·3	1	[illegible]
663	... ...	8·7	1	12	33	50·41	145	38	46·6	1	0·38
664	Taylor 5580	7·0	1	12	34	30·15	144	1	13·9	1	0·30
665	29 Virginis γ¹	2·8	...	12	34	40·15	90	42	29·3	8	0·36

636.—Groombridge 1850.
648.—Groombridge 1871.

Observed with the Madras Meridian Circle in that Year.

Number.	Star.	In Right Ascension. Annual Precession.	In Right Ascension. Secular Variation.	In Right Ascension. Proper Motion.	In Polar Distance. Annual Precession.	In Polar Distance. Secular Variation.	In Polar Distance. Proper Motion.	Number in B. A. C.
		s	s	s	"	"	"	
631	... .	+ 3·0485	+ 0·0388	...	+ 20·018	+ 0·008	...	...
632	... .	+ 3·0555	+ 0·0368	...	+ 20·058	− 0·002	...	...
633	Taylor 5334	+ 3·0471	+ 0·0421	...	+ 20·053	− 0·003	...	...
634	Lacaille 4995 ...	+ 3·0487	+ 0·0404	...	+ 20·053	− 0·006	...	...
635	Taylor 5535	+ 3·0781	− 0·0080	...	+ 20·054	− 0·008	...	...
636	R. P. L. 88	+ 3·2823	− 0·5821	...	+ 20·054	− 0·004	...	4070
637	... ...	+ 3·0679	+ 0·0285	...	+ 20·056	− 0·007	...	...
638	... ...	+ 3·0678	+ 0·0542	...	+ 20·056	− 0·008	...	...
639	... ...	+ 3·0942	+ 0·0540	...	+ 20·054	− 0·011	...	...
640	Lacaille 5041 ..	+ 3·0011	+ 0·0400	...	+ 20·054	− 0·014	...	...
641	2 Corvi ε	+ 3·0705	+ 0·0142	− 0·005	+ 20·054	− 0·016	− 0·01	4097
642	... ...	+ 3·1032	+ 0·0318	...	+ 20·049	− 0·021	...	...
643	... ...	+ 3·1334	+ 0·0568	...	+ 20·048	− 0·021	...	...
644	... ...	+ 3·1136	+ 0·0369	...	+ 20·043	− 0·022	...	...
645	... ...	+ 3·1443	+ 0·0460	...	+ 20·040	− 0·027	...	...
646	... ...	+ 3·1044	+ 0·0602	...	+ 20·038	− 0·033	...	...
647	Taylor 5048	+ 3·2106	+ 0·0640	...	+ 20·034	− 0·035	...	...
648	R. P. L. 92	+ 1·5412	+ 0·0012	+ 0·415	+ 20·028	− 0·022	+ 0·06	4130
649	15 Virginis η ...	+ 3·0719	+ 0·0027	− 0·007	+ 20·023	− 0·036	+ 0·03	4145
650	Lacaille 5119 ..	+ 3·1739	+ 0·0398	...	+ 20·009	− 0·040	...	...
651	... ...	+ 3·1994	+ 0·0496	...	+ 20·007	− 0·042	...	...
652	... ...	+ 3·2210	+ 0·0635	...	+ 20·000	− 0·044	...	...
653	α¹ Crucis	+ 3·2944	+ 0·0050	− 0·006	+ 19·986	− 0·050	+ 0·04	4167
654	... ...	+ 3·2460	+ 0·0646	...	+ 19·985	− 0·040	...	...
655	... .	+ 3·3314	+ 0·0666	...	+ 19·980	− 0·061	...	...
656	.. ..	+ 2·7881	− 0·0480	...	+ 19·980	− 0·056	...	...
657	9 Corvi β	+ 3·1398	+ 0·0164	− 0·009	+ 19·913	− 0·064	+ 0·07	4234
658	Taylor 5795... ...	+ 3·3643	+ 0·0070	...	+ 19·907	− 0·068	...	...
659	... .	+ 3·2778	+ 0·0460	...	+ 19·907	− 0·067	...	..
660	.. ..	+ 3·2720	+ 0·0447	...	+ 19·907	− 0·067	...	...
661	... .	+ 3·0518	+ 0·0008	...	+ 19·861	− 0·071	...	.
662	... ...	+ 3·3209	+ 0·0196	...	+ 19·849	− 0·079	.	
663	... .	+ 3·3600	+ 0·0848	..	+ 19·837	− 0·081	..	..
664	Taylor 5980	+ 3·3153	+ 0·0519	...	+ 19·835	− 0·082	...	4301
665	29 Virginis γ¹ ..	+ 3·0715	+ 0·0043	− 0·037	+ 19·825	− 0·073	+ 0·06	4268

648.—Proper motion in Polar Distance from "*British Association Catalogue.*"
653.—Proper motions from "*Stone's Cape Catalogue.*"
665.—Proper motions from "*Greenwich Catalogue*, 1872."

Mean Positions of Stars for 1865 January 1st.

Number.	Star.	Magnitude.	Estimations.	Mean Right Ascension.			Mean Polar Distance.			Observations.	Fraction of Year.
				h.	m.	s.	°	′	″		
666	29 Virginis γ^1	2·8	...	12	34	40·22	90	42	34·0	1	0·30
667	28 Virginis	6·5	...	12	34	58·92	96	46	27·6	1	0·27
668	W. B. E. XII. 592 ..	8·0	2	12	36	4·37	98	18	8·7	2	0·36
669	... ...	8·0	3	12	38	19·99	146	38	41·8	3	0·38
670	... ...	8·0	1	12	39	29·17	143	50	37·6	1	0·42
671	... ...	9·3	1	12	39	31·39	141	54	33·0	1	0·37
672	... ...	8·0	1	12	40	52·01	141	53	12·4	1	0·43
673	... ...	8·2	1	12	41	48·17	141	49	53·8	1	0·20
674	... ...	9·4	1	12	42	6·67	147	16	47·5	1	0·40
675	... ...	8·8	1	12	42	27·91	147	19	5·5	1	0·40
676	... ...	9·3	1	12	42	50·73	142	52	17·4	1	0·40
677	... ...	8·5	...	12	43	20·77	120	8	11·3	1	0·31
678	U Virginis Var. 3 ...	8·5	1	12	44	14·71	83	42	40·9	1	0·30
679	Radcliffe 2922	7·0	...	12	46	9·16	90	16	44·0	1	0·39
680	40 Virginis ψ	5·0	...	12	47	30·03	98	48	19·6	1	0·27
681	B. P. L. 99	5·6	...	12	48	11·26	5	51	9·0	2	0·34
682	... ...	9·0	...	12	48	32·02	126	26	50·2	1	0·39
683	... ...	8·2	2	12	48	50·29	149	34	38·0	2	0·39
684	12 Canum Venaticorum...	3·0	...	12	49	42·41	50	57	7·3	12	0·33
685	... ...	9·5	1	12	51	13·39	127	5	34·0	1	0·40
686	... ...	8·6	1	12	52	30·21	143	40	51·4	1	0·37
687	... ...	8·0	...	12	53	38·57	136	44	47·0	1	0·41
688	... .	9·0	1	12	56	6·17	104	28	22·0	1	0·39
689	... ...	9·0	2	12	57	2·51	123	36	31·2	2	0·40
690	... .	9·1	...	12	58	0·40	104	30	1·6	1	0·32
691	... ...	8·0	2	12	58	19·57	104	33	15·5	2	0·34
692	Taylor 6025	7·8	1	12	59	30·90	128	34	44·0	1	0·40
693	51 Virginis θ	4·4	...	13	2	57·08	94	49	3·8	12	0·31
694	... ...	9·0	1	13	4	[illegible]	148	12	42·2	1	0·40
695	Lacaille 5494	8·3	1	13	6	1·59	132	51	33·6	1	0·11
696	... ...	8·8	3	13	7	42·98	130	40	32·2	3	0·34
697	... ...	9·3	2	13	8	9·00	130	46	3·0	2	0·35
698	... ...	8·7	1	13	9	48·99	123	56	36·0	1	[illegible]
699	... ...	7·8	2	13	12	50·36	122	55	54·2	2	0·38
700	Lacaille 5508	6·0	2	13	14	12·30	136	21	11·6	2	0·30

668.—Comparison star for Hestia in 1861.
678.—U Virginis Var. 3.—Period 207 days. Range, 8th to 13th magnitude.
679.—Comparison star for Comet 2, 1861.
681.—Groombridge 1910.

Observed with the Madras Meridian Circle in that Year.

Number.	Star.	In Right Ascension.			In Polar Distance.			Number in B. A. C.
		Annual Precession.	Secular Variation.	Proper Motion.	Annual Precession.	Secular Variation.	Proper Motion.	
		s	s	s	"	"	"	
666	39 Virginis γ² ...	+ 3·0746	+ 0·0043	− 0·087	+ 19·886	− 0·078	+ 0·06	4368
667	28 Virginis	+ 3·0862	+ 0·0074	+ 0·001	+ 19·838	− 0·078	+ 0·04	4360
668	W. B. E. XII 502 ...	+ 3·0842	+ 0·0060	...	+ 19·807	− 0·080	...	...
669	... ...	+ 3·4081	+ 0·0581	...	+ 19·775	− 0·092	...	...
670	... ...	+ 3·3858	+ 0·0526	...	+ 19·758	− 0·098	...	...
671	... ...	+ 3·3649	+ 0·0467	...	+ 19·757	− 0·092	...	...
672	... ...	+ 3·3745	+ 0·0490	...	+ 19·737	− 0·096	...	...
673	... ...	+ 3·3900	+ 0·0490	...	+ 19·724	− 0·097	...	...
674	... ...	+ 3·4523	+ 0·0011	...	+ 19·717	− 0·100	...	...
675	... ...	+ 3·4600	+ 0·0618	...	+ 19·712	− 0·101	...	...
676	... ...	+ 3·4008	+ 0·0512	...	+ 19·706	− 0·100	...	...
677	... ...	+ 3·2766	+ 0·0313	...	+ 19·698	− 0·098	...	...
678	U Virginis Var. 3 ...	+ 3·0138	+ 0·0012	...	+ 19·682	− 0·098	...	...
679	Radcliffe 2982 ...	+ 2·5480	− 0·0844	...	+ 19·697	− 0·080	...	...
680	40 Virginis ψ ...	+ 3·1146	+ 0·0092	− 0·002	+ 19·630	− 0·101	+ 0·04	4380
681	R. P. L. 90	+ 0·3606	+ 0·2869	− 0·017	+ 19·618	− 0·019	− 0·04	4342
682	... ...	+ 3·2730	+ 0·0279	...	+ 19·607	− 0·108	...	...
683	... ...	+ 3·3614	+ 0·0492	...	+ 19·590	− 0·118	...	...
684	12 Canum Venaticorum	+ 2·8996	− 0·0162	− 0·023	+ 19·590	− 0·086	− 0·06	4395
685	... .	+ 3·2961	+ 0·0308	...	+ 19·560	− 0·115	...	...
686	... ...	+ 3·4915	+ 0·0640	...	+ 19·514	− 0·124		...
687	... ...	+ 3·3995	+ 0·0407	...	+ 19·511	− 0·128	...	...
688	... ...	+ 3·2900	+ 0·0275	...	+ 19·478	− 0·128	...	...
689	... ...	+ 3·2494	+ 0·0268	...	+ 19·437	− 0·120	...	...
690	.. ..	+ 3·3020	+ 0·0278	...	+ 19·412	− 0·129		...
691	... ...	+ 3·3024	+ 0·0279	...	+ 19·406	− 0·130	...	...
692	Taylor 6025	+ 3·2980	+ 0·0260	...	+ 19·338	− 0·132	...	...
693	51 Virginis θ ...	+ 3·1026	+ 0·0078	− 0·004	+ 19·308	− 0·133	+ 0·04	4401
694	... .	+ 3·5408	+ 0·0660	...	+ 19·308	− 0·158	...	...
695	Lacaille 5434 ..	+ 3·8133	+ 0·0980	...	+ 19·226	− 0·166	...	...
696	... ...	+ 3·5324	+ 0·0498	...	+ 19·186	− 0·163	...	...
697	... .	+ 3·3852	+ 0·0498	...	+ 19·174	− 0·159	...	...
698	. .. .	+ 3·4070	+ 0·0846	.	+ 19·132	− 0·157	...	...
699	... .	+ 3·3132	+ 0·0278	...	+ 19·049	− 0·161	...	...
700	Lacaille 5500 .	+ 3·8745	+ 0·0898	...	+ 19·013	− 0·164	...	...

Mean Positions of Stars for 1865 January 1st.

Number.	Star.	Magnitude.	Estimations.	Mean Right Ascension.			Mean Polar Distance.			Observations.	Fraction of Year.
				h.	m.	s.	°	′	″		
701	Taylor 6148	7·0	1	13	14	15·04	128	8	38·3	1	0·40
702	O. A. N. 13608	8·5	...	13	16	26·61	27	53	32·4	1	0·40
703	... ...	9·5	1	13	17	10·48	128	9	46·4	1	0·40
704	67 Virginis α (Spica) ...	1·2	...	13	18	5·00	100	27	21·4	17	0·33
705	... ...	8·3	1	13	19	5·02	143	27	5·4	1	0·43
706	Radcliffe 3011	7·5	...	13	19	28·42	34	28	47·6	1	0·30
707	R. P. L. 103	7·3	...	13	20	12·44	4	32	38·1	1	0·40
708	R Hydræ Var. 1	6·2	2	13	22	20·41	112	34	58·3	2	0·37
709	... ...	8·0	1	13	24	51·17	128	8	56·1	1	0·42
710	... ...	9·0	...	13	24	52·96	124	9	26·7	1	0·39
711	... ...	8·0	1	13	25	34·06	129	10	36·7	1	0·40
712	76 Virginis h	5·5	...	13	25	51·58	99	28	6·3	2	0·27
713	S Virginis Var. 6 ...	7·1	1	13	25	57·21	96	30	0·2	1	0·48
714	... ...	8·0	1	13	26	41·54	131	35	30·0	1	0·38
715	79 Virginis ζ	3·5	...	13	27	48·00	89	54	17·3	11	0·35
716	Taylor 7163	7·2	3	13	28	27·28	131	43	38·0	3	0·36
717	Lacaille 5614	8·0	1	13	30	0·37	138	18	36·0	2	0·40
718	... ...	9·5	1	13	34	6·46	130	10	10·1	1	0·38
719	... ...	8·6	2	13	35	46·72	136	8	46·1	2	0·43
720	... ...	8·9	1	13	36	17·96	123	5	40·1	1	0·41
721	Taylor 6368	7·7	1	13	36	48·17	147	38	40·0	1	0·43
722	Taylor 6386	7·0	1	13	36	57·04	151	46	21·1	1	0·42
723	Lacaille 5689	8·7	1	13	37	12·47	152	13	36·1	1	0·37
724	80 Virginis	6·0	...	13	38	44·81	101	44	56·7	2	0·30
725	85 Ursæ Majoris η ...	2·0	...	13	42	18·05	40	0	43·8	3	0·41
726	89 Virginis ...	5·2	...	13	42	32·50	107	27	35·1	1	0·43
727	... ...	8·0	1	13	43	17·74	128	6	30·0	1	0·39
728	... ...	8·5	1	13	43	27·64	128	18	39·7	2	0·33
729	... ...	8·3	2	13	44	15·64	127	57	2·0	2	0·41
730	... ...	8·6	1	13	46	46·80	122	54	51·0	1	0·34
731	Lacaille 5794	7·1	3	13	47	47·76	161	40	36·5	3	0·43
732	8 Bootis η	2·9	...	13	48	15·30	70	35	38·6	10	0·38
733	... ...	8·0	1	13	50	15·04	128	44	2·8	1	0·30
734	... ...	8·5	1	13	52	18·55	151	30	36·0	1	0·35
735	... ...	9·5	1	13	52	30·67	128	1	43·2	1	0·40

702.—Comparison star for Comet 2, 1861.
707.—Groombridge 3937.
708.—R Hydræ Var. 1. Period about 15 months. Range, 4th to 10th magnitude.
713.—S Virginis Var. 6. Period 371 days. Range, 6th to 12.5 magnitude.

Observed with the Madras Meridian Circle in that Year.

Number.	Star.	In Right Ascension. Annual Precession.	In Right Ascension. Secular Variation.	In Right Ascension. Proper Motion.	In Polar Distance. Annual Precession.	In Polar Distance. Secular Variation.	In Polar Distance. Proper Motion.	Number in B. A. C.
		s	s	s	"	"	"	
701	Taylor 6148	+ 3·4063	+ 0·0828	...	+ 19·012	− 0·166	...	...
702	O. A. N. 13663 ...	+ 2·3555	− 0·0180	...	+ 18·978	− 0·114	...	...
703	... ...	+ 3·4198	+ 0·0881	...	+ 18·960	− 0·173	...	...
704	67 Virginis α (*Spica.*)	+ 3·1544	+ 0·0116	− 0·006	+ 18·902	− 0·163	+ 0·04	4490
705	... ...	+ 3·6822	+ 0·0697	...	+ 18·873	− 0·190	...	...
706	Radcliffe 3011 ...	+ 2·7084	− 0·0170	...	+ 18·861	− 0·127	...	...
707	R. P. L. 108 ...	− 2·7014	+ 0·9586	...	+ 18·889	+ 0·127	...	4496
708	R Hydræ Var. 1 ...	+ 3·3676	+ 0·0192	− 0·002	+ 18·778	− 0·176	− 0·01	4501
709	... ...	+ 3·4621	+ 0·0884	...	+ 18·690	− 0·190	...	...
710	... ...	+ 3·4005	+ 0·0891	...	+ 18·696	− 0·188	...	...
711	... ...	+ 3·4360	+ 0·0884	...	+ 18·678	− 0·192	...	...
712	76 Virginis h ...	+ 3·1537	+ 0·0113	− 0·004	+ 18·664	− 0·170	+ 0·02	4521
713	S Virginis Var. 6 ...	+ 3·1279	+ 0·0096	...	+ 18·661	− 0·176	...	...
714	... ...	+ 3·5103	+ 0·0379	...	+ 18·637	− 0·197	...	—
715	79 Virginis ζ ...	+ 3·0712	+ 0·0064	− 0·010	+ 18·603	− 0·176	− 0·06	4582
716	Taylor 7188	+ 3·5209	+ 0·0931	...	+ 18·560	− 0·201	...	...
717	Lacaille 5614 ...	+ 3·4749	+ 0·0837	...	+ 18·398	− 0·202	...	...
718	... ...	+ 3·5060	+ 0·0951	...	+ 18·388	− 0·212	...	...
719	... ...	+ 3·4970	+ 0·0898	...	+ 18·369	− 0·215	...	...
720	... ...	+ 3·4997	+ 0·0889	...	+ 18·310	− 0·216	...	...
721	Taylor 6868	+ 3·9867	+ 0·0732	...	+ 18·296	− 0·243	...	...
722	Taylor 6896	+ 4·0946	+ 0·0909	...	+ 18·267	− 0·268	...	...
723	Lacaille 5680 ...	+ 4·1169	+ 0·0988	...	+ 18·278	− 0·265	...	...
724	80 Virginis	+ 3·1866	+ 0·0130	− 0·004	+ 18·222	− 0·208	− 0·01	4605
725	86 Ursæ Majoris η ...	+ 2·3849	− 0·0108	− 0·012	+ 18·004	− 0·159	+ 0·08	4607
726	89 Virginis	+ 3·2540	+ 0·0164	− 0·099	+ 18·081	− 0·213	+ 0·09	4608
727	... ...	+ 3·4690	+ 0·0867	...	+ 18·062	− 0·227	...	...
728	... ...	+ 3·4641	+ 0·0888	...	+ 18·046	− 0·223	...	...
729	... ...	+ 3·5804	+ 0·0941	...	+ 18·014	− 0·256	...	..
730	... ...	+ 3·4674	+ 0·0990	...	+ 17·937	− 0·253	...	...
731	Lacaille 5794 ..	+ 4·2088	+ 0·0960	...	+ 17·877	− 0·290	...	...
732	8 Bootis η	+ 2·8617	− 0·0006	− 0·004	+ 17·860	− 0·199	+ 0·36	4646
733	... .	+ 3·4852	+ 0·0896	...	+ 17·779	− 0·243	...	...
734	... .	+ 4·2319	+ 0·0920	...	+ 17·685	− 0·279	...	
735	... .	+ 3·3850	+ 0·0843	...	+ 17·687	− 0·268	.	...

708.—Proper motions from "*Greenwich Catalogue 1872.*"

Mean Positions of Stars for 1865 January 1st.

Number.	Star.	Magnitude.	Estimations.	Mean Right Ascension.			Mean Polar Distance.			Observations.	Fraction of Year.
				h.	m.	s.	°	′	″		
736	... ...	8·8	1	13	53	11·46	135	41	8·5	1	0·43
737	β Centauri	1·2	...	13	54	10·57	140	40	12·5	1	0·39
738	93 Virginis τ	4·4	...	13	54	46·82	87	46	2·8	6	0·30
739	94 Virginis	6·6	...	13	58	8·78	98	14	46·7	1	0·30
740	Lalande 26660	7·4	1	13	59	56·00	67	11	12·6	1	0·34
741	Taylor 6570	8·6	2	14	0	5·72	140	56	19·1	2	0·31
742	... ...	9·3	1	14	0	26·14	150	51	20·8	1	0·32
743	Taylor 6595	7·0	1	14	1	20·07	124	14	21·8	1	0·46
744	... ...	8·4	1	14	2	5·64	124	16	30·3	1	0·42
745	R. P. L. 108	7·3	...	14	3	47·71	3	35	45·4	1	0·94
746	Lacaille 5944	7·8	3	14	5	9·01	151	4	25·3	2	0·38
747	Taylor 6610	5·0	...	14	5	31·50	146	27	7·7	2	0·44
748	98 Virginis κ	4·3	...	14	5	41·72	99	38	36·1	2	0·35
749	16 Bootis α (*Arcturus*) ...	0·0	...	14	9	30·86	70	6	46·0	13	0·40
750	100 Virginis λ	4·6	...	14	11	46·50	102	44	53·7	3	0·30
751	... .	7·8	2	14	14	30·57	150	46	22·0	2	0·42
752	.. ..	8·5	1	14	14	37·72	122	36	8·3	1	0·30
753	2 Libræ	6·8	...	14	16	10·04	101	5	45·9	3	0·30
754	Taylor 6721	7·0	1	14	17	25·52	101	8	18·0	1	0·44
755	S Bootis Var. 2	9·2	1	14	18	21·54	35	34	27·0	1	0·17
756	... ...	8·2	1	14	20	1·30	127	0	9·7	2	0·40
757	Lacaille 5962	7·8	1	14	22	46·12	120	47	1·8	1	0·32
758	... ...	9·3	1	14	23	0·17	120	46	17·3	1	0·32
759	O. A. N. 14684	7·0	...	14	25	52·25	20	8	30·3	1	0·66
760	25 Bootis ρ	3·0	...	14	26	0·68	59	2	4·9	15	0·40
761	O. A. N. 14802	6·6	...	14	27	2·17	30	7	18·5	1	0·48
762	α¹ Centauri	3·5	...	14	30	27·11	150	16	40·4	5	0·41
763	α² Centauri	1·0	...	14	30	27·42	150	16	22·0	2	0·40
764	Lacaille 6027	7·8	1	14	31	7·49	122	47	38·8	1	0·40
765	... ...	8·6	1	14	32	40·51	121	44	34·1	1	0·33
766	... ...	8·5	1	14	32	50·36	126	18	2·4	1	0·41
767	... ...	9·5	1	14	34	35·96	150	17	39·4	1	0·47
768	5 Libræ	6·8	...	14	38	31·31	104	53	17·8	2	0·31
769	36 Bootis ε (*Mirac*) ...	3·0	...	14	39	5·41	62	21	18·4	11	0·41
770	... ...	7·8	2	14	39	24·23	134	7	33·5	2	0·39

745.—Groombridge 2090.
755.—S Bootis Var. 2. Period 272 days. Range, 8th to 13th magnitude.
750.—761.—Comparison stars for Comet 2, 1862.

Observed with the Madras Meridian Circle in that Year.

Number.	Star.	In Right Ascension. Annual Precession.	Secular Variation.	Proper Motion.	In Polar Distance. Annual Precession.	Secular Variation.	Proper Motion.	Number in B. A. C.
		s	s	s	"	"	"	
736	... ...	+ 3·7213	+ 0·0458	...	+ 17·688	− 0·265	...	...
737	β Centauri	+ 4·1675	+ 0·0841	− 0·010	+ 17·611	− 0·301	+ 0·06	4660
738	93 Virginis τ ...	+ 3·0475	+ 0·0084	+ 0·001	+ 17·492	− 0·221	+ 0·07	4672
739	94 Virginis	+ 3·1694	+ 0·0115	− 0·002	+ 17·406	− 0·207	+ 0·01	4686
740	Lalande 26696 ...	+ 2·7911	− 0·0023	...	+ 17·370	− 0·210	...	...
741	Taylor 6670	+ 4·2270	+ 0·0863	...	+ 17·364	− 0·315	...	...
742	... ...	+ 4·2749	+ 0·0807	...	+ 17·319	− 0·320	...	...
743	Taylor 6386	+ 3·6330	+ 0·0302	...	+ 17·305	− 0·268	...	...
744	... ...	+ 3·6340	+ 0·0308	...	+ 17·270	− 0·260	...	...
745	R. P. L. 108... ...	− 7·8687	+ 2·5079	...	+ 17·200	+ 0·883	...	...
746	Lacaille 5844 ..	+ 4·3286	+ 0·0012	...	+ 17·188	− 0·335	...	...
747	Taylor 6616	+ 4·1294	+ 0·0686	...	+ 17·119	− 0·330	...	4709
748	98 Virginis κ ...	+ 3·1206	+ 0·0122	+ 0·001	+ 17·114	− 0·250	− 0·02	4716
749	16 Bootis α (*Arcturus*)	+ 2·8132	+ 0·0004	− 0·070	+ 16·960	− 0·227	+ 1·98	4729
750	100 Virginis λ ...	+ 3·2366	+ 0·0140	− 0·002	+ 16·880	− 0·264	− 0·02	4748
751	... ...	+ 4·3965	+ 0·0900	...	+ 16·689	− 0·362	...	...
752	.. ..	+ 3·5401	+ 0·0234	...	+ 16·678	− 0·298	...	...
753	2 Libræ	+ 3·2159	+ 0·0132	− 0·004	+ 16·618	− 0·270	+ 0·09	4766
754	Taylor 6721	+ 3·2105	+ 0·0132	...	+ 16·856	− 0·272	...	4772
755	8 Bootis Var. 2 ...	+ 2·0108	− 0·0082	...	+ 16·510	− 0·174	...	...
756	... ...	+ 3·0882	+ 0·0834	...	+ 16·427	− 0·313	...	...
757	Lacaille 5968 ...	+ 3·7216	+ 0·0806	...	+ 16·298	− 0·324	...	...
758	... .	+ 3·7223	+ 0·0864	...	+ 16·270	− 0·324	...	...
759	O. A. N. 14634 ...	+ 0·9057	+ 0·0360	...	+ 16·129	− 0·086	...	...
760	25 Bootis ρ	+ 2·5348	− 0·0015	− 0·008	+ 16·121	− 0·283	− 0·14	4808
761	O A. N. 14652 ..	+ 0·8880	+ 0·0366	...	+ 16·067	− 0·084	...	...
762	α¹ Centauri	+ 4·5004	+ 0·0578	− 0·476	+ 15·890	− 0·410	− 0·81	4681
763	α² Centauri	+ 4·5004	+ 0·0578	− 0·476	+ 15·890	− 0·410	− 0·81	4682
764	Lacaille 6027 ...	+ 3·5698	+ 0·0284	...	+ 15·860	− 0·329	...	...
765	... ...	+ 3·8985	+ 0·0274	...	+ 15·762	− 0·330	...	...
766	... .	+ 3·6801	+ 0·0819	...	+ 15·780	− 0·320	.	...
767	. . ..	+ 4·8618	+ 0·0874	...	+ 15·552	− 0·498	...	..
768	5 Libræ . ..	+ 3·2860	+ 0·0152	− 0·003	+ 15·446	− 0·314	+ 0·01	4895
769	36 Bootis ε (*Mirac*)	+ 2·6240	− 0·0001	− 0·006	+ 15·414	− 0·262	− 0·01	4876
770	. ... ..	+ 3·6335	+ 0·0204	...	+ 15·396	− 0·340	.	.

737, 762.–763. Proper motions from "*Stone's Cape Catalogue.*"
739.—Proper motions from "*Greenwich Catalogue* 1872."

Mean Positions of Stars for 1865 January 1st.

Number.	Star.	Magnitude.	Estimations.	Mean Right Ascension			Mean Polar Distance.			Observations.	Fraction of Year.
				h.	m.	s.	°	′	″		
771	.. ..	9·0	..	14	40	34·11	127	3	39·7	1	0·41
772	Brisbane 5060	8·9	1	14	41	28·01	131	16	59·2	1	0·47
773	9 Libræ α²	3·0	.	14	43	[illegible]	105	28	40·1	11	0·39
774	Lalande 27123	7·8	1	14	47	20·67	100	27	36·2	1	0·36
775	... .	9·4	1	14	47	36·06	150	41	29·5	1	0·30
776	... ...	9·0	1	14	50	27·36	130	32	35·9	1	0·47
777	7 Ursæ Minoris β Var. 1...	2·1	..	14	51	8·30	15	17	32·6	1	0·46
778	Taylor 7017	7·3	2	14	57	19·46	150	36	13·2	2	0·36
779	... ...	8·5	1	14	57	46·26	101	30	54·6	1	0·41
780	46 Bootis φ	4·5	...	14	58	39·05	62	31	29·3	5	0·41
781	47 Bootis ε	6·6	...	15	0	57·56	41	10	35·1	1	0·41
782	Taylor 7079	6·0	1	15	3	28·79	128	7	30·1	1	0·49
783	... ... 2nd ..	9·0	1	15	3	37·68	123	18	59·1	1	0·37
784	24 Libræ ι¹	4·0	...	15	4	31·76	109	16	43·7	2	0·32
785	B. P. L. 111	6·9	...	15	5	37·16	5	31	38·0	1	0·46
786	27 Libræ β ...	2·7	...	15	9	44·66	98	52	57·8	9	0·44
787	... ...	8·4	2	15	11	54·94	130	21	14·6	2	0·38
788	... ...	9·0	1	15	15	4·48	124	15	23·5	1	0·37
789	... ...	9·2	1	15	17	8·01	130	3	51·4	1	0·40
790	Lacaille 6377 ...	7·8	1	15	19	24·01	130	11	6·3	1	0·47
791	32 Libræ ζ¹	6·2	..	15	20	34·72	106	14	37·2	3	0·36
792	W. B. E. XV. 396 ...	8·8	1	15	22	1·57	101	15	40·8	1	0·40
793	B. P. L. 114	7·1	..	15	22	3·00	2	15	14·0	1	0·47
794	.. ..	8·5	1	15	22	39·39	125	10	22·6	1	0·37
795	.. ..	8·1	1	15	22	32·43	151	37	14·4	1	0·41
796	W. B. E. XV. 469 ...	9·4	2	15	24	5·60	101	38	11·0	2	0·33
797	. .. .	9·3	1	15	24	37·32	130	9	9·0	1	0·47
798	38 Libræ γ ...	4·0	..	15	27	08·37	104	30	13·0	1	0·33
799	... ...	8·2	1	15	28	45·44	[illegible]	35	35·3	1	0·34
800	5 Coronæ Bor. α (*Alphera*)	2·4	.	15	28	28·33	62	40	44·7	7	0·44
801	... ...	8·4	1	15	30	13·48	129	33	39·7	1	0·37
802	... ...	9·7	1	15	32	22·40	116	31	60·3	1	[illegible]
803	... .	9·5	1	15	33	[illegible]	136	25	34·1	1	[illegible]
804	W. B. B. XV. 636 ..	8·2	1	15	34	39·08	102	19	37·3	1	0·41
805	W. B. E. XV. 676 ..	8·8	1	15	35	59·74	109	41	[illegible]	1	0·42

774.—Comparison star for Iris in 1861.
777.—β Ursæ Minoris Var. 1.—(Kochab.) Supposed to vary irregularly from 2·0 to [illegible] [illegible].
781.—Comparison star for Comet [illegible], 1861.
786.—Groombridge 2213.
792.—796.—804.—805.—Comparison stars for [illegible] in 1864.
783.—Groombridge [illegible].

Observed with the Madras Meridian Circle in that Year.

Number	Star	In Right Ascension.			In Polar Distance.			Number in B. A. C.
		Annual Precession	Secular Variation.	Proper Motion.	Annual Precession.	Secular Variation.	Proper Motion.	
		s	s	s	"	"	"	
771	... ...	+ 3·7220	+ 0·0366	...	+ 15·390	− 0·387	...	..
772	Brisbane 5089 ..	+ 3·8821	+ 0·0379	...	+ 15·380	− 0·389	...	..
773	9 Libræ α[1] .	+ 3·3142	+ 0·0154	− 0·007	+ 15·162	− 0·334	+ 0·06	4995
774	Lalande 27123 ..	+ 3·3674	+ 0·0178	.	+ 14·086	− 0·395	...	..
775	...	+ 3·8012	+ 0·0630		+ 14·086	− 0·381	...	...
776	... ...	+ 3·8464	+ 0·0363	...	+ 14·750	− 0·386	...	...
777	7 Ursæ Min. β Var 1.	− 0·2490	+ 0·1022	− 0·006	+ 14·719	+ 0·018	+ 0·03	4996
778	Taylor 7017	+ 4·7292	+ 0·0466	...	+ 14·365	− 0·486	...	...
779	... .	+ 3·9007	+ 0·0371	...	+ 14·318	− 0·406	...	...
780	43 Bootis ψ	+ 2·9833	+ 0·0010	− 0·013	+ 14·260	− 0·281	0·00	4969
781	47 Bootis κ	+ 1·9925	+ 0·0018	− 0·010	+ 14·121	− 0·212	...	4980
782	Taylor 7070...	+ 3·6950	+ 0·0273	...	+ 13·070	− 0·398	...	...
783	... . 2nd .	+ 3·6794	+ 0·0264	..	+ 13·985	− 0·302	...	...
784	24 Libræ ι[1] . .	+ 3·4002	+ 0·0171	− 0·002	+ 13·898	− 0·364	+ 0·01	4985
785	R. P. L. 111	− 6·0837	+ 1·1969	...	+ 13·830	+ 0·736	...	5082
786	27 Libræ β	+ 3·2380	+ 0·0117	− 0·009	+ 13·365	− 0·358	+ 0·01	5034
787	... ...	+ 3·0176	+ 0·0346	...	+ 13·464	− 0·481	...	...
788	... ...	+ 3·7469	+ 0·0274	...	+ 13·218	− 0·418	...	...
789	... ...	+ 3·9844	+ 0·0384	...	+ 13·082	− 0·440	...	..
790	Lacaille 6377 ...	+ 3·9380	+ 0·0334	.	+ 12·960	− 0·444	...	...
791	32 Libræ ζ[1]	+ 3·8712	+ 0·0149	+ 0·002	+ 12·948	− 0·394	+ 0·05	5080
792	W. B. E. XV. 385 ...	+ 3·2776	+ 0·0124	...	+ 12·754	− 0·378	...	..
793	R. P. L. 114 ...	− 28·1469	+ 7·7715	...	+ 12·751	+ 2·001		5140
794	.. ..	+ 3·8004	+ 0·0270	...	+ 12·724	− 0·434		..
795	.. ..	+ 4·0855	+ 0·0482	...	+ 12·720	− 0·567		...
796	W. B. E. XV. 429 ..	+ 3·2962	+ 0·0125		+ 12·615	− 0·377	...	.
797	... ...	+ 3·9606	+ 0·0327	...	+ 12·670	− 0·454	.	..
798	38 Libræ γ	+ 3·3411	+ 0·0186	+ 0·008	+ 12·344	− 0·389	− 0·03	5131
799	... .	+ 3·4564	+ 0·0286	...	+ 12·395	− 0·440	...	..
800	5 Cor. Bor. α (*Alphecca*)	+ 2·5338	+ 0·0023	+ 0·009	+ 12·340	− 0·297	+ 0·07	5148
801	... .	+ 3·9490	+ 0·0314		+ 12·108	− 0·469		.
802	.. ..	+ 3·6078	+ 0·0201		+ 12·018	− 0·120	...	.
803	. ...	+ 3·9891	+ 0·0274		+ 11·983	− 0·459	...	..
804	W. B. E. XV. 646 ...	+ 3·3072	+ 0·0135	..	+ 11·980	− 0·308	...	...
805	W. B. E. XV. 675 .	+ 3·3157	+ 0·0126	.	+ 11·795	− 0·306	..	

781.—Proper motions, in Right Ascension only, from "*Greenwich Catalogue* 1865."
792.—Proper motions from "*Greenwich Catalogue* 1872."

Mean Positions of Stars for 1865 January 1st.

Number.	Star.	Magnitude.	Estimations.	Mean Right Ascension.			Mean Polar Distance.			Observations.	Fraction of Year.
				h	m	s	°	′	″		
806	24 Serpentis α	2·7	...	15	37	37·20	83	6	50·7	5	0·42
807	Lalande 28787	8·4	...	15	42	9·00	92	40	6·2	1	0·43
808	Lalande 28970	8·0	1	15	48	0·68	70	40	14·7	1	0·42
809	7 Scorpii δ	2·5	...	15	52	21·34	112	14	5·2	2	0·30
810	W. B. E. XV. 1047 ...	8·0	1	15	55	59·39	91	10	20·2	1	0·41
811	8 Scorpii β¹	2·9	...	15	57	36·40	109	25	59·6	4	0·45
812	R. P. L. 116	6·0	...	16	4	30·87	4	18	54·2	2	0·57
813	W. B. E. XVI. 88 ...	8·0	1	16	6	6·00	102	41	14·8	1	0·68
814	1 Ophiuchi δ	2·8	...	16	7	16·28	93	20	40·1	3	0·40
815	... ...	7·8	3	16	9	12·26	112	35	3·2	3	0·60
816	... ...	9·7	1	16	9	46·72	112	33	44·8	1	0·41
817	O. A. S. 15504	8·8	1	16	11	24·95	106	41	35·5	1	0·57
818	20 Scorpii σ	3·0	...	16	12	59·31	115	15	56·5	1	0·43
819	... ...	9·4	1	16	15	49·00	152	17	11·8	1	0·53
820	... ...	9·2	1	16	19	3·44	129	30	44·4	1	0·41
821	21 Scorpii α (*Antares*) ...	1·1	...	16	21	8·02	116	7	45·5	2	0·52
822	... ...	8·5	1	16	22	46·33	49	20	57·5	1	0·57
823	... ...	9·3	1	16	28	48·00	152	16	9·8	1	0·56
824	30 Herculis g Var. 5 ...	5·1	...	16	24	12·42	47	40	11·0	2	0·83
825	... ...	9·2	1	16	26	29·30	130	56	0·1	1	0·48
826	... ...	9·2	1	16	28	27·91	182	10	58·0	1	0·55
827	... ...	9·3	1	16	29	19·80	130	54	1·2	1	0·46
828	α Trianguli Australis ...	2·2	...	16	34	24·24	158	46	29·5	1	0·56
829	... ...	8·0	1	16	34	41·30	134	7	0·8	1	0·57
830	40 Herculis ζ	3·1	.	16	36	11·91	58	9	3·6	4	0·83
831	... ...	8·8	1	16	36	46·18	130	58	5·5	1	0·46
832	O. A. S. 15992	9·0	1	16	39	25·89	111	55	39·2	1	0·36
833	20 Ophiuchi	4·7	.	16	42	32·01	100	32	29·4	2	0·38
834	... ...	8·2	1	16	44	29·70	131	1	36·3	1	0·46
835	... ...	8·6	1	16	44	41·86	130	18	29·1	1	0·33
836	Taylor 7815	8·4	2	16	46	38·97	130	17	39·6	2	0·42
837	. .. .	9·3	1	16	47	7·66	130	59	49·1	1	0·46
838	Taylor 7822	8·6	3	16	47	37·80	130	17	36·9	3	0·34
839	27 Ophiuchi κ	3·4	...	16	51	16·65	80	24	46·3	4	0·34
840	... ...	8·4	1	16	51	43·63	132	53	49·6	1	0·36

807.—Comparison star for Donati's Comet of 1858. Double: the n. p. star observed.
808—810.—Comparison stars for Comet 2, 1862.
812.—Carrington 3468.
822.—Comparison star for Comet 2, 1861.
824.—30 Herculis g Var. 5.—Changes irregularly from 5th to 6·3 magnitude.

Observed with the Madras Meridian Circle in that Year.

Number.	Star.	In Right Ascension. Annual Precession.	Secular Variation.	Proper Motion.	In Polar Distance. Annual Precession.	Secular Variation.	Proper Motion.	Number in B. A. C.
		s	s	s	"	"	"	
806	24 Serpentis α ...	+ 2·9414	+ 0·0062	+ 0·000	+ 11·678	− 0·354	− 0·06	5196
807	Lalande 23757 ..	+ 3·1904	+ 0·0088	...	+ 11·346	− 0·361	...	...
808	Lalande 28970 ..	+ 2·6821	+ 0·0039	...	+ 10·022	− 0·333	...	...
809	7 Scorpii δ	+ 3·5361	+ 0·0150	− 0·001	+ 10·601	− 0·448	+ 0·01	5303
810	W. B. E. XV. 1047..	+ 3·0976	+ 0·0079	...	+ 10·380	− 0·392	...	...
811	8 Scorpii β¹	+ 3·4780	+ 0·0142	− 0·002	+ 10·209	− 0·441	+ 0·02	5330
812	R. P. L. 116 .. .	+ 12·4435	+ 1·7408	...	+ 9·684	+ 1·595	...	...
813	W. B. E. XVI. 86 ..	+ 3·3367	+ 0·0110	...	+ 9·561	− 0·431	...	...
814	1 Ophiuchi δ ..	+ 3·1400	+ 0·0081	− 0·006	+ 9·472	− 0·408	+ 0·13	5414
815	... .	+ 3·5045	+ 0·0147	...	+ 9·322	− 0·464	...	...
816	... .	+ 3·3646	+ 0·0147	...	+ 9·277	− 0·485	...	...
817	O. A. S. 15504 ..	+ 3·4980	+ 0·0121	...	+ 9·150	− 0·440	...	...
818	20 Scorpii σ	+ 3·6356	+ 0·0156	− 0·003	+ 9·088	− 0·478	+ 0·01	5447
819	... .	+ 5·3588	+ 0·0692	...	+ 8·806	− 0·706	...	...
820	... .	+ 4·0675	+ 0·0240	...	+ 8·639	− 0·540	...	...
821	21 Scorpii α (*Antares*)	+ 3·6679	+ 0·0150	− 0·001	+ 8·396	− 0·491	+ 0·03	5498
822	... ...	+ 1·9921	+ 0·0040	...	+ 8·265	− 0·260	...	...
823	... .	+ 5·3980	+ 0·0699	...	+ 8·172	− 0·722	...	...
824	30 Herculis g Var. 5.	+ 1·9640	+ 0·0042	+ 0·006	+ 8·141	− 0·265	− 0·07	5568
825	... .	+ 4·1363	+ 0·0235	...	+ 7·950	− 0·557	...	...
826	... .	+ 5·4164	+ 0·0614	...	+ 7·790	− 0·731	...	...
827	... ...	+ 4·1408	+ 0·0239	...	+ 7·730	− 0·560	...	...
828	α Trianguli Australis	+ 6·2771	+ 0·0907	0·000	+ 7·318	− 0·866	+ 0·06	5578
829	... .	+ 4·2790	+ 0·0247	...	+ 7·294	− 0·584	...	...
830	40 Herculis ζ ...	+ 2·3964	+ 0·0083	− 0·084	+ 7·172	− 0·316	− 0·45	5604
831	... .	+ 4·1574	+ 0·0215	...	+ 7·121	− 0·570	...	...
832	O. A. S. 15952 ...	+ 3·5774	+ 0·0114	...	+ 6·998	− 0·488	...	...
833	20 Ophiuchi	+ 3·8068	+ 0·0090	+ 0·004	+ 6·944	− 0·498	+ 0·06	5687
834	... ...	+ 4·1728	+ 0·0198	...	+ 6·490	− 0·570	...	...
835	... .	+ 4·1456	+ 0·0192		+ 6·471	− 0·575	...	..
836	Taylor 7815	+ 4·1467	+ 0·0191		+ 6·400	− 0·576	...	...
837	.. ..	+ 4·1730	+ 0·0192		+ 6·370	− 0·591	...	...
838	Taylor 7882	+ 4·1496	+ 0·0190	...	+ 6·236	− 0·578	...	...
839	27 Ophiuchi κ ...	+ 2·8568	+ 0·0044	− 0·023	+ 5·984	− 0·402	− 0·02	5765
840	... .	+ 3·8089	+ 0·0130	...	+ 5·890	− 0·547	...	...

828.—Proper motions from "*Stone's Cape Catalogue.*"
833.—Proper motions from "*Greenwich Catalogue* 1864."
839.—Proper motions from "*Greenwich Catalogue* 1872."

Mean Positions of Stars for 1865 January 1st.

Number.	Star.	Magnitude.	Estimations.	Mean Right Ascension.			Mean Polar Distance.			Observations.	Fraction of Year.
				h.	m.	s.	°	′	″		
841	... ...	8·6	1	16	52	9·46	122	44	57·0	1	0·89
842	O. A. S. 16263	8·0	1	16	54	2·24	110	28	39·1	1	0·57
843	... ...	8·5	2	16	55	58·52	130	54	46·4	2	0·51
844	O. A. S. 16288	8·0	1	16	56	31·41	119	50	18·0	1	0·67
845	Taylor 7926	7·0	1	16	58	50·75	136	51	8·5	1	0·57
846	22 Ursæ Minoris ε ...	4·5	...	16	59	54·36	7	41	47·0	2	0·06
847	35 Ophiuchi η	2·6	...	17	2	33·62	106	38	17·0	1	0·36
848	Lacaille 7103	7·0	2	17	4	40·01	128	7	46·8	2	0·62
849	... ...	9·4	1	17	5	40·02	130	58	47·4	1	0·56
850	... ...	8·0	1	17	5	48·39	130	50	39·5	1	0·57
851	... ...	8·5	1	17	6	15·38	137	25	6·5	1	0·30
852	... ...	8·2	1	17	7	13·00	130	42	6·4	1	0·45
853	64 Herculis α Var. 1 ...	3·2	...	17	8	30·51	75	27	12·6	8	0·54
854	... ...	8·4	1	17	9	5·35	124	4	21·4	1	0·58
855	Taylor 8017	7·7	1	17	13	25·08	114	46	58·3	1	0·57
856	42 Ophiuchi θ	3·4	...	17	13	43·18	114	51	41·0	8	0·54
857	8 Aræ ... ... 2nd...	7·0	2	17	18	56·02	130	38	30·8	2	0·57
858	... ...	9·2	1	17	21	17·42	130	45	42·7	1	0·47
859	Brisbane 6091	8·8	1	17	21	34·89	149	27	5·0	1	0·55
860	... ...	8·8	1	17	21	36·76	130	32	50·7	1	0·58
861	Lacaille 7315	7·0	1	17	22	12·66	130	55	58·1	1	0·57
862	... ...	9·1	1	17	27	4·13	150	36	41·8	1	0·58
863	... ...	8·0	2	17	28	5·24	130	48	30·9	2	0·38
864	55 Ophiuchi α	2·2	...	17	28	40·02	77	21	21·6	4	0·67
865	55 Serpentis ξ	3·7	...	17	29	51·37	105	18	37·5	4	0·44
866	λ Scorpii	2·6	...	17	33	8·90	126	5	22·0	2	0·36
867	56 Serpentis ο	4·4	...	17	33	49·69	102	47	57·4	1	0·36
868	... ...	9·0	1	17	34	10·30	128	57	31·1	1	0·35
869	Lacaille 7406	7·9	1	17	34	46·53	128	44	13·2	1	0·57
870	... ...	9·2	1	17	36	8·27	150	36	6·0	1	0·58
871	... ...	8·5	1	17	37	48·96	126	30	30·6	1	0·47
872	... ...	8·0	1	17	39	37·60	127	21	38·2	1	0·48
873	86 Herculis μ	3·5	...	17	41	10·47	62	11	54·0	7	0·37
874	... ...	8·5	1	17	43	8·92	119	27	29·5	1	0·68
875	... ...	8·2	1	17	46	24·30	128	38	13·5	2	0·66

853.—α Herculis Var. 1.—Changes irregularly between 3rd and 4th magnitudes.
871.—872.—875.—Comparison stars for Donati's Comet of 1864.

Observed with the Madras Meridian Circle in that Year.

Number.	Star.	In Right Ascension.			In Polar Distance.			Number in B. A. C.
		Annual Precession.	Secular Variation.	Proper Motion.	Annual Precession.	Secular Variation.	Proper Motion.	
		s	s	s	"	"	"	
841	... ...	+ 3·8968	+ 0·0137	...	+ 5·851	− 0·547	...	...
842	O. A. S. 16283 ...	+ 3·5497	+ 0·0083	...	+ 5·693	− 0·488	...	...
843	... ...	+ 4·1850	+ 0·0171	...	+ 5·587	− 0·589	...	...
844	O. A. S. 16288 ...	+ 3·8036	+ 0·0110	...	+ 5·463	− 0·537	...	..
845	Taylor 7990	+ 4·4196	+ 0·0203	...	+ 5·204	− 0·620	...	...
846	22 Ursæ Minoris ε ...	− 0·4215	+ 0·3041	+ 0·009	+ 5·107	+ 0·002	− 0·01	5780
847	35 Ophiuchi η ...	+ 3·4320	+ 0·0074	+ 0·001	+ 4·967	− 0·487	− 0·12	5781
848	Lacaille 7166 ...	+ 4·0911	+ 0·0120	...	+ 4·795	− 0·581	...	...
849	... ...	+ 4·1977	+ 0·0140	...	+ 4·709	− 0·597	...	...
850	... ...	+ 4·1957	+ 0·0148	...	+ 4·698	− 0·597	...	...
851	... ...	+ 4·4372	+ 0·0187	...	+ 4·660	− 0·633	...	...
852	... .	+ 4·1921	+ 0·0144	...	+ 4·580	− 0·597	...	...
853	64 Herculis α Var. 1.	+ 2·7383	+ 0·0086	− 0·008	+ 4·469	− 0·391	− 0·04	5821
854	... ...	+ 3·9641	+ 0·0113	...	+ 4·418	− 0·565	...	...
855	Taylor 8017	+ 3·0762	+ 0·0080	...	+ 4·046	− 0·527	...	5846
856	42 Ophiuchi θ ...	+ 3·6780	+ 0·0080	− 0·008	+ 4·028	− 0·523	− 0·02	5851
857	δ Aræ ... 2nd...	+ 5·4037	+ 0·0269	− 0·009	+ 3·576	− 0·778	+ 0·09	...
858	... ...	+ 4·2082	+ 0·0111	...	+ 3·371	− 0·606	...	...
859	Brisbane 6001 ..	+ 5·2189	+ 0·0227	...	+ 3·360	− 0·753	...	...
860	... .	+ 4·1996	+ 0·0109	...	+ 3·383	− 0·606	...	...
861	Lacaille 7315 ...	+ 4·2153	+ 0·0109	...	+ 3·393	− 0·608	...	...
862	... .	+ 5·4109	+ 0·0219	...	+ 2·871	− 0·788	...	...
863	... ...	+ 4·2130	+ 0·0094	...	+ 2·770	− 0·609	...	...
864	55 Ophiuchi α ...	+ 2·7746	+ 0·0080	+ 0·004	+ 2·738	− 0·403	+ 0·20	5941
865	55 Serpentis ξ .	+ 3·4950	+ 0·0047	− 0·004	+ 2·680	− 0·498	+ 0·04	5949
866	ε Scorpii	+ 4·1451	+ 0·0070	0·000	+ 2·844	− 0·601	+ 0·08	5970
867	56 Serpentis o ...	+ 3·8739	+ 0·0041	− 0·008	+ 2·386	− 0·490	+ 0·04	5976
868	... .	+ 4·1464	+ 0·0075	...	+ 2·246	− 0·612	...	...
869	Lacaille 7406 ...	+ 4·1898	+ 0·0072	...	+ 2·208	− 0·601	...	...
870	... .	+ 5·4822	+ 0·0162	...	+ 2·084	− 0·707	...	..
871	... . ..	+ 4·0801	+ 0·0062	...	+ 1·980	− 0·590	...	...
872	... .. .	+ 4·0988	+ 0·0060	...	+ 1·760	− 0·595	...	...
873	86 Herculis μ ..	+ 2·3804	+ 0·0026	− 0·026	+ 1·645	− 0·346	+ 0·74	6021
874	.. ..	+ 3·7945	+ 0·0045	...	+ 1·381	− 0·563	.	..
875	..	+ 4·1360	+ 0·0052	...	+ 1·450	− 0·606	...	...

857 – 866. Proper motions from "*Stone's Cape Catalogue.*"
865.—Proper motions from "*Greenwich Catalogue 1872*"

Mean Positions of Stars for 1865 January 1st.

Number.	Star.	Magnitude.	Estimations.	Mean Right Ascension.			Mean Polar Distance.			Observations	Fraction of Year.
				h.	m.	s.	°	′	″		
876	... ...	8·8	1	17	50	46·87	132	7	38·1	1	0·50
877	Lacaille 7517	8·0	1	17	52	30·06	140	10	23·6	1	0·57
878	Lacaille 7518	7·8	2	17	52	48·50	140	12	15·7	2	0·56
879	33 Draconis γ (*Etanin*) ...	2·4	...	17	53	28·08	38	29	40·0	1	0·61
880	Taylor 8655	7·0	1	17	57	4·10	133	25	40·8	1	0·65
881	... ...	7·7	1	17	59	38·97	130	26	12·2	1	0·64
882	... ...	8·7	1	18	1	15·07	131	43	35·0	1	0·57
883	... ...	8·2	1	18	2	53·60	131	44	37·6	1	0·67
884	13 Sagittarii μ¹	4·1	...	18	5	41·36	111	6	27·7	7	0·57
885	Lalande 33818	8·0	3	18	15	4·73	101	56	21·9	3	0·57
886	Lalande 33946	6·9	3	18	15	38·80	102	4	38·2	3	0·67
887	23 Ursæ Minoris δ ...	4·8	...	18	16	33·61	3	28	45·8	5	0·30
888	θ Coronæ Australis ...	5·1	...	18	23	51·60	132	24	20·8	1	0·60
889	Taylor 8651...	7·3	1	18	27	38·26	140	18	40·1	1	0·56
890	3 Lyræ α (*Vega*)	0·2	...	18	32	32·02	51	20	36·5	5	0·60
891	... ...	9·0	1	18	34	43·07	136	44	54·6	1	0·63
892	Lacaille 7882	7·0	...	18	37	40·03	140	5	27·1	1	0·50
893	10 Lyræ β Var. 1... ...	3·6	...	18	45	5·03	56	47	32·8	7	0·64
894	32 Sagittarii ν¹	5·0	...	18	46	1·11	112	54	36·3	1	0·66
895	37 Sagittarii ξ²	3·5	...	18	49	40·47	111	10	52·6	1	0·30
896	39 Sagittarii ο	3·9	...	18	56	36·38	111	56	9·3	1	0·67
897	R. P. L. 131	6·6	...	18	58	33·56	3	17	54·2	1	0·17
898	17 Aquilæ ζ	3·1	...	18	59	12·21	76	20	5·8	8	0·63
899	... ...	8·5	1	19	0	25·86	120	47	3·0	1	0·64
900	... ...	6·1	1	19	10	3·34	107	0	31·5	1	0·70
901	25 Aquilæ ω ...	5·1	...	19	11	35·69	78	36	44·6	7	0·62
902	... .	8·2	2	19	16	38·86	120	32	39·1	2	0·60
903	30 Aquilæ δ ..	3·5	...	19	18	41·38	87	9	7·0	8	0·62
904	Taylor 8960 ...	6·1	...	19	22	13·87	140	27	35·2	1	0·63
905	... .	8·5	1	19	25	0·13	120	56	47·4	1	0·67
906	51 Sagittarii h¹	5·3	...	19	27	49·59	115	0	41·5	1	0·30
907	52 Sagittarii h²	4·6	...	19	28	39·29	115	10	42·7	5	0·67
908	56 Sagittarii e¹	5·0	...	19	34	47·40	106	26	14·3	2	0·67
909	50 Aquilæ γ	2·8	...	19	39	50·32	79	42	49·0	5	0·61
910	8 Vulpeculæ Var. 2.	9·0	1	19	42	51·64	62	2	52·6	1	0·67

[illegible].—Comparison stars for Donati's Comet of 1858.
885.—886.—Comparison stars for Ada in 1863.
893.—β Lyræ Var. 1.—Period 12.91 days. Range, 3.5 to 4.5 magnitude
[illegible].—Carrington [illegible].
910.—8 Vulpeculæ Var. 2.—Period 67.5 days. Range, 8.5 to 9.5 magnitude

Observed with the Madras Meridian Circle in that Year.

Number.	Star.	In Right Ascension.			In Polar Distance.			Number in B. A. C.
		Annual Precession.	Secular Variation.	Proper Motion.	Annual Precession.	Secular Variation.	Proper Motion.	
		s	s	s	"	"	"	
876	... .	+ 5·5080	+ 0·0073		+ 0·807	− 0·810	...	...
877	Lacaille 7517 ...	+ 5·3144	+ 0·0058	...	+ 0·657	− 0·774	...	...
878	Lacaille 7518 ...	+ 5·3143	+ 0·0052	...	+ 0·689	− 0·775	...	...
879	33 Draconis γ (*Etanin*)	+ 1·3915	+ 0·0030	0·000	+ 0·571	− 0·208	+ 0·04	6091
880	Taylor 8365	+ 4·8375	+ 0·0084	− 0·006	+ 0·286	− 0·682	+ 0·12	6112
881	... ..	+ 5·4292	+ 0·0012	...	+ 0·080	− 0·702	...	...
882	... ..	+ 4·2044	+ 0·0011	...	− 0·114	− 0·022	...	...
883	... .	+ 4·2650	+ 0·0007	...	− 0·263	− 0·022	...	...
884	13 Sagittarii μ^1 ..	+ 3·5875	+ 0·0000	− 0·004	+ 0·498	− 0·528	+ 0·01	6168
885	Lalande 33818 ..	+ 3·3538	+ 0·0004	...	− 1·310	− 0·467	...	...
886	Lalande 33945 ..	+ 3·3575	+ 0·0004	...	− 1·368	− 0·468	...	...
887	23 Ursæ Minoris δ ...	− 10·4015	− 0·4618	+ 0·048	− 1·390	+ 2·886	− 0·08	6251
888	θ Coronæ Australis .	+ 4·2860	− 0·0040	0·000	− 2·084	− 0·680	− 0·03	6296
889	Taylor 8551 ...	+ 5·3014	− 0·0183	...	− 2·201	− 0·767	...	...
890	3 Lyræ α (*Vega*) ...	+ 2·0130	+ 0·0016	+ 0·017	− 2·868	− 0·280	− 0·28	6365
891	... ...	+ 4·4770	− 0·0098	...	− 3·086	− 0·644	...	...
892	Lacaille 7832 ...	+ 5·2752	− 0·0210	...	− 3·248	− 0·758	...	...
893	10 Lyræ β Var. 1 ..	+ 2·2137	+ 0·0015	− 0·002	− 3·120	− 0·315	+ 0·08	6429
894	32 Sagittarii ν ..	+ 3·6257	− 0·0018	− 0·004	− 4·000	− 0·516	+ 0·01	6404
895	37 Sagittarii ξ² ...	+ 3·3807	− 0·0043	− 0·001	− 4·213	− 0·508	+ 0·08	6461
896	39 Sagittarii ο ..	+ 3·5943	− 0·0063	+ 0·001	− 4·102	− 0·506	+ 0·06	6507
897	R. P. L. 131 ...	− 18·2908	− 1·6014	...	− 5·069	+ 2·882	...	...
898	17 Aquilæ ζ	+ 2·7578	+ 0·0008	− 0·006	− 5·123	− 0·397	+ 0·07	6598
899	... .	+ 4·1347	− 0·0140	...	− 5·960	− 0·573	...	...
900	... . ..	+ 3·4056	− 0·0055	...	− 6·085	− 0·479	...	..
901	25 Aquilæ ω .. .	+ 2·9165	− 0·0003	− 0·003	− 6·154	− 0·398	− 0·08	6606
902	... .	+ 4·1272	− 0·0164	.	− 6·582	− 0·566	...	..
903	30 Aquilæ δ .. .	+ 3·0094	− 0·0015	+ 0·014	− 6·782	− 0·410	− 0·10	6646
904	Taylor 8980	+ 4·7618	− 0·0327	...	− 7·042	− 0·647	...	6669
905	... .	+ 4·1151	− 0·0181	. .	− 7·269	− 0·567	...	...
906	51 Sagittarii h^1 .	+ 3·6606	− 0·0100	− 0·002	− 7·498	− 0·491	0·00	6704
907	52 Sagittarii h^2 .	+ 3·6542	− 0·0102	+ 0·002	− 7·538	− 0·490	− 0·02	6706
908	56 Sagittarii e^1	+ 3·4394	− 0·0075	+ 0·001	− 8·080	− 0·436	− 0·02	6742
909	50 Aquilæ γ ..	+ 2·8820	− 0·0011	+ 0·001	− 8·463	− 0·373	0·00	6772
910	8 Vulpeculæ Var. 3	+ 2·4606	+ 0·0011		− 8·700	− 0·310	...	.

[illegible] —[illegible].—Proper motions from "*Stone's Cape Catalogue.*"

29

Mean Positions of Stars for 1865 January 1st.

Number.	Star.	Magnitude.	Estimations.	Mean Right Ascension. h.	m.	s.	Mean Polar Distance. °	′	″	Observations.	Fraction of Year.
911	53 Aquilæ α (*Altair*) ...	1·0	...	19	44	11·67	81	20	8·9	6	0·64
912	60 Aquilæ β	4·0	...	19	48	40·76	83	55	41·5	3	0·67
913	... ...	9·1	1	19	53	5·44	147	10	40·0	1	0·67
914	... ...	9·0	...	19	55	41·11	151	51	30·7	1	0·61
915	Taylor 9208	5·1	...	19	56	46·70	122	35	56·3	1	0·60
916	... ...	8·7	3	19	56	50·64	130	21	22·0	3	0·67
917	λ Ursæ Minoris	6·5	...	19	59	9·25	1	5	44·2	2	0·10
918	O. A. N. 20046	9·3	1	20	2	40·74	82	23	22·0	1	0·65
919	R Capricorni Var. 1 ...	9·7	1	20	3	44·00	104	39	54·0	1	0·74
920	R Delphini Var. 2 ...	9·7	1	20	8	24·18	81	19	7·6	1	0·74
921	O. A. S. 20356	7·0	1	20	8	24·96	110	25	57·4	1	0·70
922	6 Capricorni α¹	3·8	...	20	10	30·62	102	57	30·2	5	0·68
923	... ...	7·6	1	20	10	38·90	140	8	50·2	1	0·64
924	... ...	8·0	1	20	11	13·74	106	16	33·2	1	0·58
925	Lalande 39095	8·3	2	20	14	44·30	105	15	18·3	2	0·61
926	α Pavonis	2·1	...	20	14	50·32	147	9	22·3	2	0·65
927	X Capricorni Var. 7 ...	9·9	1	20	15	1·81	106	26	36·5	1	0·75
928	Lalande 39195	8·8	1	20	15	34·65	105	18	10·6	1	0·63
929	Lacaille 8441	8·0	2	20	19	16·87	121	6	40·5	2	0·69
930	11 Capricorni ρ	5·0	...	20	21	0·30	108	15	26·9	5	0·70
931	... ...	...	...	20	23	46·14	80	1	50·3	1	0·67
932	Lalande 39625	8·0	1	20	24	33·84	86	2	14·9	1	0·66
933	... ...	9·2	1	20	27	16·98	131	5	40·2	1	0·67
934	... ...	9·5	1	20	30	57·56	140	56	10·7	1	0·61
935	5 Capricorni Var. 2 ..	9·0	1	20	34	0·33	109	32	10·1	1	0·66
936	. ...	8·0	1	20	35	30·24	123	58	17·8	1	0·73
937	50 Cygni α (*Deneb*) ...	1·5	...	20	36	40·60	45	13	2·8	5	0·70
938	2 Aquarii ε	3·8	...	20	40	21·10	99	50	15·9	1	0·67
939	W. B. H. XX. 1094 ..	9·3	1	20	40	36·30	106	24	5·0	1	0·73
940	.. ..	10·5	1	20	41	12·62	105	18	5·8	1	0·73
941	T Aquarii Var. 4	9·5	1	20	42	48·61	95	28	44·6	1	0·67
942	Lacaille 8571 ..	7·7	...	20	42	59·06	130	12	46·6	1	0·64
943	.. .	9·7	1	20	43	40·77	124	57	56·9	2	0·70
944	Taylor 9238 .	7·4	3	20	44	37·41	101	56	24·0	2	0·73
945	. - . .	6·7	.	20	47	40·06	169	1	3·4	1	0·73

919.—R Capricorni Var. 1.—Period 347 days. Range, 9th to below 14th magnitude.
920.—R Delphini Var. 2.—Period 284 days. Range, 8th to 13th magnitude.
921.—Comparison star for Parthenope in 1862.
924.—[illegible]—[illegible]—Comparison stars for Hestia in 1862.
927.—X Capricorni Var. 7.—Period unknown. Range, 10·4 to below 13th magnitude.
935.—5 Capricorni Var. 2.—Supposed to change from 9th to 11th magnitude.
941.—T Aquarii Var. 4.—Period 203 days. Range, 7·0 to 12·5 magnitude.

Observed with the Madras Meridian Circle in that Year.

Number.	Star.	In Right Ascension. Annual Precession.	In Right Ascension. Secular Variation.	In Right Ascension. Proper Motion.	In Polar Distance. Annual Precession.	In Polar Distance. Secular Variation.	In Polar Distance. Proper Motion.	Number in B.A.C.
		s	s	s	"	"	"	
911	53 Aquilæ α (*Altair.*)	+ 2·8921	− 0·0014	+ 0·036	− 8·808	− 0·374	− 0·38	6802
912	60 Aquilæ β	+ 2·9456	− 0·0020	+ 0·002	− 9·158	− 0·378	+ 0·47	6883
913	... ...	+ 4·8978	− 0·0528	...	− 9·499	− 0·620	...	...
914	... ...	+ 5·2600	− 0·0700	...	− 9·698	− 0·668	...	...
915	Taylor 9208	+ 3·8156	− 0·0175	...	− 9·706	− 0·488	...	6877
916	... ...	+ 4·0637	− 0·0244	...	− 9·788	− 0·518	...	...
917	λ Ursæ Minoris ...	− 57·0046	− 29·8498	− 0·086	− 9·964	+ 7·296	− 0·01	6899
918	O. A. N. 20046 ...	+ 1·2603	− 0·0074	...	− 10·280	− 0·154	...	...
919	R Capricorni Var. 1...	+ 3·3722	− 0·0087	...	− 10·309	− 0·418	...	...
920	R Delphini Var. 3 ...	+ 2·8998	− 0·0017	...	− 10·657	− 0·358	...	...
921	O. A. S. 20866 ...	+ 3·4040	− 0·0116	...	− 10·659	− 0·427	...	...
922	6 Capricorni α[1] ...	+ 3·3311	− 0·0064	+ 0·001	− 10·818	− 0·408	0·00	6974
923	... ...	+ 4·9569	− 0·0640	...	− 10·817	− 0·604	...	...
924	... .	+ 3·4001	− 0·0098	...	+ 10·872	− 0·412	...	...
925	Lalande 39086 ...	+ 3·3065	− 0·0100	...	− 11·128	− 0·408	...	...
926	α Pavonis	+ 4·7906	− 0·0594	0·000	− 11·189	− 0·594	+ 0·10	7004
927	X Capricorni Var. 7...	+ 3·4001	− 0·0101	...	− 11·144	− 0·407	...	...
928	Lalande 39196 ...	+ 3·3940	− 0·0101	...	− 11·184	− 0·408	...	...
929	Lacaille 8441 ...	+ 3·7396	− 0·0192	...	− 11·380	− 0·444	...	...
930	11 Capricorni ρ ...	+ 3·4321	− 0·0115	− 0·006	− 11·536	− 0·408	+ 0·01	7032
931	... ...	+ 2·9970	− 0·0031		− 11·771	− 0·340	...	...
932	Lalande 39525 ...	+ 2·9074	− 0·0031		− 11·885	− 0·347	...	...
933	... ...	+ 3·7170	− 0·0200		− 12·010	− 0·439	...	...
934	... ...	+ 4·8974	− 0·0742		− 12·275	− 0·560	...	...
935	S Capricorni Var. 2...	+ 3·4488	− 0·0128		− 12·486	− 0·395	...	...
936	... ...	+ 3·7722	− 0·0231	...	− 12·621	− 0·428	...	..
937	50 Cygni α (*Deneb*)...	+ 2·0482	+ 0·0021	− 0·002	− 12·677	− 0·236	0·00	7171
938	2 Aquarii ε	+ 2·3622	− 0·0084	− 0·001	− 12·915	− 0·367	+ 0·01	7196
939	W. B. E. XX. 1024 ..	+ 3·3638	− 0·0109	...	− 12·964	− 0·367	...	...
940	... ...	+ 3·3611	− 0·0109	...	− 12·971	− 0·367	...	...
941	T Aquarii Var. 4 ..	+ 3·1723	− 0·0066		− 13·079	− 0·316		...
942	Lacaille 8571 .	+ 4·6413	− 0·0787		− 13·089	− 0·529		...
943	. .	+ 3·7786	− 0·0246		− 13·137	− 0·410		...
944	Taylor 9688 .	+ 3·2617	− 0·0093	...	− 13·158	− 0·356		7232
945	.	+ 4·7380	− 0·0744	.	− 13·300	− 0·507		

926. Proper motions from "*Stone's Cape Catalogue*."

Mean Positions of Stars for 1865 January 1st.

Number.	Star.	Magnitude.	Estimations.	Mean Right Ascension.			Mean Polar Distance.			Observations.	Fraction of Year.
				h.	m.	s.	°	′	″		
946	32 Vulpeculæ ...	5·1	...	20	48	46·29	62	27	15·5	6	0·71
947	... ...	9·7	1	20	50	45·82	140	46	30·4	1	0·72
948	Lacaille 8680 ...	9·2	1	20	51	40·11	136	37	48·0	2	0·74
949	Lacaille 8686 ...	7·8	2	20	52	22·67	126	34	50·3	2	0·67
950	... ...	9·5	1	20	54	1·91	142	59	0·6	1	0·73
951	Taylor 9772	8·0	1	21	0	31·56	145	7	8·2	1	0·69
952	61 Cygni ... ... 1st...	5·5	...	21	0	50·71	51	54	49·7	3	0·73
953	... ...	9·5	1	21	2	58·89	145	6	29·5	1	0·72
954	Lacaille 8712	8·2	2	21	4	15·03	140	48	18·2	3	0·73
955	64 Cygni 3	3·5	...	21	7	11·38	60	19	32·4	5	0·74
956	... ...	10·5	1	21	8	20·18	110	40	43·3	1	0·73
957	... ...	9·6	2	21	11	5·07	120	31	42·1	2	0·72
958	Brisbane 7012	7·0	1	21	14	9·52	151	46	46·1	1	0·75
959	T Capricorni Var. 3 ...	8·8	2	21	14	38·80	105	39	56·0	2	0·69
960	... ...	9·5	3	21	15	3·85	130	15	59·2	3	0·74
961	5 Cephei α (*Alderamin*)...	2·6	...	21	15	21·19	27	38	6·6	2	0·74
962	... ...	9·3	1	21	18	33·01	163	32	19·1	1	0·75
963	Taylor 9981	6·0	2	21	18	46·05	142	53	8·0	2	0·70
964	... ...	8·6	2	21	20	10·30	150	47	35·5	2	0·76
965	... ...	9·7	2	21	23	1·14	110	7	13·9	2	0·70
966	22 Aquarii β ...	3·1	...	21	24	27·02	96	0	49·3	7	0·73
967	... ...	8·2	1	21	26	53·37	140	33	11·1	1	0·76
968	8 Cephei β ...	3·4	...	21	26	51·68	20	1	58·4	1	0·76
969	... ...	9·0	1	21	29	34·90	164	2	16·0	1	0·78
970	.. ..	9·4	2	21	33	56·38	103	0	7·8	2	0·78
971	Taylor 10068	7·6	1	21	34	21·78	134	6	27·0	1	0·76
972	Taylor 10096	6·4	1	21	34	32·49	115	6	51·4	1	0·70
973	.. ..	9·0	1	21	34	46·65	134	0	11·5	1	0·77
974	T. Cephei Var. 3	8·0	3	21	36	30·64	11	89	1·8	3	0·78
975	8 Pegasi ε	2·4	..	21	37	33·27	80	44	34·0	4	0·72
976	μ Cephei Var. 3 ...	4·0	...	21	39	22·67	31	50	19·5	1	0·75
977	.. ... .	10·0	2	21	40	51·92	102	32	9·3	3	0·70
978	Taylor 10136 . .	7·0	1	21	41	3·40	137	11	9·7	1	0·74
979	. .	9·1	1	21	42	30·67	132	31	9·2	1	0·76
980	16 Pegasi .	5·0	...	21	46	55·34	61	42	33·3	3	0·74

959.—T Capricorni Var. 3.—Period 269 days. Range, 9th to below 14th magnitude.
965.—Observed by mistake for Doris.
974.—δ Cephei Var. 3.—Period 385 days. Range, 5th to 11·5 magnitude.
976.—μ Cephei Var. 1.—Changes irregularly from 4th to 6th magnitude.
977.—Observed by mistake for Urania.

Observed with the Madras Meridian Circle in that Year.

Number.	Star.	In Right Ascension. Annual Precession.	In Right Ascension. Secular Variation.	In Right Ascension. Proper Motion.	In Polar Distance. Annual Precession.	In Polar Distance. Secular Variation.	In Polar Distance. Proper Motion.	Number in B. A. C.
		s	s	s	"	"	"	
946	32 Vulpeculae ..	+ 2·5554	+ 0·0026	− 0·002	− 13·472	− 0·270	0·00	7286
947	... .	+ 4·6934	− 0·0730	...	− 13·396	− 0·407	...	...
948	Lacaille 8530 ..	+ 3·8001	− 0·0272	...	− 13·656	− 0·400	...	...
949	Lacaille 8633 ..	+ 3·7908	− 0·0272	.	− 13·701	− 0·396	...	...
950	... .	+ 4·3882	− 0·0553	...	− 13·607	− 0·456	...	...
951	Taylor 9772	+ 4·4086	− 0·0621	...	− 14·214	− 0·440	...	...
952	61 Cygni, 1st ..	+ 2·3338	+ 0·0044	+ 0·869	− 14·287	− 0·388	− 3·23	7336
953	... .	+ 4·4057	− 0·0626	...	− 14·364	− 0·448	...	...
954	Lacaille 8712 ..	+ 4·4900	− 0·0685	...	− 14·442	− 0·446	...	...
955	64 Cygni 3	+ 2·5505	+ 0·0038	− 0·003	− 14·610	− 0·348	+ 0·07	7369
956	... .	+ 3·4191	− 0·0140		− 14·687	− 0·338		..
957	... .	+ 3·8136	− 0·0320		− 14·850	− 0·368		...
958	Brisbane 7012 ...	+ 4·7206	− 0·0017		− 15·089	− 0·450		...
959	T Capricorni Var. 3	+ 3·3200	− 0·0120		− 15·046	− 0·314		...
960	... ...	+ 3·8196	− 0·0831		− 15·062	− 0·361		...
961	5 Cephei α (*Alderamin*)	+ 1·4161	− 0·0071	+ 0·021	− 15·090	− 0·130	− 0·01	7416
962	... ...	+ 4·8365	− 0·1061	...	− 15·367	− 0·449	...	...
963	Taylor 9961	+ 4·2151	− 0·0675	...	− 15·294	− 0·391	...	7448
964	... ...	+ 4·6091	− 0·0871	...	− 15·372	− 0·425	...	...
965	... ...	+ 3·3420	− 0·0147	...	− 15·531	− 0·305	...	...
966	22 Aquarii β	+ 3·1626	− 0·0071	− 0·001	− 15·611	− 0·332	0·00	7475
967	... ...	+ 4·0788	− 0·0510	...	− 15·698	− 0·363	...	...
968	8 Cephei β	+ 0·8007	− 0·0845	0·000	− 15·744	− 0·066	+ 0·04	7493
969	... .	+ 3·8611	− 0·0894	...	− 15·897	− 0·337	...	...
970	... .	+ 3·2559	− 0·0106	...	− 16·118	− 0·276	...	...
971	Taylor 10049 ...	+ 3·8411	− 0·0696	...	− 16·146	− 0·386	...	7538
972	Taylor 10066 ..	+ 4·2090	− 0·0949	...	− 16·149	− 0·367	...	7540
973	... .	+ 3·8363	− 0·0894	...	− 16·101	− 0·394	...	...
974	8 Cephei Var. 2 ...	− 0·6115	− 0·1623	...	− 16·268	+ 0·089	...	...
975	8 Pegasi ε	+ 2·9462	− 0·0005	+ 0·003	− 16·306	− 0·292	0·00	7561
976	μ Cephei Var. 2 .	+ 1·8281	+ 0·0080	...	− 16·207	− 0·147	...	[illegible]
977	... ...	+ 3·3417	− 0·0104	...	− 16·471	− 0·363	...	...
978	Taylor 10196	+ 3·8988	− 0·0451	...	− 16·481	− 0·317	...	7601
979	... .	+ 3·7623	− 0·0872	...	− 16·574	− 0·302	...	...
980	16 Pegasi	+ 2·7366	+ 0·0062	+ 0·001	− 16·789	− 0·210	+ 0·01	7627

Mean Positions of Stars for 1863 January 1st.

Number.	Star.	Magnitude.	Estimations.	Mean Right Ascension.			Mean Polar Distance.			Observations.	Fraction of Year.
				h.	m.	s.	°	′	″		
981	Lacaille 8968 ...	7·6	3	21	47	16·02	135	53	3·5	3	0·72
982	... ...	9·0	2	21	47	28·79	133	12	13·7	2	0·77
983	Taylor 10190 ...	6·3	...	21	51	10·18	146	31	30·0	2	0·77
984	.. ..	9·8	1	21	52	50·47	166	37	55·9	1	0·72
985	... ...	9·3	1	21	53	54·17	150	40	0·7	1	0·73
986	α¹ Indi	6·0	...	21	56	20·63	150	17	16·4	4	0·73
987	... ...	7·0	2	21	56	15·89	136	2	17·3	2	0·78
988	34 Aquarii α	3·2	...	21	58	50·80	90	58	30·0	6	0·77
989	α Gruis	1·9	...	21	59	42·69	137	36	48·7	1	0·75
990	... ...	9·5	2	22	3	22·81	101	8	37·2	2	0·75
991	... ...	8·0	3	22	5	27·52	101	5	20·6	3	0·74
992	W. B. R. XXII. 99 ...	7·0	2	22	6	24·63	90	26	29·8	2	0·70
993	... ...	8·5	2	22	9	12·07	146	27	1·4	2	0·73
994	43 Aquarii θ	4·3	...	22	9	42·46	98	27	16·2	10	0·76
995	... ...	7·7	2	22	12	31·16	150	37	32·4	2	0·78
996	... ...	9·0	1	22	13	13·22	146	28	18·1	1	0·78
997	... ...	8·8	2	22	15	23·86	82	47	7·3	2	0·71
998	... ...	10·2	1	22	16	2·54	82	42	16·1	1	0·70
999	... ...	9·5	2	22	16	55·12	135	58	6·7	2	0·77
1000	... ...	9·3	1	22	18	54·56	140	46	27·7	1	0·76
1001	56 Aquarii 3 ... 1st...	4·7	...	22	21	52·66	90	42	34·8	1	0·88
1002	56 Aquarii 3 ... 2nd...	4·5	...	22	21	52·81	90	42	38·3	3	0·80
1003	... ...	9·0	1	22	21	54·89	100	37	20·7	1	0·76
1004	57 Aquarii σ	4·8	...	22	23	30·03	101	32	5·9	2	0·68
1005	R. P. L. 180	5·4	...	22	26	24·73	4	34	33·8	3	0·57
1006	. ...	9·8	1	22	24	21·46	135	41	31·5	1	0·77
1007	. .. .	8·0	1	22	25	58·75	141	29	56·5	3	0·74
1008	62 Aquarii η . .	4·2	...	22	28	25·06	90	49	46·6	9	0·74
1009	T Aquarii Var. 3 ..	10·8	2	22	28	48·85	95	18	13·9	2	0·70
1010	Lacaille 9188	6·6	...	22	30	57·42	130	33	24·6	1	0·76
1011	Taylor 10487 .. .	6·1	...	22	32	11·34	145	7	31·4	3	0·80
1012	. . .. .	9·0	3	22	34	16·12	135	31	1·2	3	0·76
1013	42 Pegasi ζ .	3·6	...	22	34	40·74	70	58	21·7	11	0·76
1014	. .	9·1	1	22	36	31·00	130	36	28·8	1	0·48
1015	Lacaille 9236 . -	6·5		22	37	42·80	146	46	16·9	1	0·67

992.—Comparison star for Comet 2, 1862.
1005.—Groombridge 3950.
1009.—T Aquarii Var. 3. Period unknown. Supposed to change between 9th and 11th magnitudes.

Observed with the Madras Meridian Circle in that Year.

Number.	Star.	In Right Ascension. Annual Precession.	In Right Ascension. Secular Variation.	In Right Ascension. Proper Motion.	In Polar Distance. Annual Precession.	In Polar Distance. Secular Variation.	In Polar Distance. Proper Motion.	Number in B. A. C.
		s	s	s	"	"	"	
981	Lacaille 8968 ...	+ 3·8369	− 0·0428	...	− 16·765	− 0·300	...	...
982	... ...	+ 3·7676	− 0·0368	...	− 16·808	− 0·292	...	...
983	Taylor 10120 ...	+ 4·1489	− 0·0696	...	− 16·980	− 0·316	...	7645
984	... .	+ 3·8177	− 0·0441	...	− 17·040	− 0·296	...	...
985	... .	+ 4·2837	− 0·0872	...	− 17·102	− 0·328	...	...
986	ϵ^2 Indi	+ 4·2758	− 0·0945	...	− 17·206	− 0·314	...	7669
987	... ...	+ 3·7748	− 0·0480	...	− 17·298	− 0·272	...	...
988	34 Aquarii α	+ 3·0886	− 0·0041	− 0·008	− 17·318	− 0·219	+ 0·02	7688
989	α Gruis	+ 3·8061	− 0·0457	+ 0·011	− 17·366	− 0·270	+ 0·16	7692
990	... ...	+ 3·2004	− 0·0088	...	− 17·514	− 0·219	...	...
991	... .	+ 3·1977	− 0·0092	...	− 17·602	− 0·216	...	...
992	W. B. E. XXII. 98...	+ 3·0768	− 0·0037	...	− 17·642	− 0·205	...	...
993	... .	+ 4·0098	− 0·0681	...	− 17·756	− 0·264	...	...
994	43 Aquarii θ	+ 3·1640	− 0·0075	+ 0·006	− 17·777	− 0·206	+ 0·08	7773
995	... .	+ 4·1457	− 0·0848	...	− 17·890	− 0·265	...	...
996	... ...	+ 3·0796	− 0·0077	...	− 17·917	− 0·263	...	...
997	... ...	+ 2·9975	0·0000	...	− 18·008	− 0·186	...	...
998	... ..	+ 2·9971	+ 0·0001	...	− 18·027	− 0·184	...	...
999	.. ...	+ 3·6731	− 0·0422	...	− 18·060	− 0·226	...	...
1000	.. ...	+ 3·7710	− 0·0516	...	− 18·136	− 0·237	...	...
1001	55 Aquarii ζ 1st...	+ 3·0700	− 0·0033	+ 0·009	− 18·245	− 0·176	− 0·08	7832
1002	55 Aquarii ζ 2nd..	+ 3·0700	− 0·0033	+ 0·009	− 18·245	− 0·176	− 0·08	7833
1003	... ...	+ 3·1762	− 0·0085	...	− 18·246	− 0·180	...	...
1004	57 Aquarii σ... ...	+ 3·1880	− 0·0089	− 0·004	− 18·308	− 0·182	− 0·06	7850
1005	R. P. L. 150 ...	+ 3·7344	− 1·1741	+ 0·048	− 18·306	+ 0·232	− 0·06	7851
1006	... ...	+ 3·6973	− 0·0412	...	− 18·384	− 0·206	...	...
1007	... .	+ 3·7486	− 0·0527	...	− 18·380	− 0·210	...	...
1008	63 Aquarii κ	+ 3·0794	− 0·0031	+ 0·008	− 18·473	− 0·166	+ 0·06	7869
1009	T Aquarii Var. 3 ...	+ 3·1477	− 0·0072	...	− 18·498	− 0·170	...	...
1010	Lacaille 9188 ...	+ 3·5108	− 0·0388	...	− 18·527	− 0·188	...	...
1011	Taylor 10477 ..	+ 3·6780	− 0·0709	...	− 18·601	− 0·208	...	7889
1012	... .	+ 4·1150	− 0·1067	...	− 18·668	− 0·218	...	..
1013	42 Pegasi ζ	+ 2·9862	+ 0·0028	+ 0·001	− 18·688	− 0·149	0·00	7898
1014	.. .. .	+ 3·4779	− 0·0327	...	− 18·741	− 0·173	.	.
1015	Lacaille 9236	+ 3·7886	− 0·0622	.	− 18·776	− 0·183	.	

989. Proper motions from "*Stone's Cape Catalogue*"

Mean Positions of Stars for 1865 January 1st.

Number.	Star.	Magnitude.	Estimations.	Mean Right Ascension. h	m	s	Mean Polar Distance. °	′	″	Observations.	Fraction of Year.
1016	... ...	8·9	1	22	40	55·80	142	37	44·4	1	0·79
1017	... ...	8·3	3	22	44	46·02	130	0	38·7	3	0·75
1018	... ..	9·6	2	22	44	47·72	146	32	44·2	2	0·75
1019	... ..	9·0	1	22	44	54·10	[illegible]	[illegible]	13·9	1	0·78
1020	... ...	9·0	1	22	49	16·76	135	27	36·6	1	0·75
1021	24 Pis. Aust. α (*Fomalhaut*)	1·3	...	22	50	11·14	120	20	13·9	3	0·82
1022	... ...	8·1	2	22	50	22·70	110	50	46·2	2	0·84
1023	.. ..	9·3	2	22	51	30·08	151	32	36·1	2	0·75
1024	Lacaille 9353	6·6	...	22	53	30·79	144	41	15·1	2	0·76
1025	54 Pegasi α (*Markab*) ...	2·6	...	22	58	2·22	76	31	15·3	12	0·77
1026	... ...	8·6	2	22	59	23·90	150	21	47·8	2	0·82
1027	R Pegasi Var. 2	10·2	3	22	59	52·33	80	11	6·3	3	0·76
1028	... ...	9·6	2	23	4	21·85	130	48	57·6	2	0·77
1029	Lacaille 9394	8·0	2	23	5	13·40	145	50	20·2	2	0·81
1030	Lacaille 9405	8·0	2	23	7	30·10	150	26	46·9	2	0·80
1031	... ...	8·3	1	23	8	11·74	150	30	09·3	1	0·87
1032	Lacaille 9428	7·0	1	23	9	58·87	151	44	15·2	1	0·87
1033	6 Piscium γ	3·8	...	23	10	9·96	87	27	17·7	12	0·78
1034	... ...	9·8	2	23	11	0·47	137	26	16·5	2	0·80
1035	... ...	9·6	2	23	12	16·51	127	24	32·9	2	0·80
1036	... ...	8·0	2	23	15	20·80	130	46	06·2	3	0·80
1037	Taylor 10768	5·7	1	23	17	36·38	147	35	25·4	1	0·87
1038	8 Piscium κ	5·0	...	23	20	0·80	[illegible]	[illegible]	59·0	13	0·79
1039	... ...	7·0	2	23	20	46·01	[illegible]	51	[illegible]	2	0·87
1040	... ...	9·6	1	23	21	6·06	137	27	39·8	1	0·82
1041	... .	8·6	2	23	23	40·31	[illegible]	57	16·2	2	0·75
1042	... .	9·6	1	23	25	32·92	[illegible]	51	40·3	1	0·87
1043	Taylor 10804 ...	6·9	1	23	27	33·08	[illegible]	34	17·0	1	0·86
1044	... ...	8·7	2	23	27	49·24	[illegible]	14	27·2	2	0·73
1045	.. ... —	8·5	3	23	29	53·13	[illegible]	51	39·1	3	0·77
1046	.. ..	9·1	1	23	29	57·98	137	19	40·3	1	0·87
1047	... .	8·5	2	23	30	29·20	[illegible]	08	22·7	2	0·77
1048	17 Piscium ι	4·3	...	23	33	0·38	[illegible]	6	19·1	11	0·78
1049	Lacaille [illegible]	5·7	...	23	34	[illegible]	[illegible]	44	38·1	1	0·79
1050	19 Piscium	5·2		23	39	39·86	87	15	40·6	2	0·85

1027.—R Pegasi Var. 2.—Period 382 days. Range, 7th to 13th magnitude.

Observed with the Madras Meridian Circle in that Year.

Number.	Star.	In Right Ascension. Annual Precession	In Right Ascension. Secular Variation.	In Right Ascension. Proper Motion.	In Polar Distance. Annual Precession.	In Polar Distance. Secular Variation.	In Polar Distance. Proper Motion.	Number in B. A. C.
		s	s	s	"	"	"	
1016	...	+ 3·0840	− 0·0591	..	− 18·873	− 0·162	...	...
1017	...	+ 3·4911	− 0·0817	...	− 18·984	− 0·164	...	...
1018	...	+ 3·7001	− 0·0804	...	− 18·984	− 0·166	...	...
1019	...	+ 3·7762	− 0·0807	...	− 18·980	− 0·160	...	..
1020	..	+ 3·4817	− 0·0948	...	− 19·107	− 0·147	...	...
1021	21 Pis. Aust. α (*Fom.*)	+ 3·3067	− 0·0210	+ 0·022	− 19·132	− 0·135	+ 0·18	7992
1022	.. .. .	+ 3·2250	− 0·0140	...	− 19·137	− 0·138	...	...
1023	... .	+ 3·7987	− 0·0796	...	− 19·166	− 0·165	...	...
1024	Lacaille 9353 ..	+ 3·5872	− 0·0669	...	− 19·294	− 0·185	...	8089
1025	54 Pegasi α (*Markab*)	+ 2·9798	+ 0·0056	+ 0·008	− 19·306	− 0·107	+ 0·02	8094
1026	... ...	+ 3·6683	− 0·0728		− 19·368	− 0·130		...
1027	R Pegasi Var. 2 ..	+ 3·0121	+ 0·0031		− 19·369	− 0·106		...
1028	... .	+ 3·3496	− 0·0805		− 19·467	− 0·109		...
1029	Lacaille 9394 .	+ 3·5395	− 0·0571		− 19·494	− 0·114		...
1030	Lacaille 9405 ...	+ 3·6072	− 0·0703		− 19·531	− 0·111		...
1031	... .	+ 3·6021	− 0·0704	...	− 19·545	− 0·110	...	...
1032	Lacaille 9423 ..	+ 3·6103	− 0·0743	...	− 19·579	− 0·106	...	8101
1033	6 Piscium γ	+ 3·0501	+ 0·0005	+ 0·047	− 19·588	− 0·087	+ 0·01	8106
1034	... .	+ 3·2985	− 0·0264	...	− 19·601	− 0·093	...	...
1035	... .	+ 3·2986	− 0·0263	...	− 19·622	− 0·087	...	...
1036	... ...	+ 3·2958	− 0·0296	...	− 19·675	− 0·086	...	...
1037	Taylor 10748	+ 3·4591	− 0·0392	...	− 19·713	− 0·086	...	8157
1038	8 Piscium κ	+ 3·0989	0·0000	+ 0·005	− 19·731	− 0·069	+ 0·12	8169
1039	...	+ 3·0670	+ 0·0003	...	− 19·762	− 0·068	...	...
1040	... ..	+ 3·3182	− 0·0375	...	− 19·766	− 0·074	...	...
1041	... ...	+ 3·4230	− 0·0604		− 19·803	− 0·070		...
1042	. ...	+ 3·2303	− 0·0275		− 19·809	− 0·062		...
1043	Taylor 10904 ..	+ 3·3070	− 0·0656		− 19·864	− 0·060		8208
1044	... .	+ 3·8744	− 0·0672		− 19·867	− 0·060		...
1045	... .	+ 3·8627	− 0·0598		− 19·[illegible]	− 0·065		.
1046	. ... ..	+ 3·2617	− 0·0360	...	− 19·[illegible]	− 0·053	...	..
1047	... .	+ 3·3573	− 0·0398	...	− 19·[illegible]	− 0·051	...	.
1048	17 Piscium ι	+ 3·0385	+ 0·0030	+ 0·025	− 19·917	− 0·042	+ 0·45	[illegible]
1049	Lacaille 9583 .	+ 3·1707	− 0·0349	...	− 19·970	− 0·034	...	
1050	19 Piscium ...	+ 3·0064	+ 0·0021	− 0·004	− 19·975	− 0·031	+ 0·03	8263

Mean Positions of Stars for 1866 January 1st.

Number.	Star.	Magnitude.	Estimations.	Mean Right Ascension.			Mean Polar Distance.			Observations.	Fraction of Year.
				h	m	s	°	′	″		
1051	... ...	9·7	...	23	41	7·84	126	46	20·3	1	0·79
1052	δ Sculptoris	4·6	...	23	41	53·37	118	52	26·0	5	0·82
1053	... ...	8·7	1	23	42	6·82	150	49	37·0	1	0·56
1054	Lalande 46660	9·2	4	23	42	0·12	88	18	54·5	4	0·79
1055	... ...	9·6	1	23	47	46·66	126	50	37·6	1	0·74
1056	... ...	9·1	2	23	50	0·41	148	53	5·1	2	0·78
1057	... ...	8·8	1	23	51	46·98	152	20	18·6	1	0·76
1058	28 Piscium ω ...	4·2	...	23	52	22·70	83	58	3·3	0	0·79
1059	Lacaille 9696 ...	9·8	...	23	53	35·67	148	50	56·2	1	0·75
1060	... ...	9·3	2	23	56	1·55	130	16	41·6	2	0·79
1061	... ...	8·0	1	23	56	10·82	124	7	26·1	2	0·58
1062	Taylor 10994	8·4	3	23	57	50·51	157	26	41·0	3	0·78
1063	Lacaille 9721	5·7	...	23	59	10·01	139	40	24·1	2	0·75

Observed with the Madras Meridian Circle in that Year.

Number	Star.	In Right Ascension.			In Polar Distance.			Number in B. A. C.
		Annual Precession.	Secular Variation.	Proper Motion.	Annual Precession.	Secular Variation.	Proper Motion.	
		s	s	s	"	"	"	
1051	.. ..	+ 3·1004	− 0·0344	..	− 19·982	− 0·029	...	...
1052	δ Sculptoris	+ 3·1202	− 0·0161	+ 0·000	− 19·982	− 0·026	+ 0·10	8275
1053	. ... ..	+ 3·2563	− 0·0569	..	− 19·984	− 0·033		...
1054	Lalande 46950 ..	+ 3·0980	+ 0·0018	.	− 19·984	− 0·030		...
1055	... ...	+ 3·1395	− 0·0267	...	− 20·006	− 0·015		...
1056	.. .. ..	+ 3·1067	− 0·0516		− 20·006	− 0·011	.	...
1057	. ... ..	+ 3·1063	− 0·0660	...	− 20·042	− 0·008	..	...
1058	28 Piscium ω ..	+ 3·0672	+ 0·0047	+ 0·010	− 20·044	− 0·005	+ 0·13	8031
1059	Lacaille 9693 ..	+ 3·1283	− 0·0407	...	− 20·047	− 0·004	...	...
1060	. . ..	+ 3·0918	− 0·0340	...	− 20·052	+ 0·001		...
1061	. .	+ 3·0872	− 0·0185	...	− 20·052	+ 0·001		...
1062	Taylor 10994 . .	+ 3·0920	− 0·0495	..	− 20·054	+ 0·005		.
1063	Lacaille 9721 ..	+ 3·0769	− 0·0586	..	− 20·056	+ 0·008	..	..

1052.—Proper motions from "*Stone's Cape Catalogue.*"

SEPARATE RESULTS

OF

OBSERVATIONS

OF THE FIXED STARS,

MADE WITH THE

MADRAS MERIDIAN CIRCLE

IN THE YEAR

1866.

Separate Results of Madras Meridian Circle Observations in 1866.

Number and Date.	Magnitude.	Mean Right Ascension 1866. h.	m.	s.	No. of Wires.	Mean Polar Distance 1866. °	′	″	Observer.
1		*Taylor* 11010.							
Nov. 12	8·0	0	0	31·89	...	147	31	58·6	R
2		21 *Andromedae* α *Alpherat.*							
Sep. 22	—	0	1	27·90	...	61	38	59·3	R
Oct. 10	...		1	27·94	...		38	59·6	M
11	...		1	27·86	...		38	59·5	R
27	...		1	27·80	...		38	58·5	R
29	...		1	27·90	...		38	59·3	R
31	...		1	27·86	5		38	58·7	R
Nov. 8	...		1	27·92	...		38	59·3	R
14	...		1	27·91	...		38	59·5	R
3		*Lacaille* 9739.							
Nov. 5	8·0	0	2	10·16	...	130	28	50·1	R
4		*Lacaille* 9757.							
Nov. 6	8·0	0	4	31·16	...	131	7	7·8	R
5		88 *Pegasi* γ, *Algenib.*							
Oct. 30	...	0	6	20·26	...	75	33	41·3	R
Nov. 3	...		6	20·19	...		33	41·5	R
7	...		6	20·20	...		33	41·9	R
9	...		6	20·23	...		33	41·8	R
10	...		6	20·23	...		33	42·2	R
14	...		6	20·25	...		33	41·9	R
19	...		6	20·22	...		33	39·8	R
6		*Anon.*							
Nov. 6	9·2	0	6	44·46	...	131	6	22·3	R
7		*Lalande* 163.							
Oct. 9	7·7	0	7	44·39	...	89	36	54·2	M
10	7·9		7	44·33	...		36	54·1	M
12	7·1		7	44·60	...		36	54·7	M
8		*Anon.*							
Nov. 5	9·0	0	9	39·83	5	109	31	11·3	R
9		*Anon.*							
Nov. 8	9·9	0	9	39·06	...	163	54	27·9	R
10		*Anon.*							
Oct. 31	9·9	0	10	40·71	...	132	0	11·7	R
Nov. 12	9·2		10	40·51	...		0	13·6	R
11		*Lacaille* 41.							
Sep. 26	8·0	0	12	39·73	...	130	51	28·4	M
Oct. 27	8·2		12	39·65	...		51	23·4	R
12		*Anon.*							
Oct. 30	7·5	0	15	54·62	...	51	59	24·9	R
13		*Lacaille* 61.							
Nov. 3	7·5	0	16	7·58	...	180	0	18·7	R
6	7·0		16	7·56	...		0	19·8	R
14		*Anon.*							
Oct. 27	9·3	0	18	26·29	5	102	56	39·3	R
Nov. 5	9·0		18	26·81	...		56	36·4	R
15		*Lacaille* 81.							
Nov. 6	7·1	0	18	44·20	...	120	0	2·3	R
16		10 *Ceti.*							
Aug. 27	—	0	19	44·90	...	90	47	32·1	M
28	...		19	45·04	—		47	33·2	M
17		12 *Ceti.*							
Oct. 11	—	0	23	11·98	...	94	41	55·1	M
31	...		23	12·00	...		41	54·1	R
Nov. 10	...		23	11·96			41	53·6	R
12	...		23	11·98	...		41	53·7	R
14	...		23	11·97	..		41	52·6	R
19	. .		23	11·91	...		41	51·6	R
22	. .		23	12·04	...		41	51·0	R

Separate Results of Madras Meridian Circle Observations in 1866.

Number and Date.	Magnitude.	Mean Right Ascension 1866. h. m. s.	No. of Wires.	Mean Polar Distance 1866. ° ′ ″	Observer.
18		*Lacaille* 132.			
Oct. 9	8·0	0 27 23·90	5	151 53 16·8	M
12	7·8	27 24·28	...	53 17·8	M
Nov. 5	9·0	27 23·10	...	53 16·3	R
6	8·9	27 23·93	5	53 13·8	R
19		*Anon.*			
Oct. 5	8·5	0 31 32·34	...	98 27 41·6	M
Nov. 3	8·5	31 32·38	...	27 41·1	R
20		*Lalande* 1010.			
Oct. 30	9·5	0 32 21·73	...	92 31 46·8	R
Nov. 7	9·5	32 21·64	...	31 46·5	R
21		18 *Cassiopeae* α *Var.* 2, *Shedir.*			
Nov. 8	...	0 32 25·14	...	34 11 32·9	R
9	...	32 26·19		11 33·6	R
22		16 *Ceti* β.			
Oct. 31	...	0 36 51·86	..	108 40 22·6	R
Nov. 10	...	36 51·64	...	40 21·8	R
13	...	36 51·73	...	40 22·5	R
19	...	36 51·88	...	43 19·9	R
Dec. 6	...	36 51·38	...	40 22·8	M
23		*W. B. E.* 0·628.			
Nov. 22	9·3	0 37 0·17	...	95 04 46·3	R
24		*Anon.*			
Nov. 26	9·5	0 38 39·46	5	139 41 13·0	R
25		*W. B. E.* 0·697.			
Oct. 9	9·1	0 40 40·57	...	95 13 51·4	M
12	9·6	40 40·65	...	13 51·1	M
27	9·5	40 40·46	...	13 52·1	R
26		*W. B. E.* 0·705.			
Nov. 3	9·0	0 41 14·24	...	91 36 39·5	R
7	9·0	41 14·38	...	36 40·1	R
27		63 *Piscium* δ			
Sep. 24	...	0 41 44·83	...	83 8 41·0	M
26	...	41 44·04	...	8 41·5	M
Nov. 19	...	41 45·36	4	8 40·0	R
28		*W. B. E.* 0·716.			
Nov. 27	9·5	0 41 47·30	...	94 36 7·8	R
29		20 *Ceti.*			
Aug. 27	5·6	0 46 9·38	..	91 52 22·2	M
30		*Anon.*			
Oct. 26	...	0 47 41·06	6	163 30 41·0	R
Nov. 8	10·0	47 40·94	5	30 39·6	R
22	9·7	47 40·98	5	30 34·8	R
31		λ¹ *Toucani.*			
Nov. 3	6·5	0 48 2·02	...	153 36 0·1	R
32		*Anon.*			
Nov. 9	9·6	0 49 6·15	4	163 40 10·1	R
26	9·5	49 6·32	5	40 7·8	R
33		*Anon.*			
Nov. 7	9·0	0 49 9·60	5	138 48 26·4	R
34		*Lacaille* 264.			
Nov. 27	...	0 50 40·06	6	154 41 26·0	R
35		2 *Ursae Minoris*—*s.p.*			
May 3	..	0 50 57·68	3	4 27 49·6	W
4	...	50 58·30	3	27 47·4	W
36		*R. P. L.* 14—*s.p.*			
Apl 19		0 51 13·98	3	3 34 2·1	R
20	...	51 14·94	3	34 13·0	R

Separate Results of Madras Meridian Circle Observations in 1866.

Number and Date.	Magnitude.	Mean Right Ascension 1866. h. m. s.	No. of Wires.	Mean Polar Distance 1866. ° ′ ″	Observer.
37		*71 Piscium ε*			
Sep. 24	.	0 56 50·30	...	82 40 54·7	M
25	...	56 50·42	...	40 56·2	M
Oct. 10	...	56 50·34	...	40 51·7	M
Nov. 12	...	56 50·41	...	40 55·5	R
20	...	56 50·51	...	40 59·5	R
38		*20 Ceti.*			
Nov. 27	...	1 1 5·11	...	88 43 27·6	R
39		*Anon.*			
Oct. 9	7·0	1 5 50·81	...	120 52 26·1	M
29	9·0	5 51·05	...	52 26·3	R
40		*86 Piscium ζ (1st)*			
Aug. 28	...	1 6 48·79	...	83 8 4·1	M
41		*S Cassiopeae Var. 4*			
Nov. 27	...	1 9 50·73	3	19 3 48·0	R
42		*1 Ursae Minoris α, Polaris.*			
Oct. 25	...	1 9 56·92	3	1 34 19·4	R
Nov. 8	...	9 57·30	1	34 17·6	R
11	...	9 57·04	1	34 17·6	R
26	.	9 57·81	1	34 15·6	R
		1 Ursae Minoris α, Polaris.—s.p.			
Apl 14	...	1 9 57·21	3	1 34 15·3	M
43		*Anon.*			
Nov. 22	9·2	1 10 11·78	...	81 49 4·0	R
44		*Anon.*			
Aug. 28	...	1 17 6·38	4	86 30 47·1	M

Number and Date.	Magnitude.	Mean Right Ascension 1866. h. m. s.	No. of Wires.	Mean Polar Distance 1866. ° ′ ″	Observer.
45		*45 Ceti θ¹*			
Nov. 5	.	1 17 19·50	...	98 52 33·1	R
6	..	17 19·30	...	52 34·3	R
26	..	17 19·46	...	52 31·5	R
27		17 19·40	...	52 34·3	R
30		17 19·52	...	52 34·8	R
Dec. 6		17 19·40	...	52 34·3	M
12	...	17 19·61	..	52 32·3	M
46		*Taylor 465.*			
Oct. 10	7·1	1 19 36·26	...	91 5 46·4	M
47		*R Piscium Var. 1.*			
Oct. 30	9·6	1 23 46·46	...	87 45 41·3	R
Nov. 22	8·3	23 46·33	...	45 42·2	R
26	...	23 46·36		45 42·2	R
48		*90 Piscium υ*			
Nov. 6	...	1 24 15·98	—	75 20 46·0	R
7	...	24 15·92	...	20 46·3	R
8	...	24 15·91	...	20 45·6	R
27	...	24 15·98	...	20 46·2	R
30	...	24 15·97	...	20 46·3	R
49		*Anon.*			
Dec. 8	7·0	1 29 5·24	4	109 41 54·6	M
50		*Taylor 526.*			
Oct. 30	7·0	1 30 13·51	.	149 40 30·2	R
Dec. 10	6·0	30 13·48	.	40 30·6	M
51		*Anon.*			
Nov. 8	9·0	1 31 29·14		130 51 40·5	R
52		*α Eridani, Achernar.*			
Nov. 7	...	1 32 49·21	.	147 56 7·8	R
28		32 48·29	—	56 5·0	R

Separate Results of Madras Meridian Circle Observations in 1866.

Number and Date.	Magnitude.	Mean Right Ascension 1860. h. m. s.	No. of Wires.	Mean Polar Distance 1860. ° ′ ″	Observer.
53		106 *Piscium* ν			
Aug 28	...	1 34 27·53	...	85 11 30·6	M
Oct. 23		34 27·48	...	11 30·8	M
Nov. 6	..	34 27·57	...	11 30·1	R
9	...	34 27·36	...	11 30·5	R
22	...	34 27·50	...	11 29·8	R
27	...	34 27·51	...	11 29·4	R
28		34 27·62	...	11 30·0	R
30	...	34 27·54	...	11 30·0	R
Dec. 6		34 27·50	...	11 30·5	M
54		110 *Piscium* ο			
Sep. 25	...	1 38 19·23	...	81 31 4·9	M
28	.	38 19·16	...	31 8·2	M
55		*Anon.*			
Oct. 30	9·0	1 39 57·76	..	140 26 43·5	R
56		*Lacaille* 516.			
Oct. 31	...	1 40 2·34	...	151 41 30·3	R
Nov. 26	8·0	40 2·34	..	41 28·9	R
57		*Anon.*			
Oct. 29	9·5	1 42 2·08	...	130 14 38·6	R
Nov. 10	9·3	42 1·99	.	14 37·4	R
22	9·3	42 2·05	...	14 34·8	R
58		6 *Arietis* β			
Nov. 3	...	1 47 14·54	...	69 50 55·5	R
6	...	47 14·48	...	50 56·4	R
7	...	47 14·53	..	50 55·1	R
9	...	47 14·46	...	50 55·3	R
28	...	47 14·53	...	50 54·1	R
Dec. 12	...	47 14·63	...	50 56·8	M
59		*V Piscium Var.* 5.			
Oct. 25	7·8	1 47 17·96	...	81 52 49·5	R
27	7·7	47 17·72	5	52 47·6	R
Dec. 6	7·7	47 17·90	...	52 47·9	M
8	7·2	47 17·64	...	52 47·2	M
10	7·6	47 17·87	.	52 46·0	M
60		*Anon.*			
Oct. 30	9·3	1 48 26·99	...	150 4 37·7	R
61		*Lacaille* 508.			
Oct. 29	9·2	1 52 6·70	...	160 7 40·1	R
62		*Anon.*			
Oct. 27	9·0	1 54 57·51	...	139 56 5·7	R
30	8·5	54 57·67	5	56 5·9	R
63		*Taylor* 673.			
Oct. 31	7·5	1 56 21·89	6	73 28 31·8	R
Nov. 10	7·0	56 21·88	...	28 31·5	R
64		13 *Arietis* α			
Oct. 25	...	1 59 37·49	...	67 10 28·6	R
Nov. 7	...	59 37·48	...	10 24·0	R
13	...	59 37·35	...	10 23·5	R
20	...	59 37·46	...	10 20·1	R
28	...	59 37·34	...	10 29·6	R
Dec. 8	...	59 37·45	...	10 28·1	M
10	...	59 37·39	...	10 28·4	M
12	...	59 37·30	...	10 28·6	M
17	...	59 37·47	...	10 25·6	R
65		*Anon.*			
Nov. 22	9·5	2 1 7·28	...	148 49 28·4	R
66		*Anon.*			
Oct. 29	9·5	2 2 0·14	...	160 1 35·3	R
67		17 *Arietis* η			
Oct. 24	...	2 5 10·45	6	69 26 14·0	M
68		65 *Ceti* ξ¹			
Sep. 25	...	2 5 54·05	...	81 48 29·6	M
Dec. 17	...	5 53·90	...	47 0·5	R
18	.	5 54·06	...	47 1·3	R

Separate Results of Madras Meridian Circle Observations in 1866.

Number and Date.	Magnitude.	Mean Right Ascension 1866. h.	m.	s.	No. of Wires.	Mean Polar Distance 1866. °	′	″	Observer.
69				*Anon.*					
Oct. 30	9·5	2	7	2·81	...	148	39	55·3	R
70				*67 Ceti.*					
Oct. 25	...	2	10	18·00	5	97	2	30·0	R
29	...		10	17·93	...		2	29·6	R
Nov. 13	..		10	18·06	...		2	27·9	R
Dec. 8			10	18·01	...		2	28·5	M
10	...		10	18·01	3		2	29·1	M
18	...		10	18·08	5		2	30·9	R
71				*68 Ceti o, Var. 1, Mira.*					
Oct. 27	9·8	2	12	34·31	5	93	35	17·3	R
31	10·0		12	34·82	5		35	16·7	R
Dec. 17	...		12	34·74	...		35	17·9	R
72				*Anon.*					
Dec. 12	8·4	2	12	42·41	...	98	35	3·9	M
73				*Anon.*					
Oct. 30	9·0	2	15	48·84	...	152	38	33·5	R
74				*Anon.*					
Nov. 3	8·3	2	16	26·73	5	151	17	48·4	R
Dec. 13	7·9		16	28·09	6		17	50·6	M
75				*Anon.*					
Nov. 10	..	2	20	14·56	5	146	32	10·1	R
Dec. 8	8·0		20	14·51	5		32	10·1	M
76				*73 Ceti. ξ²*					
Oct. 24	–	2	21	2·18		89	8	32·5	M
25			21	2·17	...		8	33·2	R
Nov. 13			21	2·20	—		8	31·5	R
Dec. 10	...		21	2·25			8	31·9	R
17	...		21	2·20	5		8	32·4	R
18	...		21	2·20	.		8	33·1	R
77				*Anon.*					
Nov. 3	9·5	2	24	18·21	...	152	25	7·1	R
8	9·5		24	18·23	...		25	6·5	R
78				*Anon.*					
Oct. 30	9·0	2	24	31·81	...	117	3	13·4	R
79				*Anon.*					
Dec. 6	7·9	2	27	28·97	4	147	11	43·5	M
80				*Anon.*					
Nov. 13	9·5	2	30	49·02	..	117	31	23·7	R
81				*Anon.*					
Dec. 18	9·3	2	31	19·08	...	151	34	53·4	R
82				*Anon.*					
Nov. 3	9·3	2	34	5·36	...	74	26	2·1	R
6	9·3		34	5·62	..		26	3·3	R
83				*Lacaille 840 (1st).*					
Nov. 8	8·0	2	36	3·50	...	130	8	29·0	R
84				*Lacaille 849 (2nd).*					
Jan. 3	8·3	2	36	8·71	...	130	8	42·2	M
Dec. 8	7·9		36	8·65	...		8	44·5	M
85				*86 Ceti γ.*					
Dec. 12	..	2	36	21·26	5	87	19	51·7	M
13	...		36	21·30	..		19	51·7	M
86				*μ Ceti.*					
Sep. 28		2	37	41·97	—	80	27	13·3	M
Dec. 17			37	42·13			27	13·6	R

Separate Results of Madras Meridian Circle Observations in 1866.

Number and Date.	Magnitude.	Mean Right Ascension 1866. h. m. s.	No. of Wires.	Mean Polar Distance 1866. ° ′ ″	Observer.
87		*Lacaille* 868.			
Nov. 13	8·9	2 38 36·[illegible]	..	147 12 40·3	M
88		*W. B. N. II.* 676.			
Nov. 10	8·0	2 40 15·[illegible]	.	75 19 53·3	M
89		*Anon.*			
Jan. 6	9·4	2 41 45·85	...	151 2 11·3	M
90		*Anon.*			
Jan. 4	7·9	2 43 21·34	6	149 0 6·3	M
91		*Anon.*			
Nov. 3	9·2	2 44 32·16	5	149 13 19·2	M
8	9·3	44 32·46	...	13 19·2	M
Dec. 8	8·0	44 32·28	6	13 20·4	M
92		*Anon.*			
Dec. 17	9·2	2 46 22·20		76 27 25·2	M
93		*Lacaille* 941.			
Nov. 8	...	2 50 [illegible]	...	146 28 26·0	M
94		*Anon.*			
Jan. 6	8·0	2 52 25·08		150 16 28·6	M
95		91 *Ceti* λ			
Sep. 26	...	2 52 32·07	...	81 37 42·6	M
96		*Anon.*			
Nov. 27	9·2	2 53 10·[illegible]	5	146 48 57·8	M
97		*Lacaille* 969.			
Dec. 17	8·0	2 54 37·[illegible]		141 13 29·7	M
98		92 *Ceti* α, *Menkar.*			
Dec. 10	...	2 55 16·51	3	86 26 17·1	M
13	...	55 16·[illegible]	5	26 18·1	M
99		25 *Persei* ρ *Var.* 2.			
Nov. 13		2 56 35·92	.	51 40 [illegible]·8	M
100		*Taylor* 1037.			
Dec. 8	7·0	2 56 36·94	...	150 21 7·8	M
12	7·0	56 36·77	...	21 6·4	M
101		*Anon.*			
Jan. 8	..	2 [illegible] 34·15	...	151 19 32·6	M
Nov. 27	8·0	[illegible] 33·96	5	19 31·7	M
102		57 *Arietis* δ.			
Dec. 13	..	3 3 [illegible]	...	70 46 59·1	M
17	...	3 [illegible] 12	...	46 [illegible]·5	M
18	...	3 [illegible]·14	...	[illegible] [illegible]·5	M
19	..	3 [illegible] 19	4	46 [illegible]·5	M
103		*Taylor* 1081.			
Jan. 10	7·8	3 5 15·21	5	151 30 40·8	M
11	7·6	5 16·37	4	30 40·9	M
104		*Taylor* 1092.			
Nov. 13	7·0	3 7 18·42	...	148 19 1·7	M
27	7·8	7 18·35	5	18 [illegible]·9	M
105		*Anon.*			
Dec. 7	8·0	3 10 22·69	...	131 48 44·6	M
18	8·7	10 22·60	5	48 41·4	M
106		*Taylor* 1127.			
Dec. 16	8·6	3 11 31·76	5	131 46 38·9	M
107		*Anon.*			
Oct. 24	8·5	3 12 27·91	.	130 57 46·0	M

Separate Results of Madras Meridian Circle Observations in 1866.

Number and Date.	Magnitude.	Mean Right Ascension 1866. h.	m.	s.	No. of Wires.	Mean Polar Distance 1866. °	′	″	Observer.
108				*Anon.*					
Jan. 5	8·2	3	12	46·64	...	130	49	43·5	M
12	8·4		12	46·70	5		49	49·7	M
109				33 *Persei* α.					
Dec 13	...	3	14	46·68	5	40	37	7·6	M
110				*Anon.*					
Nov. 13	...	3	14	53·76	5	150	5	53·7	R
Dec. 17	9·5		14	53·92	5		5	53·8	R
111				ζ² *Reticuli.*					
Nov. 27	...	3	13	13·58	5	153	1	7·3	R
112				*Anon.*					
Jan. 15	7·3	3	20	37·21	...	140	28	5·3	M
Dec. 7	7·0		20	36·96	6		28	6·8	R
113				*Anon.*					
Nov. 27	9·5	3	20	47·06	5	54	47	20·9	M
114				*R Persei Var.* 3.					
Dec. 22	9·7	3	21	31·88	6	54	47	36·8	R
115				*R. P. L.* 34.—*s.p.*					
May 1	...	3	22	52·92	2	3	46	55·9	R
3	...		22	58·14	3		46	54·2	R
116				5 *Tauri f*					
Oct. 26	...	3	23	29·64	...	77	31	30·0	M
28	...		23	29·55	...		31	31·3	R
117				*Lacaille* 1148.					
Jan 5	5·7	3	27	3·08	5	126	24	20·7	M
118				*Lacaille* 1149.					
Jan. 6	7·0	3	28	40·12		120	14	41·8	M
8	...		28	39·95	...		13	43·1	M
Nov. 27	8·0		28	39·91	5		13	40·7	R
Dec. 18	8·0		28	39·90	...		14	45·8	R
119				*Anon.*					
Dec. 7	6·8	3	30	15·71	...	131	50	36·1	M
22	7·0		30	15·80	5		50	38·3	R
120				*Lacaille* 1159.					
Jan. 15	6·5	3	30	18·78	5	151	28	12·0	M
121				*Lacaille* 1192.					
Nov. 20	9·0	3	36	1·42	...	167	48	21·1	R
122				*Anon.*					
Jan. 10	9·0	3	35	22·98	...	120	12	55·6	M
11	8·8		35	22·94	...		12	54·5	M
123				*Taylor* 1256.					
Jan. 5	7·9	3	35	29·61	5	130	12	22·0	M
124				*Lacaille* 1200.					
Oct. 24	6·9	3	36	27·19	...	166	40	8·7	M
125				*Anon.*					
Jan 6	9·0	3	38	9·92	...	128	13	30·7	M
Dec. 8	9·0		38	9·96			13	31·6	M
126				*Anon.*					
Nov. 27	8·0	3	39	27·28	...	66	30	8·1	R
127				25 *Tauri* η, *Alcyone.*					
Dec. 7		3	39	31·27		66	16	44·1	M
18			39	31·25			16	44·2	R
22			39	31·30			16	43·4	R

Separate Results of Madras Meridian Circle Observations in 1866.

Number and Date.	Magnitude.	Mean Right Ascension 1866. h. m. s.	No. of Wires.	Mean Polar Distance 1866. ° ′ ″	Observer.
128		*Anon.*			
Jan. 3	8.6	3 45 18.38	..	76 27 38.9	M
129		*Anon.*			
Jan. 8	8.0	3 46 26.09	4	146 38 17.2	M
130		*Anon.*			
Jan. 6	7.7	3 48 5.30	4	160 40 51.2	M
10	8.0	48 6.03	...	40 55.7	M
11	8.0	48 5.90	...	40 51.7	M
131		34 *Eridani* γ^1			
Jan. 12	...	3 51 46.73	...	103 53 30.9	M
Nov. 26	...	51 46.73	...	53 28.7	R
Dec 7	...	51 46.73	...	53 30.9	M
8	...	51 46.61	...	53 30.6	M
132		*Anon.*			
Nov. 27	9.9	3 53 9.39	...	128 55 1.1	R
133		35 *Tauri* λ *Var.* 1.			
Oct. 24	...	3 53 15.42		77 53 27.2	M
25	...	53 15.46		53 28.1	R
134		*Lacaille* 1327.			
Dec. 10	6.0	3 51 20.29	.	158 51 8.9	M
135		*R. P. L.* 35.			
Dec 6	..	3 53 27.24	2	4 48 12.9	M
22	...	53 27.30	3	48 13.7	R
		R. P. L. 35—*s.p.*			
May 28		3 53 28.90	3	4 48 12.1	R
136		*Lacaille* 1347.			
Jan 6	7.1	3 58 9.32		150 2 13.5	R
137		*Lalande* 7561.			
Jan. 5	8.0	3 58 20.57	...	71 52 0.8	M
8	8.0	58 20.52		52 1.8	M
138		*Lacaille* 1359.			
Jan. 15	7.9	4 0 3.62	...	147 40 44.8	M
139		*Lacaille* 1375.			
Jan. 13		4 2 57.10	...	109 50 29.8	R
140		*Lalande* 7764.			
Dec. 7	7.9	4 3 36.08	...	74 46 32.6	M
141		38 *Eridani* o^1			
Jan. 4	.	4 5 10.59	...	97 11 21.1	M
10	...	5 10.58	...	11 22.0	M
11	...	5 10.66	...	11 21.7	M
Nov. 23	...	5 10.43	.	11 19.7	R
Dec. 8	—	5 10.66	...	11 22.5	M
142		*Anon.*			
Jan. 5	8.0	4 9 20.94	...	140 30 51.0	M
6	8.0	9 20.08		30 51.5	M
15	7.7	9 20.90	...	30 53.4	M
143		*Anon.*			
Jan. 8	8.1	4 9 30.73		120 16 28.2	M
16	9.0	9 30.81	.	16 28.9	R
144		*Lacaille* 1425.			
Nov 27	6.2	4 13 2.08		162 31 45.8	R
145		*Anon.*			
Dec. 10	8.9	4 13 54.70	..	70 51 13.1	M
146		*U Tauri Var.* 7.			
Jan 19	10.0	4 14 0.70	—	70 30 28.3	R
Dec 28	10.0	14 0.68		30 28.3	R

Separate Results of Madras Meridian Circle Observations in 1866.

Number and Date.	Magnitude.	Mean Right Ascension 1866. h.	m.	s.	No. of Wires.	Mean Polar Distance 1866. °	′	″	Observer.
147		61 *Tauri* δ¹							
Nov. 22	...	4	15	12·56	5	72	46	26·5	R
148		*Anon.*							
Jan. 13	...	4	15	44·32	...	128	30	33·4	R
149		*Anon.*							
Jan. 11	8·0	4	16	48·11	...	149	4	8·9	M
150		74 *Tauri* ε							
Jan. 6	...	4	20	47·60	...	71	7	11·6	M
10	...		20	47·67	...		7	11·5	M
12	...		20	47·73	...		7	11·2	M
20	...		20	47·61	...		7	12·9	R
26	...		20	47·64	5		7	13·5	R
Dec. 7	...		20	47·64	..		7	12·3	M
20	...		20	47·68	...		7	10·5	R
22	...		20	47·61	...		7	12·2	R
151		*R Tauri Var.* 2.							
Jan. 16	9·3	4	20	57·34	4	80	8	22·3	M
19	9·3		20	57·27	...		8	22·5	R
152		*Anon.*							
Jan. 19	10·2	4	23	8·78	5	80	27	48·8	R
153		*Anon.*							
Jan. 17	10·2	4	22	29·65	5	80	20	50·1	R
154		*Lacaille* 1519.							
Jan. 16	7·0	4	25	37·01	...	156	5	51·5	M
155		*Lacaille* 1520.							
Jan. 24	8·6	4	26	46·66	...	117	28	41·9	R
156		87 *Tauri* α, *Aldebaran.*							
Jan. 5	...	4	28	13·98	...	73	46	47·2	M
10	...		28	14·01	...		46	49·2	M
11	...		28	13·98	...		46	46·6	M
12	...		28	14·01	...		46	48·1	M
18	...		28	14·09	...		46	47·8	R
26	...		28	13·99	...		45	49·5	R
Dec. 7	...		28	13·90	...		46	49·4	M
20	...		28	14·00	...		46	46·8	R
157		*Anon.*							
Jan. 28	5·2	4	29	30·89		140	14	2·8	R
158		*Anon.*							
Jan. 16	9·1	4	31	46·76		112	50	20·4	M
159		*Anon.*							
Jan. 18	9·7	4	32	0·99	.	100	45	50·2	R
26	9·5		33	1·04	...		45	49·2	R
160		*Anon.*							
Dec. 17	8·5	4	33	41·76	5	67	31	51·3	R
161		*Anon.*							
Jan. 5	9·0	4	33	36·67	5	139	51	27·4	M
162		*Anon.*							
Jan. 15	8·7	4	34	32·69	5	101	26	30·0	R
163		96 *Tauri.*							
Jan. 22	7·0	4	35	7·14	...	66	10	6·6	R
Dec. 19	7·4		35	7·01	...		10	5·8	M
164		*Lacaille* 1567.							
Jan. 8	8·6	4	35	12·84		156	20	28·3	M
165		*Lacaille* 1566.							
Jan. 24	7·8	4	35	[illegible]		168	28	9·4	R

Separate Results of Madras Meridian Circle Observations in 1866.

Number and Date.	Magnitude.	Mean Right Ascension 1866. h.	m.	s.	No. of Wires.	Mean Polar Distance 1866. °	′	″	Observer.
166			*Anon.*						
Jan. 28	9·7	4	36	26·88	...	64	19	0·9	n
167			*Lacaille* 1582.						
Jan. 6	7·6	4	37	34·19	...	152	38	29·3	n
17	7·2		37	30·37	5		38	28·8	n
19	7·8		37	30·06	..		38	27·7	n
168			*Anon.*						
Jan. 10	9·3	4	39	34·53	...	128	44	49·6	n
169			*Anon.*						
Jan. 16	8·9	4	40	21·05	..	151	30	41·8	n
170			*Lacaille* 1596.						
Jan. 18	7·6	4	41	41·30	..	128	21	22·9	n
22	7·8		41	41·73	5		21	25·4	n
25	7·9		41	41·69	...		21	24·7	n
171			*ε Doradûs.*						
Jan. 24	6·5	4	42	39·41	...	149	58	44·3	n
172			*Anon.*						
Jan. 17	9·7	4	43	34·81		100	41	0·4	n
28	9·8		43	34·67			41	0·0	n
173			*Anon.*						
Jan. 19	9·7	4	45	33·92	...	129	34	49·9	n
174			*Anon.*						
Jan. 15	7·7	4	46	57·41	...	133	3	51·4	M
175			*Lacaille* 1656.						
Jan. 8	7·9	4	47	58·08	4	149	1	49·1	M
Feb. 6	7·3		47	57·90	.		1	49·3	M
176			*3 Aurigae ι.*						
Jan. 4	...	4	48	16·47	..	57	2	37·6	M
5	...		48	16·13	...		2	38·0	M
6	...		48	16·33	...		2	38·3	M
12	...		48	16·09	..		2	39·2	M
27	...		48	16·16	...		2	39·6	n
177			*Taylor* 1761.						
Jan. 17	7·8	4	50	3·81	4	129	18	29·2	n
Feb. 5	8·0		50	3·65	4		18	26·0	M
Dec. 16	6·9		50	3·51	...		18	27·3	M
178			*Anon.*						
Jan. 16	10·5	4	51	32·72	4	92	8	36·3	n
18	10·7		51	32·60	5		8	32·5	n
179			*7 Aurigae ε Var. 1.*						
Jan. 23	3·5	4	52	21·42	...	46	22	44·8	n
24	3·0		52	21·38	...		22	48·2	n
Dec. 17	4·0		52	21·35	5		22	44·5	n
180			*R Leporis Var. 1.*						
Jan. 25	8·2	4	53	39·45	...	105	0	38·1	n
28	8·4		53	39·62	...		0	37·9	n
30	8·5		53	39·44	...		0	38·3	n
181			*Anon.*						
Jan. 18	8·9	4	53	51·04	3	158	0	36·3	M
19	8·5		53	50·74	...		0	32·6	n
Dec. 8	7·8		53	50·71	..		0	30·8	M
182			*Taylor* 1797.						
Dec. 7	6·9	4	54	53·12	...	148	14	46·9	M
183			*Anon.*						
Jan. 17	9·3	4	55	0·56	5	130	17	31·8	n
184			*Lacaille* 1697.						
Feb. 8	7·7	4	56	35·16		129	6	39·3	M

Separate Results of Madras Meridian Circle Observations in 1866.

Number and Date.	Magnitude.	Mean Right Ascension 1866. h. m. s.	No. of Wires.	Mean Polar Distance 1866. ° ′ ″	Observer.
185		*Taylor* 1811.			
Dec. 18	5·8	4 57 6·90	...	120 54 54·8	R
186		2 *Leporis* ε			
Jan. 6	...	4 50 47·26	...	112 38 12·2	M
8	...	50 47·36	...	38 12·9	M
26	...	50 47·46	...	38 12·6	R
27	...	50 47·45	...	38 12·7	R
Feb. 1	...	50 47·42	...	38 11·9	R
6	...	50 47·30	...	38 12·7	M
Dec. 22	...	50 47·54	...	38 12·8	R
187		*Anon.*			
Jan. 19	9·0	5 0 3·52	...	138 23 44·4	R
24	8·8	0 3·42	6	23 46·0	R
Feb. 7	8·0	0 3·54	...	23 46·1	M
188		*Anon.*			
Jan. 17	9·7	5 1 45·81	5	151 36 59·9	R
23	9·7	1 45·74	5	36 59·6	R
189		*Lacaille* 1756.			
Jan. 19	8·0	5 4 0·46	5	151 43 33·0	R
30	8·5	4 0·22	...	43 33·4	R
190		*Anon.*			
Jan. 23	9·6	5 4 37·64	...	138 34 30·3	R
191		*Lacaille* 1757.			
Jan. 29	8·2	5 4 54·14	...	160 3 12·3	R
Feb. 3	7·7	4 54·27		3 12·7	M
192		13 *Aurigae* α, *Capella*.			
Jan. 26	...	5 6 47·58	...	44 8 32·8	R
27		6 47·60	...	8 33·4	R
193		*Anon.*			
Jan. 19	9·0	5 6 36·03	...	130 5 56·3	R
Feb. 8	8·0	6 36·17	...	5 54·1	M
194		19 *Orionis* β, *Rigel*.			
Jan. 8	...	5 8 5·91	...	98 21 36·4	M
13	...	8 5·96	...	21 34·7	R
Dec. 19	...	8 5·99	...	21 33·4	R
195		*Anon.*			
Jan. 10	9·8	5 11 0·04	...	139 49 29·9	R
17	9·8	10 59·98	...	49 22·1	R
23	9·5	10 59·98	...	49 28·1	R
Feb. 10	9·0	10 59·92	5	49 28·8	M
196		*Anon.*			
Jan. 22	9·7	5 12 1·38	...	139 46 39·9	R
23	9·5	12 1·39	4	46 39·7	R
24	9·7	12 1·30	...	46 42·4	R
197		*Anon.*			
Jan. 18	9·7	5 12 59·92	...	137 4 37·7	R
19	9·3	12 59·91		4 37·5	R
198		*Anon.*			
Jan. 29	8·7	5 13 28·44	4	138 41 35·6	R
Feb. 5	8·0	13 28·75	...	41 34·2	M
199		*Anon.*			
Jan. 27	8·3	5 14 51·86	...	163 29 13·0	R
Feb. 12	8·0	14 51·88	5	29 11·9	M
200		*Lacaille* 1822.			
Jan. 30	8·0	5 15 46·73		141 43 6·5	R
201		*Anon.*			
Jan. 25	8·0	5 17 29·94	...	136 7 14·4	R

Separate Results of Madras Meridian Circle Observations in 1866.

Number and Date.	Magnitude.	Mean Right Ascension 1866. h.	m.	s.	No. of Wires.	Mean Polar Distance 1866. °	′	″	Observer.
202			*112 Tauri β*						
Jan. 8	...	5	17	49·40	...	61	30	31·9	M
11	...		17	49·48	...		30	33·9	M
15	...		17	49·51	...		30	31·2	M
16	...		17	49·32	...		30	35·0	R
17	...		17	49·34	...		30	35·3	R
20	...		17	49·34	...		30	34·0	R
25	...		17	49·32	...		30	34·8	R
Feb. 1	...		17	49·31	...		30	33·6	R
Dec. 19	...		17	49·29	5		30	34·5	R
203			*R. P. L. 40—s.p.*						
June 26	...	5	19	21·52	5	4	53	50·8	R
204			*Lacaille 1854.*						
Jan. 18	8·9	5	21	42·08	...	137	12	46·9	R
205			*Anon.*						
Jan. 10	10·3	5	21	50·13	5	50	40	53·6	R
206			*Anon.*						
Jan. 28	7·0	5	23	34·02	6	152	43	6·0	R
207			*Anon.*						
Jan. 22	7·7	5	23	31·47	...	151	13	20·9	R
Feb. 7	7·6		23	31·36	5		13	20·8	M
208			*λ Doradûs.*						
Jan. 30	6·0	5	24	22·00	...	149	1	36·1	R
209			*34 Orionis δ Var. 1.*						
Jan. 11		5	25	9·72	...	90	24	3·9	M
13			25	9·67	...		24	6·4	M
15			25	9·68	...		24	6·8	M
16			25	9·73			24	6·7	M
17			25	9·73			24	6·1	R
24			25	9·73			24	5·9	R
Feb. 1	...		25	9·69			24	4·4	R
210			*Anon.*						
Feb. 5	8·8	5	26	15·56	...	160	35	15·9	M
16	8·3		26	15·59	...		35	16·2	M
211			*Anon.*						
Jan. 29	8·9	5	26	26·07	...	166	51	13·4	R
212			*11 Leporis α*						
Jan. 6	...	5	26	40·14	...	107	56	14·6	M
20	...		26	40·30	...		56	15·4	R
23	...		26	40·26	6		56	14·9	R
Feb. 14	...		26	40·16	...		56	15·3	R
213			*Taylor 2057.*						
Feb. 15	7·1	5	28	2·65	...	161	56	31·1	M
214			*Lalande 10582.*						
Jan. 22	8·3	5	29	16·29	...	78	16	8·3	R
31	9·0		29	16·16	...		16	7·9	R
Feb. 3	8·0		29	16·28	...		16	6·5	M
215			*46 Orionis ε*						
Jan. 16	...	5	29	24·96	...	91	17	26·0	R
17	...		29	25·01	5		17	26·8	R
Feb. 1	...		29	24·98	...		17	25·6	R
216			*123 Tauri ζ*						
Jan. 26	...	5	29	39·66	...	68	14	28·8	R
27	...		29	39·27	...		14	28·6	R
Nov. 23	..		29	39·19	...		14	29·9	R
217			*Anon.*						
Jan. 23	9·7	5	29	49·91	5	125	21	59·6	R
Feb. 6	9·8		29	49·84			21	57·9	M
218			*Anon.*						
Jan. 19	7·9	5	31	1·08		139	12	58·2	R
Dec. 18	8·0		31	0·99	5		12	59·1	R

Separate Results of Madras Meridian Circle Observations in 1866.

Number and Date.	Magnitude.	Mean Right Ascension 1866. h. m. s.	No. of Wires.	Mean Polar Distance 1866. ° ' "	Observer.
279		*Anon.*			
Jan. 18	9·3	5 31 41·90	6	128 42 10·0	R
280		*Lacaille* 1949.			
Feb. 7	5·9	5 32 16·40	...	154 19 1·5	M
281		*Anon.*			
Jan. 10	8·6	5 32 44·12	5	150 11 20·6	R
30	8·3	33 46·07	...	11 30·2	R
282		*Anon.*			
Jan. 18	9·2	5 33 46·00	4	128 41 10·8	R
283		*Anon.*			
Jan. 26	9·3	5 34 21·51	...	152 7 55·7	R
Feb. 5	8·9	34 21·54	...	7 53·1	M
8	8·0	34 21·51	...	7 55·7	M
284		α *Columbae.*			
Jan. 4	...	5 34 47·81	...	124 8 50·0	M
15	...	34 47·96	...	8 51·7	M
17	...	34 47·88	...	8 52·3	R
24	...	34 47·80	...	8 52·2	R
285		*Lacaille* 1971.			
Feb. 16	7·1	5 36 23·44	...	140 11 28·0	M
286		*Anon.*			
Jan. 16	9·3	5 36 47·44	...	129 57 50·1	R
287		*Anon.*			
Feb. 3	9·0	5 38 25·71	.	130 5 21·5	M
288		*Anon.*			
Jan. 21	8·7	5 39 17·81	..	79 0 11·3	R
289		*Taylor* 2145.			
Jan. 23	6·5	5 39 53·16	...	135 53 44·3	R
290		*Anon.*			
Jan. 20	9·0	5 39 56·87	...	135 48 5·0	R
291		*W. B. E. V.* 1011.			
Jan. 22	8·0	5 40 35·61	6	78 57 41·1	R
25	8·3	40 35·62	...	57 40·1	R
Feb. 6	8·5	40 35·61	4	57 39·1	M
292		*Lacaille* 1984.			
Feb. 10	7·6	5 40 45·06	...	130 15 18·7	M
293		*Anon.*			
Feb. 5	7·0	5 42 10·19	...	135 3 30·9	M
7	7·6	42 10·31	5	3 31·2	M
294		*Anon.*			
Jan. 19	9·1	5 43 29·67	5	130 6 32·3	R
295		54 *Orionis* χ^1			
Jan. 26	...	5 46 20·91	...	69 45 3·1	R
296		*Anon.*			
Feb. 13	9·0	5 47 4·94	5	135 46 46·6	M
297		*Lalande* 11166.			
Jan. 23	8·0	5 47 8·21	...	77 32 27·2	R
24	8·0	47 8·18	...	32 29·8	R
30	7·9	47 8·16	..	32 27·9	R
298		58 *Orionis* = *Var.* 2, *Betelgeux.*			
Feb. 1	.	5 47 35·04	...	82 37 15·1	R
8		47 34·98	...	37 16·1	R

Separate Results of Madras Meridian Circle Observations in 1866.

Number and Date.	Magnitude.	Mean Right Ascension 1866. h.	m.	s.	No. of Wires.	Mean Polar Distance 1866. °	′	″	Observer.
239 *Anon.*									
Feb 10	7·9	5	40	57·28	...	136	46	6·9	M
240 *Lacaille* 2073.									
Feb. 13	7·7	5	50	34·02	...	137	12	37·4	R
241 *Anon.*									
Feb. 5	9·5	5	51	0·87	...	130	40	58·4	M
242 *Anon.*									
Jan. 25	9·0	5	52	46·07	...	129	32	33·5	R
Feb. 15	8·6		52	46·05	5		32	33·2	M
243 *R. P. L.* 43.									
Feb. 3	...	5	52	51·67	3	3	11	19·5	M
244 *Anon.*									
Feb. 7	7·8	5	56	20·74	...	131	7	14·1	M
245 *Lalande* 11455.									
Jan. 20	.	5	55	56·17	...	78	19	12·3	R
23	7·3		55	56·10	...		19	13·1	R
24	7·7		55	56·20	...		19	12·5	R
246 *Anon.*									
Feb. 6	8·7	5	56	13·48	5	129	57	13·1	M
247 *Taylor* 2301.									
Feb. 8	6·9	5	58	31·73	...	149	6	15·0	M
248 *Taylor* 2310.									
Feb. 12	6·9	5	59	39·60	.	130	29	3·6	M
249 *Anon.*									
Feb. 5	8·0	5	59	44·92	...	129	49	48·7	M
10	7·9		59	44·73	...		49	49·5	M
25	8·0		59	44·90	...		49	48·8	M
250 67 *Orionis* ν									
Jan. 8	...	5	59	56·21	...	75	13	7·9	M
20	...		59	56·20	...		13	7·8	R
Feb. 24	...		59	55·30	...		13	6·2	M
251 *Anon.*									
Feb. 7	7·1	6	0	51·27	...	137	28	38·0	M
14	7·6		0	51·08	...		28	38·5	M
252 *Anon.*									
Feb. 15	8·1	6	2	28·03	...	128	44	41·5	M
253 *Lalande* 11732.									
Jan. 4	8·5	6	3	29·30	...	77	58	56·1	M
6	8·4		3	29·36	..		58	55·1	M
Feb. 3	8·0		3	29·47	..		58	54·8	M
254 *Anon.*									
Feb. 16	7·1	6	4	26·88	...	128	2	38·3	M
255 *Anon.*									
Jan. 24	...	6	5	44·83	...	77	51	30·5	R
25	9·2		5	44·86	...		51	31·9	R
256 *Anon.*									
Jan. 27	...	6	7	29·05	...	131	19	28·8	R
257 *Anon.*									
Feb. 7	8·6	6	8	58·73	...	131	54	47·3	M
258 *Anon.*									
Feb. 5	9·0	6	8	37·28	...	130	31	34·9	M
259 *Anon.*									
Feb. 8	9·6	6	9	31·01	...	131	20	52·9	M
12	9·0		9	31·00	4		20	50·9	M

Separate Results of Madras Meridian Circle Observations in 1866.

Number and Date.	Magnitude.	Mean Right Ascension 1860. h. m. s.	No. of Wires.	Mean Polar Distance 1860. ° ′ ″	Observer.
260		*Lalande* 12053.			
Feb. 14	8·8	6 12 31·20	...	68 51 21·2	M
16	8·0	12 31·15	...	51 18·6	R
20	8·8	12 31·20	6	51 18·8	R
261		*Lalande* 12094.			
Feb. 15	8·8	6 13 42·31	...	68 42 3·7	M
17	9·0	13 42·22	...	42 3·8	R
19	8·9	13 42·16	...	42 3·6	R
262		*Lalande* 12120.			
Jan. 3	7·0	6 14 12·89	...	77 4 40·7	M
263		13 *Geminorum* μ			
Jan. 15	...	6 14 51·15	...	67 25 14·2	M
25	...	14 51·25	...	25 17·4	R
30	...	14 51·22	...	25 16·9	R
Feb. 6	...	14 51·18	...	25 17·2	M
13	...	14 51·17	...	25 16·5	R
264		*Lalande* 12155.			
Jan. 6	7·8	6 15 5·21	...	77 22 2·1	M
12	7·6	15 5·41	5	22 2·8	M
265		*Lacaille* 2278.			
Feb. 5	8·0	6 17 4·49	5	153 56 27·7	M
266		*Anon.*			
Feb. 7	8·7	6 18 34·06	...	151 28 28·8	M
267		*Taylor* 2485.			
Feb. 12	7·9	6 18 34·25	...	151 16 10·9	M
268		*Anon.*			
Feb. 17	9·5	6 19 47·34	5	65 30 19·0	R
269		α *Argûs, Canopus.*			
Jan. 30		6 20 39·98		142 38 25·4	R
270		18 *Geminorum* ν			
Jan. 27	...	6 21 0·35	...	69 42 23·6	R
271		*Lacaille* 2312.			
Feb. 8	7·0	6 23 9·89	..	138 36 49·8	M
272		*Anon.*			
Feb. 11	8·4	6 22 10·61	...	129 48 47·6	M
273		*Lacaille* 2329.			
Feb. 9	6·9	6 24 6·06	...	136 44 38·0	M
20	7·0	24 6·02	5	44 35·4	R
274		*Taylor* 2541.			
Feb. 5	6·1	6 24 57·21	...	147 55 4·0	M
275		*Lacaille* 2348.			
Feb. 26	7·3	6 26 38·08	...	132 3 48·2	M
276		*Anon.*			
Feb. 7	8·7	6 28 29·30	...	130 55 53·1	R
277		*Taylor* 2580.			
Jan. 12	6·9	6 29 50·72	...	151 46 56·7	R
Feb. 12	6·7	29 50·68		46 56·4	M
13	6·7	29 50·73	...	46 54·7	M
278		24 *Geminorum* γ.			
Jan. 3	...	6 29 58·39	...	73 30 23·2	M
5	...	29 59·35	...	30 23·2	R
16	...	29 59·16	...	30 25·0	R
22	...	29 58·17	...	30 24·1	R
23	...	29 58·30	...	30 24·2	M
25	..	29 58·23	...	30 24·9	R
27	...	29 59·16	...	30 24·0	R
30	.	29 58·30		29 23·9	R
Feb. 3		29 58·14	...	30 23·7	R
6		29 58·26	...	30 24·9	M
Dec. 28	—	29 58·21	5	30 26·6	R

Separate Results of Madras Meridian Circle Observations in 1860.

Number and Date.	Magnitude.	Mean Right Ascension 1860. h. m. s.	No. of Wires.	Mean Polar Distance 1860. ° ′ ″	Observer.
279		*Anon.*			
Feb. 3	8·9	6 31 29·08	5	140 0 17·2	M
16	8·9	31 28·84	...	0 19·1	M
280		*R Monocerotis Var. 1.*			
Feb. 13	10·4	6 31 50·98	6	81 9 0·0	R
17	10·5	31 50·98	4	8 59·4	R
19	10·6	31 50·93	5	8 53·3	R
20	10·4	31 50·86	5	8 56·9	R
22	...	31 51·19	5	8 56·8	R
281		*Anon.*			
Feb. 8	7·7	6 31 31·64	...	130 28 1·0	M
9	7·7	34 34·53	...	28 2·6	M
282		*Anon.*			
Feb. 21	9·2	6 35 51·82	...	130 38 9·6	R
26	8·4	35 51·42	5	38 10·3	M
283		*Taylor 2652.*			
Feb. 7	6·2	6 39 31·86	5	151 24 57·0	M
284		51 *Cephei*			
Jan. 5	...	6 36 39·52	3	2 45 25·2	M
13	...	35 33·08	2	45 24·3	M
23	...	36 39·61	3	45 25·8	R
Feb. 10	...	36 40·18	2	45 26·8	M
		51 *Cephei—s.p.*			
July 6	...	6 36 39·62	3	2 46 24·5	M
Aug. 9	...	36 40·85	3	46 25·9	R
285		*Lacaille 2451.*			
Feb. 3	8·0	6 39 11·31	...	126 57 48·9	M
286		*Anon.*			
Feb 15	8·4	6 39 11·69	5	131 8 31·3	M
287		9 *Canis Majoris* α, *Sirius.*			
Feb. 22	...	6 39 14·24	...	106 32 8·9	R
288		*Anon.*			
Jan. 3	8·0	6 40 28·22	5	131 1 25·7	M
Feb. 12	8·0	40 28·36	...	1 23·0	M
289		*Anon.*			
Feb. 26	9·2	6 40 36·41	5	151 56 55·3	R
290		*Anon.*			
Feb. 6	8·7	6 42 27·39	...	100 57 4·2	M
8	8·1	42 27·39	...	57 4·0	M
9	8·1	42 27·46	...	57 4·8	M
291		*Anon.*			
Feb. 14	9·0	6 42 41·01	5	109 56 14·2	M
16	9·0	42 41·17	...	56 14·9	M
292		*Anon.*			
Feb. 24	8·7	6 42 41·27	...	109 36 29·7	R
293		*Anon.*			
Feb. 7	7·6	6 43 46·01	...	128 30 32·1	M
294		*Anon.*			
Feb. 23	9·3	6 44 53·54	...	128 30 26·7	M
295		*Taylor 2724.*			
Jan. 12	6·6	6 44 56·98	...	141 36 11·2	M
296		*Anon.*			
Feb. 17	9·6	6 45 27·09	5	138 10 15·7	R
24	9·6	45 26·97	5	10 13·0	R

Separate Results of Madras Meridian Circle Observations in 1866.

Number and Date.	Magnitude.	Mean Right Ascension 1866. h. m. s.	No. of Wires.	Mean Polar Distance 1866. ° ′ ″	Observer.
297		*α Pictoris.*			
Jan. 4	5·6	6 46 48·98	...	151 47 54·9	M
Feb. 15	5·2	46 49·24	5	47 55·1	M
298		*Lacaille* 2532.			
Mar. 7	6·3	6 48 14·75	...	150 5 41·7	M
299		*Anon.*			
Feb. 17	9·5	6 48 54·62	6	130 10 28·0	M
25	9·3	48 54·55	6	10 28·8	R
300		*Anon.*			
Feb. 12	9·0	6 48 58·24	5	123 46 9·2	M
21	9·3	48 58·19	...	46 11·1	R
301		*Lacaille* 2538.			
Feb. 22	8·6	6 51 32·71	...	114 47 41·4	R
302		21 *Canis Majoris ε*			
Jan. 18	...	6 53 21·59	...	118 47 31·5	R
19	...	53 21·62	...	47 30·2	R
22	...	53 21·73	6	47 32·1	R
23	...	53 21·64	...	47 31·3	R
24	...	53 21·68	...	47 31·0	R
Feb. 3	...	53 21·61	...	47 29·9	M
5	...	53 21·73	...	47 28·1	M
6	...	53 21·61	...	47 30·6	M
13	...	53 21·59	...	47 31·7	R
303		*Anon.*			
Feb. 8	8·6	6 53 51·60	...	120 47 42·6	M
304		*Anon.*			
Jan. 12	8·0	6 55 29·61	.	120 36 4·7	M
Feb. 23	7·7	55 29·48	5	34 5·7	M
305		*Anon.*			
Jan. 20	9·3	6 56 22·28	—	120 17 31·6	R
306		23 *Canis Majoris γ*			
Jan. 4	...	6 57 41·71	..	105 26 15·4	M
5	...	57 41·69	...	26 15·3	M
18	...	57 41·64	5	26 16·6	R
19	...	57 41·77	—	26 15·9	R
24	...	57 41·81	...	26 17·1	R
29	—	57 41·66	...	26 16·0	R
Feb. 26	...	57 41·82	...	26 18·4	R
27	...	57 41·72	...	26 15·0	R
307		*Lalande* 13707.			
Jan. 27	8·2	6 59 13·35	...	67 6 50·8	R
Feb. 7	7·9	59 13·40	...	6 49·3	M
308		*R Geminorum Var.* 2.			
Jan. 26	7·6	6 59 17·17	...	67 5 37·8	R
31	7·7	59 17·17	...	5 36·5	R
Feb. 8	7·7	59 17·19	—	5 37·3	M
9	7·5	59 17·17	—	5 37·3	M
10	7·5	59 17·15	—	5 37·2	M
12	7·6	59 17·20	...	5 36·1	M
309		*Anon.*			
Feb. 13	9·3	7 0 20·54	...	80 30 16·6	R
21	8·8	0 20·65	...	30 16·3	R
310		*Taylor* 2851.			
Feb. 21	7·1	7 0 22·98	5	125 44 57·3	M
Mar. 10	7·6	0 22·20	5	44 59·2	M
311		*Anon.*			
Feb. 15	8·9	7 1 18·98	—	61 4 17·3	M
17	9·3	1 18·74	—	4 19·6	R
23	9·3	1 18·66	5	4 19·7	R
312		*R Canis Majoris Var.* 1.			
Feb. 19	9·5	7 1 20·28	6	79 36 2·5	R
20	9·5	1 20·25	—	36 3·7	R

Separate Results of Madras Meridian Circle Observations in 1866.

Number and Date.	Magnitude.	Mean Right Ascension 1866. h.	m.	s.	No. of Wires.	Mean Polar Distance 1866. °	′	″	Observer.
313		*W. B. E. VII.* 1920.							
Feb. 28	8.2	7	1	34·32	...	79	44	47·9	n
Mar. 1	8·3		1	31·48	...		44	46·3	n
314		*Anon.*							
Jan. 20	9·6	7	3	27·76	5	130	14	41·2	n
Feb. 5	8·5		3	27·70	...		14	39·3	m
315		*Anon.*							
Feb. 9	9·4	7	5	46·24	...	69	58	47·4	m
10	9·0		5	46·24	...		58	47·5	m
23	9·9		5	46·26	...		58	47·3	n
316		*Anon.*							
Feb. 27	...	7	6	48·30	5	109	2	36·8	n
317		*Taylor* 2923.							
Mar. 5	7·9	7	7	18·69	5	150	21	29·7	m
318		*Anon.*							
Jan. 29	9·1	7	8	57·54	5	139	17	31·8	n
Mar. 1	9·6		8	57·65	5		17	31·1	n
319		*Taylor* 2940.							
Mar. 6	7·9	7	9	32·27	5	120	57	34·6	m
320		*Anon.*							
Jan. 26	9·6	7	9	50·37	4	130	18	44·9	n
Mar. 1	9·8		9	50·31	5		18	46·6	n
321		54 *Geminorum* λ							
Jan. 29	.	7	10	28·44	...	73	13	16·1	n
322		55 *Geminorum* δ							
Jan. 13	...	7	12	7·15	5	67	46	29·9	n
29	...		12	7·11	...		46	27·6	n
30	...		12	7·11	...		46	29·5	n
31	...		12	7·06	...		46	28·5	n
Feb. 7	...		12	7·14	...		46	29·0	m
13	...		12	7·10	...		46	28·0	m
15	...		12	7·06	...		46	29·0	m
16	...		12	7·06	...		46	28·5	m
17	...		12	7·08	...		46	29·1	n
23	...		12	7·10	...		46	28·5	m
26	...		12	6·98	...		46	28·2	m
323		*Anon.*							
Mar. 12	8·0	7	12	37·26	5	152	48	8·9	m
324		*Anon.*							
Jan. 16	9·5	7	13	5·26	...	129	16	15·2	n
21	9·3		13	5·26	...		16	12·9	n
325		*Anon.*							
Jan. 12	8·0	7	14	33·26	...	123	49	49·4	m
Feb. 24	7·7		14	33·09	4		49	47·4	m
326		*Lacaille* 2805.							
Feb. 27	8·9	7	17	20·55	...	168	8	15·2	n
Mar. 5	7.9		17	20·60	...		8	18·4	m
327		*Anon.*							
Jan. 16	9·0	7	17	29·04	4	129	13	42·8	n
20	9·0		17	29·03	...		13	41·2	n
Feb. 24	8·3		17	28·96	...		13	39·6	n
Mar. 8	8·1		17	29·03	...		13	40·4	m
328		*Anon.*							
Jan. 17	9·8	7	17	41·24	...	129	46	24·1	n
Feb. 23	9·7		17	41·01	...		46	24·9	n

Separate Results of Madras Meridian Circle Observations in 1866.

Number and Date.	Magnitude.	Mean Right Ascension 1866. h. m. s.	No. of Wires.	Mean Polar Distance 1866. ° ′ ″	Observer.
329		*R. P. L.* 45.			
Jan. 25	...	7 17 49·56	7	0 50 17·5	R
Feb. 20	...	17 49·50	7	50 20·1	R
Mar. 1	...	17 48·82	7	50 21·3	R
330		*Taylor* 3043.			
Jan. 3	7·3	7 19 17·88	...	129 16 30·3	M
20	7·0	19 17·73	...	16 29·8	R
331		*Lacaille* 2807.			
Feb. 13	8·2	7 19 35·80	...	142 15 36·7	R
Mar. 7	7·0	19 36·60	5	15 36·6	M
332		*Anon.*			
Mar. 6	9·1	7 19 50·60	...	152 36 50·1	M
333		*Anon.*			
Feb. 17	7·9	7 21 38·02	...	131 50 39·7	M
Mar. 15	7·0	21 37·82	...	50 41·1	M
334		*Anon.*			
Jan. 17	9·1	7 23 31·56	...	120 45 32·8	R
Feb. 22	9·2	23 31·63	...	45 32·1	R
Dec. 23	8·9	23 31·49	5	45 33·6	R
335		*Anon.*			
Jan. 16	8·1	7 23 57·78	...	120 11 30·7	R
336		*S Canis Minoris Var.* 2.			
Jan. 26	7·7	7 25 38·64	...	81 38 54·9	R
27	7·8	25 38·76	...	38 57·5	R
Feb. 21	7·8	25 38·70	...	38 56·3	R
22	7·5	25 38·80	...	38 57·5	M
337		*Anon.*			
Feb. 24	8·0	7 26 20·98	5	120 16 30·7	M
338		68 *Geminorum.*			
Jan. 29	...	7 26 57·61	.	73 58 19·0	R
339		66 *Geminorum* α², *Castor.*			
Feb. 5	...	7 26 2·78	8	57 49 16·8	M
12	...	26 2·75	5	49 15·0	M
14	...	26 3·08	...	49 17·9	M
15	...	26 2·79	...	49 18·4	M
26	...	26 2·84	...	49 16·7	R
27	...	26 2·62	...	49 17·6	R
340		*Anon.*			
Jan. 17	9·7	7 26 41·53	5	129 45 16·7	R
341		*Anon.*			
Jan. 20	8·3	7 28 0·25	6	120 42 59·7	R
Feb. 13	8·0	28 0·02	...	43 0·4	R
17	8·5	28 0·26	6	42 59·2	R
Mar. 5	8·0	28 0·22	...	42 57·7	M
342		*Taylor* 3126.			
Feb. 23	9·8	7 29 37·08	...	148 16 58·9	R
Mar. 6	8·9	29 37·06	...	16 1·5	M
7	7·0	29 37·08	...	15 59·9	M
343		*Taylor* 3133.			
Mar. 8	8·7	7 31 8·94	4	131 10 49·9	M
9	8·5	31 8·74	—	10 46·0	M
344		*Anon.*			
Jan. 17	8·9	7 31 58·80	...	129 41 8·9	R
Feb. 17	9·2	31 58·84	..	41 9·5	R
345		10 *Canis Minoris* α, *Procyon.*			
Jan. 3	—	7 32 17·13	...	84 26 4·5	M
23	...	32 17·17	..	26 5·7	R
31	...	32 17·21	.	26 3·7	R
Feb. 3	...	32 17·04		26 3·1	M

Separate Results of Madras Meridian Circle Observations in 1866.

Number and Date.	Magnitude.	Mean Right Ascension 1866. h.	m.	s.	No. of Wires.	Mean Polar Distance 1866. °	′	″	Observer.
Feb. 5	...	7	32	17·07	...	64	26	4·5	M
7	...		32	17·06	...		26	5·0	M
8	...		32	17·02	...		26	3·7	M
9	...		32	17·13	...		26	3·7	M
12	...		32	17·07	...		26	2·8	M
14	...		32	17·17	...		26	5·5	M
15	...		32	17·11	...		26	6·0	M
16	...		32	17·20	...		26	4·8	M
23	...		32	17·12	...		26	6·4	M
26	...		32	17·20	...		26	6·3	N

346 *S Geminorum Var.* 3.

Number and Date.	Magnitude.	h.	m.	s.	No. of Wires.	°	′	″	Observer.
Jan. 24	10·6	7	34	41·04	5	66	12	42·5	N
26	10·6		34	41·00	6		12	40·6	N
27	10·6		34	40·75	5		12	44·8	N

347 *Anon.*

Number and Date.	Magnitude.	h.	m.	s.	No. of Wires.	°	′	″	Observer.
Jan. 16	8·9	7	35	10·87	...	120	38	7·7	N
18	9·0		35	10·90	5		38	7·9	N

348 *Anon.*

Number and Date.	Magnitude.	h.	m.	s.	No. of Wires.	°	′	″	Observer.
Feb. 13	9·9	7	35	17·13	4	66	16	13·8	N

349 *Taylor* 3195.

Number and Date.	Magnitude.	h.	m.	s.	No. of Wires.	°	′	″	Observer.
Mar. 13	7·9	7	36	34·65	5	130	19	22·3	M

350 *Anon.*

Number and Date.	Magnitude.	h.	m.	s.	No. of Wires.	°	′	″	Observer.
Mar. 5	7·7	7	36	34·82	...	123	56	28·1	M
6	8·0		36	34·76	5		56	29·3	M

351 *Anon.*

Number and Date.	Magnitude.	h.	m.	s.	No. of Wires.	°	′	″	Observer.
Jan. 16	7·2	7	36	49·31	5	120	57	34·8	N
18	8·0		36	49·30	...		57	38·7	N

352 78 *Geminorum β, Pollux.*

Number and Date.	Magnitude.	h.	m.	s.	No. of Wires.	°	′	″	Observer.
Jan. 23	...	7	37	6·68	...	61	30	12·2	N
29	...		37	6·70	...		30	11·4	N
31	...		37	6·69	...		30	13·4	N
Feb. 1	...	7	37	6·68	...	61	30	12·6	P
3	...		37	6·67	...		30	11·9	M
5	...		37	6·66	...		30	12·3	M
7	...		37	6·82	...		30	13·0	M
8	...		37	6·69	...		30	13·6	M
9	...		37	6·79	...		30	15·6	M
10	...		37	6·77	...		30	13·6	M
14	...		37	6·90	...		30	14·0	M
15	...		37	6·77	...		30	14·0	M
16	...		37	6·82	...		30	13·6	M
19	...		37	6·92	6		30	13·3	N
23	...		37	6·75	...		30	14·5	M
24	...		37	6·69	...		30	13·0	M

353 *Anon.*

Number and Date.	Magnitude.	h.	m.	s.	No. of Wires.	°	′	″	Observer.
Mar. 7	7·0	7	39	9·76	...	131	8	54·3	M
8	7·0		39	9·67	...		8	54·6	M
10	7·0		39	9·61	5		8	53·6	M

354 *Anon.*

Number and Date.	Magnitude.	h.	m.	s.	No. of Wires.	°	′	″	Observer.
Mar. 9	8·3	7	40	57·20	4	133	9	54·9	M
14	8·6		40	57·46	...		9	55·7	M

355 *T Geminorum Var.* 4.

Number and Date.	Magnitude.	h.	m.	s.	No. of Wires.	°	′	″	Observer.
Jan. 26	10·1	7	41	15·46	5	66	56	8·9	N
27	10·0		41	15·38	...		56	8·4	N
30	9·9		41	15·35	...		56	9·0	N
Feb. 17	9·6		41	15·29	...		56	8·6	N
23	9·6		41	14·95	6		56	9·4	N
27	...		41	15·28	...		56	7·4	N

356 *Anon.*

Number and Date.	Magnitude.	h.	m.	s.	No. of Wires.	°	′	″	Observer.
Mar. 12	7·6	7	41	34·20	...	144	18	59·0	M

357 *Lacaille* 3031.

Number and Date.	Magnitude.	h.	m.	s.	No. of Wires.	°	′	″	Observer.
Jan. 17	7·2	7	45	7·88	5	144	23	46·4	N
Mar. 6	6·9		45	7·96	5		23	45·1	W

358 *Anon.*

Number and Date.	Magnitude.	h.	m.	s.	No. of Wires.	°	′	″	Observer.
Mar. 5	8·0	7	46	10·64	...	129	25	9·4	W
13	8·0		46	10·66	...		25	6·6	W

Separate Results of Madras Meridian Circle Observations in 1866.

Number and Date.	Magnitude.	Mean Right Ascension, 1866. h.	m.	s.	No. of Wires.	Mean Polar Distance 1866. °	′	″	Observer.
859		*U. Geminorum Var. 5.*							
Jan. 16	9·5	7	47	9·31	..	67	38	59·0	R
19	9·6		47	9·26	...		38	58·1	R
20	9·8		47	9·24	...		39	0·3	R
22	9·9		47	9·24	...		38	59·6	R
23	9·7		47	9·14	...		39	0·4	R
24	9·8		47	9·28	...		38	59·5	R
25	9·9		47	9·37	...		39	0·5	R
27	10·1		47	9·11	...		39	1·4	R
30	10·5		47	9·19	5		39	2·3	R
860		*Taylor* 3310.							
Feb. 21	...	7	48	26·01	5	149	18	11·2	R
26	7·7		48	26·08	5		18	9·6	R
861		*Anon.*							
Feb. 22	9·3	7	48	25·76	...	67	46	26·0	R
862		*Anon.*							
Mar. 1	8·4	7	49	3·28	5	130	26	22·5	R
863		1 *Cancri.*							
Dec. 22	...	7	49	28·71	...	73	51	17·5	R
864		*Anon.*							
Feb. 20	.	7	49	56·39	...	129	17	40·6	R
Mar. 7	8·1		49	56·67	...		17	41·6	M
865		*Anon.*							
Feb. 5	7·6	7	49	57·77	...	149	8	45·0	M
866		*Anon.*							
Feb. 24	7·7	7	50	9·24	—	129	24	14·5	M
867		*Taylor* 3339.							
Mar. 9	7·9	7	51	28·65		144	17	13·7	M
868		*Anon.*							
Mar. 10	8·0	7	52	8·09	5	104	22	42·8	M
12	8·1		52	8·08	5		22	44·5	M
869		5 *Cancri.*							
Dec. 22	...	7	53	51·90	5	73	10	48·6	R
870		*Taylor* 3373.							
Mar. 6	7·6	7	54	16·73	...	141	12	13·2	M
871		6 *Cancri.*							
Jan. 3	...	7	55	17·05	...	61	49	59·1	M
22	...		55	17·02	...		50	0·5	R
29	...		55	17·05	...		50	1·7	R
Feb. 1	...		56	17·18	...		49	59·7	P
7	...		55	17·09	...		50	0·1	M
8	...		55	17·13	...		50	1·1	M
9	...		55	17·09	..		49	59·5	M
10	...		55	16·99	...		50	0·9	M
12	...		55	17·20	...		49	58·0	M
17	...		55	17·03	...		50	1·7	R
19	...		55	17·05	4		50	0·2	R
21	...		55	16·99	...		50	0·6	R
22	...		55	16·91	...		50	0·3	R
24			55	17·08	...		50	1·3	R
Mar. 15	.		55	17·09	...		50	2·9	M
872		*Anon.*							
Mar. 14	7·9	7	55	24·41	...	124	30	33·0	M
873		*Brisbane* 1955.							
Mar. 5	6·7	7	55	39·99	5	132	56	7·1	M
13	6·9		55	39·16	...		56	6·0	M
874		*Anon.*							
Jan. 24	9·2	8	1	24·95	...	130	31	44·2	R
Mar. 10	8·6		1	24·64	...		31	42·9	M

Separate Results of Madras Meridian Circle Observations in 1866.

Number and Date	Magnitude	Mean Right Ascension 1866. h. m. s.	No. of Wires	Mean Polar Distance 1866. ° ′ ″	Observer
875		*Lacaille* 3174.			
Mar 7	7·8	8 1 27·12	...	136 34 16·8	M
876		15 *Argûs.*			
Feb. 1	...	8 1 50·43	...	113 55 13·2	P
13	...	1 50·33	...	55 13·1	M
17	...	1 50·41	...	55 13·9	M
19	...	1 50·38	...	55 13·0	M
21	...	1 50·28	...	55 12·1	M
22	...	1 50·23	...	55 11·6	M
23	...	1 50·34	...	55 13·9	M
24	...	1 50·27	...	56 13·1	M
Mar. 1	...	1 50·42	...	55 13·1	M
16	...	1 50·22	...	56 12·0	M
17	...	1 50·36	...	55 13·0	M
877		*Anon.*			
Mar. 16	9·5	8 2 7·91	...	118 47 0·3	M
878		*Anon.*			
Mar. 9	7·9	8 2 16·15	3	128 30 45·2	M
12	8·3	2 16·20	..	30 46·7	M
879		16 *Cancri* ζ			
Feb. 26	...	8 4 31·44	.	71 57 8·5	M
880		*Anon.*			
Feb. 2	8·0	8 6 8·16	...	129 40 54·1	M
Mar. 6	7·7	6 8·05	...	40 55·7	M
14	7·0	6 8·07	..	40 54·1	M
881		*Anon.*			
Jan. 29	9·0	8 8 38·14	...	124 38 46·3	M
882		*R Cancri Var.* 1			
Jan. 24	7·3	8 9 10·47	4	77 51 51·3	M
26	8·2	9 10·49	...	51 56·1	M
27	8·2	9 10·60	..	51 55·2	M

Number and Date	Magnitude	Mean Right Ascension 1866. h. m. s.	No. of Wires	Mean Polar Distance 1866. ° ′ ″	Observer
883		*Anon.*			
Jan. 30	9·6	8 10 20·20	...	151 25 20·7	M
884		*Anon.*			
Mar. 17	9·0	8 10 25·50	4	77 38 8·8	M
885		*Lalande* 16224.			
Feb. 24	7·8	8 10 40·25	...	73 54 36·3	M
Mar. 9	7·0	10 40·37	...	54 30·6	M
10	6·8	10 40·20	...	54 34·5	M
886		*Anon.*			
Feb. 14	8·4	8 11 20·20	3	152 5 2·4	M
887		*Anon.*			
Mar. 8	8·3	8 12 21·73	...	120 44 3·4	M
15	...	12 21·67	4	44 5·4	M
888		*Anon.*			
Jan. 30	9·2	8 12 40·65	...	120 41 20·1	M
Feb. 28	8·5	12 40·70	...	41 16·9	M
Mar. 12	8·0	12 40·69	...	41 17·1	M
13	8·0	12 40·72	...	41 16·4	M
19	9·0	12 40·84	5	41 18·3	M
889		*Anon.*			
Mar. 20	10·0	8 13 1·76	5	130 45 55·2	M
890		*Anon.*			
Mar. 6	7·7	8 14 53·88	...	148 17 16·4	M
7	7·9	14 53·91	...	17 17·7	M
891		*Lacaille* 3207.			
Feb. 13	8·6	8 16 2·64	4	163 30 18·7	M
16	8·0	16 2·74	5	30 14·1	M
Mar. 5	8·1	16 3·08	...	30 12·2	M

Separate Results of Madras Meridian Circle Observations in 1866.

Number and Date.	Magnitude.	Mean Right Ascension 1866. h. m. s.	No. of Wires.	Mean Polar Distance. 1866. ° ′ ″	Observer.
392		*20 Cancri d*¹			
Feb. 26	...	8 16 41·34	...	71 14 26·3	R
393		*Anon.*			
Feb. 26	9·0	8 17 26·80	...	141 16 16·2	R
394		*Anon.*			
Feb. 12	8.5	8 17 29·15	...	77 40 30·6	M
Mar. 14	8·6	17 29·13	...	40 35·3	M
395		*Anon.*			
Mar. 23	10·3	8 18 49·94	4	151 1 8·5	R
396		*Taylor* 3599.			
Feb. 7	7·8	8 20 19·92	...	144 56 9·1	M
8	7·7	20 19·74	...	56 10·6	M
Mar. 9	7·1	20 19·81	...	53 8·2	M
16	...	20 19·98	5	56 10·1	R
397		*W. B. N. VIII.* 459.			
Feb. 14	8·6	8 20 23·00	...	74 27 44·7	M
398		29 *Cancri*			
Jan. 29	...	8 21 8·72	...	75 30 54·6	R
30	...	21 8·86	...	30 54·8	R
399		*Taylor* 3607.			
Mar. 10	6·9	8 21 15·48	...	144 56 46·5	M
16	..	21 15·36	4	55 46·4	R
400		*Anon.*			
Feb. 3	9·0	8 21 31·41	5	131 41 59·6	M
401		*Anon.*			
Mar. 19	..	8 23 14·60	...	73 35 40·0	R
21	9·3	23 14·85	...	35 40·9	R
402		*Taylor* 3620			
Feb. 6	8·0	8 23 14·98	...	130 49 11·7	M
403		*Anon.*			
Mar. 1	8·9	8 23 36·65	...	133 33 50·9	R
404		*Anon.*			
Mar. 8	9·0	8 24 51·70	...	73 46 4·8	M
405		33 *Cancri* η.			
Feb. 9	...	8 24 57·30	...	69 6 24·7	M
10	...	24 57·40	...	6 24·2	M
17	...	24 57·29	...	6 23·1	R
19	...	24 57·34	...	6 24·2	R
20	...	24 57·31	...	6 23·5	R
23	...	24 57·26	...	6 23·1	R
Mar. 5	...	24 57·27	...	6 20·4	M
7	...	24 57·34	...	6 23·8	M
17	...	24 57·30	...	6 24·3	R
406		*Taylor* 3651.			
Mar. 6	7·6	8 25 44·01	...	130 3 40·2	M
407		*Taylor* 3652.			
Mar. 16	7·9	8 25 46·28	...	130 3 2·6	M
408		*Lacaille* 3593.			
Mar. 20	8·7	8 25 50·62	6·7	140 40 35·2	R
409		*W. B. N. VIII.* 635.			
Mar. 13	8·4	8 27 46·63	...	73 41 42·5	M
14	8·6	27 46·64	...	41 46·3	M
410		*Taylor.*			
Feb. 12	7·1	8 28 36·62	...	74 13 30·8	M
Mar. 15	7·0	28 36·66		13 31·7	M

Separate Results of Madras Meridian Circle Observations in 1866.

Number and Date.	Magnitude.	Mean Right Ascension 1866. h.	m.	s.	No. of Wires.	Mean Polar Distance 1866. °	′	″	Observer.
411 *Anon.*									
Feb. 24	9·8	8	28	54·27	...	70	41	5·1	n
412 *W. B. N. VIII.* 684.									
Mar. 22	9·2	8	29	7·86	...	70	30	22·5	n
413 *Lacaille* 3430.									
Mar. 9	7·3	8	29	15·00	5	151	38	3·3	M
414 *Anon.*									
Mar. 19	8·8	8	30	19·37	...	128	47	36·6	n
28	8·0		30	19·44	5		47	33·8	n
415 *Anon.*									
Feb. 6	8·6	8	33	13·90	...	129	28	51·9	M
416 *W. B. N. VIII.* 852.									
Mar. 6	8·0	8	34	8·54	5	74	8	4·9	M
7	8·4		34	8·69	...		8	2·9	M
417 *Anon.*									
Mar. 10	8·3	8	34	10·28	5	164	30	49·3	M
418 *Anon.*									
Mar. 17	9·2	8	34	48·98	...	160	46	37·8	n
419 45 *Cancri* A[1]									
Jan. 29	...	8	35	49·24	...	76	50	28·9	n
30	...		35	49·26	...		50	28·7	n
Mar. 23	.		35	49·12	6		50	28·1	n
420 *Lacaille* 3491.									
Mar 20	8·4	8	36	3·38	5	132	22	15·3	n
421 *b Velorum.*									
Mar 10	...	8	36	19·98	.	136	10	25·5	n
422 *S. Cancri Var.* 2.									
Feb. 22	8·5	8	36	16·71	...	70	29	11·1	n
Mar. 22	8·9		36	16·83	...		29	14·7	n
23	9·0		36	16·68	...		30	11·3	n
423 *Lalande* 17231.									
Mar. 13	7·8	8	37	50·84	...	74	28	7·3	M
14	7·8		37	50·94	...		28	9·0	M
424 *W. B. N. VIII.* 977.									
Feb. 12	9·7	8	38	21·00	5	74	49	28·7	M
Mar. 27	9·7		38	21·88	..		49	29·3	n
425 11 *Hydrae* ε									
Feb. 1	...	8	39	40·73	...	96	5	31·6	r
13	...		39	40·70	...		5	32·3	n
16	...		39	40·69	...		5	31·3	M
21	...		39	40·77	...		5	30·8	n
Mar. 6	...		39	40·76	...		5	31·9	M
8	...		39	40·73	...		5	33·0	M
19	...		39	40·72	...		5	31·2	n
22	...		39	40·76	5		5	34·3	n
28	...		39	40·66	...		5	33·3	n
426 *Anon.*									
Mar. 12	7·6	8	40	38·80	...	129	15	36·4	M
427 *Lacaille* 3534.									
Mar. 17	8·6	8	42	21·65	...	129	18	22·4	n
428 *W. B. N. VIII.* 1043.									
Mar. 15	7·9	8	42	29·12	6	74	40	23·5	M
26	8·7		42	29·20	...		40	19·4	n
429 *Anon.*									
Feb. 22	9·2	8	46	30·47	3	166	10	54·2	n

Separate Results of Madras Meridian Circle Observations in 1866.

Number and Date.	Magnitude.	Mean Right Ascension 1866. h. m. s.	No. of Wires.	Mean Polar Distance 1866. ° ′ ″	Observer.
430		*Lacaille* 3673.			
Feb. 2	9·0	8 44 15·34	5	152 41 54·6	M
6	7·7	44 14·95	...	41 52·5	M
9	7·7	44 15·25	...	41 51·1	M
431		*S Hydrae Var.* 3.			
Feb. 24	8·2	8 46 34·60	6	86 26 40·7	R
Mar. 9	9·0	46 34·47	...	26 40·1	M
10	7·8	46 34·46	...	26 41·7	M
16	9·1	46 34·63	...	26 41·0	R
19	8·0	46 34·62	...	26 40·4	R
432		*Anon.*			
Mar. 20	9·6	8 46 49·43	...	133 56 42·3	R
433		*R. P. L.* 60.			
Mar. 9	...	8 46 50·28	3	5 17 22·6	M
434		*Anon.*			
Mar. 5	7·9	8 46 59·80	...	136 6 27·3	M
Apl. 5	9·1	47 0·00	4	6 23·0	M
435		*Anon.*			
Mar. 24	9·0	8 47 44·25	5	138 56 21·9	R
Apl. 6	9·8	47 44·20	5	56 21·6	M
436		*T Cancri Var.* 3.			
Feb. 22	9·0	8 49 0·69	...	[illegible] 38 36·2	R
Mar. 13	8·9	49 0·75	...	38 27·7	M
22		49 0·57	...	38 29·0	R
28	8·5	49 0·73	...	38 27·7	R
437		*T Hydrae Var.* 4.			
Feb. 13	9·0	8 49 8·61	5	[illegible] 37 36·3	R
15	7·7	49 8·62	..	37 26·0	M
20	8·3	49 8·69	...	37 31·7	R
23	7·8	49 8·60	5	37 36·3	M
24	7·7	49 9·39	3	37 31·6	M
Mar. 6	9·2	49 8·33	..	37 35·9	M
14	9·0	49 9·62	4	37 37·4	M
438		*Anon.*			
Mar. 24	8·0	8 49 17·87	...	132 54 49·1	R
27	9·6	49 17·81	...	54 47·6	R
439		9 *Ursae Majoris* ι.			
Feb. 1	...	8 50 1·21	...	41 25 6·4	R
440		*Anon.*			
Mar. 12	8·2	8 50 19·60	5	132 57 28·9	M
441		65 *Cancri* α.			
Jan. 3	...	8 51 9·32	...	77 37 53·2	M
Feb. 26	...	51 9·33	6	37 54·4	R
27	...	51 9·30	5	37 57·4	R
Mar. 26	...	51 9·49	...	37 56·3	R
442		*Anon.*			
Mar. 27	9·2	8 54 5·96	6	132 54 10·5	R
443		*Anon.*			
Mar. 16	9·7	8 54 24·93	5	130 38 21·7	M
444		*Anon.*			
Mar. 7	9·8	8 54 51·98	.	130 36 36·3	M
10	8·1	54 51·97		36 30·4	M
445		*Anon.*			
Feb. 20	8·0	8 55 1·66	4	102 29 24·6	R
446		*Anon.*			
Feb. 24	9·5	8 56 40·61	5	106 48 29·2	R
447		*Anon.*			
Feb. 5	9·5	8 56 47·26	...	129 46 29·9	M

Separate Results of Madras Meridian Circle Observations in 1866.

Number and Date.	Magnitude.	Mean Right Ascension 1866. h.	m.	s.	No. of Wires.	Mean Polar Distance 1866. °	′	″	Observer.
448				*Anon.*					
Mar. 30	..	8	50	59·10	4	140	36	4·1	R
449				*Anon.*					
Feb. 28	9·0	8	53	0·29	5	146	50	10·6	R
450				*Anon.*					
Feb. 2	9·1	8	58	12·73	...	146	19	56·9	M
451				*Anon.*					
Mar. 9	8·0	8	59	9·40	5	145	38	37·4	M
24	8·5		59	9·45	5		38	38·3	R
Apl. 10	8·1		59	9·54	...		38	36·0	M
452				*76 Cancri* κ					
Jan. 3	..	9	0	29·26	...	78	47	41·0	M
Feb. 26	...		0	29·30	...		47	48·1	R
27	.		0	29·23	..		47	42·9	R
453				*Anon.*					
Mar. 14	..	9	1	51·87	...	123	57	41·6	M
454				*Lacaille* 3705.					
Mar 8	7·0	9	2	14·00	...	151	17	33·8	M
19			2	16·09	...		17	34·0	R
455				*Anon.*					
Jan. 30	9·3	9	4	1·65	..	139	48	13·6	R
Mar. 17	9·0		4	1·31	4		48	14·9	M
456				*Anon.*					
Feb. 22	8·5	9	4	28·27	...	130	30	8·4	R
Mar 22	9·0		4	28·41	...		30	10·8	R
457				*Anon.*					
Mar. 10	8·0	9	4	29·14	...	138	46	57·6	M
12	7·7		4	29·36	4		46	57·4	M
15	8·0		4	29·25	...		46	56·0	M
17	9·0		4	29·38	...		46	58·3	M
458				*Lacaille* 3713.					
Mar. 26	7·9	9	4	38·26	...	148	40	41·1	R
27	8·2		4	38·27	...		40	42·4	R
459				*Taylor* 4021.					
Mar. 13	7·0	9	5	36·38	...	138	44	38·5	M
Apl. 4	7·8		5	35·47	5		44	39·7	M
460				*Anon.*					
Apl. 11	8·0	9	6	30·91	...	102	29	54·2	M
461				*Anon.*					
Mar. 20	9·3	9	8	17·66	...	148	14	46·8	R
462				*Lacaille* 3774.					
Feb. 28	8·8	9	9	56·15	...	157	10	14·2	R
Mar. 24	9·9		9	56·00	5		10	16·2	R
463				*83 Cancri.*					
Feb. 2	...	9	11	30·64	...	71	48	48·3	M
21	...		11	29·88	...		48	44·9	R
23	...		11	30·00	...		48	43·9	R
24	...		11	29·81	...		43	44·1	M
Mar. 5	...		11	29·73	...		48	41·6	M
9	...		11	29·86	...		43	43·9	M
10	...		11	29·95	...		48	46·2	M
13	...		11	29·97	..		43	41·7	M
14	...		11	29·66	...		48	48·9	M
27	.		11	29·67	...		48	46·4	R
464				*Anon.*					
Jan. 30	8·2	9	13	10·02	...	78	19	57·7	R

Separate Results of Madras Meridian Circle Observations in 1866.

Number and Date.	Magnitude.	Mean Right Ascension 1866. h.	m.	s.	No. of Wires.	Mean Polar Distance 1866. °	′	″	Observer.
465				*ι Argûs.*					
Mar. 15	...	9	13	30·12	...	149	38	54·0	M
19	...		13	30·23	...		38	56·4	R
466				*Anon.*					
Mar. 21	9·0	9	15	19·26	...	143	49	11·5	R
Apl. 5	7·0		15	19·33	...		49	10·7	M
7	7·0		15	19·38	...		49	13·4	M
467				*Anon.*					
Mar. 20	9·3	9	16	10·30	...	140	8	5·0	R
23	9·3		16	10·38	5		8	4·9	R
Apl. 6	8·0		16	10·60	5		8	5·3	M
468				*Anon.*					
Feb. 23	9·5	9	17	39·89	...	75	5	59·2	R
Mar. 26	...		17	39·17	5		6	0·9	R
469				*Anon.*					
Feb. 19	7·0	9	19	36·60	4	75	7	3·5	R
25	6·9		19	36·70	...		7	3·5	R
470				*Anon.*					
Mar. 17	9·8	9	20	38·06	...	137	28	27·7	R
19	9·0		20	38·06	...		28	27·5	R
471				*Anon.*					
Mar. 26	9·5	9	20	46·30	...	158	28	46·1	R
472				30 *Hydrae α, Var. 2.*					
Feb. 1	...	9	21	0·01	...	98	4	47·6	M
2	...		21	0·09	...		4	49·4	M
16	...		21	0·06	...		4	46·3	M
20	...		21	0·12	...		4	46·3	R
24	...		21	0·11	...		4	46·8	M
Mar. 5	...		21	0·19	...		4	46·7	M
6	...		21	0·17	...		4	49·4	M
7	...		21	0·08	...		4	46·0	M
12	...		21	0·06	...		4	47·7	M
Mar. 13	...	9	21	0·27	...	98	4	46·5	M
14	...		21	0·02	...		4	47·4	M
16	...		21	0·14	...		4	47·1	M
27	...		21	0·15	...		4	47·6	R
Apl. 9	...		21	0·07	...		4	47·2	M
473				25 *Ursae Majoris θ*					
Mar. 22	...	9	23	52·80	5	37	42	56·3	R
23	...		23	54·86	...		42	56·0	R
474				*Anon.*					
Apl. 6	9·5	9	26	26·62	4	158	41	16·9	M
475				*Anon.*					
Mar. 26	...	9	24	37·73	...	130	26	44·0	R
476				6 *Leonis h*					
Jan. 30	...	9	24	46·31	...	79	41	42·8	R
31	...		24	46·45	6		41	42·6	R
477				*Anon.*					
Mar. 9	8·0	9	26	50·21	...	144	53	40·4	M
478				*Taylor* 4282.					
Mar. 6	7·3	9	27	43·84	...	146	33	51·9	M
8	7·3		27	43·69	...		33	54·7	M
479				*Anon.*					
Feb. 27	9·6	9	28	3·08	5	129	46	24·1	R
28	9·0		28	3·31	5		46	24·2	R
Mar. 16	9·0		28	3·36	5		46	24·9	R
19	8·3		28	3·43	5		46	25·6	R
20	9·0		28	3·35	5		46	25·1	R
480				*Anon.*					
Apl. 7	7·9	9	28	31·71	...	144	0	18·5	R

Separate Results of Madras Meridian Circle Observations in 1866.

Number and Date.	Magnitude.	Mean Right Ascension 1866. h. m. s.	No. of Wires.	Mean Polar Distance 1866. ° ′ ″	Observer.
481		*Anon.*			
Feb. 27	9·6	9 38 30·87	4	128 47 27·3	R
Mar. 16	9·3	9 28 30·34	6	47 24·6	R
482		*Anon.*			
Mar. 27	9·9	9 29 6·06	...	128 50 6·1	R
483		*Anon.*			
Mar. 7	9·1	9 30 36·86	5	146 34 6·8	M
Apl. 5	9·4	29 36·94	4	34 7·8	M
484		*Taylor* 4259.			
Feb. 10	5·9	9 32 2·04	...	138 46 21·6	M
Mar. 13	5·5	33 1·90	...	46 23·5	M
Apl. 4	5·7	32 1·99	...	46 18·6	M
485		*Anon.*			
Feb 20	9·0	9 32 33·48	...	130 51 34·8	R
486		*R. P. L.* 69—*s.p.*			
Sep. 19	...	9 33 39·13	1	3 47 22·1	R
487		14 *Leonis* o			
Jan. 30	...	9 34 39·79	...	79 30 1·1	R
Mar. 26	..	35 50·01	...	30 0·0	R
27	...	34 50·86	5	29 59·9	R
488		*Anon.*			
Apl. 12	8·8	9 35 37·75	...	151 56 52·9	M
489		*Anon.*			
Apl. 10	8·2	9 35 53·07	...	148 40 40·2	M
11	9·0	35 53·26	...	40 38·3	M
490		*Anon.*			
Apl. 6	8·5	9 36 52·80	...	151 54 28·2	M
491		*R Leonis Minoris Var.* 1.			
Feb. 17	8·8	9 37 31·70	...	54 56 30·1	R
23	7·5	37 31·61	...	56 30·0	M
27	...	37 31·74	5	56 29·5	R
28	8·0	37 31·68	...	56 28·5	R
Mar. 5	7·6	37 31·56	...	56 27·4	M
6	7·7	37 31·84	...	56 28·6	M
7	7·5	37 31·79	...	56 27·9	M
492		17 *Leonis* ε			
Feb. 1	...	9 38 14·15	...	65 34 39·6	M
3	...	38 14·30	...	34 41·3	M
20	...	38 14·31	4	34 38·6	R
Mar. 2	...	38 14·08	...	34 40·0	R
9	...	38 14·50	...	34 39·1	M
10	...	38 14·36	...	34 40·9	M
19	...	38 14·51	4	36 40·7	R
22	...	38 14·26	...	34 42·3	R
26	...	38 14·40	...	34 40·9	R
493		*B. F.* 1383.			
Apl. 9	...	9 39 8·70	...	82 40 33·9	M
494		18 *Leonis.*			
Mar. 26	...	9 39 10·11	...	77 34 33·1	R
27	...	39 10·08	...	34 30·0	R
495		*l Carinae Var.* 1.			
Apl. 6	5·7	9 41 34·00	...	151 53 38·7	M
496		*Anon.*			
Apl. 4	8·0	9 41 34·86	5	130 46 33·2	M
5	8·0	41 34·80	5	46 34·2	M
7	8·0	41 34·75	5	46 33·8	M
18	8·0	41 34·74	...	46 34·3	M
497		*Anon.*			
Mar. 8	8·0	9 44 9·48	...	147 3 10·3	M
9	8·0	44 9·46	5	3 9·6	M

Separate Results of Madras Meridian Circle Observations in 1866.

Number and Date.	Magnitude.	Mean Right Ascension 1866. h. m. s.	No. of Wires.	Mean Polar Distance 1866. ° ′ ″	Observer.
498		*Anon.*			
Mar. 2	9·2	9 46 0·78	...	139 3 25·9	R
499		*Anon.*			
Feb. 20	9·4	9 46 31·72	5	129 7 29·2	R
500		*R. P. L.* 70.			
Mar. 28	...	9 46 46·20	3	5 26 26·3	R
501		*Anon.*			
Apl. 7	9·5	9 49 37·11	4	152 8 16·0	M
502		*W. B. E. IX.* 1057.			
Jan. 31	7·8	9 49 51·01	...	85 7 17·3	R
Feb. 16	7·4	49 51·05	...	7 16·0	M
Mar. 28	7·3	49 50·95	...	7 19·2	R
503		*Anon.*			
Mar. 26	9·2	9 50 51·52	5	145 39 46·8	R
Apl. 14	8·9	50 51·87	...	39 46·3	M
28	7·9	50 51·68	...	39 46·1	M
504		29 *Leonis* π.			
Jan. 3	—	9 53 7·78	...	81 18 51·3	M
Feb. 1	...	53 7·72	...	18 52·2	M
20	...	53 7·79	...	18 50·7	R
27	...	53 7·86	...	18 52·7	R
28	...	53 7·80	...	18 53·6	R
Mar. 1	—	53 7·79	...	18 53·8	R
5	...	53 7·84	...	18 52·5	M
6	...	53 7·68	...	18 54·4	M
7	...	53 7·74	...	18 54·3	M
8	...	53 7·88	...	18 51·3	M
9	...	53 7·84	...	18 52·6	M
10	...	53 7·81	...	18 52·0	M
13	...	53 7·74	...	18 53·2	M
14	...	53 7·86	...	18 54·3	M
Apl. 6	...	53 7·94	...	18 51·6	M
505		*Anon.*			
Mar. 24	9·3	9 53 56·88	5	152 7 28·1	R
Apl. 6	9·0	53 56·74	...	7 22·0	M
506		*Anon.*			
Feb. 14	8·0	9 54 13·11	...	143 53 50·5	M
Mar. 3	8·0	54 13·00	...	53 50·1	M
Apl. 5	8·1	54 13·05	...	53 49·3	M
507		*Taylor* 4445.			
Apl. 10	7·9	9 54 47·21	...	147 29 13·7	M
508		*Anon.*			
Apl. 13	8·0	9 56 56·17	..	147 24 50·3	M
509		*Anon.*			
Mar. 21	9·3	9 56 8·35	...	129 57 50·3	R
510		*Anon.*			
Mar. 17	9·5	9 57 12·30	...	129 57 13·0	R
21	9·0	57 12·37	4	57 14·7	R
511		*Taylor* 4476.			
Mar. 28	8·8	9 57 56·68	...	145 36 29·5	R
512		*Anon.*			
Apl. 18	9·2	9 59 17·84	5	148 54 41·5	R
513		*Taylor* 4484.			
Apl. 12	7·4	9 59 40·41	...	151 30 38·6	M
514		32 *Leonis* α, *Regulus.*			
Jan. 3	...	10 1 14·02	...	77 22 45·4	M
Feb. 1	...	1 13·88	...	22 46·3	M
27	...	1 14·07	5	22 46·2	R
28	...	1 13·94	-	22 46·9	R

Separate Results of Madras Meridian Circle Observations in 1866.

Number and Date.	Magnitude.	Mean Right Ascension 1866. h.	m.	s.	No. of Wires.	Mean Polar Distance 1866. °	'	"	Observer.
Mar. 1	...	10	1	13·96	...	77	22	46·6	R
2	...		1	13·96	...		22	46·3	R
6	...		1	13·99	—		22	47·1	M
7	...		1	14·06	...		23	47·0	M
8	...		1	13·96	...		23	47·6	M
9	...		1	13·93	...		23	44·7	M
10	...		1	13·87	...		22	46·8	M
13	...		1	13·96	...		23	44·7	M
Apl. 9	...		1	13·91	...		22	46·3	M
515				*Anon.*					
Mar. 24	9·0	10	1	26·37	3	130	2	33·6	R
Apl. 14	8·5		1	26·12	...		2	34·9	M
17	9·2		1	26·10	...		2	33·3	R
516				*Lacaille* 4164.					
Mar. 5	7·9	10	2	17·06	3	143	54	36·2	M
15	...		2	16·89	...		54	39·6	M
Apl. 3	7·9		2	16·87	3		54	37·3	M
6	7·1		2	16·76	5		54	37·5	M
517				*Anon.*					
Jan. 31	9·3	10	2	51·70	...	129	53	11·1	R
518				*Anon.*					
Mar. 26	9·3	10	5	18·08	...	140	30	21·4	R
Apl. 7	8·9		5	18·40	...		30	20·3	M
519				*Anon.*					
Jan. 31	9·1	10	7	12·40	5	129	59	46·9	R
Mar. 3	8·3		7	12·36	...		56	41·4	M
Apl. 10	8·0		7	12·32	...		59	41·5	M
520				*Anon.*					
Apl. 11	9·1	10	9	6·35	...	139	53	14·9	M
13	8·7		9	6·40	...		53	17·0	M
521				*Taylor* 4577.					
Mar. 2	9·0	10	9	54·95	6	125	37	32·7	R
522				*Anon.*					
Apl. 19	6·2	10	9	86·67	4	146	34	57·5	R
523				*Anon.*					
Apl. 19	9·5	10	10	30·55	...	129	51	46·6	R
524				41 *Leonis* γ^1					
Mar. 12	...	10	12	34·81	...	69	28	56·0	M
13	...		12	34·86	...		28	57·6	M
14	...		12	34·89	...		28	56·5	M
Apl. 4	...		12	34·88	...		28	58·6	M
5	...		12	34·73	...		28	55·4	M
525				*Anon.*					
Mar. 2	9·6	10	12	86·57	...	126	37	31·3	R
526				*Anon.*					
Apl. 6	8·9	10	14	46·27	...	109	36	11·3	M
7	8·3		14	42·30	...		36	12·7	M
527				*Anon.*					
Apl. 17	9·3	10	16	14·86	...	75	25	6·8	R
528				*Anon.*					
Apl. 12	8·1	10	16	17·15	...	129	16	49·6	M
21	8·3		16	17·05	4		16	52·6	R
529				*Taylor* 4653.					
Mar. 3	8·0	10	18	6·11	...	151	28	49·7	M
Apl. 9	8·6		18	6·27	—		28	50·6	M
530				44 *Leonis.*					
Feb. 1	...	10	18	11·29	...	80	32	7·4	M
Mar. 27	...		18	11·30	5		32	10·7	R
28	...		18	11·36	...		32	11·4	R

Separate Results of Madras Meridian Circle Observations in 1866.

Number and Date.	Magnitude.	Mean Right Ascension 1900. h. m. s.	No. of Wires.	Mean Polar Distance 1900. ° ′ ″	Observer.
531		*Anon.*			
Apl. 19	9·5	10 16 49·75	...	146 9 4·4	n
532		*Anon.*			
Mar. 21	8·8	10 21 56·89	...	146 55 30·5	n
Apl. 13	7·9	21 57·00	...	55 30·7	m
16	7·8	21 56·81	5	55 31·7	n
533		*Anon.*			
Mar. 2	9·5	10 22 0·34	...	146 59 37·0	n
Apl. 11	9·2	22 0·07	...	59 36·6	m
17	9·8	22 0·22	5	59 36·5	n
534		*Anon.*			
Apl. 19	10·2	10 23 29·43	...	76 5 56·5	n
535		*Anon.*			
Mar. 9	8·4	10 24 9·96	...	146 55 43·1	m
Apl. 13	8·1	24 10·11	4	55 42·7	m
14	8·4	24 10·17	...	56 40·6	m
536		*Anon.*			
Apl. 23	8·6	10 24 29·96	5	146 09 31·0	m
537		*Anon.*			
Mar. 21	9·8	10 25 44·89	...	143 51 46·6	n
Apl. 10	8·4	25 44·72	...	51 46·3	m
12	9·0	25 44·75	...	51 48·0	m
16	8·3	25 44·64	5	51 46·3	n
538		47 *Leonis* ρ			
Jan. 31		10 25 46·26	...	80 0 17·7	n
Mar. 8	...	25 46·15	...	0 19·3	m
12	...	25 46·20	...	0 14·0	m
23	.	25 46·26	...	0 16·1	n
24		25 46·19	5	0 19·1	n
Apl. 4		25 46·13	...	0 18·7	m

Number and Date.	Magnitude.	Mean Right Ascension 1900. h. m. s.	No. of Wires.	Mean Polar Distance 1900. ° ′ ″	Observer.
539		*Anon.*			
Apl. 6	9·2	10 26 34·02	5	152 26 22·7	n
7	8·9	26 34·07	5	26 25·4	m
21	9·3	26 33·98	5	26 26·6	n
540		*p Carinae.*			
Apl. 9	5·0	10 27 16·02	...	150 50 30·6	m
541		*Anon.*			
Apl. 17	9·7	10 28 9·13	4	130 50 32·6	n
18	9·5	28 8·98	...	50 32·0	n
542		*Anon.*			
Apl. 19	10·1	10 29 10·79	5	137 56 12·9	n
543		*Taylor* 4700.			
Mar. 2	5·5	10 30 20·91	...	144 51 32·5	n
544		*Anon.*			
Apl. 17	9·5	10 31 57·27	4	130 17 15·6	n
545		*Anon.*			
Apl. 5	8·0	10 35 7·52		140 6 17·0	m
23	..	35 7·69	5	6 17·1	m
546		*Taylor* 4804.			
Apl. 4	8·3	10 35 10·72	...	145 28 37·9	m
6	8·0	35 10·68	4	28 40·2	m
9	7·6	35 10·70	...	28 40·6	m
13	7·7	35 10·80	4	28 37·4	m
24	7·4	35 10·67	...	28 38·8	m
547		*Anon.*			
Apl. 18	9·0	10 37 34·09	...	131 29 57·8	n
20	9·0	37 34·17	5	29 57·4	n

Separate Results of Madras Meridian Circle Observations in 1866.

Number and Date.	Magnitude.	Mean Right Ascension 1866. h. m. s.	No. of Wires.	Mean Polar Distance 1866. ° ′ ″	Observer.
548		*Taylor 4840.*			
Mar. 2	8·0	10 38 44·15	...	149 34 54·7	R
Apl. 13	6·8	38 44·21	5	34 56·9	M
14	6·7	38 44·28	...	34 53·7	M
549		*Anon.*			
Apl. 11	9·0	10 39 33·76	...	148 51 40·1	M
550		*η Argûs Var. 1.*			
Mar. 22	...	10 39 52·22	...	149 58 53·0	R
551		*Anon.*			
Apl. 10	9·0	10 41 2·21	4	137 1 36·0	M
552		*Anon.*			
Mar. 3	9·5	10 42 0·30	...	149 23 29·6	R
Apl. 7	8·6	42 0·13	...	23 31·3	M
553		*58 Leonis l*			
Feb. 1	...	10 42 12·64	...	73 44 49·5	R
Mar. 16	...	42 12·73	...	44 48·3	M
24	...	42 12·70	...	44 49·1	R
Apl. 19	...	42 12·72	...	44 49·2	R
554		*Anon.*			
Apl. 16	9·5	10 42 29·95	...	75 5 29·1	R
555		*Anon.*			
Apl. 29	9·0	10 42 41·45	...	111 5 1·4	R
556		*Taylor 4896.*			
Apl. 10	6·9	10 43 53·76	5	137 2 26·2	M
557		*Anon.*			
Apl. 21	9·2	10 43 57·95	6	137 3 26·5	R
558		*Anon.*			
Apl. 8	7·9	10 45 9·22	...	141 46 46·9	M
17	8·0	45 9·16	...	46 46·5	R
27	7·0	45 9·04	5	46 42·3	R
559		*Lacaille 4502.*			
Apl. 9	7·8	10 46 36·17	...	141 5 28·4	M
26	7·0	46 36·34	...	5 23·3	M
560		*Taylor 4915.*			
Apl. 12	8·6	10 46 37·78	...	135 30 19·5	M
23	8·0	46 37·75	...	30 23·8	R
561		*Anon.*			
Mar. 30	9·4	10 48 4·57	...	129 29 51·3	R
Apl. 19	9·5	48 4·27	...	29 51·6	R
562		*Anon.*			
Apl. 5	8·2	10 49 21·26	...	141 45 31·0	M
563		*Taylor 4945.*			
Apl. 13	7·0	10 50 0·94	...	144 54 27·6	M
14	7·0	50 0·91	...	54 26·9	M
564		*Taylor 4955.*			
Apl. 7	7·0	10 50 45·80	5	147 20 16·6	R
565		*Anon.*			
Apl. 13	9·2	10 52 0·40	6	148 46 7·7	M
566		*59 Leonis c*			
Feb. 28	...	10 53 47·99	...	83 10 47·7	R
Mar. 1	...	53 48·09	...	10 48·9	R
567		*Anon.*			
Apl. 14	8·8	10 53 53·71	...	126 32 29·1	R
27	9·0	53 53·08	..	32 28·8	R
29	8·9	53 52·60	...	32 30·3	R

Separate Results of Madras Meridian Circle Observations in 1866.

Number and Date.	Magnitude.	Mean Right Ascension, 1866. h. m. s.	No. of Wires.	Mean Polar Distance 1866. ° ′ ″	Observer.
568		30 *Ursae Majoris* α, *Dubhe.*			
Apl. 26	..	10 55 26·08	...	27 31 36·9	M
569		*Anon.*			
Apl. 11	9.1	10 56 42·81	...	140 17 23·4	M
23	8·2	56 42·66	...	17 26·4	M
570		*Anon.*			
Apl. 13	9·0	10 57 4·66	...	146 38 5·9	M
571		*Anon.*			
Apl. 17	9·8	10 57 6·06	...	146 36 30·4	R
572		*Lacaille* 4576.			
May 3	7·7	10 57 54·24	...	129 35 8·2	M
11	7·9	57 54·41	5	35 12·8	M
573		63 *Leonis* χ			
Feb. 23	...	10 58 6·22	5	81 56 27·6	R
Mar. 16	...	58 6·17	...	56 23·7	R
17	...	58 6·16	...	56 23·8	R
21	..	58 6·18	...	56 27·7	R
23	...	58 6·25	...	56 24·4	R
24	...	58 6·20	...	56 25·1	R
26	...	58 6·19	...	56 23·7	R
Apl. 19		58 6·20	...	56 24·8	R
574		*Anon.*			
Apl. 16	8·6	10 58 17·03	5	140 39 54·9	R
May 9	9·8	58 17·44	6	39 50·9	M
10	8·6	58 17·22	5	39 52·4	M
575		*Lacaille* 4595.			
Apl. 9	8·1	10 59 28·28	5	144 59 29·9	M
10	9·8	59 27·95	...	59 27·6	M
16	8·8	59 27·28	6	59 28·1	M
576		*Anon.*			
May. 4	8·3	11 1 6·09	...	135 34 12·4	M
577		67 *Leonis.*			
Apl. 21	6·0	11 1 37·21	...	64 37 5·6	R
578		*Lalande* 21371.			
Apl. 27	7·8	11 3 36·66	...	77 58 15·0	R
579		*Lalande* 21416.			
Mar. 3	8·8	11 5 5·02	...	67 13 1·8	R
580		*Anon.*			
Apl. 17	10·3	11 5 47·33	4	98 51 4·8	R
581		*Anon.*			
Mar. 13	8·4	11 5 54·81	...	148 59 32·6	M
582		68 *Leonis* δ.			
Mar. 23	...	11 6 30·66	...	68 44 36·7	R
24	...	6 29·93	—	44 38·8	R
Apl. 7	...	6 29·97	...	44 39·0	M
10	...	6 29·84	...	44 39·3	M
11	...	6 29·90	...	44 36·6	M
12	...	6 29·42	...	44 39·3	M
14	...	6 29·83	...	44 38·6	M
583		*Anon.*			
Apl. 21	8·0	11 7 12·23	5	146 40 53·0	R
23	7·0	7 12·60		40 55·2	M
584		*Anon.*			
Apl. 18	8·9	11 8 28·63	...	130 51 29·0	R
585		*Anon.*			
Apl. 25	9·3	11 9 33·58	.	145 15 33·7	R

Separate Results of Madras Meridian Circle Observations in 1866.

Number and Date.	Magnitude.	Mean Right Ascension 1866. h. m. s.	No. of Wires.	Mean Polar Distance 1866. ° ′ ″	Observer.
586		74 *Leonis* φ			
Mar. 28	...	11 9 50·98	...	92 56 13·1	n
587		*Anon.*			
May 3	6·0	11 11 13·70	...	127 35 37·6	n
588		12 *Crateris* δ			
Mar. 16	...	11 12 38·04	...	104 3 16·2	n
21	...	12 38·58	...	3 14·5	n
22	...	12 38·60	...	3 14·3	n
26	...	12 38·68	5	3 14·1	n
Apl. 6	—	12 38·89	...	3 12·5	M
11	—	12 39·03	...	3 12·1	M
13	...	12 38·85	...	3 14·4	M
14	...	12 39·56	...	3 11·7	M
16	...	12 38·54	...	3 14·1	n
19	...	12 38·40	...	3 13·6	n
21	...	12 39·51	...	3 13·7	n
589		*Anon.*			
Apl. 27	8·3	11 12 54·20	...	150 38 47·0	n
May 4	7·7	12 54·90	...	38 46·9	M
590		*Anon.*			
Mar. 23	6·0	11 16 56·65	...	128 22 16·7	n
Apl. 17	7·5	16 55·48	...	22 16·9	n
28	7·8	16 56·65	...	22 15·6	n
591		*Lacaille 4706.*			
Apl. 28	7·0	11 16 10·78	...	146 52 10·0	n
592		*Taylor 3920.*			
Mar. 2	7·8	11 19 6·10	3	131 40 10·5	M
593		*Anon.*			
Apl. 27	7·8	11 19 39·07	...	120 31 38·2	n
594		84 *Leonis* τ			
Feb. 1	...	11 21 2·80	5	86 24 22·0	n
595		*Anon.*			
Mar. 23	9·8	11 21 48·06	5	163 58 27·1	n
596		*Taylor 5345—2nd.*			
Apl. 13	8·3	11 22 8·28	...	131 56 26·9	n
21	8·5	22 8·30	...	56 26·6	n
597		*Anon.*			
Mar. 16	8·0	11 22 29·39	6	150 4 54·5	n
598		*Anon.*			
Apl. 28	8·9	11 22 59·98	...	146 54 59·2	n
599		*Anon.*			
May 3	9·6	11 23 29·45	6	98 21 52·3	M
600		*Anon.*			
Apl. 16	9·0	11 23 40·63	...	146 9 20·0	n
601		*Anon.*			
Mar 6	9·3	11 24 46·91	...	94 46 8·4	n
602		*Anon.*			
Apl. 27	9·4	11 25 26·40	...	129 33 25·6	n
603		*Anon.*			
May 6	8·8	11 26 20·20	6	148 51 36·8	n
604		*Anon.*			
Apl 27	9·0	11 28 25·65		125 30 47·5	n
28	9·6	28 25·71	...	30 47·7	n

Separate Results of Madras Meridian Circle Observations in 1866.

Number and Date.	Magnitude.	Mean Right Ascension 1866. h. m. s.	No. of Wires.	Mean Polar Distance. 1866. ° ′ ″	Observer.
605		*λ Centauri.*			
Feb. 1	5·0	11 29 37·89	...	152 16 45·4	R
Apl. 24	5·0	29 37·06	...	16 40·8	M
May 4	5·6	29 37·03	...	16 41·4	M
606		*Anon.*			
Apl. 9	7·0	11 29 56·62	5	140 16 22·7	M
607		91 *Leonis υ*			
Mar. 1	...	11 30 5·19	...	90 5 5·6	R
2	...	30 5·26	...	5 5·0	R
20	...	30 5·24	...	5 4·7	R
28	...	30 5·33	...	5 5·7	M
Apl. 6	...	30 5·32	...	5 3·5	M
7	...	30 5·24	...	5 5·2	M
18	...	30 5·30	...	5 4·6	M
19	...	30 5·19	...	5 4·4	R
21	...	30 5·21	...	5 4·4	R
23	...	30 5·25	...	5 3·5	M
608		*Anon.*			
May 16	9·3	11 32 14·26	6	141 15 11·4	M
609		*Anon.*			
May 11	8·2	11 33 13·21	...	141 15 15·7	M
610		*W. B. E. XI.* 571.			
Feb. 2	7·9	11 33 32·07	...	98 18 20·7	M
Mar. 17	8·0	33 31·78	...	18 22·5	R
19	8·0	33 31·80	...	18 22·2	R
611		*Anon.*			
Apl. 29	8·0	11 34 3·28	...	127 49 56·4	R
612		*Anon.*			
May 3	8·6	11 34 28·14	...	144 21 31·9	R
613		*Anon.*			
Apl. 16	9·3	11 36 9·02	...	130 40 56·2	R
18	9·0	36 8·84	...	40 57·1	R
614		*Taylor* 5384.			
Apl. 24	6·0	11 37 8·55	...	151 44 44·0	M
615		*Anon.*			
Apl. 27	8·9	11 39 49·00	...	129 34 43·1	R
616		*Anon.*			
Mar. 20	9·5	11 41 2·21	5	149 22 44·0	M
617		*Anon.*			
Mar. 2	9·2	11 41 14·86	...	128 31 5·6	R
618		94 *Leonis β*			
Mar. 3	...	11 42 13·35	...	74 40 44·5	R
Apl. 5	...	42 13·29	...	40 49·2	M
6	...	42 13·25	...	40 49·3	R
7	...	42 13·37	...	40 46·4	M
10	...	42 13·31	...	40 46·9	M
11	...	42 13·18	...	40 46·3	M
19	...	42 13·25	...	40 46·9	R
20	...	42 13·25	...	40 46·0	R
21	...	42 13·34	.	40 47·7	R
619		*Taylor* 5421.			
Mar. 24	8·0	11 46 16·98	...	129 31 50·0	R
Apl. 27	7·5	46 16·92	...	31 54·1	R
May 4	7·7	45 16·96	...	31 51·9	M
620		5 *Virginis β*			
Mar. 1	...	11 43 42·98		87 29 22·4	R
621		*Taylor* 5427.			
Feb. 2	6·0	11 44 11·23	...	94 05 20·9	M

Separate Results of Madras Meridian Circle Observations in 1866.

Number and Date.	Magnitude.	Mean Right Ascension 1860. h.	m.	s.	No. of Wires.	Mean Polar Distance 1866. °	′	″	Observer.
672 *Taylor* 5433.									
Mar. 24	8·2	11	44	57·39	6	120	38	41·7	n
673 *Groombridge* 1830.									
Mar. 22	8·4	11	45	14·86	...	51	19	18·2	n
May 3	7·2		45	14·75	...		19	16·0	M
7	7·0		45	14·70	...		19	15·4	M
674 64 *Ursae Majoris* γ									
Apl. 12	...	11	46	44·81	...	35	33	37·7	M
675 *Lacaille* 4037.									
Mar. 23	7·7	11	48	21·34	...	168	38	3·1	n
Apl. 13	7·0		48	21·38	...		38	3·9	M
May 10	7·0		48	21·43	...		39	0·4	M
676 *Lacaille* 4046.									
Apl. 24	...	11	50	15·62	5	162	30	10·8	M
May 5	...		50	15·76	4		30	14·2	M
10	8·0		50	15·98	3		30	12·5	M
11	8·3		50	15·88	6		30	14·1	M
14	8·0		50	15·85	...		30	14·3	M
677 *Anon.*									
Apl. 10	7·9	11	50	48·00	...	150	22	39·0	M
11	7·9		50	48·19	...		22	38·1	M
678 *Lacaille* 4056.									
Apl. 9	8·1	11	51	15·99	...	154	34	36·0	M
679 *Anon.*									
Apl. 27	9·5	11	52	45·36	4	154	38	59·3	n
28	9·5		52	45·47	...		38	57·8	n
680 *Taylor* 5534.									
Mar. 21	...	11	56	55·62		148	58	1·5	n
681 *R. P. L* 89.									
Apl. 28	...	11	57	57·35	5	3	40	17·3	n
May 3	...		57	57·61	3		40	15·0	M
4	...		57	58·44	3		40	16·8	M
13	...		57	57·22	3		40	16·3	M
682 *Anon.*									
Apl. 14	8·0	11	59	7·57	...	128	38	30·7	M
683 *Anon.*									
Mar. 27	9·5	11	59	37·39	6	160	20	16·7	n
Apl. 28	9·0		59	37·65	...		20	17·4	n
May 9	9·6		59	37·52	..		20	13·7	M
684 *Anon.*									
May 5	...	11	59	50·64	...	144	16	51·8	M
15	8·0		59	50·68	6		16	51·7	M
685 *Anon.*									
Apl. 9	8·3	12	1	13·94	...	169	22	9·4	M
686 *Anon.*									
May 14	9·4	12	1	40·48	4	130	3	16·6	M
687 *Lacaille* 5041.									
May 11	7·9	12	2	30·11		161	33	54·2	M
688 10 *Virginis.*									
Feb. 2	6·0	12	2	49·35	...	87	20	59·6	M
Mar. 30	...		2	49·58	.		21	1·5	n

Separate Results of Madras Meridian Circle Observations in 1866.

Number and Date.	Magnitude.	Mean Right Ascension 1866. h. m. s.	No. of Wires.	Mean Polar Distance 1866. ° ′ ″	Observer.
639		*2 Corvi ε*			
Apl. 6	...	12 3 14·18	...	111 53 27·2	M
10	...	3 14·20	...	52 25·5	M
11	...	3 14·27	...	52 26·3	M
12	...	3 14·40	...	52 24·7	M
13	...	3 14·34	...	52 24·1	M
14	...	3 14·26	4	52 25·4	M
16	...	3 14·22	...	52 23·3	R
17	...	3 14·13	...	52 23·9	R
19	...	3 14·09	...	52 23·2	R
24	...	3 14·07	...	52 26·3	M
May 8	...	3 14·30	4	52 28·2	M
640		*Anon.*			
Mar. 26	8·3	12 3 41·96	5	146 57 45·7	R
641		*Anon.*			
May 10	8·0	12 5 54·49	...	134 8 46·3	M
642		*Anon.*			
Apl. 5	8·0	12 6 13·80	5	136 26 12·5	M
643		*Anon.*			
May 17	9·2	12 6 36·49	...	148 51 20·6	R
644		*Anon.*			
May 4	8·7	12 11 59·60	...	100 23 29·9	M
11	8·6	11 59·30	6	23 31·8	M
645		*Taylor 5643.*			
May 3	6·9	12 12 57·69	...	152 6 38·5	M
646		*R. P. L. 92.*			
Apl. 25	...	12 13 50·31	3	3 40 8·6	R
647		*15 Virginis η*			
Mar. 2	...	12 13 3·00	6	89 55 20·6	R
3	...	13 2·99	...	55 20·7	M
Apl. 13	...	13 3·05	...	55 19·6	M
17	...	13 2·96	...	55 19·8	R
18	...	13 3·05	...	55 19·0	R
May 5	...	13 2·97	...	55 19·5	M
8	...	13 3·02	...	55 19·7	M
648		*Anon.*			
May 14	9·3	12 14 10·41	...	143 45 30·9	M
15	9·0	14 10·26	...	45 29·5	M
17	9·4	14 10·11	6	45 27·2	M
649		*Lacaille 5119.*			
Mar. 27	8·9	12 15 27·96	...	139 34 57·2	R
650		*Anon.*			
Apl. 7	7·9	12 15 39·59	...	141 40 36·7	M
12	8·0	15 39·56	...	40 35·7	M
651		*Anon.*			
May 10	8·3	12 16 52·76	...	147 10 26·3	M
652		*α Crucis—1st.*			
Mar. 23	...	12 19 10·01	5	152 21 24·3	R
May 19	.	19 10·49		21 22·3	R
653		*Anon.*			
May 22	8·8	12 19 53·96		141 4 20·0	R
654		*Anon.*			
Apl. 6	9·0	12 24 46·94	..	139 50 19·0	M
May 9	8·8	24 46·64	..	50 16·7	M

Separate Results of Madras Meridian Circle Observations in 1866.

Number and Date.	Magnitude.	Mean Right Ascension 1860. h.	m.	s.	No. of Wires.	Mean Polar Distance 1860. °	'	"	Observer.
655			*9 Corvi β*						
Apl. 5	...	12	27	21·24	...	112	30	19·8	M
17	...		27	21·07	...		30	20·8	R
21	...		27	21·01	...		30	20·7	R
23	..		27	21·19	..		30	19·6	M
May 5	...		27	21·15	...		30	19·4	M
19	...		27	21·21	...		30	18·8	R
21	...		27	21·13	..		30	19·6	R
26	...		27	21·17	...		30	20·6	R
656			*Anon.*						
May 10	8·4	12	27	58·70	..	141	40	33·6	M
657			*25 Virginis f*						
Apl. 27	..	12	29	58·47	...	96	5	34·3	R
658			*Anon.*						
May 4	8·8	12	30	57·38	...	142	20	19·5	M
659			*R Virginis Var. 2.*						
May 8	7·5	12	31	42·01	...	83	16	26·5	M
660			*Taylor 5830.*						
May 15	7·0	12	31	38·67	...	141	1	33·3	M
661			*29 Virginis γ¹*						
Mar. 2	...	12	34	52·20	...	90	42	30·5	R
31	...		34	52·25	...		42	40·6	R
Apl. 27	...		34	52·19	...		42	40·6	R
May 18	...		34	52·31	...		42	47·5	R
662			*S Ursae Majoris Var. 2.*						
May 26	9·7	12	38	8·37	4	28	10	21·3	R
663			*Taylor 5863.*						
Apl. 13	7·7	12	38	39·05	...	143	52	42·3	M
May 10	6·9		38	39·25	...		52	48·4	M
21	8·0		38	39·30			52	48·3	R
23	7·4		38	39·33	5		52	46·1	R
664			*Anon.*						
May 22	9·0	12	38	32·57	4	116	50	24·3	R
665			*Anon.*						
May 16	8·0	12	42	31·26	..	147	19	28·3	R
28	9·5		42	31·37	..		19	23·1	R
666			*Anon.*						
May 14	8·0	12	44	30·65	6	141	36	17·7	M
15	8·0		44	30·38	...		36	16·8	M
667			*R. P. L. 90.*						
Apl. 5	...	12	48	10·65	2	5	51	33·0	M
14	...		48	10·26	3		51	32·0	M
668			*Anon.*						
May 11	9·0	12	48	26·11	...	126	26	39·5	M
669			*Anon.*						
May 23	9·0	12	46	50·02	...	149	04	46·6	R
670			*12 Canis Venaticorum.*						
Apl. 21	...	12	49	46·33	...	50	37	27·1	M
671			*Taylor 5974.*						
May 9	8·0	12	53	1·39	...	146	30	10·6	M
672			*Anon.*						
May 14	9·0	12	55	0·56	5	121	23	41·6	M
16	8·3		55	0·65	5		23	41·5	R
17	9·0		55	0·70			23	41·7	R
673			*48 Virginis.*						
Mar. 30	...	12	57	0·22		97	58	31·3	R
31	.		57	0·31			58	31·2	R

Separate Results of Madras Meridian Circle Observations in 1866.

Number and Date.	Magnitude.	Mean Right Ascension 1866. h.	m.	s.	No. of Wires.	Mean Polar Distance 1866. °	′	″	Observer.
674 *Anon.*									
May 22	9.3	12	57	5.07	...	123	25	50.0	R
675 *Lacaille* 5381.									
Apl. 11	7.9	12	57	14.68	...	129	57	43.1	M
676 *Anon.*									
May 28	9.3	12	58	12.73	...	121	29	21.4	R
677 *Anon.*									
May 11	8.0	12	58	22.57	4	134	23	31.8	M
14	7.0		58	22.26	...		23	34.7	M
678 *Taylor* 6025.									
Apl. 10	8.0	12	59	31.13	...	123	34	0.8	M
May 22	7.3		59	31.27	...		34	4.1	R
23	7.7		59	31.34	5		34	2.5	R
679 51 *Virginis θ*									
Mar. 8	...	13	3	0.73	...	94	40	22.8	M
Apl. 27	...		3	0.86	...		40	22.3	R
May 4	...		3	0.71	...		40	21.0	M
9	...		3	0.61	...		40	20.6	M
25	..		3	0.76	...		40	23.6	R
29	...		3	0.78	.		40	21.2	R
680 *W Virginis Var.* 1.									
Apl. 19	8.0	13	7	0.60	...	106	30	33.7	R
May 23	8.3		7	0.67	...		30	33.1	R
681 *R. P. L.* 101—*s.p.*									
Oct. 25	.	13	10	3.63	3	1	37	56.3	R
Nov. 3	.		10	3.80	1		37	54.3	R
11			10	4.04	2		37	57.3	R
682 *Anon.*									
May 16	8.3	13	12	39.08	5	122	57	12.3	R
683 *Taylor* 6148.									
May 11	7.0	13	14	19.63	...	128	8	57.6	M
17	7.5		14	19.40	...		8	57.2	R
22	7.7		14	18.51	...		8	57.7	R
684 *O. A. N.* 13563.									
May 26	9.1	13	16	29.01	...	27	53	51.8	R
685 *Anon.*									
Apl. 21	9.5	13	17	18.06	6	126	10	5.2	R
May 17	9.7		17	18.90	..		10	3.1	R
686 67 *Virginis α, Spica.*									
Mar. 3	...	13	18	8.14	...	100	27	39.2	M
31	...		18	8.17	...		27	39.8	R
Apl. 27	...		18	8.12	...		27	40.9	R
28	...		18	8.20	...		27	40.8	R
May 21	...		18	8.08	...		27	40.4	R
687 *Lacaille* 5546.									
May 28	9.2	13	19	48.77	...	116	23	6.4	R
688 *R. P. L.* 103.									
Apl. 19	...	13	20	11.27	3	4	33	40.7	R
689 *Anon.*									
May 26	10.7	13	23	34.21	5	88	23	30.4	R
690 *Anon.*									
Apl. 21	8.9	13	24	54.30	.	125	9	19.6	R
May 16	9.0		24	54.36	.		9	16.2	R
691 *Anon.*									
June 9	9.0	13	25	3.06	6	131	9	43.9	M

Separate Results of Madras Meridian Circle Observations in 1866.

Number and Date.	Magnitude.	Mean Right Ascension 1866. h. m. s.	No. of Wires.	Mean Polar Distance 1866. ° ′ ″	Observer.
692		*76 Virginis h*			
Mar. 31	...	13 25 51·80	...	99 25 24·7	R
Apl. 27	...	25 51·79	...	25 24·2	R
28	..	25 51·72	...	25 26·1	R
693		*S Virginis Var.* 6.			
May 1	8·0	13 26 0·03	...	96 30 21·6	R
13	7·5	26 0·17	...	30 19·3	M
June 7	6·9	26 0·15	...	30 21·1	M
694		79 *Virginis* ζ			
May 5	...	13 27 51·81	...	99 51 26·6	M
22	...	27 52·02	...	51 25·4	R
23	...	27 52·10	...	54 26·1	R
29	...	27 52·09	...	54 24·1	R
695		*Taylor* 7183.			
Apl. 19	7·0	13 29 30·80	...	131 48 46·4	R
30	7·0	29 30.56	...	48 46·5	R
696		*Anon.*			
June 8	7·7	13 33 5·21	5	129 2 11·1	M
697		*Anon.*			
May 23	9·5	13 34 10·06	...	129 10 26·4	R
698		82 *Virginis m.*			
Mar. 31	..	13 31 34·96	..	98 1 26·5	R
699		*Anon.*			
May 1	8·2	13 36 40·90	...	123 4 4·6	R
700		*Anon.*			
Apl. 30	9·2	13 36 21·51	...	128 5 57·1	R
701		*Taylor* 6966.			
June 5	7·5	13 37 1·26	4	151 46 42·2	M
702		*Anon.*			
Apl. 10	7·8	13 37 26·97	5	128 40 54·2	R
703		*Anon.*			
June 9	8·6	13 44 21·36	...	123 7 7·8	M
704		*Anon.*			
June 7	8·3	13 43 31·30	[illegible]	128 18 40·9	M
8	8.6	43 30·97	[illegible]	18 46·0	M
705		*Anon.*			
Apl. 19	8·3	13 44 23·04	5	137 57 29·9	R
706		8 *Bootis* η			
Apl. 26	...	13 48 13·22	...	70 55 44·4	R
May 1	...	48 13·26	...	55 46·9	R
3	...	48 13·30	...	55 45·8	M
4	...	48 13·31	..	55 46·5	M
June 12	...	48 13·14	...	55 47·6	M
707		*Anon.*			
June 9	8·8	13 52 28·74	...	161 31 34·6	M
708		β *Centauri.*			
Apl. 27	...	13 54 28·72	4	149 48 29·4	R
May 16	...	54 24·18	4	48 30·3	R
709		93 *Virginis* τ			
May 1	...	13 54 49·68	...	87 46 21·1	R
4	...	54 49·56	...	46 24·9	M
7	...	54 49·70	...	46 20·6	M
8	...	54 49·61	4	46 22·4	M
11	...	54 49·67	...	46 24·2	M
14	...	54 49·74	...	46 21·0	M
16	...	54 49·69	...	46 19·8	R
17	..	54 49·77	..	46 20·5	R
25		54 49·65	...	46 21·4	R
June 1	.	54 49·72	...	46 19·6	M
6	...	54 49·64	...	46 21·4	M

Separate Results of Madras Meridian Circle Observations in 1866.

Number and Date.	Magnitude.	Mean Right Ascension 1866. h.	m.	s.	No. of Wires.	Mean Polar Distance 1866. °	′	″	Observer.
710			*Lacaille* 5794.						
June 6	6·9	13	57	13·97	5	153	46	8·2	M
711			94 *Virginis.*						
May 26	...	13	59	12·18	...	98	15	8·1	R
712			*Taylor* 6570.						
June 5	8·4	14	0	9·90	...	149	56	38·8	M
713			*R. P. L.* 108—*s.p.*						
Dec. 6	...	14	3	39·57	2	3	36	3·2	M
714			*Taylor* 6616.						
June 9	5·9	14	5	39·86	...	146	27	24·1	M
715			98 *Virginis κ*						
Apr. 28	...	14	5	46·09	...	99	38	54·3	R
May 26	...		5	46·05	...		38	54·7	R
716			16 *Bootis α, Arcturus.*						
May 1	...	14	9	33·06	...	70	7	7·9	R
3	...		9	32·90	...		7	10·0	M
11	...		9	33·04	...		7	8·6	M
12	...		9	33·18	...		7	7·7	M
14	...		9	32·96	...		7	8·8	M
16	...		9	33·02	...		7	7·1	R
17	...		9	32·97	...		7	7·4	R
June 5	...		9	33·01	...		7	8·7	M
6	...		9	33·00	...		7	9·8	M
8	...		9	33·00	...		7	10·2	M
12	...		9	33·00	...		7	8·6	M
717			*Anon.*						
June 7	6·2	14	14	36·45	...	150	46	39·8	M
9	7·9		14	36·46	...		46	39·3	M
718			*Taylor* 6721.						
June 8	7·9	14	17	39·09	.	101	3	22·4	M
719			*Taylor* 6740.						
May 17	8·0	14	19	12·77	...	138	48	39·0	R
720			*Anon.*						
May 16	8·5	14	20	4·88	...	127	9	26·0	R
721			25 *Bootis ρ*						
May 3	...	14	26	3·14	...	59	2	23·4	M
14	...		26	3·28	...		2	22·6	M
26	...		26	3·30	5		2	21·4	R
June 6	...		26	3·17	5		2	20·4	M
8	...		26	3·18	3		2	24·1	M
722			*α Centauri*—1*st.*						
May 16	...	14	30	31·29	...	150	16	56·2	R
21	...		30	31·34	...		16	56·4	R
June 26	...		30	31·00	...		16	54·8	R
723			36 *Bootis ε* (*Mirac.*)						
May 14	...	14	39	8·02	...	62	21	36·1	M
June 5	...		39	7·78	...		21	36·7	M
6	...		39	8·06	...		21	34·4	M
13	...		39	8·00	...		21	35·7	M
724			9 *Librae α²*						
Apl. 30	...	14	43	26·02	...	105	28	59·9	R
May 2	...		43	25·95	...		28	60·2	R
28	...		43	25·13	...		28	60·1	R
June 12	...		43	26·18	5		28	59·5	M
725			*O. A. S.* 14112.						
June 5	7·8	14	50	50·78	...	109	11	13·5	M
6	...		50	50·95	...		11	12·6	M
7	7·9		50	50·90			11	13·9	M
726			*Anon.*						
June 6	9·0	14	51	36·94	5	29	30	59·6	M

Separate Results of Madras Meridian Circle Observations in 1866.

Number and Date.	Magnitude.	Mean Right Ascension 1866. h.	m.	s.	No. of Wires.	Mean Polar Distance 1866. °	′	″	Observer.
727 *O. A. N.* 15004.									
June 8	7·8	14	58	54·66	5	80	21	32·0	M
728 *Taylor* 7017.									
June 9	7·6	14	57	14·77	...	150	34	36·0	M
729 43 *Bootis* ψ									
May 16	...	14	58	48·22	...	62	31	45·5	R
26	...		58	48·22	...		31	42·0	R
June 16	...		58	48·19	5		31	46·2	R
26	...		58	48·20	—		31	46·6	R
29	...		58	48·21	...		31	46·2	R
730 21 *Librae* ν¹									
Apl. 30	...	14	59	9·15	...	106	44	6·8	R
731 *R. P. L.* 111.									
May 28	...	15	5	29·67	3	5	31	38·9	R
732 27 *Librae* β									
May 18	...	15	9	47·93	...	98	53	10·8	R
June 5	...		9	47·87	...		53	11·6	M
8	...		9	47·91	3		53	8·3	M
9	...		9	47·92	—		53	10·4	M
16	...		9	47·86	...		53	13·0	R
22	...		9	47·98	...		53	13·0	R
25	...		9	47·84	...		53	13·1	R
26	...		9	47·85	...		53	11·0	R
733 *Anon.*									
July 17	9·0	15	14	19·77	3	128	7	36·6	R
734 *S. Serpentis Var* 3.									
May 21	10·2	15	13	58·04	6	75	12	12·9	R
735 *S Coronae Borealis Var* 2.									
June 29	9·5	15	15	56·32	5	58	8	28·5	R
736 *Lacaille* 6877.									
May 26	8·0	15	19	1·84	5	130	11	29·4	R
737 *Anon.*									
May 23	9·3	15	20	21·65	5	129	9	1·0	R
738 *W. B. E. XV.* 305.									
May 22	8·9	15	22	5·17	...	101	16	55·5	R
739 *Taylor* 7220.									
June 26	...	15	22	14·02	...	123	6	59·2	R
July 17	...		22	14·13	4		7	1·1	R
740 38 *Librae* γ									
May 28	...	15	28	2·08	...	104	20	24·3	R
741 5 *Coronae Borealis* α, *Alphecca.*									
June 7	...	15	29	0·95	...	62	49	59·7	M
16	...		29	0·96	6		50	1·1	R
29	...		29	0·92	...		49	59·0	R
July 17	...		29	0·74	...		49	57·8	R
742 *W. B. E. XV.* 587.									
June 5	7·8	15	31	30·60	3	106	27	35·9	M
743 *W. B. E. XV.* 645.									
May 22	9·1	15	34	20·23	3	102	19	46·2	R
744 *Anon.*									
May 21	8·5	15	34	30·03	...	129	1	50·3	R
745 24 *Serpentis* α									
May 2	...	15	37	40·06	...	83	9	2·6	R
June 23	...		37	40·04	..		9	4·6	R

Separate Results of Madras Meridian Circle Observations in 1866.

Number and Date.	Magnitude.	Mean Right Ascension 1866. h. m. s.	No. of Wires.	Mean Polar Distance 1866. ° ′ ″	Observer.
746		*O. A. S.* 14674.			
June 16	..	15 39 31·98	...	104 49 3·6	R
747		*O. A. S.* 14934.			
July 7	9·6	15 42 56·11	...	107 52 48·1	M
24	9·3	42 56·23	5	52 47·3	M
748		*W. B. E. XV.* 838.			
June 5	7·7	15 44 6·09	...	101 27 21·5	M
749		*O. A. S.* 14963.			
July 23	7·0	15 44 35·18	...	108 1 54·7	M
750		46 *Librae* θ			
Apl. 30	...	15 46 11·79	...	106 30 1·7	R
May 1	...	46 11·86	...	30 0·6	R
751		*Radcliffe* 3402.			
May 23	8·0	15 46 34·70	...	47 1 35·2	R
752		16 *Ursae Minoris* ζ			
July 16	...	15 48 [illegible]	4	11 47 40·3	R
753		49 *Librae.*			
May 31	...	15 53 [illegible]	...	106 8 9·1	R
754		8 *Scorpii* β¹			
Apl. 30	.	15 57 [illegible]	...	109 26 10·9	R
May 1	...	57 [illegible]	...	26 10·2	R
2		57 [illegible]	...	26 10·0	R
19		57 [illegible]	...	26 9·8	R
June 16	.	57 [illegible]	6	26 10·2	R
755		*O. A. S.* 15237.			
June 3	8·4	15 59 [illegible]		109 31 48·3	M
July 16	8·0	59 [illegible]	3	34 43·7	R
756		*Lalande* 29306.			
June 7	8·2	15 59 33·00	...	107 34 [illegible]	M
9	8·6	59 33·19	—	31 19·7	M
757		*R Herculis Var.* 2			
May 17	9·5	16 0 12·28	...	71 16 3·7	R
758		*Lalande* 29301.			
June 1	7·0	16 1 56·35	5	102 41 39·8	M
759		*O. A. S.* 15342.			
June 6	8·0	16 3 39·22	...	107 45 50·4	M
7	7·8	3 38·12	...	45 50·5	M
9	8·3	3 38·16	...	45 49·5	M
760		*R. P. L.* 116.			
May 26	—	16 4 16·[illegible]	8	4 19 5·9	R
		R. P. L. 116—*s.p.*			
Dec. 22	...	16 4 17·37	8	4 19 5·4	R
761		*Anon.*			
June 3	7·6	16 4 18·61	...	107 [illegible] 53·9	M
July 19	7·8	4 [illegible]·61		53 54·1	R
762		*Weisse XVI.* 83.			
May 13	9·0	16 4 9·00		108 41 34·6	R
763		1 *Ophiuchi* δ.			
June 19	..	16 7 19·44	...	93 30 51·3	R
27	.	7 19·33		30 49·9	R
July 6		7 19·60	5	30 49·6	M
16		7 19·30	...	30 50·4	R
30	.	7 19·47	—	30 51·0	R

Separate Results of Madras Meridian Circle Observations in 1866.

Number and Date.	Magnitude.	Mean Right Ascension 1866. h.	m.	s.	No. of Wires.	Mean Polar Distance 1866. °	′	″	Observer.
764				*Anon.*					
May 19	10·5	16	12	31·55	6	107	27	3·2	n
21	10·4		12	31·13	...		27	2·7	n
22	10·2		12	31·20	...		27	1·8	n
765				*Anon.*					
May 17	9·5	16	16	42·85	...	107	25	58·9	n
19	9·7		16	42·81	5		25	58·7	n
July 16	10·0		16	42·60	...		25	52·9	n
19	9·3		16	42·60	...		25	53·6	n
766				*Taylor* 8521.					
July 20	9·0	16	17	35·84	...	113	5	37·9	n
767				7 *Ophiuchi* ψ					
May 28	..	16	19	15·72	...	108	8	57·6	n
768				21 *Scorpii* α, *Antares.*					
June 5	...	16	21	11·69	5	116	7	54·5	m
19	...		21	11·74	5		7	53·6	n
July 6	..		21	11·67	6		7	58·9	m
7	...		21	11·60	...		7	53·1	m
769				14 *Draconis* η					
July 19	...	16	23	10·92	4	28	10	54·8	n
770				9 *Ophiuchi* ω					
May 28		16	24	11·86	...	111	10	30·7	n
29	...		24	11·82	5		10	35·4	n
771				30 *Herculis* g *Var.* 5.					
June 6	6·0	16	26	14·54	5	47	40	20·6	m
July 20	6·0		26	14·87	4		40	20·1	n
772				*Taylor* 7723.					
May 1		16	33	49·60	...	107	25	47·3	n
2	.		33	49·46	..		25	46·9	n
773				α *Trianguli Australis.*					
July 16	...	16	34	30·29	...	158	46	39·1	n
774				40 *Herculis* ζ					
July 5	...	16	36	14·12	...	58	9	8·7	m
13	...		36	14·09	...		9	11·6	m
19	...		36	16·07	...		9	8·8	n
20	...		36	14·02	4		9	9·9	n
24	...		36	14·11	...		9	10·3	m
775				*Anon.*					
July 16	9·0	16	44	32·88	4	131	1	44·3	n
776				*Taylor* 7842.					
May 1	...	16	48	17·64	...	108	35	28·1	n
2	...		48	17·84	5		35	28·1	n
July 19	6·8		48	17·78	...		35	28·6	n
777				27 *Ophiuchi* κ					
July 5	...	16	51	19·62	...	80	24	31·4	m
13	...		51	19·86	4		24	32·5	m
17	...		51	19·80	...		24	32·2	n
20	..		51	19·76	...		24	31·3	n
778				*O. A. S.* 16292.					
July 19	9·0	16	54	4·72	5	116	14	22·8	n
779				*Anon.*					
July 16	9·0	16	56	28·86	..	100	36	47·3	n
780				22 *Ursae Minoris* ε—s.p.					
Jan. 5		16	59	46·76	5	7	44	51·1	m
15	..		59	49·04	2		44	39·5	m
22	...		59	46·81	5		44	36·1	n
781				35 *Ophiuchi* η.					
May 30	..	17	2	41·96		105	33	22·1	n

Separate Results of Madras Meridian Circle Observations in 1866.

Number and Date.	Magnitude.	Mean Right Ascension 1866. h.	m.	s.	No. of Wires.	Mean Polar Distance 1866. °	′	″	Observer.
782		*Anon.*							
July 19	.	17	5	52·61	5	130	58	34·2	n
783		64 *Herculis* α *Var.* 1.							
July 16	...	17	8	32·27		75	27	16·0	n
20	...		8	32·25	...		27	17·0	n
784		42 *Ophiuchi* θ.							
June 10	.	17	13	46·86	...	114	51	46·0	n
July 16	...		13	46·80	...		51	44·9	n
17	..		13	47·07	...		51	45·0	n
19	...		13	46·91	...		51	46·3	n
20	...		13	46·91	...		51	46·7	n
785		*Anon.*							
Aug. 17	7·9	17	25	12·21	...	130	48	55·1	n
786		*Anon.*							
Aug. 19	7·9	17	28	38·11	.	128	14	45·6	n
787		55 *Ophiuchi* α.							
July 16		17	28	46·81	..	77	20	24·5	n
19	.		28	46·83	4		20	24·0	n
Aug. 7	.		28	46·[illegible]	5		20	25·1	n
9			28	46·80	...		20	25·6	n
16	.		28	46·[illegible]	...		20	22·7	n
20	.		28	46·[illegible]	...		20	23·8	n
21			28	46·[illegible]	...		20	25·3	n
22			28	46·79	4		20	24·2	n
788		55 *Serpentis* ξ.							
May 2	...	17	30	54·80		105	16	40·3	n
20			30	55·[illegible]	5		16	39·2	n
789		*Lacaille* 7406.							
June 25	...	17	34	39·73	5	128	44	16·7	n
Aug. 10	7·7		34	39·[illegible]	6		44	14·3	n
790		58 *Ophiuchi*.							
June 25	...	17	35	[illegible]	5	111	[illegible]	54·0	n
791		*Anon.*							
Aug. 20	7·9	17	37	47·86	6	126	29	22·7	n
21	7·8		37	48·10	..		29	24·2	n
792		86 *Herculis* μ							
July 5	...	17	41	13·02	...	62	11	58·6	n
Aug. 10	...		41	12·97			11	57·4	n
11	...		41	12·[illegible]	5		11	57·7	n
22	..		41	12·[illegible]	...		11	56·3	n
793		*Lacaille* 7504.							
Aug. 9	7·8	17	48	49·46	5	129	6	32·9	n
794		*Lacaille* 7517.							
Aug. 17	7·7	17	52	34·31	.	140	10	25·4	n
795		*Lacaille* 7518.							
Aug. 16	7·9	17	52	58·80	5	140	12	16·4	n
796		33 *Draconis* γ							
Aug. 10	.	17	53	29·64	...	38	29	29·5	n
11	...		53	29·67	3		29	29·4	n
22			53	29·84	...		29	25·6	n
797		*Anon.*							
Aug. 14	7·9	18	5	8·15	...	80	1	12·3	n
798		*Anon.*							
Aug. 9	10·0	18	4	1·[illegible]		80	9	[illegible]	n

Separate Results of Madras Meridian Circle Observations in 1866.

Number and Date.	Magnitude.	Mean Right Ascension 1866. h. m. s.	No. of Wires.	Mean Polar Distance 1866. ° ′ ″	Observer.
799		*13 Sagittarii μ¹*			
May 30	...	18 5 46·08	...	111 5 28·0	R
31	...	5 44·98	...	5 26·7	R
June 26	...	5 44·96	...	5 26·1	R
July 13	...	5 45·09	...	5 27·2	M
19	..	5 44·98	...	5 28·0	R
Aug. 7		5 44·96	...	5 28·0	R
10	..	5 45·06	...	5 26·5	R
11	.	5 44·90	...	5 27·3	R
17	...	5 44·84	...	5 27·2	M
21	...	5 44·98	..	5 27·7	M
800		*Taylor* 8461.			
Aug. 7	...	18 14 33·17	...	134 10 22·7	R
801		23 *Ursae Minoris δ*			
July 5	...	18 15 38·96	5	3 23 44·6	M
Aug. 9	...	15 31·61	3	23 42·8	R
		23 *Ursae Minoris δ—s.p.*			
Feb. 3	...	18 16 34·81	2	3 23 46·7	M
10	...	15 34·72	3	23 42·6	M
802		*Lalande* 33845.			
Aug. 10	...	18 16 42·37	...	108 4 37·7	R
18	7·0	15 42·26	...	4 38·1	M
803		*δ¹ Telescopii.*			
Aug. 17	6·0	18 22 6·97	...	135 30 48·5	M
804		*Taylor* 8551.			
Aug. 16	7·2	18 27 38·74	...	109 13 28·4	M
805		*Anon.*			
Aug. 20	7·8	18 28 36·16	..	135 34 27·6	M
806		3 *Lyrae α, Vega.*			
Aug. 7	...	18 32 33·97	..	51 30 28·4	R
10	...	32 34·01	...	30 23·2	R
11	...	32 33·98	...	30 26·1	R
17	...	32 33·97	5	30 20·9	M
21	...	32 34·01	4	30 23·2	M
22		32 34·04	...	30 21·6	M
23	...	32 33·92	...	30 23·1	M
25	...	32 34·04	...	30 22·6	M
807		*Lacaille* 7832.			
Aug. 16	7·1	18 37 49·15	...	149 5 28·8	M
808		*R Scuti Var.* 1.			
Aug. 10	7·0	18 40 19·90	...	95 50 48·6	R
13	7·3	40 19·84	...	50 46·9	M
20	7·0	40 19·85	...	50 47·6	M
Sep. 8	6·9	40 19·81	6	50 46·4	M
809		*Lacaille* 7872.			
Aug. 22	6·5	18 46 29·28	5	138 44 35·5	M
29	6·5	46 29·17	...	44 35·8	R
810		*Lacaille* 7878.			
Aug. 25	6·7	18 48 2·30	...	138 44 33·4	M
30	7·5	48 2·35	...	44 31·8	R
811		10 *Lyrae β Var.* 1.			
Aug. 7	...	18 45 7·98	5	56 47 27·0	R
11	...	45 7·89	...	47 25·9	R
17	...	45 7·86	...	47 28·5	M
28	...	45 7·92	...	47 27·9	M
Sep. 7	...	45 7·96	..	47 29·9	M
812		*Anon.*			
Sep. 8	6·6	18 47 2·97	5	137 44 46·0	M
10	7·8	47 3·01	5	43 45·8	M

Separate Results of Madras Meridian Circle Observations in 1866.

Number and Date.	Magnitude.	Mean Right Ascension 1866. h.	m.	s.	No. of Wires.	Mean Polar Distance 1866. °	′	″	Observer.
813		37 *Sagittarii* ξ²							
June 27	...	18	40	44·34	...	111	16	48·3	R
814		13 *Lyrae Var.* 2.							
Aug. 7		18	51	15·36	4	46	13	48·9	R
9	5·3		51	15·29	5		13	44·5	R
16	4·9		51	15·28	...		13	44·1	M
20	5·6		51	15·46	...		13	43·7	M
815		*O. A. S.* 18960.							
Aug. 18	7·4	18	53	46·14	5	121	7	29·9	M
29	8·0		53	46·40	...		7	28·8	R
Sep. 7	7·4		53	46·06	6		7	30·4	M
816		*Anon.*							
Aug. 11	8·2	18	54	18·53	...	122	56	5·5	R
Sep. 12	7·7		54	18·80	5		56	5·5	M
13	...		54	18·65	5		56	4·4	M
817		39 *Sagittarii* o							
Aug 21	...	18	56	30·09	3	111	56	7·1	M
28	...		56	30·06	...		56	4·7	M
818		17 *Aquilae* ζ.							
Aug 3	...	18	59	14·96	...	76	20	0·9	R
17			59	15·02	...		20	1·2	M
23	...		59	14·97	...		20	0·8	M
25	...		59	14·94	...		20	0·1	M
Sep. 8	...		59	14·96	...		20	0·6	M
819		*Anon.*							
Aug 11	9·4	19	0	47·88		81	1	21·2	R
820		41 *Sagittarii* π							
Aug 21	...	19	1	47·42	3	111	14	2·0	M
23			1	47·41	...		14	1·1	M
821		*Anon.*							
Aug. 29	8·2	19	3	15·54	5	122	22	39·4	R
Sep. 10	7·7		3	15·40	.		22	38·6	M
12	7·9		3	16·26			22	34·1	M
822		*Anon.*							
Aug. 9	7·8	19	3	21·27	...	122	50	59·1	R
18	7·9		3	21·18	4		50	56·3	M
20	7·9		3	21·36	4		50	58·5	M
823		*Anon.*							
Sep. 14	9·0	19	7	14·86	.	120	47	29·4	M
824		*Anon.*							
July 29	7·0	19	8	2·66	6	122	7	0·7	R
31	7·1		8	2·65	—		7	1·3	M
Aug. 16	7·3		8	2·61	...		7	0·3	M
28	7·0		8	2·40	...		7	0·6	M
825		*Anon.*							
Aug. 11	9·8	19	8	36·72	5	120	[illegible]	30·2	R
826		*Anon.*							
Aug. 11	7·5	19	9	30·26	6	120	46	56·7	R
Sep. 14	7·0		9	30·06	4		46	56·3	M
827		*Anon.*							
Sep. 13	.	19	10	6·83	...	107	9	29·3	M
828		*Anon.*							
Aug 25	7·6	19	10	15·29	..	122	20	47·6	M
829		23 *Aquilae* ω							
Aug. 3	...	19	11	31·33	5	78	[illegible]	25·5	R
Sep. 15	...		11	31·33	4		[illegible]	26·6	M

Separate Results of Madras Meridian Circle Observations in 1866.

Number and Date.	Magnitude.	Mean Right Ascension 1866. h. m. s.	No. of Wires.	Mean Polar Distance 1866. ° ′ ″	Observer.
930		*30 Aquilae δ*			
Aug. 16	...	19 18 44·42	...	87 9 0·2	M
18	...	18 44·39	...	9 0·7	M
20	...	18 44·29	...	8 59·5	M
21	...	18 44·38	...	9 1·3	M
25	...	18 44·38	...	9 0·8	M
29	...	18 44·35	...	8 59·2	R
Sep. 12	...	18 44·40	...	9 0·6	M
13	...	18 44·42	...	9 0·1	M
14	...	18 44·44	...	8 59·5	M
931		*Taylor* 8950.			
Sep. 7	6·3	19 22 18·24	...	140 27 51·1	M
15	6·2	22 18·32	...	27 50·3	M
932		52 *Sagittarii h²*			
Aug. 18	...	19 28 32·87	...	115 10 34·9	M
20	...	28 32·86	...	10 36·1	M
22	...	28 32·89	...	10 34·5	M
30	...	28 32·96	...	10 34·6	R
Sep. 8	...	28 32·94	...	10 36·0	M
933		*Lacaille* 8178.			
Sep. 7	7·7	19 31 46·06	6	143 15 16·7	M
934		*Anon.*			
Sep. 10	8·4	19 31 31·56	...	127 16 47·6	M
935		56 *Sagittarii f*			
Aug. 22	...	19 34 32·47	...	110 4 50·2	M
28	...	34 32·35	...	4 51·7	M
936		50 *Aquilae γ*			
Aug. 9	...	19 39 56·29	...	79 42 41·6	R
16	...	39 56·21	...	42 40·2	M
19	...	39 53·17	.	42 40·0	M
20	...	39 56·29	...	42 39·5	M
Sep. 12	...	39 56·31	.	42 40·6	M
937		*S Vulpeculae Var.* 3.			
Sep. 8	8·0	19 42 54·02	...	63 2 48·4	M
10	8·0	42 54·67	...	2 48·5	M
938		53 *Aquilae α, Altair.*			
Aug. 9	...	19 44 14·61	5	81 28 59·0	R
16	...	44 14·66	4	28 59·8	M
20	...	44 14·50	...	28 59·6	R
Sep. 12	...	44 14·50	...	29 0·6	M
939		57 *Sagittarii.*			
Aug. 22	...	19 44 24·73	...	109 22 37·2	M
940		60 *Aquilae β*			
Aug. 3	...	19 48 48·73	...	83 56 32·5	R
20	...	48 48·71	...	56 33·1	M
Sep. 7	...	48 48·64	...	56 34·0	M
14	...	48 48·68	...	56 33·2	M
941		*Anon.*			
Sep. 10	7·0	19 58 54·87	...	130 21 13·2	M
942		λ *Ursae Minoris—s.p.*			
Jan. 28	...	19 58 11·62	3	1 5 32·8	R
Feb. 20	...	58 11·86	7	5 32·6	R
Mar. 1	...	58 11·84	7	5 31·4	R
8	...	58 11·76	2	5 34·1	M
943		*Anon.*			
Sep. 13	8·8	20 7 47·23	...	81 22 7·0	M
14	8·9	7 47·10	...	22 7·8	M
27	8·1	7 47·20	...	22 7·4	M
944		*R Sagittarii Var.* 1.			
Sep. 19	9·3	20 7 37·46	..	73 40 29·3	R
945		*O. A. S.* 20356.			
Sep. 16	7·7	20 8 27·02	...	119 25 48·6	M

Separate Results of Madras Meridian Circle Observations in 1866.

Number and Date.	Magnitude.	Mean Right Ascension 1866. h.	m.	s.	No. of Wires.	Mean Polar Distance 1866. °	′	″	Observer.
846		*6 Capricorni α²*							
June 30	...	20	10	37·00	...	102	57	29·1	R
Aug. 29	...		10	37·00	...		57	28·0	M
Sep. 8	...		10	36·86	...		57	27·4	M
13	...		10	36·81	...		57	28·6	M
15	...		10	37·00	4		57	29·2	M
20	...		10	37·01	...		57	28·9	R
847		*Anon.*							
Sep. 28	7·5	20	10	38·87	5	149	8	41·6	M
848		*Anon.*							
Aug. 3	9·7	20	12	13·56	5	88	46	53·2	R
849		*α Pavonis.*							
Aug. 30	..	20	15	1·73	...	147	0	30·6	R
850		*Lalande* 39125.							
Sep. 18	8·9	20	15	37·36	...	106	12	39·2	R
851		*11 Capricorni ρ*							
Aug. 18	...	20	21	12·77	...	109	15	16·6	M
28	...		21	12·78	...		15	15·7	M
29	...		21	12·87	...		15	15·1	R
30	...		21	12·85	...		15	15·2	R
Sep. 8	.		21	12·71	...		15	15·4	M
10	...		21	12·72	.		15	15·9	M
13	.		21	12·84	...		15	16·0	M
14	...		21	12·84	...		15	15·9	M
23	..		21	12·68	...		15	16·4	R
28	...		21	12·88	..		15	16·4	M
29	..		21	12·75	...		15	14·1	M
Oct. 1			21	12·73	5		15	15·2	M
3	...		21	12·87			15	14·2	M
852		*Anon.*							
Aug. 11	9·9	20	26	20·40		121	12	17·8	R
853		*Anon.*							
Oct. 2	9·3	20	28	50·46	...	121	6	5·6	M
854		*R. P. L.* 143.							
Sep. 19	...	20	29	34·34	3	5	13	3·9	R
855		*Anon.*							
Sep. 22	9·0	20	31	2·36		149	35	0·4	R
856		*S Capricorni Var.* 2.							
Aug. 3	9·6	20	34	4·21	...	109	31	57·3	R
Sep. 18	9·3		34	4·26	...		31	56·8	M
24	9·0		34	4·25	4		31	56·1	M
Oct. 5	9·1		34	4·16	...		31	57·1	M
857		*Anon.*							
Aug. 9	9·2	20	35	54·12	5	134	29	57·6	R
858		*50 Cygni α, Deneb.*							
July 24	...	20	36	51·70	5	46	11	39·0	M
Sep. 7	...		36	51·48	...		11	40·9	M
10	...		36	51·40	...		11	39·6	M
Oct. 9	...		36	51·76	...		11	39·6	M
859		*W. B. E. XX.* 978.							
Aug. 11	9·0	20	36	53·36	5	73	28	37·7	R
Oct. 6	8·9		36	53·29	...		28	37·4	M
860		*S Delphini Var.* 2.							
Sep. 14	9·2	20	36	54·65	...	73	28	39·7	M
27	9·1		36	54·43	4		28	39·9	M
861		*2 Aquarii ε*							
June 30	.	20	40	25·18		99	39	4·0	R
Aug. 28			40	25·13			39	4·3	M

Separate Results of Madras Meridian Circle Observations in 1866.

Number and Date.	Magnitude.	Mean Right Ascension 1866. h.	m.	s.	No. of Wires.	Mean Polar Distance 1866. °	′	″	Observer.
862		*W. B. E. XX.* 1024.							
Sep. 18	9·5	20	40	50·74	...	105	28	50·6	R
863		*Anon.*							
Sep. 23	10·4	20	41	15·40	4	106	17	56·3	R
864		6 *Aquarii* μ							
Aug. 28	...	20	46	28·33	...	99	20	3·7	M
865		32 *Vulpeculae.*							
Sep. 10	...	20	48	50·79	...	62	27	2·3	M
19	...		48	50·86	...		27	2·2	R
24	...		48	50·90	...		27	3·7	M
28	...		48	50·83	...		27	2·4	M
29	...		48	50·80	...		27	2·5	V
Oct. 5	...		48	50·88	...		27	1·0	M
6	...		48	51·08	...		27	3·4	M
9	...		48	50·87	...		27	2·0	M
866		*Lacaille* 8630.							
Sep. 12	7·0	20	51	31·58	6	126	38	46·3	M
27	7·0		51	31·77	...		38	45·7	M
867		*Anon.*							
Oct. 11	9·0	20	51	6·41	4	142	28	47·8	M
868		*Anon.*							
Sep. 18	10·1	20	53	31·21	5	98	48	21·3	R
22	10·2		53	31·15	5		48	20·8	R
869		*Taylor* 9772—1*st.*							
Sep. 27	7·7	21	0	36·11	.	146	6	30·6	M
870		*Taylor* 9772—2*nd.*							
Oct. 5	7·9	21	0	36·64	...	146	6	30·0	M
12	8·0		0	36·34	5		6	31·4	M
15	7·9		0	34·38			6	30·3	M
871		61 *Cygni*—1*st.*							
Sep. 19	...	21	0	58·40	...	51	54	28·4	R
Oct. 9	...		0	58·38	...		54	29·8	M
10	...		0	58·38	...		51	27·6	M
872		13 *Aquarii* ν							
Sep. 20	...	21	2	17·48	...	101	34	46·2	R
873		*Anon.*							
Oct. 11	9·3	21	3	3·17	5	145	6	16·7	M
874		*Lacaille* 8712.							
Sep. 26	8·3	21	4	20·13	5	148	46	5·0	M
875		64 *Cygni* ζ							
Aug. 29	...	21	7	13·98	...	60	19	17·0	R
30	...		7	13·81	...		19	16·5	R
Sep. 10	...		7	13·98	...		19	19·6	M
12	...		7	13·94	...		19	17·3	M
14	...		7	13·89	—		19	17·9	M
16	...		7	13·98	...		19	16·9	R
18	...		7	14·01	5		19	17·6	R
19	...		7	14·01	5		19	17·5	R
22	...		7	13·92	...		19	17·6	R
25	...		7	13·97	4		19	17·1	M
29	...		7	13·91	4		19	16·3	M
Oct. 1	...		7	13·90	...		19	15·9	M
5	...		7	13·98	...		19	16·5	M
6	...		7	14·08	..		19	16·1	
876		*Anon.*							
Oct. 2	9·6	21	11	8·93	..	129	31	29·5	M
877		*Brisbane* 7012.							
Sep. 18	..	21	14	14·80	6	131	48	29·3	R
27	7·8		14	14·09	—		48	31·6	M
Oct. 12	9·0		14	14·15	5		46	31·4	M

Separate Results of Madras Meridian Circle Observations in 1866.

Number and Date.	Magnitude.	Mean Right Ascension 1866. h.	m.	s.	No. of Wires.	Mean Polar Distance 1866. °	′	″	Observer.
878			*T Capricorni Var.* 3.						
Sep. 23	9·9	21	14	32·25	...	105	30	41·2	R
879			5 *Cephei* α						
Sep. 19	...	21	15	22·60	5	27	58	52·2	R
20	...		15	22·56	5		58	53·4	R
880			*Anon.*						
Sep. 12	7·7	21	20	14·71	5	150	47	21·0	M
881			22 *Aquarii* β						
Sep. 16	...	21	24	30·22	...	96	0	33·3	M
18	...		24	30·15	...		0	33·7	R
19	...		24	30·18	...		0	33·1	R
20	...		24	30·06	...		0	33·4	R
24	...		24	30·19	...		0	33·1	M
27	...		24	30·01	5		0	35·1	M
29	...		24	30·07	...		0	32·4	M
Oct. 1	...		24	30·11	...		0	32·5	M
5	...		24	30·05	...		0	32·0	M
882			8 *Cephei* β						
Aug. 17	...	21	28	56·51	5	20	1	37·2	M
30	...		28	56·16	...		1	37·8	R
883			*Anon.*						
Sep. 14	9·0	21	28	39·15	...	134	3	49·6	M
884			23 *Aquarii* ξ						
June 30	...	21	30	36·87	..	98	27	13·3	R
Sep. 30	..		30	36·92	...		27	18·6	R
885			*Taylor* 10068.						
Sep. 29	7·5	21	34	29·51	...	134	6	11·8	M
886			*S Cephei Var.* 3.						
Aug. 29	7·0	21	36	49·34	3	11	58	25·3	R
887			8 *Pegasi* ε						
Sep. 16	...	21	37	36·15	...	81	44	17·1	R
24	...		37	36·19	.		44	17·1	M
28	...		37	36·29	...		44	16·9	M
Oct. 1	...		37	36·11	.		44	15·4	M
3	...		37	36·29	.		44	16·3	M
11	..		37	36·20	..		44	16·3	M
22	..		37	36·25	...		44	16·3	M
888			48 *Capricorni* λ						
June 30	...	21	39	19·02	3	101	39	57·4	R
889			μ *Cephei Var.* 1.						
Sep. 27	6·0	21	39	24·46	3	31	20	3·6	M
890			*W. B. E. XXI.* 975.						
Sep. 19	9·7	21	41	16·14	...	97	19	12·1	R
891			16 *Pegasi*.						
Sep. 22	...	21	46	57·87	...	61	42	17·3	R
Oct. 6	...		46	57·98	...		42	15·0	M
15	...		46	58·00	...		42	16·1	
22	..		46	57·98	...		42	16·2	M
892			*Anon.*						
Sep. 13	8·9	21	47	48·82	...	133	11	36·7	M
893			*Taylor* 10190.						
Aug. 16	6·5	21	51	14·12	...	148	31	21·9	M
27	6·2		51	14·10	...		31	22·9	M
894			*Anon.*						
Oct. 11	9·5	21	52	54·19	...	136	37	41·6	M
895			ε *Indi*.						
Aug. 29	5·5	21	53	5·07	...	147	20	5·1	R
Sep. 19	...		53	5·01	...		20	4·6	R
Oct. 6	5·9		53	5·06	6		20	6·8	M

Separate Results of Madras Meridian Circle Observations in 1866.

Number and Date.	Magnitude.	Mean Right Ascension 1866. h. m. s.	No. of Wires.	Mean Polar Distance 1866. ° ′ ″	Observer.
896		*Anon.*			
Sep. 22	9·0	21 53 58·71	...	130 49 48·9	R
897		34 *Aquarii* ι			
Aug. 27	...	21 58 56·86	...	90 53 12·8	M
Sep. 27	...	58 54·05	4	53 10·7	M
Oct. 3	...	58 54·08	...	53 11·0	M
9	...	58 54·09	...	53 11·4	M
12	...	58 54·06	...	53 11·0	M
15	...	58 53·92	...	53 11·5	M
16	...	58 53·97	...	53 11·1	M
17	...	58 54·08	...	53 10·0	M
24	...	58 53·93	...	53 11·2	M
898		α *Gruis.*			
Sep. 15	...	21 59 46·52	...	137 36 31·1	M
899		*Anon.*			
Sep. 28	7·9	22 9 11·76	...	98 21 30·6	M
Oct. 6	8·0	9 11·77	...	21 31·0	M
900		*Anon.*			
Oct. 12	8·9	22 9 16·00	...	146 34 48·6	M
901		43 *Aquarii* θ			
Sep. 15	.	22 9 46·43	...	98 24 57·9	M
22	...	9 45·76	4	24 59·5	R
25	...	9 46·39	5	24 57·6	M
Oct. 3	...	9 46·31	3	24 59·2	M
15	...	9 46·51	—	24 58·9	M
16	...	9 46·04	...	24 59·3	M
17	...	9 46·24	...	24 58·0	M
902		*Anon.*			
Oct. 11	9·0	22 13 17·02	.	144 28 1·1	M
903		*Anon.*			
Oct. 16	9·0	22 16 58·44	4	135 57 49·2	M
904		55 *Aquarii* ζ—*1st.*			
Aug. 27	6·2	22 21 56·04	...	90 42 15·6	M
Sep. 28	6·6	21 56·67	...	42 14·6	M
Oct. 10	5·0	21 55·67	...	42 14·6	M
905		55 *Aquarii* ζ—*2nd.*			
Oct. 17	6·2	22 21 56·67	...	90 42 19·1	M
906		*Anon.*			
Sep. 26	7·0	22 21 57·91	...	109 37 9·6	M
Oct. 12	7·8	21 57·61	...	37 12·0	M
907		27 *Cephei* δ *Var* 2.			
Sep. 15	...	22 24 12·08	3	32 14 18·1	M
908		*Anon.*			
Oct. 15	7·9	22 25 59·60	5	141 29 36·7	M
909		62 *Aquarii* η			
Aug. 27	..	22 28 28·18	...	90 46 27·9	M
Oct. 6	...	28 28·10	...	46 28·6	M
17	...	28 28·10	...	46 28·9	M
22	...	28 28·19	..	46 27·6	M
24	...	28 28·14	...	46 27·6	M
Nov. 5	...	28 28·11	...	46 28·6	M
910		*Lacaille* 9188.			
Oct. 13	6·9	22 30 0·67	5	139 33 7·4	M
16	7·1	30 0·54	5	33 6·6	M
911		48 *Pegasi* μ			
Aug. 27	...	22 34 46·76	...	79 52 4·0	M
28	...	34 46·64	...	52 2·6	M
Oct. 2	...	34 46·66	2	52 1·7	M
24	..	34 46·68	...	52 2·6	M
Nov. 5	.	34 46·71	...	52 2·8	M

Separate Results of Madras Meridian Circle Observations in 1866.

Number and Date.	Magnitude.	Mean Right Ascension 1866. h.	m.	s.	No. of Wires.	Mean Polar Distance 1866. °	′	″	Observer.
912		*W. B. E. XXII.* 844.							
Oct. 11	8.8	22	40	49.32	...	87	48	4.0	M
913		*Anon.*							
Nov. 5	...	22	48	38.73	...	130	36	2.1	R
914		*O. A. S.* 22500.							
Oct. 5	8.0	22	49	7.09	...	119	19	13.9	M
11	8.0		49	7.20	...		19	14.6	M
12	8.1		49	7.17	...		19	14.2	M
915		*S Aquarii Var.* 2.							
Oct. 29	9.2	22	49	56.89	...	111	3	27.6	R
916		24 *Piscis Australis* α, *Fomalhaut.*							
Aug. 28	...	22	50	14.81	...	120	19	56.1	M
Sep. 26	...		50	14.87	...		19	54.9	M
Oct. 15	...		50	14.88	...		19	55.3	M
16	...		50	14.83	...		19	56.1	M
17	...		50	14.83	5		19	54.8	M
23	...		50	14.86	...		19	56.9	M
917		*Anon.*							
Sep. 6	7.1	22	50	36.73	...	110	39	36.6	M
Oct. 24	7.3		50	36.91	..		39	39.2	M
918		*Anon.*							
Oct. 29	8.6	22	51	54.96	...	96	22	30.1	R
31	8.3		51	54.70	.		22	30.6	R
919		54 *Pegasi* α, *Markab.*							
Sep. 27	...	22	58	5.19	...	75	30	56.3	M
Oct. 12	...		58	5.13	...		30	56.2	M
30	...		58	5.32	.		30	54.5	R
920		*Lacaille* 9372.							
Oct. 9	6.0	23	0	39.09	...	140	27	34.4	M
921		*Anon.*							
Oct. 11	9.2	23	4	26.22	...	130	45	37.2	M
922		*Anon.*							
Oct. 12	8.5	23	8	15.26	5	160	30	39.3	M
27	...		8	15.23	6		30	39.2	R
Nov. 3	8.7		8	15.10	5		30	37.7	R
923		*Lacaille* 9423.							
Sep. 8	7.0	23	10	2.53	..	151	43	37.0	M
924		6 *Piscium* γ							
Aug. 28	...	23	10	12.97	...	87	26	58.2	M
Sep. 13	...		10	13.10	.		26	59.4	R
19	...		10	13.10	5		26	56.0	R
25	...		10	13.16	—		26	59.7	M
27	...		10	13.11	...		26	57.8	M
Oct. 24	...		10	13.11	...		26	58.5	M
30	...		10	13.13	...		26	56.2	R
Nov. 1	...		10	13.19	.		26	57.2	R
5	...		10	13.06	.		26	56.6	R
925		*Anon.*							
Oct. 31	8.5	23	10	42.96	...	130	45	56.5	R
926		96 *Aquarii.*							
Sep. 22	...	23	12	27.67	...	96	51	38.3	R
927		*Anon.*							
Nov. 5	9.0	23	15	49.15	5	130	29	9.4	R
928		*Taylor* 10748.							
Aug. 29	6.9	23	17	39.58		147	33	4.6	M
Oct. 24	6.9		17	39.41	—		33	4.9	M
929		*Anon.*							
Nov. 3	9.3	23	19	49.57	5	151	37	26.3	R

Separate Results of Madras Meridian Circle Observations in 1866.

Number and Date.	Magnitude.	Mean Right Ascension 1866. h. m. s.	No. of Wires.	Mean Polar Distance 1866. ° ′ ″	Observer.
930		8 *Piscium* κ			
Aug. 27	...	23 30 3·75	...	89 23 41·3	M
Sep. 26	...	30 3·88	...	23 39·8	M
Oct. 11	...	30 3·85	...	23 40·1	M
12	...	30 3·66	...	23 40·4	M
27	...	30 3·80	5	23 39·6	R
29	...	30 3·78	...	23 39·7	R
Nov. 1	...	30 3·78	...	23 37·8	R
931		*Anon.*			
Oct. 30	8·5	23 30 51·70	...	88 51 2·3	R
Nov. 6	8·6	30 51·66	...	51 2·1	R
932		*Anon.*			
Oct. 31	9·3	23 21 9·87	...	137 27 5·3	R
933		10 *Piscium* θ			
Sep. 25	...	23 31 16·20	...	84 21 23·3	M
934		*Anon.*			
Nov. 3	10·2	23 35 30·82	5	129 51 20·3	R
5	9·8	35 30·27	...	51 19·7	R
935		*Taylor* 10804.			
Aug. 23	6·5	23 27 34·19	..	147 33 37·4	M
936		*R. P. L.* 158.			
Nov. 7	...	23 27 30·15	3	3 25 54·1	R
		R. P. L. 158—*s.p.*			
Apl. 5	...	23 27 50·02	5	3 25 55·8	M
937		*Anon.*			
Oct. 31	9·6	23 30 1·13	...	137 19 24·5	R
Nov. 4	9·7	30 1·01	...	19 26·6	R
938		*Anon.*			
Oct. 27	...	23 30 31·[illegible]	3	119 26 1·7	R

Number and Date.	Magnitude.	Mean Right Ascension 1866. h. m. s.	No. of Wires.	Mean Polar Distance 1866. ° ′ ″	Observer.
939		17 *Piscium* ι			
Aug. 27	...	23 33 3·48	...	85 5 20·9	M
Sep. 26	...	33 3·61	...	5 22·8	M
Oct. 29	...	33 3·50	...	5 23·6	R
30	...	33 3·38	...	5 20·5	R
Nov. 1	...	33 3·65	...	5 25·0	R
8	...	33 3·61	...	5 20·3	R
940		35 *Cephei* γ			
Sep. 19	...	23 33 52·65	5	13 6 52·3	R
23	...	33 52·31	5	6 54·1	R
941		*Anon.*			
Nov. 5	9·0	23 34 26·95	...	146 26 49·6	R
942		*Anon.*			
Nov. 7	9·7	23 36 53·16	5	100 1 20·4	R
9	9·2	36 52·90	...	1 23·1	R
943		*Anon.*			
Nov. 5	9·3	23 41 10·75	...	130 46 0·3	R
6	9·2	41 10·78	...	46 0·5	R
944		δ *Sculptoris.*			
Sep. 26	...	23 41 56·47	...	118 52 16·6	M
Oct. 27	...	41 56·50	...	52 17·2	R
31	...	41 56·64	...	52 16·7	R
Nov. 3	...	41 56·64	...	52 16·5	R
945		*Lalande* 46650.			
Oct. 2	8·4	23 42 12·22	5	84 18 25·8	M
946		*Anon.*			
Nov. 8	9·7	23 42 49·52	3	150 53 34·9	R
947		22 *Piscium.*			
Sep. 24	...	23 45 6·36	...	87 48 51·9	M

Separate Results of Madras Meridian Circle Observations in 1866.

Number and Date.	Magnitude.	Mean Right Ascension 1866. h. m. s.	No. of Wires.	Mean Polar Distance 1866. ° ′ ″	Observer.
958		*Anon.*			
Oct. 11	9·3	23 47 3·10	5	130 48 2·0	M
959		*Lacaille* 9638.			
Nov. 7	7·7	23 47 8·04	4	130 17 19·7	R
960		*Anon.*			
Nov. 9	...	23 47 37·20	...	130 46 12·8	R
961		*Anon.*			
Nov. 5	9·1	23 47 49·06	5	128 50 13·8	R
6	9·8	47 49·97	...	50 19·2	R
962		*Lacaille* 9641.			
Oct. 10	7·9	23 48 8·21	3	128 6 36·1	M
27	8·5	48 8·28	...	6 36·6	R
963		28 *Piscium* ω			
Sep. 26	...	23 52 23·87	...	83 52 41·9	M
Oct. 13	...	52 23·90		52 43·3	M
964		*Lacaille* 9696.			
Oct. 27	7·0	23 53 38·61	.	143 50 36·2	R
965		*Anon.*			
Nov. 5	8·9	23 56 4·61	...	130 16 22·3	R
966		*Taylor* 10990.			
Nov. 6	9·2	23 57 0·23	..	148 34 31·6	R
967		*Lacaille* 9721.			
Aug. 27	6·9	23 59 22·11	5	139 49 14·4	M

MEAN POSITIONS OF STARS

OBSERVED WITH THE

MADRAS MERIDIAN CIRCLE

IN THE YEAR

1866

REDUCED TO JANUARY 1, OF THAT YEAR.

Mean Positions of Stars for 1866 January 1st.

Number.	Star.	Magnitude.	Estimations.	Mean Right Ascension. h.	m.	s.	Mean Polar Distance. °	′	″	Observations.	Fraction of Year.
1	Taylor 11010	8·0	1	0	0	34·89	147	34	33·6	1	0·86
2	21 Androm. α (*Alpheratz*)..	2·0	...	0	1	27·01	61	38	50·0	8	0·81
3	Lacaille 9789	8·0	1	0	2	10·16	180	23	60·1	1	0·84
4	Lacaille 9787	8·0	1	0	4	31·16	131	7	7·8	1	0·86
5	88 Pegasi γ (*Algenib*) ...	3·0	...	0	6	20·22	75	33	41·5	7	0·85
6	... ...	9·2	1	0	6	44·46	131	6	22·3	1	0·85
7	Lalande 163 ...	7·7	3	0	7	41·82	80	26	54·3	3	0·77
8	... ...	9·0	1	0	9	38·63	140	31	11·3	1	0·94
9	... ...	9·9	1	0	9	39·06	153	54	27·9	1	0·86
10	... ...	9·6	2	0	10	40·08	162	0	12·7	2	0·86
11	Lacaille 41 ...	8·1	2	0	12	39·60	160	51	23·4	2	0·78
12	... ...	7·5	1	0	15	54·02	61	59	21·9	1	0·98
13	Lacaille 61 ...	7·3	2	0	16	7·57	180	0	19·1	2	0·54
14	... ...	9·2	2	0	18	30·90	162	53	37·9	2	0·88
15	Lacaille 81 ...	7·1	1	0	18	44·20	180	0	2·3	1	0·95
16	10 Ceti	6·7	...	0	19	41·97	90	47	32·7	2	0·65
17	12 Ceti	6·2	...	0	23	11·90	94	41	53·1	7	0·85
18	Lacaille 132	8·4	4	0	27	24·09	161	58	16·1	4	0·51
19	... ...	8·5	2	0	31	32·30	82	27	41·4	2	0·80
20	Lalande 1010	9·5	2	0	32	21·08	82	31	46·7	2	0·94
21	18 Cassiopeæ α Var. 2 ...	2·2	...	0	32	56·17	34	11	53·3	2	0·85
22	16 Ceti β	2·1	...	0	36	51·68	108	48	21·9	5	0·97
23	W. B. E. 0.693	9·3	1	0	37	0·17	98	48	46·3	1	0·99
24	... ...	9·5	1	0	39	59·46	150	44	13·0	1	0·90
25	W. B. E. 0.697	9·2	3	0	40	40·69	86	13	51·5	3	0·79
26	W. B. E. 0.705	9·0	2	0	41	14·32	94	36	39·8	2	0·84
27	63 Piscium δ	4·6	...	0	41	43·97	83	8	40·9	3	0·78
28	W. B. E. 0.716	9·5	1	0	41	47·39	94	36	7·8	1	0·90
29	20 Ceti	5·0	...	0	46	0·88	91	52	22·2	1	0·65
30	... ...	9·0	2	0	47	40·99	153	39	39·8	3	0·86
31	λ¹ Toucani	6·5	1	0	48	2·62	153	36	0·1	1	0·84
32	... ...	9·6	2	0	49	0·34	153	49	9·0	2	0·88
33	... ...	9·0	1	0	49	0·80	133	48	36·3	1	0·85
34	Lacaille 294	7·9	...	0	50	40·06	154	41	38·0	1	0·90
35	R. P. L. 12	4·5	...	0	50	39·08	4	27	48·1	2	0·84

7.—Star occulted by the moon, when totally eclipsed, on 1866 Sep. 24.
20.—Comparison star for Ariadne in 1861
21.—α Cassiopeæ Var. 2 (*Schedir*). Changes irregularly between 2·2 and 2·8 magnitude.
23.—25—26.—28.—Comparison stars for Europa in 1861
35.—2 Ursæ Minoris.

Observed with the Madras Meridian Circle in that Year.

Number.	Star.	In Right Ascension.			In Polar Distance.			Number in B. A. C.
		Annual Precession.	Secular Variation.	Proper Motion.	Annual Precession.	Secular Variation.	Proper Motion.	
		s	s	s	"	"	"	
1	Taylor 11010 ...	+ 3·0668	− 0·0452	...	− 20·055	+ 0·010	...	...
2	21 Androm. α (*Alpheratz*)	+ 3·0767	+ 0·0182	+ 0·000	− 20·065	+ 0·018	+ 0·15	4
3	Lacaille 9789 ...	+ 3·0613	− 0·0283	...	− 20·064	+ 0·013	...	...
4	Lacaille 9757 ...	+ 3·0491	− 0·0280	...	− 20·061	+ 0·018	...	22
5	88 Pegasi γ (*Algenib*).	+ 3·0815	+ 0·0100	0·000	− 20·048	+ 0·022	+ 0·02	26
6	... ...	− 3·0878	− 0·0282		− 20·046	+ 0·022		..
7	Lalande 163	+ 3·0724	+ 0·0026		− 20·044	+ 0·024		...
8	... ...	+ 2·9781	− 0·0453		− 20·038	+ 0·027		...
9	... ...	+ 2·9572	− 0·0540		− 20·036	+ 0·027		...
10	... .	+ 2·9550	− 0·0498		− 20·034	+ 0·029		...
11	Lacaille 41	+ 3·0083	− 0·0221	...	− 20·024	+ 0·038	...	...
12	... ...	+ 3·1446	+ 0·0871	...	− 20·007	+ 0·041	...	...
13	Lacaille 61	+ 2·9982	− 0·0800	...	− 20·006	+ 0·040	...	...
14	... ...	+ 2·8597	− 0·0472	...	− 19·988	+ 0·048	...	...
15	Lacaille 81	+ 2·9805	− 0·0205	...	− 19·989	+ 0·044	...	...
16	10 Ceti	+ 3·0705	+ 0·0026	+ 0·006	− 19·981	+ 0·047	+ 0·03	96
17	12 Ceti	+ 3·0609	+ 0·0008	− 0·002	− 19·953	+ 0·056	+ 0·01	113
18	Lacaille 132	+ 2·7737	− 0·0412	...	− 19·912	+ 0·067	...	...
19	... ...	+ 3·0064	+ 0·0070	...	− 19·865	+ 0·071	...	...
20	Lalande 1010 ...	+ 3·0900	+ 0·0076	...	− 19·856	+ 0·072	...	...
21	18 Cassiopeæ α Var. 2	+ 3·3536	+ 0·0553	+ 0·006	− 19·848	+ 0·080	+ 0·04	169
22	16 Ceti β	+ 2·9996	− 0·0065	+ 0·018	− 19·796	+ 0·080	− 0·02	196
23	W. B. E. 0.696 ..	+ 3·0573	+ 0·0020	...	− 19·794	+ 0·080	...	...
24	... .	+ 2·6370	− 0·0840	...	− 19·780	+ 0·076	...	...
25	W. B. E. 0.697 ...	+ 3·0503	+ 0·0015	...	− 19·733	+ 0·057	...	...
26	W. B. E. 0.706 ...	+ 3·0534	+ 0·0019	...	− 19·732	+ 0·087	..	...
27	63 Piscium δ ...	+ 3·1012	+ 0·0077	+ 0·003	− 19·728	+ 0·090	+ 0·05	222
28	W. B. E. 0.716 ...	+ 3·0526	+ 0·0019	...	− 19·722	+ 0·088	...	...
29	20 Ceti	+ 3·0682	+ 0·0034	− 0·004	− 19·650	+ 0·097	+ 0·01	242
30	... .	+ 2·6148	− 0·0838	...	− 19·622	+ 0·094	...	...
31	λ¹ Toucani	+ 2·5110	− 0·0827	...	− 19·615	+ 0·084	...	261
32	... .	+ 2·4951	− 0·0328	...	− 19·599	+ 0·096	...	...
33	... .	+ 2·8002	− 0·0183	...	− 19·598	+ 0·096	...	..
34	Lacaille 264	+ 2·4602	− 0·0818	...	− 19·564	+ 0·097	...	...
35	B. P. L. 12	+ 6·9400	+ 1·2042	+ 0·036	− 19·561	+ 0·223	+ 0·01	262

17.—Proper motions from "*Greenwich Catalogue 1872*"

Mean Positions of Stars for 1866 January 1st.

Number.	Star.	Magnitude.	Estimations.	Mean Right Ascension.			Mean Polar Distance.			Observations.	Fraction of Year.
				h.	m.	s.	°	′	″		
36	R. P. L. 14	6·2	...	0	54	13·70	3	34	11·2	2	0·30
37	71 Piscium ε	4·5	...	0	55	59·41	82	49	54·5	5	0·90
38	20 Ceti	6·3	...	1	1	5·11	89	42	27·6	1	0·90
39	... ...	8·5	2	1	5	50·98	129	52	26·2	2	0·80
40	56 Piscium ζ ... 1st ...	5·4	...	1	6	43·79	83	8	4·1	1	0·80
41	S Cassiopeæ Var. 4 ...	...	...	1	9	50·73	18	5	48·0	1	0·90
42	1 Ursæ Minoris α (*Polaris*)	2·2	...	1	9	57·44	1	24	17·1	5	0·74
43	... ...	9·2	1	1	10	14·74	81	49	4·0	1	[illegible]
44	... ...	7·7	1	1	17	6·18	96	30	47·4	1	0·66
45	45 Ceti θ¹	3·8	...	1	17	10·51	98	52	33·2	7	0·90
46	Taylor 465...	7·1	1	1	19	35·95	91	5	40·4	1	0·77
47	R Piscium Var. 1... ...	9·2	2	1	23	48·52	87	46	41·9	3	0·87
48	99 Piscium η	4·5	...	1	24	18·91	75	20	46·9	5	0·87
49	... ...	7·9	1	1	29	5·88	150	41	58·6	1	[illegible]
50	Taylor 595...	6·5	2	1	30	13·40	148	49	30·4	2	0·88
51	... ...	9·0	1	1	31	28·11	130	51	40·5	1	0·85
52	α Eridani (*Achernar*) ...	1·0	...	1	32	48·25	147	55	6·4	2	0·88
53	106 Piscium ν	4·7	...	1	34	27·55	85	11	29·9	9	0·86
54	110 Piscium ο	4·4	...	1	38	19·21	81	31	4·1	2	0·78
55	... ...	9·0	1	1	39	57·76	149	30	48·5	1	0·88
56	Lacaille 516	8·0	1	1	40	2·30	151	41	38·7	2	0·87
57	... ...	9·8	3	1	42	2·02	130	14	36·9	3	[illegible]
58	6 Arietis β	2·8	...	1	47	14·62	69	[illegible]	56·1	6	0·87
59	Y Piscium Var. 5... ...	7·6	5	1	47	17·86	81	52	45·8	5	[illegible]
60	... ...	9·8	1	1	48	36·98	130	4	37·7	1	[illegible]
61	Lacaille 588 ...	9·2	1	1	52	6·70	149	7	40·1	1	0·62
62	... .	8·8	2	1	54	57·89	130	55	5·8	2	0·62
63	Taylor 678 ...	7·8	3	1	56	21·89	72	30	31·7	2	0·84
64	13 Arietis α ...	2·0	...	1	59	37·41	67	10	32·8	9	0·93
65	... .	9·5	1	1	1	7·98	109	45	30·4	1	[illegible]
66	... ...	9·5	1	2	2	0·14	130	1	54·3	1	0·82
67	17 Arietis η	5·4	...	2	5	19·46	69	26	14·0	1	0·81
68	65 Ceti ξ¹	4·4	...	2	5	34·04	81	47	0·5	3	[illegible]
69	... .	9·5	1	2	7	2·84	148	35	55·3	1	[illegible]
70	67 Ceti	5·5	...	2	10	18·05	97	2	30·2	6	[illegible]

36.—Groombridge 195.
41.—S Cassiopeæ Var. 1.—Period 615 days.—Range, 7th to 13th magnitude.
47.—R Piscium Var. 1.—Period 345 days.—Range, 7·5 to 12th magnitude.
59.—Y Piscium Var. 5.—Supposed to vary between the 8th and 9th magnitude.

Observed with the Madras Meridian Circle in that Year.

Number.	Star.	In Right Ascension. Annual Precession.	In Right Ascension. Secular Variation.	In Right Ascension. Proper Motion.	In Polar Distance. Annual Precession.	In Polar Distance. Secular Variation.	In Polar Distance. Proper Motion.	Number in B. A. C.
		s	s	s	"	"	"	
36	R. P. L. 14... ...	+ 3·0964	+ 1·0602	− 0·171	− 19·407	+ 0·396	− 0·02	723
37	71 Piscium ε ..	+ 3·1127	+ 0·0057	− 0·002	− 19·460	+ 0·119	0·00	308
38	29 Ceti	+ 3·0901	+ 0·0058	+ 0·010	− 19·340	+ 0·126	+ 0·46	334
39	... ...	+ 2·7556	− 0·0135	...	− 19·233	+ 0·122	...	...
40	86 Piscium ζ... 1st ..	+ 3·1153	+ 0·0090	+ 0·008	− 19·211	+ 0·130	+ 0·07	368
41	S Cassiopeæ Var. 4...	+ 4·2998	+ 0·1621	...	− 19·131	+ 0·196	...	...
42	1 Urs. Min. α (*Polaris*)	+ 19·4564	+ 13·5606	+ 0·065	− 19·128	+ 0·861	0·00	360
43	... ...	+ 3·1301	+ 0·0090	...	− 19·120	+ 0·146	...	...
44	... ...	+ 3·0218	+ 0·0098	...	− 18·990	+ 0·153	...	...
45	46 Ceti θ[1]	+ 3·0029	+ 0·0018	− 0·007	− 18·925	+ 0·154	+ 0·22	420
46	Taylor 465	+ 3·0083	+ 0·0086	...	− 18·863	+ 0·160	...	433
47	R. Piscium Var. 1 ...	+ 3·0904	+ 0·0073	..	− 18·731	+ 0·169	...	...
48	99 Piscium η... ...	+ 3·1978	+ 0·0142	0·000	− 18·713	+ 0·176	0·00	463
49	... ...	+ 2·1691	− 0·0135	...	− 18·536	+ 0·128	...	...
50	Taylor 525	+ 2·2244	− 0·0185	...	− 18·521	+ 0·133	...	...
51	... .	+ 2·6227	− 0·0101	...	− 18·479	+ 0·137	...	...
52	α Eridani (*Achernar*)..	+ 2·2390	− 0·0128	+ 0·008	− 18·436	+ 0·137	+ 0·07	507
53	106 Piscium ν ..	+ 3·1171	+ 0·0091	− 0·004	− 18·376	+ 0·191	− 0·04	518
54	110 Piscium ο ..	+ 3·1550	+ 0·0111	+ 0·006	− 18·237	+ 0·199	− 0·01	537
55	... ...	+ 2·1132	− 0·0100	...	− 18·177	+ 0·138	...	...
56	Lacaille 510	+ 2·0226	− 0·0095	...	− 18·174	+ 0·138	...	548
57	... ...	+ 2·5848	− 0·0084	...	− 18·100	+ 0·170	...	...
58	6 Arietis β	+ 3·2984	+ 0·0183	+ 0·002	− 17·990	+ 0·286	+ 0·11	577
59	V Piscium Var. 5 ...	+ 3·1581	+ 0·0111	...	− 17·897	+ 0·210	...	...
60	... ...	+ 2·0119	− 0·0067	...	− 17·845	+ 0·142	...	...
61	Lacaille 593	+ 2·0212	− 0·0061	...	− 17·703	+ 0·147	...	...
62	... ...	+ 2·5130	− 0·0080	...	− 17·594	+ 0·184	...	...
63	Taylor 678	+ 3·2794	+ 0·0167	...	− 17·526	+ 0·269	...	632
64	13 Arietis α	+ 3·3527	+ 0·0203	+ 0·012	− 17·396	+ 0·262	+ 0·15	649
65	... ...	+ 1·9136	− 0·0093	...	− 17·310	+ 0·144	...	...
66	... .	+ 2·3021	− 0·0088	...	− 17·279	+ 0·192	...	...
67	17 Arietis η	+ 3·3331	+ 0·0188	+ 0·000	− 17·131	+ 0·290	− 0·01	662
68	66 Ceti ξ[1]	+ 3·1790	+ 0·0116	− 0·004	− 17·104	+ 0·240	+ 0·01	664
69	... .	+ 1·9169	− 0·0021	...	− 17·052	+ 0·154	...	..
70	67 Ceti	+ 2·9832	+ 0·0040	+ 0·003	− 16·900	+ 0·242	+ 0·14	704

36 — 40. Proper motions from "*Greenwich Catalogue* 1872."
38.- Proper motions from "*Greenwich Catalogue* 1864."
52.—Proper motions from "*Stone's Cape Catalogue.*"

Mean Positions of Stars for 1866 January 1st.

Number.	Star.	Magnitude.	Estimations.	Mean Right Ascension.			Mean Polar Distance.			Observations.	Fraction of Year.
				h.	m.	s.	°	′	″		
71	68 Ceti o Var. 1 (*Mira*) ...	9·9	2	2	12	34·80	93	56	17·8	3	0·87
72	... ...	8·4	1	2	12	42·41	93	36	3·8	1	0·86
73	... ...	9·0	1	2	15	42·84	132	33	38·6	1	0·83
74	... ...	8·1	2	2	16	36·84	151	17	46·5	2	0·89
75	... ...	8·0	1	2	20	14·64	146	32	10·1	2	0·90
76	73 Ceti ξ²	4·4	...	2	21	2·22	82	8	32·6	6	0·80
77	... ...	9·5	2	2	24	18·22	152	35	7·0	2	0·84
78	... ...	9·0	1	2	24	31·81	147	2	13·4	1	0·88
79	... ...	7·9	1	2	27	29·97	147	11	43·5	1	0·83
80	... ...	9·5	1	2	30	40·02	147	34	23·7	1	0·87
81	... ...	9·8	1	2	31	19·08	151	38	33·4	1	0·96
82	... ...	9·4	2	2	34	5·59	74	56	2·7	2	0·84
83	Lacaille 849 ... 1st...	8·0	1	2	36	3·60	150	8	39·0	1	0·85
84	Lacaille 849 ... 2nd...	8·1	2	2	36	8·68	150	8	48·4	2	0·47
85	86 Ceti γ	3·6	...	2	36	21·44	87	19	51·7	2	0·85
86	μ Ceti	4·4	...	2	37	42·05	80	27	13·5	2	0·55
87	Lacaille 883	8·9	1	2	38	36·30	147	12	40·3	1	0·87
88	W. B. N. II. 676 ...	8·0	1	2	40	14·67	75	10	58·8	1	0·95
89	... ...	9·4	1	2	41	46·56	151	2	11·3	1	0·01
90	... ...	7·9	1	2	43	21·24	148	0	6·3	1	0·01
91	... ...	8·8	3	2	44	32·29	144	13	19·6	3	0·94
92	... ...	9·2	1	2	46	22·39	76	37	36·2	1	0·95
93	Lacaille 941 ...	6·6	...	2	50	29·98	146	25	36·0	1	0·85
94	... .	8·0	1	2	52	35·08	130	18	39·6	1	0·91
95	91 Ceti λ	4·6	...	2	52	32·07	81	37	42·6	1	0·78
96	... .	9·2	1	2	53	19·81	148	43	57·8	1	0·90
97	Lacaille 980	8·0	1	2	54	57·09	144	13	29·7	1	0·96
98	92 Ceti α (*Menkar*) ..	2·7	...	2	55	16·89	86	30	17·6	2	0·94
99	25 Persei ρ Var. 2 ..	3·7	...	2	56	36·92	51	40	58·8	1	0·97
100	Taylor 1037	7·0	2	2	58	56·98	150	21	7·1	2	0·94
101	Taylor 1047	8·0	1	2	59	54·06	151	19	38·2	3	0·86
102	57 Arietis δ	4·5	...	3	3	59·17	70	46	57·2	4	0·98
103	Taylor 1061 .. .	7·6	2	3	5	15·39	151	39	49·4	2	0·98
104	Taylor 1062	7·2	2	3	7	16·39	138	19	0·3	2	0·99
105	... ...	5·4	2	3	10	38·60	131	43	44·4	2	0·93

71.—68 Ceti o Var. 1 (*Mira*).—Period 331 days.—Range, 2nd to 10th magnitude.
98.—Comparison star for Victoria in 1861.
99.—25 Persei ρ Var. 2.—Changes irregularly from 3·5 to 4·3 magnitude.

Observed with the Madras Meridian Circle in that Year.

Number.	Star.	In Right Ascension.			In Polar Distance.			Number in B. A. C.
		Annual Precession.	Secular Variation.	Proper Motion.	Annual Precession.	Secular Variation.	Proper Motion.	
		s	s	s	"	"	"	
71	68 Ceti *o* Var. 1 (*Mira*)	+ 3·0268	+ 0·0064	− 0·001	− 16·792	+ 0·348	+ 0·23	720
72	... ...	+ 3·0262	+ 0·0004	...	− 16·789	+ 0·349	...	...
73	... ...	+ 1·6349	+ 0·0066	...	− 16·640	+ 0·140	...	...
74	... .	+ 1·7027	+ 0·0036	...	− 16·604	+ 0·146	...	...
75	... ...	+ 1·9102	− 0·0005	...	− 16·416	+ 0·167	...	...
76	73 Ceti ξ²	+ 3·1786	+ 0·0117	+ 0·001	− 16·377	+ 0·276	+ 0·02	760
77	... ...	+ 1·5541	+ 0·0083	...	− 16·209	+ 0·141	...	...
78	... .	+ 1·8505	+ 0·0008	...	− 16·106	+ 0·167	...	...
79	... ...	+ 1·8275	+ 0·0016	...	− 16·048	+ 0·167	...	...
80	... ...	+ 1·7840	+ 0·0027	...	− 15·867	+ 0·166	...	...
81	... ...	+ 1·5586	+ 0·0084	...	− 15·840	+ 0·146	...	...
82	... ...	+ 3·2960	+ 0·0154	...	− 15·690	+ 0·303	...	...
83	Lacaille 849 ... 1st...	+ 1·6068	+ 0·0071	...	− 15·581	+ 0·154	...	...
84	Lacaille 849 ... 2nd...	+ 1·6051	+ 0·0071	...	− 15·577	+ 0·154	...	...
85	86 Ceti γ	+ 3·1114	+ 0·0094	− 0·011	− 15·560	+ 0·294	+ 0·19	857
86	μ Ceti...	+ 3·2149	+ 0·0120		− 15·462	+ 0·305	...	846
87	Lacaille 868	+ 1·7477	+ 0·0040		− 15·443	+ 0·170	...	...
88	W. B. N. II. 676 ..	+ 3·2974	+ 0·0150		− 15·349	+ 0·315	...	...
89	... ...	+ 1·8062	+ 0·0049		− 15·264	+ 0·149	...	...
90	... ...	+ 1·6727	+ 0·0067		− 15·173	+ 0·167	...	...
91	... ...	+ 1·6623	+ 0·0062	...	− 15·106	+ 0·166	...	...
92	... .	+ 3·2949	+ 0·0144	...	− 15·067	+ 0·322	...	...
93	Lacaille 911	+ 1·7080	+ 0·0058	...	− 14·756	+ 0·176	...	...
94	... ...	+ 1·4719	+ 0·0107	...	− 14·646	+ 0·158	...	...
95	91 Ceti λ	+ 3·2006	+ 0·0117	+ 0·006	− 14·625	+ 0·336	+ 0·03	920
96	... .	+ 1·6737	+ 0·0060	...	− 14·580	+ 0·174	...	...
97	Lacaille 969	+ 1·7393	+ 0·0040	...	− 14·490	+ 0·150	...	...
98	92 Ceti α (*Menkar*) ...	+ 3·1293	+ 0·0098	− 0·002	− 14·470	+ 0·289	+ 0·11	949
99	36 Persei ρ Var. 2 ...	+ 3·8080	+ 0·0282	+ 0·010	− 14·360	+ 0·396	+ 0·11	964
100	Taylor 1037	+ 1·4884	+ 0·0116	...	− 14·265	+ 0·158	...	...
101	Taylor 1047	+ 1·3441	+ 0·0130	...	− 14·197	+ 0·145	...	981
102	57 Arietis δ	+ 3·4078	+ 0·0171	+ 0·010	− 13·988	+ 0·364	0·00	993
103	Taylor 1081	+ 1·2792	+ 0·0156	...	− 13·892	+ 0·141	...	992
104	Taylor 1092	+ 1·4922	+ 0·0100	...	− 13·721	+ 0·166	..	1002
105	... .	+ 2·1916	+ 0·0011	...	− 13·384	+ 0·243	...	...

76.—Proper motions from "*Greenwich Catalogue* 1872."

Mean Positions of Stars for 1866 January 1st.

Number.	Star.	Magnitude.	Estimations.	Mean Right Ascension.			Mean Polar Distance.			Observations.	Fraction of Year.
				h	m	s	°	′	″		
106	Taylor 1127 ...	8·4	1	3	11	51·76	131	46	58·9	1	0·93
107	... .	8·6	1	3	12	27·91	130	57	46·0	1	0·91
108	... .	8·3	2	3	12	45·67	130	49	40·1	2	0·92
109	33 Persei α ..	1·9	...	3	14	46·03	40	37	7·6	1	0·95
110	... .	9·5	1	3	14	58·81	150	5	53·5	2	0·91
111	3² Reticuli	5·7	...	3	15	18·53	153	1	7·3	1	0·90
112	... ...	7·2	2	3	20	37·07	140	26	0·1	2	0·40
113	.. ..	9·5	1	3	20	47·00	54	47	30·0	1	0·90
114	R Persei Var. 3	9·7	1	3	21	31·53	54	47	35·9	1	0·97
115	R. P. L. 34	5·8	...	3	22	52·53	3	40	38·6	2	0·83
116	5 Tauri *f*	4·8	...	3	23	38·55	77	31	30·7	2	0·81
117	Lacaille 1143 ...	5·0	...	3	27	3·03	130	24	39·7	1	0·01
118	Lacaille 1149 ...	7·7	3	3	28	40·00	150	16	42·4	4	0·47
119	... ...	6·0	2	3	30	15·80	161	30	36·2	2	0·95
120	Lacaille 1150 ..	6·5	1	3	30	18·79	161	23	12·0	1	0·04
121	Lacaille 1192 ...	9·0	1	3	35	1·42	147	46	21·1	1	0·90
122	... ...	8·9	2	3	36	22·15	160	12	54·1	2	0·03
123	Taylor 1200 ...	7·0	1	3	36	39·64	150	12	52·0	1	0·01
124	Lacaille 1200 ...	6·0	1	3	36	27·19	146	40	8·7	1	0·81
125	... ...	9·0	2	3	39	0·04	136	12	30·7	2	0·47
126	... ...	8·0	1	3	39	27·38	66	30	8·1	1	0·90
127	25 Tauri η (*Alcyone*) ...	3·0	...	3	39	31·28	66	18	46·9	3	0·85
128	... ...	8·6	1	3	45	10·88	76	27	38·9	1	0·01
129	... ...	5·0	1	3	46	35·09	140	28	17·2	1	0·02
130	... ...	7·9	3	3	46	5·94	180	40	51·9	3	0·02
131	34 Eridani γ¹	3·1	...	3	51	40·71	103	53	30·3	4	0·70
132	... ...	9·0	1	3	53	9·89	139	35	1·1	1	0·90
133	35 Tauri λ Var. 1 ..	3·8	...	3	53	15·44	77	53	27·7	2	0·81
134	Lacaille 1297	6·0	1	3	54	20·99	158	51	5·3	1	0·94
135	R. P. L. 36	6·7	...	3	55	36·01	4	48	12·9	3	0·77
136	Lacaille 1307	7·1	1	3	58	0·32	149	2	16·5	1	0·01
137	Lalande 7601	8·0	2	3	54	20·65	74	62	1·1	2	0·03
138	Lacaille 1330 ..	7·0	1	4	0	3·62	117	40	40·8	1	0·04
139	Lacaille 1375 ..	8·0	..	4	2	57·10	144	50	35·8	1	0·63
140	Lalande 7704 ..	7·0	1	4	3	36·03	74	48	32·6	1	0·93

114.— R Persei Var. 3.— Period 209 days.—Range, 8·5 to 12·5 magnitude.
115.— Groombridge 612.
133.— 35 Tauri λ Var. 1.— Period 3·95 days.—Range, 3·5 to 4·3 magnitude.
135.— Groombridge 750.
137.— 140.— Comparison stars for [illegible] in [illegible].

Observed with the Madras Meridian Circle in that Year.

Number.	Star.	In Right Ascension. Annual Precession.	Secular Variation.	Proper Motion.	In Polar Distance. Annual Precession.	Secular Variation.	Proper Motion.	Number in B. A. C.
		s	s	s	"	"	"	
106	Taylor 1127 .. .	+ 2·1653	+ 0·0012	...	− 13·420	+ 0·213	...	...
107	... .	+ 2·2080	+ 0·0011	...	− 13·300	+ 0·240	...	.
108	.. .. ..	+ 2·2110	+ 0·0012	...	− 13·370	+ 0·246	..	..
109	38 Persei o	+ 4·3430	+ 0·0468	+ 0·002	− 13·299	+ 0·472	+ 0·05	1013
110	... .	+ 1·3247	+ 0·0138	...	− 13·230	+ 0·151	...	...
111	3^2 Reticuli	+ 1·0952	+ 0·0808	+ 0·190	− 13·203	+ 0·120	− 0·65	1051
112	... .	+ 1·3316	+ 0·0183	...	− 12·400	+ 0·135	...	..
113	... ...	+ 3·7070	+ 0·0270	...	− 12·839	+ 0·461	..	...
114	R. Persei Var 3 ..	+ 3·7098	+ 0·0278	...	− 12·708	+ 0·432	...	...
115	B. P. L. 34 ...	+ 18·7270	+ 3·2080	+ 0·136	− 12·607	+ 2·119	+ 0·03	1061
116	5 Tauri f	+ 3·3015	+ 0·0130	+ 0·002	− 12·066	+ 0·379	+ 0·03	1087
117	Lacaille 1143 ...	+ 0·9710	+ 0·0287	...	− 12·418	+ 0·117	...	1103
118	Lacaille 1140 ..	+ 1·2190	+ 0·0186	...	− 12·301	+ 0·146	...	...
119	.. ..	+ 1·0995	+ 0·0100	...	− 12·190	+ 0·131	..	..
120	Lacaille 1139 ...	+ 1·1190	+ 0·0160	...	− 12·148	+ 0·185	...	..
121	Lacaille 1192 ..	+ 1·3619	+ 0·0130	...	− 11·887	+ 0·165		..
122	. ...	+ 1·1861	+ 0·0139	...	− 11·893	+ 0·145		.
123	Taylor 1256 . ..	+ 1·1854	+ 0·0159	...	− 11·825	+ 0·145		1141
124	Lacaille 1280 ..	+ 1·4960	+ 0·0107	...	− 11·765	+ 0·173		...
125	... .	+ 1·3962	+ 0·0046	...	− 11·684	+ 0·298		...
126	....	+ 3·5475	+ 0·0176	...	− 11·542	+ 0·468	...	..
127	25 Tauri η *(Alcyone)*	+ 3·5518	+ 0·0177	+ 0·001	− 11·568	+ 0·480	+ 0·06	1108
128	... ...	+ 3·3402	+ 0·0134	...	− 11·120	+ 0·410	...	...
129	... ...	+ 1·3913	+ 0·0111	...	− 11·026	+ 0·178	...	...
130	... .	+ 1·0627	+ 0·0177	...	− 10·916	+ 0·196	...	...
131	34 Eridani γ^1 ..	+ 2·7918	+ 0·0047	+ 0·002	− 10·645	+ 0·361	+ 0·12	1234
132	... .	+ 2·1700	+ 0·0080	...	− 10·542	+ 0·274	...	..
133	35 Tauri λ Var. 1	+ 3·3162	+ 0·0115	− 0·002	− 10·334	+ 0·416	+ 0·02	1241
134	Lacaille 1327 ..	+ 0·7490	+ 0·0250	...	− 10·156	+ 0·097	...	1248
135	B. P. L. 35... ..	+ 16·6808	+ 1·8176	+ 0·067	− 10·370	+ 2·084	− 0·05	1235
136	Lacaille 1347 ..	+ 1·1612	+ 0·0148	...	− 10·169	+ 0·149	..	.
137	Lalande 7891 ..	+ 2·[illegible]	+ 0·0134	...	− 10·158	+ 0·480	..	...
138	Lacaille 1350	+ 1·2012	+ 0·0131	...	− 10·098	+ 0·160	...	
139	Lacaille 1375 ..	+ 1·1431	+ 0·0144	...	− 9·804	+ 0·148	...	
140	Lalande 7764	+ 3·3012	+ 0·0121	...	− 9·751	+ 0·430	..	.

111.—Proper motions from "*Stone's Cape Catalogue.*"
115.—116.—133.—135.—Proper motions from "*Greenwich Catalogue 1872.*"

Mean Positions of Stars for 1866 January 1st.

Number.	Star.	Magnitude.	Estimations.	Mean Right Ascension. h.	m.	s.	Mean Polar Distance. °	′	″	Observations.	Fraction of Year.
141	36 Eridani e¹ ..	4·1	...	4	5	10·49	97	11	21·4	5	0·[illegible]
142	... ...	7·0	3	4	9	20·94	140	30	32·0	3	0·02
143	... ...	8·6	2	4	9	50·78	129	18	33·6	2	0·03
144	Lacaille 1495 ...	6·2	1	4	10	2·82	152	31	46·8	1	0·90
145	... ...	8·9	1	4	13	54·76	70	51	13·4	1	0·94
146	U Tauri Var. 7	10·0	2	4	14	0·69	70	30	28·8	2	0·51
147	61 Tauri δ¹	4·0	...	4	15	12·56	72	46	26·5	1	0·[illegible]
148	... ...	9·5	...	4	15	44·32	136	39	38·4	1	0·03
149	... ...	8·0	1	4	16	46·11	140	4	8·9	1	0·03
150	74 Tauri ε	3·7	...	4	20	47·66	71	7	12·0	8	0·28
151	R Tauri Var. 2	9·3	2	4	20	57·31	80	8	22·4	2	0·04
152	... ...	10·2	1	4	22	8·73	80	27	49·3	1	0·05
153	... ...	10·2	1	4	22	38·66	80	20	59·4	1	0·04
154	Lacaille 1519	7·0	1	4	25	37·01	138	5	51·5	1	0·04
155	Lacaille 1520	8·0	1	4	26	46·66	147	38	44·9	1	0·06
156	87 Tauri α (*Aldebaran*) ...	1·0	...	4	28	13·99	73	45	47·8	8	0·26
157	... ...	8·2	1	4	28	30·90	140	14	2·8	1	0·09
158	... ...	9·1	1	4	31	46·76	142	59	20·4	1	0·04
159	... ...	9·0	2	4	32	0·97	130	47	56·7	2	0·08
160	... ...	8·5	1	4	33	41·76	67	31	51·3	1	0·[illegible]
161	... ...	9·0	1	4	33	56·60	130	51	34·4	1	0·01
162	... ...	8·7	1	4	34	32·69	144	55	36·0	1	0·04
163	96 Tauri	7·3	2	4	35	7·06	68	10	6·2	2	0·80
164	Lacaille 1567 ...	6·5	...	4	35	12·84	132	50	33·3	1	0·02
165	Lacaille 1566 ...	7·3	1	4	36	48·05	145	35	9·4	1	0·06
166	... ...	9·7	1	4	36	[illegible]	64	19	0·9	1	0·0[illegible]
167	Lacaille [illegible] ...	7·5	3	4	37	30·21	132	34	[illegible]	3	0·04
168	.. .. ..	9·8	1	4	39	34·32	155	43	40·6	1	0·08
169	. . .	8·9	1	4	40	21·05	151	20	41·8	1	0·04
170	Lacaille 1593	7·8	3	4	41	41·63	139	21	24·3	3	0·05
171	α Doradûs	5·6	.	4	42	30·94	145	58	46·3	1	0·06
172	... . .	9·5	2	4	43	34·71	130	41	0·3	2	0·05
173	. .	9·7	1	4	45	32·92	130	34	59·9	1	0·05
174	. .	7·7	1	4	45	56·41	153	3	51·4	1	0·04
175	Lacaille 1636	7·7	2	4	47	57·[illegible]	140	1	[illegible]	2	0·06

146.—U Tauri Var. 7.—Period unknown.—Range, 9th to 10·5 magnitude.
151.—R Tauri Var. 2.—Period 326 days.—Range, 8th to below 13th magnitude.
168.—The Mean Polar Distance of this star is unaccountably about 30″ too small.

Observed with the Madras Meridian Circle in that Year.

Number.	Star.	In Right Ascension.			In Polar Distance.			Number in B.A.C.
		Annual Precession.	Secular Variation.	Proper Motion.	Annual Precession.	Secular Variation.	Proper Motion.	
		s	s	s	"	"	"	
141	38 Eridani o^1 ...	+ 2·0241	+ 0·0038	− 0·002	− 9·621	+ 0·379	− 0·07	1290
142	... .	+ 1·0307	+ 0·0155	...	− 9·311	+ 0·141	...	...
143	.. ..	+ 2·1016	+ 0·0085	...	− 9·272	+ 0·276	...	...
144	Lacaille 1495 ..	+ 0·7756	+ 0·0210	...	− 9·028	+ 0·105	...	...
145	... .	+ 3·4871	+ 0·0128	...	− 8·950	+ 0·469	...	...
146	U Tauri Var. 7 ...	+ 3·4959	+ 0·0129	...	− 8·948	+ 0·460	...	...
147	61 Tauri δ^1	+ 3·4440	+ 0·0119	+ 0·004	− 8·854	+ 0·465	+ 0·08	1346
148	... ...	+ 2·1112	+ 0·0036	...	− 8·812	+ 0·281	...	...
149	... .	+ 1·0633	+ 0·0146	...	− 8·739	+ 0·141	...	...
150	74 Tauri ϵ	+ 3·4871	+ 0·0120	+ 0·005	− 8·413	+ 0·460	+ 0·08	1376
151	R Tauri Var. 2 ...	+ 3·2833	+ 0·0092		− 8·400	+ 0·436	...	...
152	... .	+ 3·2765	+ 0·0090		− 8·305	+ 0·439	...	...
153	... ...	+ 3·2792	+ 0·0090		− 8·279	+ 0·440	...	...
154	Lacaille 1510 ...	+ 0·6574	+ 0·0212		+ 8·098	+ 0·081	...	...
155	Lacaille 1520 ...	+ 1·1404	+ 0·0122		− 7·989	+ 0·157	...	...
156	87 Tauri α (*Aldebaran*)	+ 3·4303	+ 0·0105	+ 0·004	− 7·818	+ 0·464	+ 0·17	1420
157	... ...	+ 1·5917	+ 0·0067	...	− 7·794	+ 0·217	...	...
158	... ...	+ 1·4285	+ 0·0082	...	− 7·662	+ 0·196	...	...
159	... ...	+ 2·0001	+ 0·0040	...	− 7·490	+ 0·274	...	...
160	... ...	+ 3·3563	+ 0·0122	...	− 7·375	+ 0·459	...	...
161	... .	+ 1·9901	+ 0·0040	...	− 7·356	+ 0·274	...	...
162	... ...	+ 1·2967	+ 0·0095	...	− 7·306	+ 0·180	...	...
163	96 Tauri	+ 3·0227	+ 0·0125	+ 0·004	− 7·280	+ 0·406	+ 0·00	1436
164	Lacaille 1567 ...	+ 0·6936	+ 0·0196	...	− 7·262	+ 0·097	...	...
165	Lacaille 1366 ...	+ 1·0850	+ 0·0128	...	− 7·204	+ 0·144	...	...
166	... .	+ 3·0728	+ 0·0130		− 7·151	+ 0·806		...
167	Lacaille 1582 ..	+ 0·6545	+ 0·0189		− 7·078	+ 0·092		1400
168	... .	+ 2·0370	+ 0·0037		− 6·894	+ 0·285		...
169	... .	+ 0·7719	+ 0·0163		− 6·831	+ 0·100		...
170	Lacaille 1595 ..	+ 2·0754	+ 0·0036		− 6·720	+ 0·288		...
171	α Doradûs	+ 0·8395	+ 0·0141	...	− 6·667	+ 0·125		1468
172	... .	+ 1·0908	+ 0·0040	...	− 6·578	+ 0·277		...
173	... .	+ 2·0308	+ 0·0037	...	− 6·401	+ 0·284		...
174	... ...	+ 0·5771	+ 0·0196	...	− 6·398	+ 0·088		...
175	Lacaille 1606... ..	+ 0·9864	+ 0·0124	..	− 6·200	+ 0·136		

161.—Proper motions from "*Greenwich Catalogue* 1904."

Mean Positions of Stars for 1866 January 1st.

Number.	Star.	Magnitude.	Estimations.	Mean Right Ascension.			Mean Polar Distance.			Observations.	Fraction of Year.
				h.	m.	s.	°	′	″		
176	8 Aurigæ ι	2·7	...	4	48	16·26	57	2	58·5	5	0·03
177	Taylor 1761	7·3	3	4	50	3·98	120	18	27·6	8	0·36
178	... ...	10·6	2	4	51	32·61	92	8	34·4	2	0·04
179	7 Aurigæ ε Var. 1 ...	3·2	...	4	52	21·37	46	22	44·0	3	[illegible]
180	R Leporis Var. 1	8·4	3	4	53	30·44	106	0	37·1	3	0·07
181	... ...	8·1	3	4	53	50·88	103	0	34·6	3	0·04
182	Taylor 1797 ...	6·9	1	4	54	33·12	148	16	46·0	1	0·05
183	... ...	10·3	1	4	56	0·56	130	17	34·8	1	0·04
184	Lacaille 1697 ...	7·7	1	4	56	55·18	129	6	50·2	1	0·10
185	Taylor 1811 ...	6·4	...	4	57	6·90	129	54	54·8	1	0·06
186	2 Leporis ε ...	3·8	...	4	59	47·37	112	33	12·5	7	0·19
187	... ...	8·6	3	5	0	3·49	106	28	44·8	8	0·07
188	... ...	9·7	2	5	1	46·73	151	26	39·8	2	0·05
189	Lacaille 1736 ...	8·8	2	5	4	0·34	154	48	38·2	2	0·06
190	... ...	9·6	1	5	4	37·84	136	34	30·2	1	0·07
191	Lacaille 1737	8·0	2	5	4	54·21	160	3	12·5	2	0·06
192	13 Aurigæ α (*Capella*) ...	0·2	...	5	6	47·80	44	8	38·1	2	0·07
193	... ...	8·5	2	5	6	56·10	130	5	58·7	2	0·06
194	19 Orionis β (*Rigel*) ...	0·3	...	5	8	5·02	98	21	33·6	3	0·34
195	... ...	9·5	4	5	10	30·02	130	48	23·1	4	0·06
196	... .	9·6	3	5	12	1·34	138	46	40·7	3	0·06
197	... ...	9·6	2	5	12	35·82	137	4	37·6	2	0·05
198	... ...	8·4	2	5	13	25·60	108	41	34·0	2	0·09
199	... ...	8·1	2	5	14	51·44	163	20	12·0	2	0·09
200	Lacaille 1822 ..	8·0	1	5	16	46·73	141	48	6·5	1	0·08
201	... .	8·0	1	5	17	38·28	168	7	14·4	1	0·07
202	112 Tauri β .	1·9	...	5	17	49·27	61	30	34·1	9	0·15
203	B. P. L. 40 ...	6·4	...	5	19	21·82	4	52	50·3	1	0·45
204	Lacaille 1881	8·0	1	5	21	42·06	137	13	45·0	1	0·06
205	... .	10·8	1	5	21	50·13	80	40	53·6	1	0·05
206	...	7·0	1	5	22	36·02	182	42	0·9	1	0·06
207	.. .. .	7·7	2	5	23	31·33	131	13	20·9	2	0·08
208	λ Doradûs — ..	5·6	...	5	24	22·60	149	1	36·1	1	0·06
209	34 Orionis δ Var. 1 .	2·4	...	5	25	9·71	90	24	5·6	7	0·05
210	... . ..	8·6	2	5	26	16·57	130	36	10·1	2	0·11

179.—ε Aurigæ Var. 1.—Supposed to be irregularly variable.
180.—R Leporis Var. 1.—Period [illegible] days. Range, 6th to 9th magnitude.
203.—Groombridge 944.
205.—Observed in mistake for [illegible] in 1864.
209.—δ Orionis Var. 1.—Supposed to vary irregularly from 2·3 to 2·7 magnitude.

Observed with the Madras Meridian Circle in that Year.

Number.	Star.	In Right Ascension. Annual Precession.	In Right Ascension. Secular Variation.	In Right Ascension. Proper Motion.	In Polar Distance. Annual Precession.	In Polar Distance. Secular Variation.	In Polar Distance. Proper Motion.	Number in B. A. C.
		s	s	s	"	"	"	
176	3 Aurigæ ι	+ 3·8966	+ 0·0144	− 0·008	− 6·175	+ 0·544	+ 0·02	1580
177	Taylor 1761	+ 2·0251	+ 0·0036	...	− 6·026	+ 0·288	...	...
178	... ...	+ 3·2486	+ 0·0098	...	− 5·901	+ 0·466	...	...
179	7 Aurigæ ε Var. 1 ...	+ 4·2012	+ 0·0199	0·000	− 5·894	+ 0·602	0·00	1540
180	R Leporis Var. 1 ..	+ 2·7287	+ 0·0038	...	− 5·739	+ 0·382	...	...
181	... ...	+ 0·5356	+ 0·0172	...	− 5·709	+ 0·080	...	...
182	Taylor 1797	+ 0·0958	+ 0·0111	...	− 5·692	+ 0·141	...	...
183	... .	+ 1·0926	+ 0·0088	...	− 5·527	+ 0·280	...	...
184	Lacaille 1697 ..	+ 2·0288	+ 0·0086	...	− 5·460	+ 0·296	...	...
185	Taylor 1811	+ 1·9965	+ 0·0089	...	− 5·434	+ 0·282	...	1861
186	2 Leporis ε	+ 2·5358	+ 0·0083	+ 0·001	− 5·308	+ 0·360	+ 0·03	1575
187	... .	+ 1·7020	+ 0·0040	...	− 5·186	+ 0·209	...	...
188	... ...	+ 0·6788	+ 0·0136	...	− 5·041	+ 0·098	...	...
189	Lacaille 1736 ..	+ 0·3245	+ 0·0158	...	− 4·881	+ 0·046	...	...
190	... .	+ 1·7476	+ 0·0045	...	− 4·793	+ 0·280	...	...
191	Lacaille 1757 ...	+ 0·8165	+ 0·0117	...	− 4·775	+ 0·116	...	...
192	13 Aurigæ α (*Capella*)	+ 4·4126	+ 0·0173	+ 0·009	− 4·615	+ 0·629	+ 0·48	1613
193	... .	+ 2·0146	+ 0·0085	...	− 4·602	+ 0·288	...	...
194	19 Orionis β (*Rigel*). .	+ 2·8806	+ 0·0040	− 0·001	− 4·608	+ 0·412	+ 0·08	1623
195	... .	+ 1·9933	+ 0·0086	...	− 4·260	+ 0·285	...	...
196	.. ..	+ 1·0983	+ 0·0036		− 4·168	+ 0·265	...	...
197	... ...	+ 1·6046	+ 0·0046		− 4·030	+ 0·280	...	...
198	... .	+ 0·4283	+ 0·0144		− 4·046	+ 0·062	...	...
199	... ...	+ 0·4130	+ 0·0138		− 3·925	+ 0·065	...	...
200	Lacaille 1822 ..	+ 1·4094	+ 0·0067		− 3·847	+ 0·204	...	...
201	... .	+ 0·4703	+ 0·0128	...	− 3·685	+ 0·070	...	...
202	112 Tauri β	+ 3·7855	+ 0·0092	+ 0·003	− 3·670	+ 0·546	+ 0·20	1651
203	R. P. L. 40	+ 18·4917	+ 0·6779	...	− 3·587	+ 2·666	...	1662
204	Lacaille 1864 ..	+ 1·0477	+ 0·0042	...	− 3·396	+ 0·298	...	...
205	... .	+ 3·8482	+ 0·0081	...	− 3·394	+ 0·554	...	..
206	. ... ..	+ 0·5193	+ 0·0118	...	− 3·258	+ 0·075	...	...
207	... .	+ 0·6687	+ 0·0098	...	− 3·179	+ 0·097	...	...
208	λ Doradûs	+ 0·8715	+ 0·0081	...	− 3·106	+ 0·127	...	1729
209	24 Orionis δ Var. 1 ..	+ 3·0689	+ 0·0038	+ 0·001	− 3·087	+ 0·443	+ 0·04	1730
210	... .	+ 1·9369	+ 0·0034	...	− 3·023	+ 0·281	...	..

Mean Positions of Stars for 1866 January 1st.

Number.	Star.	Magnitude.	Estimations.	Mean Right Ascension.			Mean Polar Distance.			Observations.	Fraction of Year.
				h.	m.	s.	°	′	″		
211	... .	8·9	1	5	26	35·87	106	5	13·4	1	0·08
212	11 Leporis α	3·7	...	5	26	49·22	107	56	15·0	4	0·06
213	Taylor 2057	7·1	1	5	28	2·68	151	56	54·1	1	0·12
214	Lalande 10382	8·4	3	5	29	16·24	78	16	7·6	3	0·07
215	46 Orionis ε	1·8	...	5	29	34·03	91	17	20·8	3	0·06
216	123 Tauri ζ ...	3·0	...	5	29	36·27	68	56	32·6	3	0·31
217	... .	9·5	2	5	30	46·88	135	21	56·3	2	0·08
218	... ...	8·0	2	5	31	0·86	130	12	57·7	2	0·51
219	... .	9·2	1	5	31	41·90	128	42	10·0	1	0·05
220	Lacaille 1949 ...	5·9	...	5	32	16·40	154	10	1·5	1	0·10
221	... ...	8·6	2	5	32	41·06	130	11	29·0	2	0·06
222	... ...	9·2	1	5	32	45·90	128	41	10·8	1	0·05
223	... .	8·7	3	5	34	21·52	132	7	54·8	3	0·09
224	α Columbæ ..	2·7	...	5	34	47·57	124	8	51·6	4	0·01
225	Lacaille 1971 ...	7·1	1	5	36	23·44	140	11	25·0	1	0·12
226	... ...	9·8	1	5	36	47·44	160	57	50·1	1	0·04
227	... ...	9·0	1	5	38	27·74	130	5	21·5	1	0·09
228	... ...	8·7	1	5	39	17·54	79	0	11·3	1	0·06
229	Taylor 2145	6·7	...	5	39	53·16	136	53	44·3	1	0·06
230	... ...	9·0	1	5	39	56·87	185	48	5·0	1	0·08
231	W. B. E. V. 1011 ...	8·3	3	5	40	25·62	78	57	30·5	3	0·07
232	Lacaille 1984	7·6	1	5	40	45·68	130	15	14·7	1	0·11
233	... ...	7·3	2	5	42	10·26	137	3	31·1	2	0·10
234	. . ...	9·1	1	5	43	24·07	130	6	52·2	1	0·05
235	54 Orionis χ^1 . ..	4·7	...	5	46	26·01	69	45	9·1	1	0·07
236	. ...	9·0	1	5	47	4·04	136	46	46·0	1	0·12
237	Lalande 11166 ...	8·0	3	5	47	8·18	78	[illegible]	30·0	3	0·07
238	58 Orionis α (*Betelgeux*) ..	0·9	...	5	47	56·06	82	37	15·3	2	0·09
239	.. ..	7·0	1	5	49	57·28	126	43	0·9	1	0·11
240	Lacaille 2073	7·7	1	5	50	31·62	137	12	37·4	1	0·12
241	. ...	9·5	1	5	51	0·67	130	42	39·4	1	0·10
242	. .	8·8	2	5	52	45·00	129	32	33·4	2	[illegible]
243	R. P. L. 68	6·6	...	5	52	51·67	3	11	19·6	1	[illegible]
244	.	7·8	1	5	53	20·71	131	7	14·1	1	0·10
245	Lalande 11486 ...	7·5	2	5	55	50·16	78	19	12·6	3	[illegible]

227.–236. – Comparison stars for Sappho in 1865.
238 – α Orionis Var. 2 (*Betelgeux*). Irregularly variable from 0·9 to 1·5 [illegible]
243 – Groombridge 1094.

Observed with the Madras Meridian Circle in that Year.

Number.	Star.	In Right Ascension. Annual Precession.	In Right Ascension. Secular Variation.	In Right Ascension. Proper Motion.	In Polar Distance. Annual Precession.	In Polar Distance. Secular Variation.	In Polar Distance. Proper Motion.	Number in B.A.C.
		s	s	s	"	"	"	
211	...	+ 0·1232	+ 0·0142	...	− 2·999	+ 0·010	...	...
212	11 Leporis α ...	+ 2·6411	+ 0·0029	+ 0·001	− 2·898	+ 0·368	0·00	1741
213	Taylor 2067 ...	+ 0·5807	+ 0·0096	..	− 2·701	+ 0·086	...	...
214	Lalande 10802 ...	+ 3·3470	+ 0·0046	...	− 2·681	+ 0·485	...	...
215	46 Orionis ε ...	+ 3·0422	+ 0·0036	− 0·002	− 2·680	+ 0·441	+ 0·01	1766
216	123 Tauri ζ ...	+ 3·5924	+ 0·0065	0·000	− 2·649	+ 0·519	+ 0·06	1767
217	... ...	+ 1·7296	+ 0·0037		− 2·633	+ 0·251	...	...
218	.. ...	+ 0·7648	+ 0·0079		− 2·530	+ 0·110	...	...
219	... ..	+ 2·0070	+ 0·0031		− 2·471	+ 0·292	..	...
220	Lacaille 1940 ..	+ 0·3123	+ 0·0106		− 2·420	+ 0·046	...	1780
221	... ...	+ 0·7548	+ 0·0076	...	− 2·380	+ 0·110		...
222	... ...	+ 2·0090	+ 0·0030	...	− 2·378	+ 0·292		...
223	... ...	+ 0·5504	+ 0·0086	...	− 2·399	+ 0·082	...	...
224	α Columbæ ...	+ 2·1707	+ 0·0027	+ 0·008	− 2·201	+ 0·816	0·00	1802
225	Lacaille 1971 ...	+ 0·8420	+ 0·0066	...	− 2·063	+ 0·130	...	...
226	... ...	+ 1·9574	+ 0·0080		− 2·027	+ 0·295		...
227	... ..	+ 1·9616	+ 0·0080		− 1·892	+ 0·304		...
228	.. ...	+ 3·3309	+ 0·0036		− 1·810	+ 0·485		...
229	Taylor 2146 ...	+ 1·6990	+ 0·0038		− 1·758	+ 0·348		1838
230	...	+ 1·7094	+ 0·0038		− 1·758	+ 0·348		...
231	W. B. H. V. 1011	+ 3·3320	+ 0·0034		− 1·696	+ 0·486	...	..
232	Lacaille 1981 ..	+ 1·9441	+ 0·0080		− 1·651	+ 0·291	...	.
233	.. ...	+ 1·6988	+ 0·0034	.	− 1·561	+ 0·289	...	...
234	.. . ..	+ 1·9190	+ 0·0096	...	− 1·404	+ 0·295	...	...
235	54 Orionis χ¹ ..	+ 3·3615	+ 0·0034	− 0·010	− 1·156	+ 0·330	+ 0·10	1876
236	... .	+ 1·7009	+ 0·0080		− 1·130	+ 0·348		...
237	Lalande 11166	+ 3·3470	+ 0·0080		− 1·126	+ 0·467		...
238	58 Orionis *Betelgeuse*	+ 3·2450	+ 0·0027	+ 0·001	− 1·057	+ 0·473	0·00	1883
239	. ..	+ 1·7025	+ 0·0030	...	− 0·879	+ 0·308		...
240	Lacaille 2073	+ 1·0390	+ 0·0030	...	− 0·898	+ 0·297		
241	..	+ 1·9823	+ 0·0027		− 0·748	+ 0·280		.
242		+ 1·1098	+ 0·0086		− 0·684	+ 0·297		
243	R. P. L. 43	+ 30·4290	+ 0·2000		− 0·621	+ 3·890		1879
244	.	+ 1·9084	+ [illegible]		− 0·392	+ 0·278		
245	Lalande 11163 ..	+ [illegible]	+ 0·0021		− 0·355	+ 0·485		

224 — 238 — Proper motions from "Greenwich Catalogue 1872."

Mean Positions of Stars for 1866 January 1st.

Number.	Star.	Magnitude.	Estimations.	Mean Right Ascension.			Mean Polar Distance.			Observations.	Fraction of Year.
				h.	m.	s.	°	′	″		
246	... ...	8·7	1	5	56	18·46	129	57	13·1	1	0·10
247	Taylor 2801	6·0	1	5	56	31·78	148	6	15·0	1	0·10
248	Taylor 2810	6·9	1	5	59	39·09	150	29	3·6	1	0·12
249	... ...	8·0	3	5	59	44·85	120	40	40·0	3	0·12
250	67 Orionis ν	4·4	...	5	59	55·26	75	13	7·8	3	0·08
251	... ...	7·4	2	6	0	54·19	137	28	37·3	2	0·11
252	... ...	8·1	1	6	2	22·02	168	44	41·5	1	0·12
253	Lalande 11732	8·3	3	6	3	39·41	77	58	54·3	3	0·04
254	... ...	7·1	1	6	4	26·32	128	2	35·3	1	0·13
255	... ...	9·2	1	6	5	44·62	77	51	30·8	2	0·08
256	... ...	9·2	...	6	7	29·06	151	18	29·8	1	0·07
257	... ...	8·6	1	6	8	53·78	131	54	47·8	1	0·10
258	... ...	9·0	1	6	8	57·22	130	31	34·8	1	0·10
259	... ...	9·0	2	6	9	31·00	181	50	51·5	2	0·11
260	Lalande 12068	8·4	3	6	12	31·20	69	51	19·6	3	0·13
261	Lalande 12094	8·7	3	6	13	46·28	68	42	3·7	3	0·13
262	Lalande 12130	7·0	1	6	14	12·30	77	4	46·7	1	0·01
263	13 Geminorum μ	3·2	...	6	14	51·13	67	26	16·6	5	0·08
264	Lalande 12156	7·2	2	6	15	5·31	77	22	2·5	2	0·02
265	Lacaille 2273	8·0	1	6	17	4·48	153	38	27·7	1	0·10
266	... ...	8·7	1	6	18	34·05	151	26	28·8	1	0·10
267	Taylor 2465	7·9	1	6	18	34·25	131	16	10·0	1	0·12
268	... .	9·5	1	6	19	47·38	65	29	39·0	1	0·13
269	α Argûs *(Canopus)* .	0·4	...	6	20	36·86	142	38	26·4	1	0·04
270	18 Geminorum ν	4·0	...	6	21	0·35	69	42	23·6	1	0·07
271	Lacaille 2312	7·0	1	6	22	9·89	166	36	46·3	1	0·10
272	.. ..	8·4	1	6	22	10·61	128	46	47·6	1	0·12
273	Lacaille 2329 .. .	7·0	2	6	24	0·04	153	44	36·7	2	0·12
274	Taylor 2541 . ..	6·1	1	6	24	57·21	147	25	4·0	1	0·18
275	Lacaille 2340	7·3	1	6	26	33·08	163	3	46·2	1	0·13
276	... ...	8·7	1	6	28	38·30	130	56	36·1	1	0·10
277	Taylor 2580	6·8	3	6	29	50·78	151	46	42·0	3	0·09
278	24 Geminorum γ	2·0		6	29	54·23	73	29	24·2	11	0·14
279	... ...	8·9	2	6	31	28·96	140	0	10·2	2	0·11
280	R Monocerotis Var. 1 ..	10·3	4	6	31	50·97	81	8	38·9	5	0·13

253.—255.—262.—264.—Comparison stars for [illegible] in [illegible].
259.—261.—Comparison stars for Ariadne in 1866.
280.—R Monocerotis Var. 1.—Irregularly variable from 9·3 to 11·5 magnitude.

Observed with the Madras Meridian Circle in that Year.

Number.	Star.	In Right Ascension.			In Polar Distance.			Number in B. A. C.
		Annual Precession.	Secular Variation.	Proper Motion.	Annual Precession.	Secular Variation.	Proper Motion.	
		s	s	s	"	"	"	
246	... .	+ 1·9522	+ 0·0026	...	− 0·381	+ 0·296	...	...
247	Taylor 2301	+ 0·9266	+ 0·0030	...	− 0·127	+ 0·136	...	1964
248	Taylor 2310	+ 0·7104	+ 0·0030	...	− 0·031	+ 0·104	...	...
249	... .	+ 1·9560	+ 0·0024	...	− 0·023	+ 0·295	...	...
250	67 Orionis ν	+ 3·4048	+ 0·0017	+ 0·001	− 0·007	+ 0·500	+ 0·02	1965
251	... .	+ 1·6143	+ 0·0024		+ 0·078	+ 0·290		...
252	... .	+ 0·3015	+ 0·0025		+ 0·300	+ 0·058		...
253	Lalande 11732 ...	+ 3·3567	+ 0·0015		+ 0·306	+ 0·460		...
254	... .	+ 2·0262	+ 0·0028		+ 0·398	+ 0·296		...
255	... .	+ 3·3607	+ 0·0013		+ 0·501	+ 0·490		...
256	... .	+ 0·6300	+ 0·0016	...	+ 0·666	+ 0·092	...	...
257	... .	+ 1·8798	+ 0·0021	...	+ 0·777	+ 0·278	...	...
258	... ...	+ 1·9300	+ 0·0021	...	+ 0·783	+ 0·281	...	...
259	... .	+ 1·8757	+ 0·0021	...	+ 0·833	+ 0·273	...	...
260	Lalande 12058 ...	+ 3·5584	+ 0·0002	...	+ 1·095	+ 0·528	—	...
261	Lalande 12094 ...	+ 3·5794	0·0000	...	+ 1·192	+ 0·522	...	...
262	Lalande 12120 ...	+ 3·3782	+ 0·0005	...	+ 1·242	+ 0·467	...	...
263	13 Geminorum μ ...	+ 3·6368	− 0·0003	+ 0·006	+ 1·290	+ 0·527	+ 0·14	2087
264	Lalande 12156 ...	+ 3·3711	+ 0·0004	...	+ 1·310	+ 0·490	...	...
265	Lacaille 2273 ...	+ 0·3416	− 0·0014	...	+ 1·492	+ 0·049	...	...
266	... .	+ 0·6146	− 0·0017	...	+ 1·633	+ 0·039	—	—
267	Taylor 2495	+ 0·6411	− 0·0007	...	+ 1·694	+ 0·092	...	...
268	... .	+ 3·6745	− 0·0011	...	+ 1·790	+ 0·538	...	...
269	α Argûs (Canopus) ..	+ 1·3292	+ 0·0010	0·000	+ 1·834	+ 0·192	0·00	2085
270	15 Geminorum ν ..	+ 3·5641	− 0·0008	− 0·002	+ 1·837	+ 0·517	+ 0·01	2090
271	Lacaille 2612 ...	+ 0·2002	− 0·0036	...	+ 1·937	+ 0·636	...	—
272	... .	+ 2·0017	+ 0·0019	...	+ 1·987	+ 0·290	...	...
273	Lacaille 2339 ..	+ 0·8700	− 0·0031	...	+ 2·106	+ 0·064	—	2121
274	Taylor 2641	+ 0·9620	− 0·0006	...	+ 2·178	+ 0·137	...	2134
275	Lacaille 2348 ..	+ 0·5678	− 0·0030	...	+ 2·318	+ 0·081	...	2142
276	... .	+ 1·0216	+ 0·0016	...	+ 2·495	+ 0·277	...	...
277	Taylor 2889	+ 0·6917	− 0·0031	...	+ 2·605	+ 0·046	.	2166
278	24 Geminorum γ ...	+ 3·4649	− 0·0015	+ 0·001	+ 2·615	+ 0·500	+ 0·04	2169
279	... ...	+ 1·4936	+ 0·0007	...	+ 2·740	+ 0·215	...	...
280	R Monocerotis Var. 1.	+ 3·2768	− 0·0007	...	+ 2·778	+ 0·473	...	...

250.—Proper motions from "*Greenwich Catalogue* 1872."
269.—Proper motions from "*Stone's Cape Catalogue.*"

Mean Positions of Stars for 1866 January 1st.

Number.	Star.	Magnitude.	Estimations.	Mean Right Ascension.			Mean Polar Distance.			Observations.	Fraction of Year.
				h.	m.	s.	°	′	″		
281	... ...	7·7	2	6	34	34·59	130	35	2·1	2	0·11
282	... ...	8·8	2	6	35	51·37	130	38	10·0	2	0·14
283	Taylor 2668	9·2	1	6	36	34·68	151	24	57·0	1	0·10
284	51 Cephei (*Hev.*)	6·3	...	6	36	30·57	2	46	25·3	6	0·22
285	Lacaille 2461	8·0	1	6	38	11·31	155	57	49·9	1	0·09
286	... ...	8·4	1	6	39	11·09	131	3	31·3	1	0·12
287	9 Canis Majoris α (*Sirius*)	− 1·4	...	6	39	14·34	106	32	5·9	1	0·11
288	... ...	8·0	2	6	40	33·29	131	1	33·9	2	0·09
289	... ...	9·2	1	6	40	38·41	151	38	55·3	1	0·15
290	... ...	8·3	3	6	42	27·48	130	57	4·2	3	0·10
291	... .	9·0	2	6	42	41·09	130	59	14·6	2	0·12
292	... ...	8·7	1	6	42	41·27	130	56	29·7	1	0·14
293	... ...	7·6	1	6	43	45·01	128	30	32·1	1	0·10
294	... ...	9·3	1	6	44	53·54	128	30	39·7	1	0·15
295	Taylor 2794 ...	8·8	1	6	44	55·80	144	36	11·2	1	0·03
296	... ...	9·6	2	6	46	37·03	130	10	18·4	2	0·11
297	α Pictoris	3·3	...	6	46	40·00	151	47	56·0	2	0·07
298	Lacaille 2682 ...	6·8	1	6	49	14·75	150	5	41·7	1	0·18
299	... ...	9·4	2	6	48	51·59	130	10	34·0	2	0·14
300	... .	9·1	2	6	48	38·22	128	46	10·2	2	0·13
301	Lacaille 2688	8·0	1	6	51	23·71	114	47	41·4	1	0·14
302	21 Canis Majoris ε ...	1·5	...	6	53	21·05	118	47	30·9	9	0·08
303	... ...	8·0	1	6	53	51·30	120	47	48·6	1	0·10
304	... ...	7·9	2	6	55	30·47	120	38	5·2	2	0·09
305	... ...	9·3	1	6	56	35·38	129	17	31·6	1	0·05
306	28 Canis Majoris γ ...	4·1	...	6	57	41·79	106	20	15·7	5	0·07
307	Lalande 13707	8·1	2	6	58	13·48	97	6	50·1	2	0·09
308	R Geminorum Var. 2. ...	7·6	6	6	59	17·19	67	5	37·7	6	0·10
309	. .. .	9·1	2	7	0	39·50	90	50	17·7	2	0·18
310	Taylor 2861	7·1	2	7	0	52·36	146	41	58·3	2	0·17
311	... ...	9·2	3	7	1	19·78	81	4	19·7	3	0·13
312	R Canis Minoris Var. 1. ...	9·7	2	7	1	30·34	79	46	3·1	2	0·14
313	W. B. R. VI. 1980 ..	8·5	2	7	1	34·38	79	41	47·1	2	0·16
314	. .	9·1	2	7	3	27·73	130	14	40·5	2	0·07
315	. ...	9·4	3	7	5	46·26	60	58	49·4	3	0·12

308.—R Geminorum Var. 2.—Period 371 days.—Range, 7th magnitude to invisibility.
309.—311.—315.—Comparison stars for [illegible] in 1868.
312.—R Canis Minoris Var. 1.—Period 336 days.—Range, 7·5 to 11th magnitude.

Observed with the Madras Meridian Circle in that Year.

Number.	Star.	In Right Ascension. Annual Precession.	In Right Ascension. Secular Variation.	In Right Ascension. Proper Motion.	In Polar Distance. Annual Precession.	In Polar Distance. Secular Variation.	In Polar Distance. Proper Motion.	Number in B. A. C.
		s	s	s	"	"	"	
281	... ...	+ 1·0415	+ 0·0015	...	+ 3·014	+ 0·279	...	...
282	... .	+ 1·0887	+ 0·0014	...	+ 3·125	+ 0·278	...	...
283	Taylor 2632	+ 0·6496	− 0·0042	...	+ 3·187	+ 0·092	...	2205
284	51 Cephei (*Hev.*) ..	+ 30·4815	− 1·8738	− 0·027	+ 3·194	+ 4·387	+ 0·08	2157
285	Lacaille 2451 ..	+ 0·1158	− 0·0092	...	+ 3·326	+ 0·016	...	...
286	... ...	+ 1·0244	+ 0·0014	...	+ 3·413	+ 0·276	...	...
287	9 Can. Maj. α (*Sirius*).	+ 2·6408	+ 0·0010	− 0·035	+ 3·416	+ 0·384	+ 1·24	2213
288	... ...	+ 1·0270	+ 0·0013	...	+ 3·380	+ 0·275	...	..
289	... .	+ 0·5986	− 0·0053	...	+ 3·599	+ 0·094	...	...
290	... ...	+ 1·9317	+ 0·0013	...	+ 3·694	+ 0·275	...	...
291	... ...	+ 1·9312	+ 0·0013		+ 3·714	+ 0·275		...
292	... ...	+ 1·9457	+ 0·0013		+ 3·714	+ 0·277		...
293	... .	+ 2·0270	+ 0·0018		+ 3·905	+ 0·298		...
294	... ...	+ 2·0296	+ 0·0014		+ 3·903	+ 0·298		...
295	Taylor 2734	+ 1·2205	− 0·0014		+ 3·907	+ 0·173		...
296	... .	+ 1·9067	+ 0·0013	...	+ 4·061	+ 0·279	...	...
297	α Pictoris	+ 0·6306	− 0·0063	− 0·010	+ 4·088	+ 0·098	− 0·18	2260
298	Lacaille 2532 ..	+ 0·7988	− 0·0050	...	+ 4·190	+ 0·112	...	...
299	... ...	+ 1·9840	+ 0·0012	...	+ 4·247	+ 0·279	...	...
300	... ...	+ 2·0227	+ 0·0013	...	+ 4·283	+ 0·296	...	...
301	Lacaille 2608 ...	+ 2·4600	+ 0·0018	...	+ 4·458	+ 0·340	...	...
302	21 Canis Majoris ε ...	+ 2·3571	+ 0·0018	0·000	+ 4·627	+ 0·332	+ 0·02	2298
303	... ...	+ 1·9890	+ 0·0012	...	+ 4·670	+ 0·290	...	...
304	... .	+ 1·9072	+ 0·0012	...	+ 4·825	+ 0·281	...	...
305	... ...	+ 2·0111	+ 0·0012	...	+ 4·898	+ 0·288	...	...
306	23 Canis Majoris γ ..	+ 2·7145	+ 0·0005	+ 0·002	+ 4·986	+ 0·351	+ 0·01	2319
307	Lalande 13707 ..	+ 3·0144	− 0·0067	...	+ 5·040	+ 0·609	...	...
308	R Geminorum Var. 2.	+ 3·0182	− 0·0060	...	+ 5·120	+ 0·608	...	...
309	... .	+ 3·7022	− 0·0081	...	+ 5·244	+ 0·582	...	...
310	Taylor 2851	+ 1·1778	− 0·0083	...	+ 5·294	+ 0·164	...	...
311	... ...	+ 3·7447	− 0·0081	...	+ 5·302	+ 0·531	...	...
312	R Canis Minoris Var. 1	+ 3·3040	− 0·0081	...	+ 5·304	+ 0·448	...	...
313	W. B. K. VI. 1980 .	+ 3·3088	− 0·0031	...	+ 5·324	+ 0·463	...	...
314	... .	+ 1·9880	+ 0·0010	...	+ 5·365	+ 0·274	...	..
315	... .	+ 3·7880	− 0·0088	...	+ 5·077	+ 0·596	...	..

284.—Proper motions from "*Greenwich Catalogue 1872.*"
297.—Proper motions from "*Stone's Cape Catalogue.*"

Mean Positions of Stars for 1866 January 1st.

Number.	Star.	Magnitude.	Estimations.	Mean Right Ascension.			Mean Polar Distance.			Observations.	Fraction of Year.
				h	m	s	°	′	″		
316	... ...	7·7	..	7	6	42·80	120	2	56·3	1	0·16
317	Taylor 2928 ...	7·0	1	7	7	16·09	150	21	29·7	1	0·17
318	... ...	9·1	2	7	8	57·61	120	17	31·5	2	0·12
319	Taylor 2940 ...	7·0	1	7	9	32·27	129	57	54·0	1	0·18
320	... ...	9·5	2	7	9	50·44	130	18	46·8	2	0·12
321	54 Geminorum λ	3·6	...	7	10	23·44	73	13	16·1	1	0·08
322	55 Geminorum δ	3·7	...	7	12	7·08	67	46	38·1	11	0·11
323	... ...	6·0	1	7	12	37·36	132	48	6·0	1	0·10
324	.. ..	9·4	2	7	13	5·31	130	16	14·1	2	0·05
325	... ...	7·9	2	7	14	33·68	138	49	49·4	2	0·09
326	Lacaille 2905	8·0	2	7	17	22·62	166	8	14·3	2	0·17
327	... ...	8·6	4	7	17	29·01	130	13	41·0	4	0·10
328	... ...	9·3	2	7	17	41·13	120	40	34·1	2	0·06
329	B. P. L. 45	7·2	...	7	17	40·32	0	59	19·6	3	0·12
330	Taylor 3043	7·2	2	7	19	17·81	139	16	39·6	2	0·08
331	Lacaille 2907	8·1	2	7	19	35·48	142	15	36·7	2	0·15
332	... ...	9·1	1	7	19	50·60	152	36	50·1	1	0·18
333	... ...	7·5	2	7	21	37·92	131	50	40·4	2	0·17
334	... ...	9·1	3	7	23	31·58	129	46	32·8	3	0·09
335	... ...	8·1	1	7	23	57·73	139	10	20·7	1	0·04
336	S Canis Minoris Var. 2 ...	7·0	4	7	25	26·78	81	23	36·3	4	0·11
337	... ...	8·0	1	7	25	30·98	129	19	30·7	1	0·15
338	65 Geminorum	5·0	...	7	25	57·61	78	38	19·0	1	0·08
339	66 Geminorum α² (*Castor*)	2·0	...	7	26	2·74	57	48	17·1	6	0·13
340	... ...	9·7	1	7	26	41·58	130	45	16·7	1	0·04
341	... ...	8·2	4	7	28	0·26	130	42	39·3	4	0·12
342	Taylor 3196	6·9	3	7	29	37·07	148	15	39·5	3	0·17
343	. ...	8·6	2	7	31	8·34	131	10	40·0	2	0·18
344	... ...	9·1	2	7	31	56·07	120	44	9·2	3	0·09
345	10 Can. Min. α (*Procyon*)...	0·8	...	7	32	17·12	84	36	4·7	14	0·10
346	S Geminorum Var. 3. ..	10·6	3	7	34	40·06	66	12	42·6	4	0·07
347	.. ..	9·0	2	7	35	10·49	120	33	7·8	2	0·04
348	... ...	9·9	1	7	36	17·13	90	16	15·9	1	0·12
349	Taylor 3166 ..	7·9	1	7	36	34·06	130	19	38·8	1	0·20
350	... ...	7·9	2	7	36	34·79	134	56	26·7	2	0·17

320.—Groombridge 1119.
336.—S Canis Minoris Var. 2.—Period 292 days.—Range, 7·3 to below 12th magnitude
346.—S Geminorum Var. 3.—Period 294 days.—Range 9·3 to below 10th magnitude

Observed with the Madras Meridian Circle in that Year.

Number.	Star.	In Right Ascension.			In Polar Distance.			Number in B. A. C.
		Annual Precession.	Secular Variation.	Proper Motion.	Annual Precession.	Secular Variation.	Proper Motion.	
		s	s	s	"	"	"	
316	... .	+ 2·0832	+ 0·0011	...	+ 5·766	+ 0·298	...	...
317	Taylor 2988	+ 0·8281	− 0·0077	...	+ 5·806	+ 0·118	...	...
318	... .	+ 1·9806	+ 0·0010	...	+ 5·944	+ 0·274	...	...
319	Taylor 2940	+ 2·0098	+ 0·0010	...	+ 5·992	+ 0·276	...	...
320	... ...	+ 1·9800	+ 0·0009	...	+ 6·017	+ 0·274	...	...
321	54 Geminorum λ ..	+ 3·4564	− 0·0055	− 0·002	+ 6·068	+ 0·478	+ 0·04	2398
322	55 Geminorum δ ...	+ 3·5015	− 0·0072	0·000	+ 6·207	+ 0·486	+ 0·02	2410
323	... ...	+ 0·5990	− 0·0119	...	+ 6·249	+ 0·080	...	...
324	... ...	+ 2·0341	+ 0·0009	...	+ 6·363	+ 0·279	...	...
325	... ...	+ 1·6234	− 0·0088	...	+ 6·410	+ 0·221	...	...
326	Lacaille 2805 ...	+ 0·5814	− 0·0132	...	+ 6·648	+ 0·077	...	...
327	... ...	+ 2·0424	+ 0·0000	...	+ 6·652	+ 0·278	...	...
328	... ...	+ 2·0226	+ 0·0000	...	+ 6·666	+ 0·275	...	...
329	R. P. L. 45 ...	+ 76·1378	− 28·0890	− 0·328	+ 6·690	+ 10·467	− 0·01	2280
330	Taylor 3043	+ 2·0486	+ 0·0009	...	+ 6·800	+ 0·277	...	...
331	Lacaille 2807 ...	+ 1·4479	− 0·0083		+ 6·826	+ 0·196		...
332	... ...	+ 0·0490	− 0·0127		+ 6·847	+ 0·086		...
333	... ...	+ 1·9800	+ 0·0006		+ 6·989	+ 0·266		...
334	... ...	+ 2·0629	+ 0·0009		+ 7·149	+ 0·274		...
335	... ...	+ 2·0551	+ 0·0009		+ 7·164	+ 0·277		...
336	S Canis Minoris Var. 2	+ 3·2006	− 0·0044	...	+ 7·300	+ 0·440	...	...
337	... ...	+ 2·0529	+ 0·0009	...	+ 7·311	+ 0·276	...	...
338	63 Geminorum ...	+ 3·4315	− 0·0066	− 0·004	+ 7·347	+ 0·468	0·00	2480
339	66 Gem. α² (*Castor*). .	+ 3·8547	− 0·0136	− 0·013	+ 7·364	+ 0·519	+ 0·08	2495
340	... ...	+ 2·0836	+ 0·0009	...	+ 7·406	+ 0·273	...	...
341	... ...	+ 2·0424	+ 0·0000	...	+ 7·518	+ 0·273	...	...
342	Taylor 3126	+ 1·4189	− 0·0032	...	+ 7·644	+ 0·168	...	2807
343	... .	+ 1·9887	+ 0·0008	...	+ 7·768	+ 0·266	...	...
344	.. ..	+ 2·0480	+ 0·0009	...	+ 7·804	+ 0·272	...	...
345	10 Can. Min. (*Procyon*)	+ 3·1019	− 0·0041	− 0·048	+ 7·880	+ 0·403	+ 1·03	2522
346	S Geminorum Var. 3	+ 3·6110	− 0·0102	...	+ 8·082	+ 0·490		
347	. .	+ 2·0467	+ 0·0008	...	+ 8·098	+ 0·270		...
348	. .. .	+ 3·6088	− 0·0131	...	+ 8·100	+ 0·479		.
349	Taylor 3195 . ..	+ 0·9312	− 0·0106	..	+ 8·208	+ 0·130	.	.
350	.. . . .	+ 2·0894	+ 0·0010	...	+ 8·204	+ 0·274	.	.

321.—329.—Proper motions from "*Greenwich Catalogue* 1872."

Mean Positions of Stars for 1866 January 1st.

Number.	Star.	Magnitude.	Estimations.	Mean Right Ascension.			Mean Polar Distance.			Observations.	Fraction of Year.
				h.	m.	s.	°	′	″		
351	... ...	7·6	2	7	36	46·31	150	57	34·3	2	0·04
352	78 Geminorum β (Pollux).	1·1	...	7	37	6·77	61	40	13·1	16	0·11
353	... ...	7·0	3	7	39	9·63	131	8	54·2	3	0·18
354	... ...	8·5	2	7	40	57·49	153	0	57·8	2	0·19
355	T Geminorum Var. 4 ...	9·8	5	7	41	15·38	66	56	8·7	6	0·21
356	... ...	7·6	1	7	41	34·90	144	18	50·0	1	0·19
357	Lacaille 3081	7·1	2	7	46	7·98	144	22	46·3	2	0·11
358	... ...	8·0	2	7	46	10·65	120	26	9·0	2	0·18
359	U Geminorum Var. 5 ...	9·9	9	7	47	0·34	67	39	0·1	9	0·05
360	Taylor 3310	7·7	1	7	48	36·02	140	18	10·4	2	0·15
361	... ...	9·3	1	7	48	35·76	67	48	35·6	1	0·14
362	... ...	8·4	1	7	49	8·36	130	36	22·5	1	0·16
363	1 Cancri	5·9	...	7	49	22·74	73	51	17·5	1	0·07
364	... ...	8·1	1	7	49	36·53	139	17	41·1	2	0·16
365	... ...	7·6	1	7	49	57·77	149	8	45·0	1	0·10
366	... ...	7·7	1	7	50	9·34	130	38	44·5	1	0·15
367	Taylor 3339 ...	7·9	1	7	51	33·06	144	17	19·7	1	0·18
368	... ...	8·1	2	7	52	2·90	148	28	43·7	2	0·19
369	5 Cancri	6·0	...	7	53	51·90	73	10	43·5	1	0·07
370	Taylor 3378 ...	7·6	1	7	56	16·73	144	12	18·2	1	0·13
371	6 Cancri	5·0	...	7	58	17·06	61	50	0·5	15	0·11
372	... ...	7·9	1	7	58	34·41	135	30	33·0	1	0·30
373	Brisbane 1886	6·8	2	7	58	30·06	132	56	7·0	2	0·18
374	... ...	8·9	2	8	1	24·80	130	31	43·6	2	0·18
375	Lacaille 3174	7·2	1	8	1	27·12	136	39	19·8	1	0·19
376	15 Argûs	2·9	...	8	1	50·32	113	55	13·0	11	0·15
377	... ...	9·5	1	8	2	7·91	113	47	9·2	1	0·30
378	... ...	8·1	3	8	2	16·24	128	39	46·0	2	0·19
379	16 Cancri ζ .	4·7	...	8	4	31·44	71	57	3·5	1	0·15
380	... .. .	7·6	3	8	8	8·10	126	40	51·0	3	0·13
381	. .. ..	9·0	1	8	8	38·14	128	38	45·3	1	0·08
382	R Cancri Var. 1 . .	7·9	3	8	9	10·46	77	51	55·2	3	0·07
383	. . ..	9·6	1	8	10	38·29	151	36	39·7	1	0·08
384	... ...	9·9	1	8	10	35·80	77	38	8·3	1	0·21
385	Lalande 16294 .	7·0	3	8	10	40·30	73	54	32·8	3	0·17

355.—T Geminorum Var. 4.—Period 288 days.—Range, 8·3 to below 14th magnitude.
359.—U Geminorum Var. 5.—Period 99 days but irregular.—Range, 9th to below 14th magnitude.
382.—R Cancri Var. 1.—Period 354 days.—Range, 6th to below 12th magnitude.
366.—Comparison star for Ariadne in 1863.

Observed with the Madras Meridian Circle in that Year.

Number.	Star.	In Right Ascension.			In Polar Distance.			Number in B.A.C.
		Annual Precession.	Secular Variation.	Proper Motion.	Annual Precession.	Secular Variation.	Proper Motion.	
		s	s	s	"	"	"	
351	... ...	+ 2·0508	+ 0·0008	...	+ 8·382	+ 0·280	...	...
352	78 Gem. β (*Pollux*) ...	+ 3·7290	− 0·0128	− 0·049	+ 8·246	+ 0·481	+ 0·06	2565
353	... ...	+ 2·0114	+ 0·0000		+ 8·400	+ 0·262	...	...
354	... ...	+ 0·0815	− 0·0161		+ 8·551	+ 0·086	...	...
355	T Geminorum Var. 4.	+ 3·0119	− 0·0110		+ 8·575	+ 0·472	...	...
356	... ...	+ 1·3802	− 0·0041		+ 8·601	+ 0·179	...	...
357	Lacaille 3081 ...	+ 1·3800	− 0·0042		+ 8·881	+ 0·179	...	...
358	... ...	+ 2·0969	+ 0·0010		+ 8·[illegible]	+ 0·269	...	...
359	U Geminorum Var. 5.	+ 3·5[illegible]	− 0·0108		+ 9·009	+ 0·460	...	...
360	Taylor 3310	+ 1·0082	− 0·0096		+ 9·150	+ 0·136	...	...
361	... ...	+ 3·5583	− 0·0109	...	+ 9·151	+ 0·464	...	...
362	... ...	+ 2·0593	+ 0·0010	...	+ 9·186	+ 0·268	...	...
363	1 Cancri ...	+ 3·4150	− 0·0094	− 0·001	+ 9·212	+ 0·480	+ 0·04	2670
364	... ...	+ 2·1015	+ 0·0011	...	+ 9·255	+ 0·268	...	...
365	... ...	+ 1·0868	− 0·0092	...	+ 9·267	+ 0·137	...	...
366	... ...	+ 2·0807	+ 0·0011	...	+ 9·272	+ 0·266	...	...
367	Taylor 3230	+ 1·4095	− 0·0041	...	+ 9·407	+ 0·180	...	...
368	... ...	+ 1·1563	− 0·0068	...	+ 9·419	+ 0·146	...	...
369	5 Cancri	+ 3·4075	− 0·0090	− 0·001	+ 9·506	+ 0·496	0·00	2664
370	Taylor 3378	+ 1·4077	− 0·0041	...	+ 9·667	+ 0·161	...	...
371	6 Cancri	+ 3·[illegible]	− 0·0148	− 0·006	+ 9·[illegible]	+ 0·468	+ 0·07	2672
372	... .	+ 2·1404	+ 0·0018	...	+ 9·677	+ 0·270	...	...
373	Brisbane 1666 ...	+ 0·7807	− 0·0165	...	+ 9·[illegible]	+ 0·086	...	2680
374	... ...	+ 1·0306	− 0·0116	...	+ 10·134	+ 0·126	...	...
375	Lacaille 3174 ...	+ 0·5245	− 0·0246	...	+ 10·137	+ 0·062	...	...
376	15 Argûs	+ 2·5[illegible]	+ 0·0009	− 0·007	+ 10·166	+ 0·318	− 0·06	2728
377	... ...	+ 2·3046	+ 0·0009	...	+ 10·168	+ 0·318	...	...
378	... ...	+ 2·1510	+ 0·0015	...	+ 10·190	+ 0·268	...	...
379	16 Cancri 3	+ 3·4461	− 0·0108	+ 0·004	+ 10·[illegible]	+ 0·496	+ 0·11	2744
380	. . .	+ 2·1807	+ 0·0016	...	+ 10·[illegible]	+ 0·261	...	..
381	... .	+ 2·1670	+ 0·0017		+ 10·675	+ 0·[illegible]		..
382	R Cancri Var. 1 ..	+ 3·3151	− 0·0080		+ 10·715	+ 0·401		. .
383	. .	+ 1·[illegible]	− 0·0183	...	+ 10·810	+ 0·118		
384	... ...	+ 3·31[illegible]	− 0·0082	...	+ 10·819	+ 0·403		...
385	Lalande 16224 ..	+ 3·3[illegible]	− 0·0097	.	+ 10·[illegible]	+ 0·412		

363.—369 — Proper motions from "*Greenwich Catalogue 1872.*"

Mean Positions of Stars for 1866 January 1st.

Number.	Star.	Magnitude.	Estimations.	Mean Right Ascension. h.	m.	s.	Mean Polar Distance. °	′	″	Observations.	Fraction of Year.
386	... ...	8·4	1	8	11	23·09	162	4	52·0	1	0·12
387	... ...	8·2	1	8	12	21·70	128	44	4·4	2	0·19
388	... ...	8·5	5	8	12	49·73	128	41	17·8	5	0·16
389	... ...	10·0	1	8	13	1·86	120	46	55·2	1	0·21
390	... ...	7·8	2	8	14	53·75	142	17	18·1	2	0·18
391	Lacaille 3297	8·0	3	8	15	2·87	53	30	18·8	3	0·14
392	20 Cancri d¹	5·9	...	8	15	41·24	71	14	26·8	1	0·15
393	... ...	9·0	1	8	17	26·20	141	16	10·2	1	0·16
394	... ...	8·6	2	8	17	29·14	77	49	33·1	2	0·16
395	... ...	10·3	1	8	18	48·91	161	1	6·5	1	0·22
396	Taylor 3600	7·5	3	8	20	19·85	144	58	9·5	4	0·15
397	W. B. N. VIII. 450 ...	8·0	1	8	20	28·00	74	27	44·7	1	0·12
398	29 Cancri	5·0	...	8	21	8·66	76	20	54·5	3	0·08
399	Taylor 3607	6·9	1	8	21	18·41	144	56	46·0	2	0·20
400	... ...	9·0	1	8	21	31·41	131	40	50·6	1	0·09
401	... ...	9·3	1	8	23	14·86	78	25	40·0	2	0·21
402	Taylor 3620 ...	8·0	1	8	23	14·98	130	48	11·7	1	0·10
403	... ...	8·9	1	8	23	36·66	128	34	50·9	1	0·16
404	... ...	9·0	1	8	24	51·70	73	46	4·3	1	0·18
405	33 Cancri η ...	5·5	...	8	24	57·38	69	6	23·6	9	0·15
406	Taylor 3651	7·6	1	8	25	44·01	130	3	46·2	1	0·18
407	Taylor 3652	7·0	1	8	25	48·23	130	3	2·6	1	0·20
408	Lacaille 3388	8·7	1	8	26	30·32	140	40	35·8	1	0·21
409	W. B. N. VIII. 635 ...	8·5	2	8	27	42·84	73	48	48·9	2	0·20
410	Taylor 3672	7·1	2	8	28	36·84	71	18	31·0	3	0·16
411	... .	9·3	1	8	28	54·27	70	41	5·1	1	0·16
412	W. B. N. VIII. 684	9·2	1	8	29	7·85	70	20	22·5	1	0·23
413	Lacaille 3409 . .	7·3	1	8	29	15·00	151	38	3·8	1	0·18
414	. . .	8·4	2	8	30	14·41	128	47	33·7	2	0·22
415	.. .. .	9·0	1	8	33	13·99	120	23	51·9	1	0·10
416	W. B. N. VIII. 693	9·3	2	8	34	8·62	74	8	3·0	2	0·18
417	.. .	8·3	1	8	34	10·23	154	30	40·3	1	0·10
418	.. .	9·3	1	8	34	43·62	129	48	37·8	1	0·21
419	36 Cancri A¹ ..	5·6	...	8	35	40·21	76	30	28·4	3	0·18
420	Lacaille 3491	8·4	1	8	36	3·53	132	22	16·3	1	0·21

397.—Comparison star for supposed new variable star, 1898 Feb. 1.
391.—392.—410.—419.—Comparison stars for [illegible] in 1898.

Observed with the Madras Meridian Circle in that Year.

Number.	Star.	In Right Ascension. Annual Precession.	In Right Ascension. Secular Variation.	In Right Ascension. Proper Motion.	In Polar Distance. Annual Precession.	In Polar Distance. Secular Variation.	In Polar Distance. Proper Motion.	Number in B. A. C.
		s	s	s	"	"	"	
386	...	+ 0·0526	− 0·0147	...	+ 10·879	+ 0·112	...	...
387	... ...	+ 2·1736	+ 0·0018	...	+ 10·940	+ 0·261	...	...
388	... ...	+ 2·1763	+ 0·0018	...	+ 10·984	+ 0·261	...	...
389	... ...	+ 2·1083	+ 0·0013	..	+ 10·998	+ 0·262	...	...
390	... ...	+ 1·6340	− 0·0021	...	+ 11·134	+ 0·193	...	...
391	Lacaille 3297 ...	+ 0·7946	− 0·0196	...	+ 11·145	+ 0·091	...	...
392	20 Cancri d¹	+ 3·4490	− 0·0114	− 0·006	+ 11·101	+ 0·413	+ 0·02	2790
393	... ...	+ 1·0961	− 0·0014	...	+ 11·310	+ 0·160	...	...
394	... ...	+ 3·3102	− 0·0084	...	+ 11·322	+ 0·394	...	...
395	... ...	+ 1·0877	− 0·0122	...	+ 11·418	+ 0·126	...	...
396	Taylor 3590	+ 1·5162	− 0·0039	...	+ 11·527	+ 0·176	...	...
397	W. B. N. VIII. 430..	+ 3·3763	− 0·0100	...	+ 11·581	+ 0·398	...	...
398	30 Cancri	+ 3·3374	− 0·0096	− 0·002	+ 11·595	+ 0·396	+ 0·01	2806
399	Taylor 3607	+ 1·5192	− 0·0098	...	+ 11·590	+ 0·176	...	...
400	... .	+ 2·1010	+ 0·0018	...	+ 11·614	+ 0·246	...	...
401	... ...	+ 3·3946	− 0·0109	...	+ 11·784	+ 0·397	...	...
402	Taylor 3620	+ 2·1361	+ 0·0020	...	+ 11·735	+ 0·248	...	—
403	... ...	+ 2·2060	+ 0·0023	...	+ 11·760	+ 0·256	...	...
404	... ...	+ 3·3962	− 0·0108	...	+ 11·849	+ 0·394	...	...
405	33 Cancri η	+ 3·4037	− 0·0120	− 0·005	+ 11·880	+ 0·404	+ 0·06	2832
406	Taylor 3651	+ 2·1675	+ 0·0022		+ 11·911	+ 0·249	...	...
407	Taylor 3662	+ 2·1681	+ 0·0022		+ 11·914	+ 0·249	...	...
408	Lacaille 3393 ...	+ 1·3346	− 0·0098		+ 11·988	+ 0·140	...	...
409	W. B. N. VIII. 686..	+ 3·3588	− 0·0107		+ 12·053	+ 0·390	...	...
410	Taylor 3672	+ 3·3732	− 0·0106		+ 12·112	+ 0·387	...	2838
411	... ...	+ 3·4452	− 0·0124		+ 12·133	+ 0·396	...	...
412	W. B. N. VIII. 694..	+ 3·4460	− 0·0124		+ 12·140	+ 0·396	...	...
413	Lacaille 3430 ...	+ 1·0819	− 0·0136		+ 12·156	+ 0·121	...	...
414	... .	+ 2·2204	+ 0·0026		+ 12·230	+ 0·252	...	...
415	.. ..	+ 2·2104	+ 0·0026		+ 12·431	+ 0·246	...	..
416	W. B. N. VIII. 883..	+ 3·3094	− 0·0100	...	+ 12·495	+ 0·370	...	...
417	.. ..	+ 0·8947	− 0·0107	...	+ 12·490	+ 0·097	...	...
418	.. ..	+ 2·2082	+ 0·0027	...	+ 12·534	+ 0·216	...	.
419	46 Cancri A¹ ..	+ 3·3152	− 0·0096	− 0·003	+ 12·609	+ 0·371	− 0·01	2846
420	Lacaille 3491 ..	+ 1·0875	− 0·0141	...	+ 12·625	+ 0·115	...	2849

392.—398.—Proper motions from "*Greenwich Catalogue* 1872."

Mean Positions of Stars for 1866 January 1st.

Number.	Star.	Magnitude.	Estimations.	Mean Right Ascension.			Mean Polar Distance.			Observations.	Fraction of Year.
				h.	m.	s.	°	′	″		
421	b Velorum	4·1	...	8	36	10·08	136	10	55·5	1	0·20
422	S Cancri Var. 2 ...	8·8	3	8	36	16·64	70	29	12·4	3	0·20
423	Lalande 17231	7·8	2	8	37	50·60	74	28	8·3	2	0·20
424	W. B. N. VIII. 977 ...	9·7	2	8	39	21·87	74	40	20·0	2	0·17
425	11 Hydræ ε ...	3·6	...	8	39	40·71	83	5	32·2	9	0·17
426	... ...	7·8	1	8	40	38·59	129	15	59·4	1	0·19
427	Lacaille 3534	8·5	1	8	42	21·65	129	18	32·4	1	0·21
428	W. B. N. VIII. 1048 ...	8·3	2	8	42	29·16	74	40	21·5	2	0·24
429	... ...	9·2	1	8	42	50·47	136	10	54·2	1	0·14
430	Lacaille 3573	7·8	3	8	44	15·19	152	41	52·7	3	0·10
431	S Hydræ Var. 3	8·0	5	8	46	34·56	86	26	40·8	5	0·19
432	... ...	8·6	1	8	46	48·48	132	53	42·3	1	0·21
433	R. P. L. 60	6·5	...	8	46	50·28	5	17	22·6	1	0·18
434	... ...	8·0	2	8	46	59·95	136	6	28·1	2	0·22
435	... ...	8·7	2	8	47	44·28	132	56	21·7	2	0·24
436	T Cancri Var. 3	8·8	3	8	48	0·66	69	39	37·7	4	0·20
437	T Hydræ Var. 4	7·9	7	8	48	8·59	98	37	36·7	7	0·16
438	... ...	8·3	2	8	49	17·84	132	51	48·4	2	0·23
439	9 Ursæ Majoris ι	3·2	...	8	50	1·21	41	28	6·4	1	0·09
440	... ...	8·2	1	8	50	10·69	132	57	23·9	1	0·19
441	65 Cancri α ...	4·3	...	8	51	0·34	77	37	36·8	4	0·14
442	... ...	9·2	1	8	54	5·95	132	54	10·5	1	0·28
443	... ...	8·7	1	8	54	34·48	130	38	21·7	1	0·20
444	... .	8·3	2	8	54	51·72	130	36	39·4	2	0·18
445	.. ..	8·0	1	8	56	1·08	142	40	24·5	1	0·14
446		8·5	1	8	56	40·61	146	46	30·2	1	0·16
447		9·5	1	8	56	47·48	129	14	40·9	1	0·10
448		8·7	...	8	56	56·10	146	35	4·1	1	0·21
449		9·0	1	8	58	9·29	147	50	10·8	1	0·16
450	...	9·1	1	8	58	12·73	140	18	36·9	1	0·09
451	. ..	8·2	3	8	59	9·49	145	38	37·2	3	0·23
452	76 Cancri κ	5·0		9	0	29·23	78	47	42·3	3	0·11
453	... - ...	8·8		9	1	54·37	128	57	41·6	1	0·20
454	Lacaille 3706	7·0	1	9	2	16·79	151	17	33·9	2	0·20
455	.. ..	9·2	2	9	4	1·38	132	48	14·4	2	0·14

422.—S Cancri Var. 2.—Period 9·48 days.—Range, 8th to 10·5 magnitude
423.—424.—428.—Comparison stars for [illegible] in 1864
431.—S Hydræ Var. 3.—Period 260 days.—Range, 8th to 12th magnitude
433.—Carrington [illegible]
436.—T Cancri Var. 3.—Period [illegible] days.—Range, 8th to 10·5 magnitude
437.—T Hydræ Var. 4.—Period 289 days.—Range, 7th to below 12th magnitude.

Observed with the Madras Meridian Circle in that Year.

Number.	Star.	In Right Ascension. Annual Precession.	In Right Ascension. Secular Variation.	In Right Ascension. Proper Motion.	In Polar Distance. Annual Precession.	In Polar Distance. Secular Variation.	In Polar Distance. Proper Motion.	Number in B. A. C.
		s	s	s	"	"	"	
421	b Velorum	+ 1·9903	+ 0·0018	...	+ 12·633	+ 0·221	...	2947
422	S Cancri Var. 2 .	+ 3·4400	− 0·0130	...	+ 12·640	+ 0·286	...	...
423	Lalande 17231 ...	+ 3·3390	− 0·0100	...	+ 12·740	+ 0·378	...	...
424	W. B. N. VIII. 977 .	+ 3·3500	− 0·0108	...	+ 12·849	+ 0·360	...	...
425	11 Hydræ ε	+ 3·1962	− 0·0071	− 0·013	+ 12·869	+ 0·351	+ 0·04	2971
426	... .	+ 2·2365	+ 0·0031		+ 12·998	+ 0·244		...
427	Lacaille 3584 ...	+ 2·2408	+ 0·0032		+ 13·048	+ 0·242		...
428	W. B. N. VIII. 1043	+ 3·3503	− 0·0109		+ 13·050	+ 0·366		...
429	... .	+ 2·0165	+ 0·0023		+ 13·091	+ 0·217		...
430	Lacaille 3573 ..	+ 1·1193	− 0·0140		+ 13·174	+ 0·117		3006
431	S Hydræ Var. 3 ...	+ 3·1345	− 0·0080		+ 13·336	+ 0·336	...	...
432	... .	+ 2·1417	+ 0·0032		+ 13·341	+ 0·233	...	...
433	R. P. L. 60... ..	+ 13·8531	− 1·7306		+ 13·341	+ 1·502	...	...
434	... .	+ 2·0363	+ 0·0026		+ 13·354	+ 0·210	...	...
435	... ...	+ 2·1466	+ 0·0032		+ 13·402	+ 0·227	...	...
436	T Cancri Var. 3 ..	+ 3·4394	− 0·0141	...	+ 13·486	+ 0·366	...	...
437	T Hydræ Var. 4 ...	+ 2·9220	− 0·0018	...	+ 13·494	+ 0·309	...	...
438	... ...	+ 2·1531	+ 0·0033	...	+ 13·503	+ 0·228	...	...
439	9 Ursæ Majoris ι ...	+ 4·1587	− 0·0446	− 0·017	+ 13·520	+ 0·448	+ 0·26	3048
440	... ...	+ 2·1560	+ 0·0034	...	+ 13·570	+ 0·226	...	...
441	65 Cancri α	+ 3·2874	− 0·0068	0·000	+ 13·623	+ 0·346	+ 0·04	3086
442	... .	+ 2·1712	+ 0·0037	...	+ 13·811	+ 0·228	...	...
443	... ...	+ 2·2427	+ 0·0039	...	+ 13·831	+ 0·231	...	...
444	... .	+ 2·3439	+ 0·0030	...	+ 13·860	+ 0·251	...	...
445	... .	+ 1·7967	+ 0·0005	...	+ 13·870	+ 0·184	...	...
446	... ...	+ 1·6079	− 0·0030		+ 13·971	+ 0·162		...
447	... ..	+ 2·2573	+ 0·0040		+ 13·961	+ 0·235		...
448	. .	+ 1·6304	− 0·0028		+ 13·992	+ 0·163		...
449	... ..	+ 1·6136	− 0·0024		+ 14·036	+ 0·162		..
450	.	+ 1·6129	− 0·0018		+ 14·070	+ 0·166		...
451	... ...	+ 1·6840	− 0·0010	...	+ 14·130	+ 0·168	..	...
452	76 Cancri κ	+ 3·2800	− 0·0091	− 0·002	+ 14·211	+ 0·330	0·00	3111
453	... . ..	+ 2·3111	+ 0·0044	...	+ 14·290	+ 0·251	...	...
454	Lacaille 3705 .	+ 1·3681	− 0·0053	...	+ 14·321	+ 0·138	...	
455	... ...	+ 2·2120	+ 0·0046	...	+ 14·427	+ 0·218	...	...

Mean Positions of Stars for 1860 January 1st.

Number.	Star.	Magnitude.	Estimations.	Mean Right Ascension.			Mean Polar Distance.			Observations.	Fraction of Year.
				h.	m.	s.	°	′	″		
456	... ...	8·0	2	9	4	28·84	130	30	9·5	2	0·18
457	... .	8·2	4	9	4	39·86	132	45	57·8	4	0·30
458	Lacaille 3713 ...	8·1	2	9	4	39·27	143	49	41·8	2	0·23
459	Taylor 4021 ...	7·2	2	9	5	36·48	138	44	39·1	2	0·23
460	... ...	8·0	1	9	6	30·01	142	29	54·2	1	0·27
461	... ...	9·8	1	9	8	17·65	146	14	44·8	1	0·21
462	Lacaille 3774 ...	8·0	2	9	9	56·08	157	10	14·7	2	0·19
463	80 Cancri	6·6	...	9	11	29·59	71	40	43·8	10	0·17
464	... ...	8·2	1	9	13	10·02	72	18	27·7	1	0·08
465	ι Argûs	2·5	...	9	13	30·90	148	48	58·7	2	0·21
466	... ..	8·3	3	9	15	19·28	148	40	11·9	3	0·26
467	... ...	8·0	3	9	16	10·21	140	8	6·3	3	0·28
468	... ...	9·5	1	9	17	29·34	76	6	0·1	2	0·20
469	... ...	7·0	2	9	19	36·66	76	7	3·5	2	0·15
470	... ...	9·2	2	9	20	28·01	137	28	27·0	2	0·21
471	... ...	9·5	1	9	20	53·20	158	39	46·1	1	0·23
472	30 Hydræ α Var. 2. ...	2·0	...	9	21	0·11	98	4	46·7	14	0·17
473	25 Ursæ Majoris θ ...	3·2	...	9	22	52·62	37	42	56·2	2	0·28
474	... ...	9·5	1	9	24	22·02	158	41	16·9	1	0·26
475	... ...	9·0	...	9	24	37·75	130	26	44·0	1	0·28
476	6 Leonis h... ...	5·4	...	9	24	46·50	79	41	42·7	2	0·08
477	... ...	8·0	1	9	26	39·21	144	58	40·4	1	0·15
478	Taylor 4282 ...	7·6	2	9	27	48·32	146	29	52·3	2	0·16
479	... ...	8·9	5	9	28	3·41	138	46	34·8	5	0·19
480	.. ..	7·0	1	9	28	31·74	144	0	13·5	1	0·26
481	... ...	9·2	2	9	28	50·71	128	47	27·9	2	0·15
482	.. ..	9·0	1	9	29	6·98	128	50	6·1	1	0·20
483	... ...	9·8	2	9	29	26·31	146	34	7·2	2	0·22
484	Taylor 4289	4·9	...	9	32	1·99	138	46	20·9	3	0·19
485	... .	9·0	1	9	32	32·48	129	54	34·3	1	0·14
486	B. P. L. 69	7·9	...	9	33	30·15	2	47	22·1	1	0·72
487	14 Leonis ο	3·8	...	9	33	39·95	79	30	0·3	3	0·18
488	... ...	8·8	1	9	34	37·42	151	56	52·9	1	0·28
489	... ...	9·1	3	9	35	53·17	148	40	29·3	2	0·27
490	... ...	8·5	1	9	35	56·29	151	54	38·2	1	0·26

472.—30 Hydræ α Var. 2.—Supposed to vary irregularly from 2·0 to 2·6 magnitude.
486.—Carrington 1418.

Observed with the Madras Meridian Circle in that Year.

Number.	Star.	In Right Ascension.			In Polar Distance.			Number in B. A. C.
		Annual Precession.	Secular Variation.	Proper Motion.	Annual Precession.	Secular Variation.	Proper Motion.	
		s	s	s	"	"	"	
456	... ...	+ 2·2306	+ 0·0047	...	+ 14·455	+ 0·296	...	...
457	... ...	+ 2·2180	+ 0·0046	...	+ 14·456	+ 0·218	...	...
458	Lacaille 3712 ...	+ 1·5064	+ 0·0010	...	+ 14·465	+ 0·176	...	...
459	Taylor 4031	+ 2·0200	+ 0·0037	...	+ 14·528	+ 0·197	...	...
460	... ...	+ 1·8756	+ 0·0022	...	+ 14·578	+ 0·182	...	...
461	... ...	+ 1·6000	− 0·0025	...	+ 14·685	+ 0·158	...	...
462	Lacaille 3774 ...	+ 0·0253	− 0·0341	...	+ 14·762	+ 0·065	...	...
463	83 Cancri	+ 3·3681	− 0·0134	− 0·012	+ 14·874	+ 0·229	+ 0·16	3171
464	... ...	+ 3·3559	− 0·0131	...	+ 14·072	+ 0·380	...	...
465	ι Argûs	+ 1·6106	− 0·0022	...	+ 14·091	+ 0·150	...	3186
466	... ...	+ 1·8686	+ 0·0080	...	+ 15·097	+ 0·174	...	...
467	... ...	+ 2·0220	+ 0·0045	...	+ 15·146	+ 0·186	...	...
468	... ...	+ 3·3036	− 0·0116	...	+ 15·229	+ 0·307	...	...
469	... ...	+ 3·3010	− 0·0116	...	+ 15·340	+ 0·308	...	...
470	... ...	+ 2·1375	+ 0·0067	...	+ 15·390	+ 0·198	...	...
471	... ...	+ 0·8641	− 0·0265	...	+ 15·413	+ 0·076	...	...
472	30 Hydræ α Var. 2 ...	+ 2·9506	− 0·0013	− 0·004	+ 15·460	+ 0·268	− 0·03	[illegible]
473	26 Ursæ Majoris θ ...	+ 4·1603	− 0·0661	− 0·111	+ 15·578	+ 0·374	+ 0·57	[illegible]
474	... ...	+ 0·0197	− 0·0575	...	+ 15·606	+ 0·077	...	...
475	... ...	+ 2·3673	+ 0·0064	...	+ 15·620	+ 0·209	...	...
476	6 Leonis h	+ 3·2945	− 0·0002	− 0·002	+ 15·688	+ 0·268	+ 0·03	[illegible]
477	... ...	+ 1·8808	+ 0·0068	...	+ 15·748	+ 0·104	...	...
478	Taylor 4222	+ 1·8819	+ 0·0029	...	+ 15·794	+ 0·153	...	...
479	... ...	+ 2·4112	+ 0·0066	...	+ 15·806	+ 0·210	...	...
480	... ...	+ 1·0428	+ 0·0017	...	+ 15·832	+ 0·167	...	...
481	... ...	+ 2·4148	+ 0·0007		+ 15·856	+ 0·209	...	...
482	... ...	+ 2·4184	+ 0·0068		+ 15·882	+ 0·208	...	...
483	... ...	+ 1·8064	+ 0·0081		+ 15·990	+ 0·186	...	...
484	Taylor 4280	+ 2·1546	+ 0·0068		+ 16·017	+ 0·182	...	3300
485	... ...	+ 2·4013	+ 0·0072		+ 10·045	+ 0·208	...	...
486	R. P. L. 60	+ 10·4440	− 5·7496	...	+ 16·096	+ 1·645	..	...
487	14 Leonis ο	+ 3·2194	− 0·0098	− 0·018	+ 16·122	+ 0·272	+ 0·04	3312
488	... .	+ 1·3989	− 0·0080	...	+ 16·265	+ 0·130	...	...
489	... ...	+ 1·7702	+ 0·0096	...	+ 16·219	+ 0·146	...	...
490	... .	+ 1·6076	− 0·0016	...	+ 16·270	+ 0·180	..	...

Mean Positions of Stars for 1866 January 1st.

Number.	Star.	Magnitude.	Estimations.	Mean Right Ascension.			Mean Polar Distance.			Observations.	Fraction of Year.
				h.	m.	s.	°	′	″		
491	R Leonis Minoris Var. 1. .	7·7	6	9	31	21·71	54	32	28·6	7	0·16
492	17 Leonis ε	3·1	...	9	36	14·26	66	36	40·9	9	0·17
493	B. F. 1388	6·0	...	9	39	5·70	52	40	32·9	1	0·27
494	18 Leonis	6·1	...	9	39	10·16	77	34	29·1	2	0·[illegible]
495	l Carinæ Var. 1	4·0	...	9	41	34·00	151	56	26·7	1	0·26
496	... ...	8·0	4	9	41	34·65	150	46	23·6	4	0·26
497	... ...	8·0	2	9	44	9·45	147	2	9·9	2	0·13
498	... ...	9·2	1	9	46	0·78	139	3	26·9	1	0·17
499	... ...	9·4	1	9	46	31·72	139	7	29·2	1	0·14
500	R. P. L. 70 ...	6·5	...	9	46	40·20	5	26	25·8	1	0·24
501	... ...	9·5	1	9	48	37·11	152	8	16·0	1	0·26
502	W. B. E. IX. 1067 ..	7·5	3	9	49	51·00	86	7	17·6	3	0·14
503	... ...	8·7	3	9	50	51·87	146	39	47·1	3	0·27
504	29 Leonis π	5·0	...	9	53	7·80	81	18	32·7	15	0·16
505	... ...	9·2	2	9	53	56·81	152	7	22·6	2	0·24
506	... ...	8·0	3	9	54	18·69	148	36	50·0	3	0·18
507	Taylor 4445 ...	7·9	1	9	54	47·21	147	29	18·7	1	0·27
508	... ...	8·0	1	9	55	56·17	147	24	30·8	1	0·23
509	... ...	9·8	1	9	56	3·56	139	57	59·8	1	0·22
510	... ...	9·3	2	9	57	13·98	139	57	13·9	2	0·21
511	Taylor 4476	8·8	1	9	57	55·68	145	35	38·5	1	0·22
512	... ...	9·2	1	9	58	13·54	143	51	41·6	1	0·29
513	Taylor 4484	7·4	1	9	58	44·41	151	30	33·5	1	0·28
514	32 Leonis α, (*Regulus*) ...	1·4	...	10	1	12·96	77	22	46·3	13	0·16
515	... ...	8·9	3	10	1	26·20	130	2	33·9	3	0·27
516	Lacaille 4164 ...	7·6	3	10	2	16·90	143	54	37·6	4	0·22
517	... .	9·3	1	10	2	51·70	139	35	11·1	1	0·08
518	... .	8·7	3	10	5	15·86	140	30	20·9	2	0·26
519	.. .. ..	8·5	3	10	7	12·37	139	56	43·3	3	0·17
520	... .. .	8·9	2	10	9	6·44	139	32	16·4	2	0·26
521	Taylor 4577 .	9·0	1	10	9	32·95	136	37	32·7	1	0·17
522	... ...	8·3	1	10	9	55·37	146	34	57·6	1	0·29
523	.. ..	9·5	1	10	10	30·56	139	51	46·8	1	0·30
524	41 Leonis γ¹	2·2	...	10	12	34·04	69	28	58·9	5	0·22
525	... ...	9·6	1	10	12	50·57	138	57	31·3	1	0·17

491.—R Leonis Minoris Var 1.—Period 376 days.—Range, 6·5 to below 11th magnitude.
495.—l Carinæ Var. 1.—Period 31 days.—Range, 3·7 to 5·2 magnitude.
500.—Carrington 1451.
502.—Comparison stars for Asia in 1864.

Observed with the Madras Meridian Circle in that Year.

Number.	Star.	In Right Ascension. Annual Precession.	In Right Ascension. Secular Variation.	In Right Ascension. Proper Motion.	In Polar Distance. Annual Precession.	In Polar Distance. Secular Variation.	In Polar Distance. Proper Motion.	Number in B. A. C.
		s	s	s	"	"	"	
491	R Leonis Min. Var. 1.	+ 3·6109	− 0·0276	...	+ 16·809	+ 0·301	...	...
492	17 Leonis ε	+ 3·4286	− 0·0180	− 0·001	+ 16·840	+ 0·288	+ 0·02	3361
493	B. F. 1388	+ 3·1712	− 0·0075	...	+ 16·388	+ 0·260	...	3386
494	18 Leonis	+ 3·2419	− 0·0108	− 0·008	+ 16·385	+ 0·266	− 0·08	3337
495	l Carinæ Var. 1 ...	+ 1·6804	− 0·0001	− 0·003	+ 16·806	+ 0·130	− 0·02	3363
496	... .	+ 2·4173	+ 0·0083	...	+ 16·507	+ 0·193		...
497	... ...	+ 1·9204	+ 0·0060	...	+ 16·634	+ 0·150		...
498	... .	+ 2·4735	+ 0·0086	...	+ 16·724	+ 0·198		..
499	... .	+ 2·4740	+ 0·0086	...	+ 16·749	+ 0·198		...
500	R. P. L. 70... ..	+ 10·7820	− 1·5841	...	+ 16·756	+ 0·857		...
501	... .	+ 1·7004	+ 0·0017	...	+ 16·849	+ 0·128	...	...
502	W. B. E. IX. 1057...	+ 3·1336	− 0·0082	...	+ 16·907	+ 0·280	...	...
503	... .	+ 2·0267	+ 0·0082	...	+ 16·955	+ 0·151	...	...
504	20 Leonis ν	+ 3·1794	− 0·0080	− 0·003	+ 17·060	+ 0·276	+ 0·03	3415
505	.. ..	+ 1·7309	+ 0·0034	...	+ 17·006	+ 0·127	...	...
506	... ...	+ 2·1128	+ 0·0097		+ 17·110	+ 0·154		...
507	Taylor 4445	+ 1·9823	+ 0·0058		+ 17·135	+ 0·140		...
508	... ...	+ 1·9943	+ 0·0086		+ 17·187	+ 0·143		...
509	... ...	+ 2·4052	+ 0·0097		+ 17·198	+ 0·180		...
510	... ...	+ 2·8008	+ 0·0089		+ 17·241	+ 0·179		...
511	Taylor 4476	+ 2·0501	+ 0·0100	...	+ 17·277	+ 0·147	...	...
512	... ...	+ 2·1427	+ 0·0105	...	+ 17·290	+ 0·151	...	...
513	Taylor 4494 . ..	+ 1·8987	+ 0·0066	...	+ 17·312	+ 0·138	...	...
514	32 Leonis α (*Regulus*)	+ 3·2208	− 0·0102	− 0·010	+ 17·422	+ 0·225	− 0·01	3480
515	... ...	+ 2·5161	+ 0·0105	...	+ 17·481	+ 0·175	...	...
516	Lacaille 4164	+ 2·1709	+ 0·0115		+ 17·490	+ 0·148		...
517	.. . ..	+ 2·5240	+ 0·0106		+ 17·492	+ 0·172		...
518	... ...	+ 2·2087	+ 0·0124		+ 17·596	+ 0·153		...
519	.. .	+ 2·5494	+ 0·0112		+ 17·675	+ 0·167		...
520	. ... ..	+ 2·3312	+ 0·0131		+ 17·753	+ 0·150		...
521	Taylor 4577 .	+ 2·5784	+ 0·0112	...	+ 17·784	+ 0·160	..	..
522	. .	+ 2·1705	+ 0·0129	.	+ 17·788	+ 0·128	.	...
523	. ..	+ 2·3420	+ 0·0134	...	+ 17·802	+ 0·140	..	...
524	41 Leonis γ[1] .	+ 3·2980	− 0·0148	+ 0·019	+ 17·882	+ 0·208	+ 0·15	3523
525	..	+ 2·8011	+ 0·0115	...	+ 17·807	+ 0·162	..	

494.—Proper motions from "*Greenwich Catalogue 1872*."
495.—Proper motions from "*Stone's Cape Catalogue.*"

Mean Positions of Stars for 1866 January 1st.

Number.	Star.	Magnitude.	Estimations.	Mean Right Ascension.			Mean Polar Distance.			Observations.	Fraction of Year.
				h	m	s	°	′	″		
526	... ...	8·6	2	10	14	[illegible]	150	[illegible]	12·0	2	[illegible]
527	... ...	9·3	1	10	16	14·56	75	25	6·8	1	0·30
528	.. ..	8·2	2	10	16	17·10	129	16	50·8	2	0·29
529	Taylor 4656 ...	8·0	2	10	18	6·19	151	23	50·2	2	0·22
530	44 Leonis	6·2	...	10	19	11·32	80	32	9·8	8	0·19
531	... ...	9·5	1	10	18	40·75	146	0	4·4	1	0·29
532	... ...	7·9	3	10	21	56·67	146	36	31·0	3	0·26
533	... .	9·4	3	10	22	0·44	146	50	37·0	3	0·24
534	... ...	10·2	1	10	23	28·48	76	5	55·5	1	0·30
535	... ...	8·4	3	10	24	10·08	146	56	42·1	3	0·25
536	... ...	8·5	1	10	24	30·00	146	30	31·0	1	0·31
537	... ...	8·8	4	10	25	41·68	146	54	44·8	4	0·28
538	47 Leonis ρ ...	4·0	...	10	25	46·20	80	0	18·0	6	0·10
539	... ...	9·1	3	10	26	34·08	152	26	23·0	3	0·28
540	p Carinae	3·6	...	10	27	15·82	150	30	50·6	1	0·27
541	.. ..	9·6	2	10	28	9·07	150	30	32·3	2	0·29
542	... ...	10·1	1	10	29	16·79	147	15	12·9	1	0·30
543	Taylor 4769 ...	6·5	1	10	30	28·91	146	41	52·5	1	0·17
544	... ...	9·5	1	10	34	37·27	139	17	15·6	1	0·29
545	... ...	8·0	1	10	35	7·86	140	6	17·1	2	0·28
546	Taylor 4821	7·6	5	10	36	10·77	143	38	36·4	2	0·27
547	... ...	9·0	2	10	37	34·09	151	29	57·5	2	0·30
548	Taylor 4848	7·0	3	10	38	44·25	149	24	56·1	3	0·24
549	... ...	9·0	1	10	39	32·76	148	51	40·1	1	0·27
550	η Argûs Var. 1	8·0	...	10	39	52·22	149	28	53·0	1	0·32
551	... ...	9·0	1	10	41	2·21	137	1	36·9	1	0·27
552	... ...	8·8	2	10	42	0·34	140	33	30·2	2	0·21
553	58 Leonis [illegible]	5·3	...	10	42	12·71	78	44	40·0	4	0·30
554	... ...	9·5	1	10	42	[illegible]	75	5	29·1	1	0·29
555	... ...	9·0	1	10	42	41·46	141	5	1·4	1	0·30
556	Taylor 4886	6·0	1	10	43	52·76	137	2	35·3	1	0·27
557	... ...	8·2	1	10	43	57·06	137	3	38·3	1	[illegible]
558	... ...	7·6	3	10	45	9·14	141	45	40·2	3	0·29
559	Lacaille 4602 ...	7·4	2	10	44	28·34	141	5	30·4	2	[illegible]
560	Taylor 4916	8·0	2	10	46	37·78	135	30	21·4	2	0·30

550.—η Argûs Var. 1.—Irregularly variable from 1st to 5th magnitude.

Observed with the Madras Meridian Circle in that Year.

Number.	Star.	In Right Ascension. Annual Precession.	In Right Ascension. Secular Variation.	In Right Ascension. Proper Motion.	In Polar Distance. Annual Precession.	In Polar Distance. Secular Variation.	In Polar Distance. Proper Motion.	Number in B. A. C.
		s	s	s	"	"	"	
526	.	+ 2·0909	+ 0·0122	...	+ 17·975	+ 0·123		.
527		+ 3·2248	− 0·0110	...	+ 18·004	+ 0·108	...	.
528	..	+ 2·5040	+ 0·0121	...	+ 18·036	+ 0·158	...	.
529	Taylor 4623 ...	+ 2·0175	+ 0·0139	...	+ 18·105	+ 0·119	...	...
530	44 Leonis	+ 3·1680	− 0·0079	− 0·007	+ 18·109	+ 0·101	+ 0·12	3601
531	...	+ 2·2205	+ 0·0152		+ 18·132	+ 0·131		...
532	...	+ 2·2204	+ 0·0160		+ 18·247	+ 0·126		..
533	..	+ 2·2186	+ 0·0160		+ 18·240	+ 0·126		.
534		+ 3·2074	− 0·0102		+ 18·302	+ 0·188		.
535	..	+ 2·2084	+ 0·0166		+ 18·327	+ 0·126		.
536	... .	+ 2·2301	+ 0·0168	...	+ 18·330	+ 0·124	...	..
537	.. .. .	+ 2·2618	+ 0·0171	...	+ 18·380	+ 0·123	...	.
538	47 Leonis ρ ...	+ 3·1602	− 0·0060	0·000	+ 18·398	+ 0·176	+ 0·08	3609
539	... .	+ 2·0565	+ 0·0154	...	+ 18·411	+ 0·111	...	
540	p Carinæ	+ 2·1227	+ 0·0165	...	+ 18·486	+ 0·114	..	3610
541	..	+ 2·1373	+ 0·0160		+ 18·466	+ 0·114		
542	... .	+ 2·2407	+ 0·0181		+ 18·304	+ 0·110		.
543	Taylor 4769	+ 2·2920	+ 0·0185		+ 18·543	+ 0·119		3635
544	... ...	+ 2·5087	+ 0·0182		+ 18·690	+ 0·126		...
545	... ...	+ 2·2634	+ 0·0180		+ 18·606	+ 0·112		.
546	Taylor 4824 .. .	+ 2·2771	+ 0·0208		+ 18·720	+ 0·111		3673
547	... ...	+ 2·2085	+ 0·0206		+ 18·772	+ 0·108		
548	Taylor 4940	+ 2·2967	+ 0·0211		+ 18·807	+ 0·108		3690
549	.. ..	+ 2·3113	+ 0·0214		+ 18·832	+ 0·108		.
550	η Argûs Var. 1 ..	+ 2·3104	+ 0·0215		+ 18·841	+ 0·107		3695
551	. ...	+ 2·5374	+ 0·0180	...	+ 18·876	+ 0·130	...	.
552	... .	+ 2·3179	+ 0·0228	...	+ 18·906	+ 0·106	...	.
553	53 Leonis l	+ 3·1606	− 0·0060	− 0·008	+ 18·911	+ 0·145	+ 0·02	3706
554	. ...	+ 3·1002	− 0·0104	...	+ 18·919	+ 0·147	...	...
555	... ...	+ 2·5941	+ 0·0206		+ 18·925	+ 0·114	...	...
556	Taylor 4996	+ 2·2080	+ 0·0190		+ 18·980	+ 0·117		.
557		+ 2·4042	+ 0·0198		+ 18·961	+ 0·116		..
558		+ 2·5278	+ 0·0213		+ 18·986	+ 0·110		.
559	Lacaille 4532 ..	+ 2·5506	+ 0·0215		+ 19·034	+ 0·109		.
560	Taylor 4915 ...	+ 2·6130	+ 0·0190	.	+ 19·036	+ 0·113		

530.—Proper motions from "Greenwich Catalogue 1864."

Mean Positions of Stars for 1866 January 1st.

Number.	Star.	Magnitude.	Estimations.	Mean Right Ascension.			Mean Polar Distance.			Observations.	Fraction of Year.
				h.	m.	s.	°	′	″		
561	... ...	9·5	2	10	46	4·42	129	29	51·2	2	0·26
562	... ...	8·2	1	10	48	21·28	141	46	31·0	1	0·26
563	Taylor 4945 ...	7·0	2	10	49	0·28	144	54	20·2	2	0·28
564	Taylor 4956 ...	7·0	1	10	50	45·50	147	20	16·5	1	0·26
565	... ...	9·2	1	10	52	0·40	143	46	7·7	1	0·30
566	59 Leonis c	5·1	...	10	53	49·04	83	10	46·3	2	0·16
567	... ...	8·9	3	10	53	58·68	135	32	26·0	3	0·31
568	50 Ursæ Majoris α (*Dubhe*)	2·0	...	10	55	26·08	27	31	23·9	1	0·31
569	... ...	8·7	2	10	55	42·88	140	17	24·4	2	0·29
570	... ...	9·0	1	10	57	4·66	145	28	5·0	1	0·28
571	... ...	9·3	1	10	57	6·95	146	36	20·4	1	0·29
572	Lacaille 4576 ...	7·8	2	10	57	54·38	129	56	10·6	2	0·26
573	63 Leonis χ ...	5·2	...	10	58	6·22	81	56	27·0	8	0·22
574	... ...	8·7	3	10	59	17·38	140	39	52·4	3	0·28
575	Lacaille 4606 ...	8·4	3	10	59	27·08	146	39	26·6	3	0·27
576	... ...	8·3	1	11	1	6·03	135	34	16·4	1	0·34
577	67 Leonis	5·8	...	11	1	37·21	64	37	6·6	1	0·30
578	Lalande 21371	7·8	1	11	3	36·63	77	23	18·0	1	0·32
579	Lalande 21416	8·2	1	11	5	5·02	67	18	1·8	1	0·17
580	... ...	10·3	1	11	5	47·23	88	51	4·8	1	0·29
581	... ...	8·4	1	11	5	54·51	149	50	32·6	1	0·19
582	68 Leonis δ ...	2·8	...	11	6	55·60	68	44	35·4	7	0·30
583	.. ..	7·5	2	11	7	12·86	145	50	56·2	2	0·30
584	... . ..	9·0	..	11	8	38·05	150	51	39·0	1	0·29
585	... ...	9·3	1	11	9	33·95	146	36	50·7	1	0·32
586	74 Leonis φ ...	4·5	...	11	9	50·98	92	55	13·1	1	0·34
587	. . ...	8·0	1	11	11	13·76	127	39	37·6	1	0·34
588	δ Crateris ...	3·9	...	11	12	39·57	104	3	13·8	11	0·28
589	.. .	8·0	2	11	12	54·22	130	32	45·5	2	0·33
590	. . .	7·8	3	11	15	58·37	134	32	13·7	3	0·25
591	Lacaille 4736 ..	7·0	1	11	16	10·78	146	52	10·0	1	0·31
592	Taylor 5080 ...	7·8	1	11	19	6·10	131	56	10·8	1	0·17
593	... .	7·8	1	11	19	30·67	129	31	36·2	1	0·28
594	81 Leonis [illegible] ...	5·1	...	11	21	2·83	95	24	22·5	1	0·30
595	.	9·3	1	11	21	49·06	138	23	27·1	1	0·32

577.—Comparison star for Thalia in 1862.

Observed with the Madras Meridian Circle in that Year.

Number.	Star.	In Right Ascension. Annual Precession.	In Right Ascension. Secular Variation.	In Right Ascension. Proper Motion.	In Polar Distance. Annual Precession.	In Polar Distance. Secular Variation.	In Polar Distance. Proper Motion.	Number in B. A. C.
		s	s	s	"	"	"	
561	... ...	+ 2·7320	+ 0·0164	...	+ 19·075	+ 0·115	...	...
562	... ...	+ 2·5801	+ 0·0222	...	+ 19·084	+ 0·106	...	...
563	Taylor 4946	+ 2·4921	+ 0·0296	...	+ 19·101	+ 0·103	...	...
564	Taylor 4955	+ 2·4615	+ 0·0260	...	+ 19·147	+ 0·097	...	...
565	... ...	+ 2·5387	+ 0·0230	...	+ 19·170	+ 0·100	...	...
566	59 Leonis ε	+ 3·1177	− 0·0062	− 0·006	+ 19·234	+ 0·122	+ 0·06	3760
567	... ...	+ 2·6815	+ 0·0204	...	+ 19·236	+ 0·108	...	...
568	50 Urs. Maj. α (*Dubhe*).	+ 3·7953	− 0·0631	− 0·017	+ 19·264	+ 0·144	+ 0·09	3777
569	... ...	+ 2·4490	+ 0·0274	...	+ 19·271	+ 0·091	...	...
570	... ...	+ 2·5457	+ 0·0262	...	+ 19·304	+ 0·093	...	...
571	... ...	+ 2·6490	+ 0·0268	...	+ 19·305	+ 0·092	...	...
572	Lacaille 4676 ...	+ 2·7762	+ 0·0179	...	+ 19·328	+ 0·100	...	...
573	63 Leonis χ	+ 3·1226	− 0·0056	− 0·024	+ 19·328	+ 0·118	+ 0·06	3798
574	... ...	+ 2·6320	+ 0·0242	...	+ 19·362	+ 0·094	...	...
575	Lacaille 4636 ...	+ 2·4013	+ 0·0287	...	+ 19·380	+ 0·087	...	...
576	... ...	+ 2·7256	+ 0·0216	...	+ 19·396	+ 0·093	...	...
577	67 Leonis	+ 3·2320	− 0·0164	+ 0·003	+ 19·403	+ 0·111	+ 0·01	3809
578	Lalande 21371 ...	+ 3·1416	− 0·0076	...	+ 19·451	+ 0·106	...	...
579	Lalande 21416 ...	+ 3·2054	− 0·0144	..	+ 19·462	+ 0·106	...	...
580	... ...	+ 3·1069	− 0·0046	...	+ 19·496	+ 0·098	...	...
581	... ...	+ 2·5580	+ 0·0311	..	+ 19·490	+ 0·070	...	...
582	68 Leonis δ	+ 3·1018	− 0·0182	+ 0·011	+ 19·521	+ 0·093	+ 0·14	[illegible]
583	... .	+ 2·6940	+ 0·0294	...	+ 19·526	+ 0·070	...	...
584	... ...	+ 2·5396	+ 0·0441	..	+ 19·568	+ 0·074	...	...
585	. ..	+ 2·6400	+ 0·0304		+ 19·571	+ 0·076	..	...
586	74 Leonis φ	+ 3·0873	+ 0·0006	− 0·000	+ 19·577	+ 0·089	+ 0·04	[illegible]
587	... ...	+ 2·8543	+ 0·0186	.	+ 19·603	+ 0·080	...	.
588	δ Crateris	+ 3·0088	+ 0·0064	− 0·009	+ 19·660	+ 0·081	− 0·14	[illegible]
589	... ...	+ 2·4436	+ 0·0280	..	+ 19·683	+ 0·077	...	...
590	. .	+ 2·8698	+ 0·0197	.	+ 19·686	+ 0·073	...	...
591	Lacaille 4726 ..	+ 2·6972	+ 0·0324	..	+ 19·690	+ 0·067	...	..
592	Taylor 5020	+ 2·6860	+ 0·0236	...	+ 19·737	+ 0·066	...	...
593	... .	+ 2·8762	+ 0·0209	.	+ 19·743	+ 0·065	...	..
594	84 Leonis τ	+ 3·0893	− 0·0080	− 0·001	+ 19·766	+ 0·072	+ 0·02	[illegible]
595	... ...	+ 2·8963	+ 0·0205	..	+ 19·777	+ 0·061	..	...

577.—Proper motions from "*Greenwich Catalogue 1872.*"

Mean Positions of Stars for 1866 January 1st.

Number.	Star.	Magnitude.	Estimations.	Mean Right Ascension.			Mean Polar Distance.			Observations.	Fraction of Year.
				h.	m.	s.	°	′	″		
596	Taylor 5246 ... 2nd ..	8·4	2	11	22	8·30	131	36	30·8	2	0·30
597	... ...	8·0	1	11	22	19·83	139	4	55·5	1	0·30
598	... ...	8·9	1	11	23	58·83	146	51	38·2	1	0·32
599	... ...	9·6	1	11	23	29·47	28	21	32·2	1	0·33
600	... ...	9·0	1	11	24	40·03	146	0	35·0	1	0·30
601	... ...	9·3	1	11	24	46·01	84	43	3·4	1	0·18
602	... ...	9·4	1	11	25	58·40	138	25	25·6	1	0·32
603	... ...	8·8	1	11	26	20·80	148	51	55·8	1	0·36
604	... ...	9·8	2	11	28	36·70	138	30	47·6	2	0·32
605	λ Centauri... ...	3·4	...	11	29	37·14	152	16	42·5	3	0·34
606	... ...	7·9	1	11	30	50·02	140	16	22·7	1	0·27
607	91 Leonis υ	4·5	...	11	30	5·38	90	5	4·7	10	0·26
608	... ...	9·3	1	11	32	14·06	144	15	11·4	1	0·35
609	... ...	8·2	1	11	33	13·21	144	15	15·7	1	0·36
610	W. B. E. XI. 671 ...	8·0	3	11	33	31·85	88	18	21·8	3	0·17
611	... ...	8·0	1	11	34	3·32	127	40	56·4	1	0·32
612	... ...	8·6	1	11	34	30·14	144	31	21·9	1	0·34
613	... ...	9·2	2	11	36	8·98	139	40	30·7	2	0·29
614	Taylor 5334 ...	6·0	1	11	37	8·86	151	44	44·0	1	0·31
615	... ...	8·9	1	11	38	48·00	139	34	43·1	1	0·32
616	... ...	9·5	1	11	41	7·81	140	32	44·0	1	0·21
617	... ...	9·2	1	11	41	14·58	136	31	5·5	1	0·17
618	94 Leonis β (*Deneb*) ...	2·2	...	11	42	13·32	74	40	46·7	7	0·27
619	Taylor 5421	7·7	3	11	43	16·94	139	31	54·0	3	0·29
620	5 Virginis β	3·7	...	11	43	46·98	87	36	52·6	1	0·16
621	Taylor 5427	6·7	...	11	44	11·33	94	36	20·9	1	0·09
622	Taylor 5438	8·3	1	11	44	57·95	139	38	44·7	1	0·23
623	Groombridge 1830 ...	7·5	3	11	45	14·79	51	19	10·7	3	0·30
624	64 Ursæ Majoris γ ...	2·6	...	11	46	40·81	35	33	37·7	1	0·36
625	Lacaille 4887	7·2	3	11	49	21·35	152	32	2·1	3	0·30
626	Lacaille 4946	8·1	3	11	50	15·81	152	30	18·1	5	0·34
627	... ...	7·9	2	11	50	48·14	130	22	30·6	2	0·27
628	Lacaille 4956	8·1	1	11	51	16·90	154	34	36·0	1	0·17
629	... ...	9·7	2	11	52	45·41	151	36	56·3	2	0·32
630	Taylor 5534 .	7·5	...	11	53	55·02	143	36	1·5	1	[illegible]

599.—Comparison star for Comet 2, 1861.
601.—Observed in mistake for Amphitrite in 1866.
610.—Comparison star for Amphitrite in 1862.

Observed with the Madras Meridian Circle in that Year.

Number.	Star.	In Right Ascension.			In Polar Distance.			Number in B. A. C.
		Annual Precession.	Secular Variation.	Proper Motion.	Annual Precession.	Secular Variation.	Proper Motion.	
		s	s	s	"	"	"	
596	Taylor 5945 2nd .	+ 2·8746	+ 0·0280	...	+ 19·782	+ 0·060	...	.
597	.. ..	+ 2·8989	+ 0·0203		+ 19·783	+ 0·060	...	...
598	... . .	+ 2·7487	+ 0·0344		+ 19·792	+ 0·066	..	...
599	.. ..	+ 3·4630	− 0·0806		+ 19·801	+ 0·074	...	...
600	... .	+ 2·7680	+ 0·0352		+ 19·818	+ 0·063	...	...
601	... ...	+ 3·0010	− 0·0028	...	+ 19·818	+ 0·060	...	..
602	... .	+ 2·9150	+ 0·0211	...	+ 19·834	+ 0·054	...	..
603	. .. ..	+ 2·9041	+ 0·0335	...	+ 19·840	+ 0·051	...	...
604	... .	+ 2·9960	+ 0·0214	...	+ 19·865	+ 0·060	...	...
605	λ Centauri .. .	+ 2·7388	+ 0·0443	0·000	+ 19·878	+ 0·044	0·00	3941
606	. ...	+ 2·7780	+ 0·0406	...	+ 19·882	+ 0·044	...	...
607	91 Leonis υ	+ 3·0718	+ 0·0003	− 0·003	+ 19·885	+ 0·049	− 0·08	3946
608	... .. .	+ 2·8478	+ 0·0366	...	+ 19·908	+ 0·041	...	...
609	.. ..	+ 2·8555	+ 0·0360	...	+ 19·918	+ 0·050	...	...
610	W. B. K. XI. 571 ..	+ 3·0767	− 0·0006	...	+ 19·922	+ 0·046	...	...
611	... ..	+ 2·9648	+ 0·0219		+ 19·926	+ 0·040	...	...
612	.. ..	+ 2·8645	+ 0·0364		+ 19·930	+ 0·037	...	...
613	... .	+ 2·9094	+ 0·0380		+ 19·940	+ 0·035	...	...
614	Taylor 5361	+ 2·8244	+ 0·0470		+ 19·955	+ 0·032	...	3978
615	... .	+ 2·9700	+ 0·0287		+ 19·969	+ 0·031	...	...
616	... .	+ 2·8818	+ 0·0469	...	+ 19·980	+ 0·030	...	...
617	... ...	+ 2·9012	+ 0·0213	...	+ 19·988	+ 0·027	...	...
618	94 Leonis β (*Deneb*).	+ 3·1065	− 0·0074	− 0·036	+ 19·996	+ 0·025	+ 0·10	3995
619	Taylor 5421	+ 2·9917	+ 0·0242	...	+ 20·001	+ 0·026	...	...
620	5 Virginis β	+ 3·0763	− 0·0008	+ 0·048	+ 20·004	+ 0·023	+ 0·28	4002
621	Taylor 5427	+ 3·0617	+ 0·0094	...	+ 20·007.	+ 0·022	...	4006
622	Taylor 5488	+ 2·9997	+ 0·0240	...	+ 20·012	+ 0·020	...	...
623	Groombridge 1830 ..	+ 3·1411	− 0·0380	+ 0·346	+ 20·013	+ 0·021	+ 5·78	4010
624	64 Ursae Majoris γ ..	+ 3·1790	− 0·0488	+ 0·011	+ 20·023	+ 0·017	0·00	4017
625	Lacaille 4887 ...	+ 2·9414	+ 0·0388	...	+ 20·029	+ 0·013	...	...
626	Lacaille 4916 .	+ 2·8880	+ 0·0546	...	+ 20·037	+ 0·009	...	...
627	. ... ..	+ 2·9709	+ 0·0306	...	+ 20·034	+ 0·009	...	..
628	Lacaille 4946 ..	+ 2·9049	+ 0·0468	...	+ 20·040	+ 0·008	...	.
629	.. ..	+ 2·9469	+ 0·0319	...	+ 20·045	+ 0·006	..	...
630	Taylor 5594	+ 3·0175	+ 0·0121	...	+ 20·038	− 0·003	...	...

605.—Proper motions from "*Stone's Cape Catalogue.*"
623.—Proper motions from "*Greenwich Catalogue 1872.*"

Mean Positions of Stars for 1866 January 1st.

Number.	Star.	Magnitude.	Estimations.	Mean Right Ascension.			Mean Polar Distance.			Observations.	Fraction of Year.
				h.	m.	s.	°	′	″		
631	R. P. L. 80 ..	6·3	...	11	57	57·66	3	40	18·0	4	0·28
632	... ...	8·0	1	11	58	7·57	128	28	33·7	1	0·24
633	... ...	9·3	3	11	59	37·59	150	20	15·9	3	0·30
634	... ...	8·0	1	11	59	50·04	144	16	51·8	2	0·35
635	... ...	8·2	1	12	1	13·04	150	22	9·4	1	0·27
636	... ...	9·4	1	12	1	48·43	130	2	16·6	1	0·36
637	Lacaille 5041 ...	7·9	1	12	2	39·11	141	23	54·2	1	0·36
638	10 Virginis ..	6·1	...	12	2	40·32	87	21	0·6	2	0·17
639	2 Corvi ε	3·1	...	12	3	14·22	111	52	26·8	11	0·29
640	... ...	8·8	1	12	3	44·06	145	57	45·7	1	0·23
641	... ...	8·0	1	12	5	54·40	134	8	40·2	1	0·35
642	... ...	8·0	1	12	6	18·80	138	38	12·5	1	0·36
643	... .	9·2	1	12	6	36·48	142	31	20·6	1	0·37
644	... ...	8·7	2	12	11	59·24	150	23	30·9	2	0·36
645	Taylor 5646 ...	6·9	1	12	12	37·69	152	6	33·5	1	0·33
646	R. P. L. 92 ...	6·7	...	12	12	55·84	2	40	8·6	1	0·31
647	15 Virginis η ...	4·1	...	12	13	2·18	89	55	19·9	7	0·37
648	... ...	9·2	3	12	14	10·30	148	46	29·2	3	0·37
649	Lacaille 5119 ..	8·9	1	12	15	27·06	138	34	57·2	1	0·28
650	.. ..	8·0	2	12	15	58·57	141	40	35·2	2	0·27
651	... ...	8·3	1	12	16	52·76	147	10	36·2	1	0·35
652	α Crucis . ... 1st..	1·5	...	12	19	10·25	152	21	28·2	2	0·31
653	.. ..	8·8	1	12	19	20·96	144	4	50·0	1	0·39
654	... ...	8·9	2	12	24	45·44	150	30	17·4	2	0·31
655	9 Corvi β	2·8	...	12	27	21·15	112	39	19·8	8	0·33
656	. ...	8·4	1	12	27	58·79	141	40	33·6	1	0·35
657	25 Virginis f	5·9	..	12	29	53·47	95	5	34·3	1	0·33
658	... ...	8·8	1	12	30	57·38	142	20	19·5	1	0·34
659	R Virginis Var. 2 ..	7·5	1	12	31	42·04	82	16	38·5	1	0·35
660	Taylor 5880	7·0	1	12	34	38·37	144	1	33·8	1	0·37
661	29 Virginis γ ... 1st.	3·5	.	12	34	52·26	90	48	40·3	4	0·26
662	S Ursæ Majoris Var. 2. ..	8·7	1	12	38	3·07	28	10	21·3	1	0·40
663	Taylor 5959	7·5	4	12	38	39·48	143	32	42·8	4	0·26
664	... .	8·0	1	12	39	32·57	143	30	58·3	1	0·40
665	... .	8·8	2	12	42	31·32	147	19	23·2	2	0·30

631.—Groombridge 1830.
646.—Groombridge 1871.
659.—R Virginis Var. 2.—Period 146 days.—Range, 6·5 to 11th magnitude.
662.—S Ursæ Majoris Var. 2.—Period 225 days.—Range, 7th to 12th magnitude.

Observed with the Madras Meridian Circle in that Year.

Number.	Star.	In Right Ascension. Annual Precession.	Secular Variation.	Proper Motion.	In Polar Distance. Annual Precession.	Secular Variation.	Proper Motion.	Number in B.A.C.
		s	s	s	"	"	"	
631	R. P. L. [illegible]	+ 3·2571	− 0·5199	...	+ 20·034	− 0·005	..	4070
632	... .	+ 3·0081	+ 0·0265	...	+ 20·055	− 0·007	.	..
633	...	+ 3·0083	+ 0·0542		+ 20·066	− 0·008	...	...
634	..	+ 3·0709	+ 0·0481		+ 20·066	− 0·009	...	...
635	..	+ 3·0817	+ 0·0640		+ 20·064	− 0·011	..	...
636	... .	+ 3·0806	+ 0·0273	...	+ 20·064	− 0·012	...	...
637	Lacaille 5041 ..	+ 3·0915	+ 0·0100	...	+ 20·064	− 0·014	...	...
638	10 Virginis ...	+ 3·0714	+ 0·0007	− 0·001	+ 20·064	− 0·013	+ 0·21	4094
639	2 Corvi ε	+ 3·0796	+ 0·0142	− 0·005	+ 20·058	− 0·016	− 0·01	4097
640	... .	+ 3·1044	+ 0·0473	...	+ 20·052	− 0·016	...	...
641	... ...	+ 3·1055	+ 0·0818	...	+ 20·048	− 0·021		...
642	... .. .	+ 3·1137	+ 0·0369	...	+ 20·047	− 0·022		...
643	... ..	+ 3·1298	+ 0·0480	...	+ 20·047	− 0·022		...
644	... .	+ 3·1050	+ 0·0692	...	+ 20·028	− 0·033		...
645	Taylor 5648 ..	+ 3·2112	+ 0·0640	...	+ 20·026	− 0·036		...
646	R. P. L. 92 .. .	+ 1·5398	+ 0·0010	+ 0·415	+ 20·023	− 0·032	+ 0·08	4180
647	15 Virginis η ..	+ 3·0720	+ 0·0027	− 0·007	+ 20·022	− 0·036	+ 0·08	4146
648	... .	+ 3·1947	+ 0·0464	..	+ 20·017	− 0·038	...	...
649	Lacaille 5119 ...	+ 3·1740	+ 0·0988	...	+ 20·009	− 0·040	...	...
650	... .	+ 3·1990	+ 0·0495	...	+ 20·006	− 0·042	...	...
651	.. .. ..	+ 3·2245	+ 0·0585	..	+ 20·000	− 0·044	...	...
652	α Crucis .. 1st..	+ 3·3461	+ 0·0680	− 0·006	+ 19·998	− 0·050	+ 0·04	4187
653	.	+ 3·2281	+ 0·0492	...	+ 19·993	− 0·049	...	...
654	... ..	+ 3·3321	+ 0·0688	..	+ 19·989	− 0·051	...	...
655	9 Corvi β ..	+ 3·1364	+ 0·0164	− 0·008	+ 19·918	− 0·054	+ 0·07	[illegible]
656	.. ..	+ 3·2778	+ 0·0480	.	+ 19·900	− 0·067	...	...
657	25 Virginis ρ... ..	+ 3·0475	+ 0·0063	− 0·004	+ 19·895	− 0·069	+ 0·35	4247
658	... .	+ 3·3053	+ 0·0476	...	+ 19·872	− 0·074	...	...
659	R Virginis, Var. 2 ..	+ 3·0171	− 0·0008	...	+ 19·868	− 0·070	...	—
660	Taylor [illegible]	+ 3·3498	+ 0·0518	...	+ 19·827	− 0·072		[illegible]
661	29 Virginis γ 1st..	+ 3·0745	+ 0·0048	− 0·037	+ 19·821	− 0·073	+ 0·05	[illegible]
662	S Ursae Majoris, Var. 2	+ 2·6894	− 0·0360	...	+ 19·779	− 0·078	.	.
663	Taylor [illegible]	+ 3·3743	+ 0·0321	...	+ 19·778	− 0·091	...	[illegible]
664	... ..	+ 3·3863	+ 0·0325	...	+ 19·757	− 0·093	—	.
665	... .	+ 3·4500	+ 0·0013	..	+ 19·711	− 0·101	...	...

646.—Proper motion in Polar distance from "*British Association Catalogue.*"
652.—Proper motions from "*Stone's Cape Catalogue.*"
661.—Proper motions from "*Greenwich Catalogue 1872.*"

Mean Positions of Stars for 1866 January 1st.

Number.	Star.	Magnitude.	Estimations.	Mean Right Ascension. h.	m.	s.	Mean Polar Distance. °	′	″	Observations.	Fraction of Year.
666	... ...	8·0	2	12	44	50·52	141	56	17·3	2	0·37
667	R. P. L. 99	5·6	...	12	48	10·52	5	51	32·6	2	0·27
668	... ...	9·0	1	12	48	35·11	125	25	59·5	1	0·35
669	... ...	9·0	1	12	49	59·02	140	34	55·6	1	0·30
670	12 Canis Venaticorum ...	3·1	...	12	49	45·38	50	57	27·1	1	0·31
671	Taylor 5974 ...	8·0	1	12	52	1·50	143	30	10·6	1	0·35
672	... .	6·8	3	12	55	0·64	124	23	41·6	3	0·37
673	48 Virginis ...	6·6	...	12	57	0·27	92	50	31·8	2	0·34
674	... ...	9·3	1	12	57	5·07	128	25	50·9	1	0·39
675	Lacaille 5381 ...	7·0	1	12	57	14·08	130	57	48·1	1	0·27
676	.. ..	9·5	1	12	58	12·73	124	29	21·4	1	0·40
677	... ...	7·5	2	12	58	22·42	124	23	34·9	2	0·36
678	Taylor 6025	7·7	3	12	59	34·21	130	34	2·5	3	0·35
679	51 Virginis θ	4·4	...	13	3	0·78	94	40	21·0	6	0·38
680	W Virginis Var. 1 ...	8·2	2	13	7	0·58	105	50	38·6	2	0·34
681	R. P. L. 101	7·5	...	13	10	4·05	1	37	50·7	3	0·34
682	... ...	8·3	1	13	12	50·46	132	57	12·3	1	0·37
683	Taylor 6148	7·4	3	13	14	18·54	128	8	57·5	3	0·37
684	O. A. N. 13563	9·1	1	13	15	26·94	27	53	51·8	1	0·40
685	... ...	9·6	2	13	17	13·78	128	10	4·2	2	0·34
686	67 Virginis α (*Spica*) ...	1·2	...	13	18	8·14	100	27	40·2	6	0·39
687	Lacaille 5546	9·2	1	13	19	48·77	143	28	0·4	1	0·40
688	R. P. L. 108	7·3	...	13	20	11·27	4	32	45·7	1	0·30
689	... ...	10·7	1	13	23	34·21	8	35	58·4	1	0·40
690	... ...	9·0	2	13	24	54·28	128	9	18·9	2	0·34
691	... ...	9·0	1	13	24	56·41	124	9	48·0	1	0·44
692	76 Virginis h	5·5	...	13	25	54·77	99	38	26·0	3	0·29
693	S Virginis Var. 6 .. .	7·5	3	13	26	0·12	96	30	20·7	3	0·38
694	79 Virginis ζ ...	3·5	..	13	27	58·01	90	54	35·6	4	0·38
695	Taylor 7190 . ..	7·0	2	13	29	30·88	131	43	46·0	2	0·31
696	... ...	7·7	1	13	33	6·21	130	2	11·1	1	0·48
697	... ...	9·5	1	13	34	10·06	129	10	38·4	1	0·40
698	82 Virginis m .	5·3	.	13	34	34·06	98	1	32·5	1	0·34
699	.. ..	8·3	1	13	35	49·90	128	4	4·0	1	0·33
700	... ...	9·3	1	13	36	21·34	128	5	57·1	1	0·33

667.—Groombridge 1940.
680.—W Virginis Var. 1.—Irregularly variable from 7th to 10·6 magnitude.
681.—Groombridge 2008.
688.—Groombridge 2007.
689.—Observed in mistake for Europa in 1864.
693.—S Virginis Var. 6.—Period 374 days.—Range, 6th to 12·5 magnitude.

Observed with the Madras Meridian Circle in that Year.

Number.	Star.	In Right Ascension. Annual Precession.	Secular Variation.	Proper Motion.	In Polar Distance. Annual Precession.	Secular Variation.	Proper Motion.	Number in B. A. C.
		s	s	s	"	"	"	
666	... ...	+ 3·4040	+ 0·0498	..	+ 19·672	+ 0·104	...	...
667	R. P. L. 99	+ 0·3583	+ 0·2860	− 0·017	+ 19·613	− 0·019	− 0·01	4342
668	... .	+ 3·2723	+ 0·0259	...	+ 19·606	− 0·108	...	...
669	... ...	+ 3·5321	+ 0·0692	...	+ 19·598	− 0·118	...	...
670	12 Canis Venaticorum	+ 2·8835	− 0·0152	− 0·023	+ 19·585	− 0·098	− 0·06	4346
671	Taylor 5974	+ 3·4810	+ 0·0546	...	+ 19·540	− 0·122	...	...
672	... ...	+ 3·2003	+ 0·0275	...	+ 19·477	− 0·123	..	...
673	48 Virginis	+ 3·0890	+ 0·0065	− 0·004	+ 19·466	− 0·119	+ 0·03	4373
674	... ...	+ 3·3607	+ 0·0298	...	+ 19·456	− 0·128	...	...
675	Lacaille 5361 ...	+ 3·3491	+ 0·0365	...	+ 19·438	− 0·128	...	...
676	... ...	+ 3·3039	+ 0·0278	...	+ 19·411	− 0·129	...	...
677	... .	+ 3·3027	+ 0·0278	...	+ 19·408	− 0·129	...	...
678	Taylor 6025	+ 3·2957	+ 0·0260	...	+ 19·392	− 0·132	...	...
679	51 Virginis θ	+ 3·1027	+ 0·0078	− 0·004	+ 19·302	− 0·132	+ 0·01	4401
680	W Virginis Var. 1 ...	+ 3·1814	+ 0·0142	...	+ 19·204	− 0·142	...	...
681	R. P. L. 101 ...	+ 11·0464	+ 8·2000		+ 19·125	+ 0·477	...	...
682	... ...	+ 3·3135	+ 0·0276		+ 19·046	− 0·161	...	...
683	Taylor 6146	+ 3·4006	+ 0·0328		+ 19·010	− 0·166	...	...
684	O. A. N. 12340 ...	+ 2·2555	− 0·0139		+ 18·978	− 0·114	...	...
685	... .	+ 3·1190	+ 0·0331		+ 18·927	− 0·173	—	...
686	67 Virginis α (*Spica*).	+ 3·1540	+ 0·0116	− 0·005	+ 18·901	− 0·160	+ 0·04	4480
687	Lacaille 5546 ...	+ 3·6390	+ 0·0599	...	+ 18·851	− 0·192	...	...
688	R. P. L. 103 ...	+ 2·0921	+ 0·0302	...	+ 18·840	+ 0·127	...	4408
689	... ...	+ 3·0609	+ 0·0065	...	+ 18·740	− 0·267	...	...
690	... ...	+ 3·4534	+ 0·0334	...	+ 18·694	− 0·190	...	...
691	... ...	+ 3·4008	+ 0·0391	...	+ 18·690	− 0·168	...	...
692	76 Virginis h ...	+ 3·1598	+ 0·0113	− 0·004	+ 18·662	− 0·176	+ 0·02	4521
693	S Virginis Var. 6 ..	+ 3·1990	+ 0·0096	...	+ 18·680	− 0·175	...	...
694	79 Virginis ζ ..	+ 3·0712	+ 0·0064	− 0·010	+ 18·600	− 0·170	− 0·06	4532
695	Taylor 7199	+ 3·5213	+ 0·0351	...	+ 18·579	− 0·201	...	...
696	... ...	+ 3·3001	+ 0·0349	...	+ 18·499	− 0·210	—	...
697	... ...	+ 3·8078	+ 0·0851	...	+ 18·398	− 0·212	...	...
698	82 Virginis m ..	+ 3·1177	+ 0·0108	− 0·010	+ 18·371	− 0·192	0·00	4566
699	... .	+ 3·4073	+ 0·0338	...	+ 18·327	− 0·215	...	...
700	... .	+ 3·3070	+ 0·0389	...	+ 18·308	− 0·216	...	...

Mean Positions of Stars for 1866 January 1st.

Number.	Star.	Magnitude.	Estimations.	Mean Right Ascension.			Mean Polar Distance.			Observations.	Fraction of Year.
				h.	m.	s.	°	′	″		
701	Taylor 6866 ...	7·5	1	13	37	1·26	151	46	42·2	1	0·43
702	... ...	7·8	1	13	37	38·07	129	20	54·2	1	0·30
703	... ...	8·0	1	13	43	21·36	123	7	7·8	1	0·44
704	... ...	8·4	2	13	43	31·00	123	13	40·5	2	0·43
705	... ...	8·8	1	13	44	22·01	127	57	30·9	1	0·30
706	8 Bootis η ...	2·9	...	13	48	18·26	70	55	46·6	5	0·36
707	... .	8·8	1	13	52	22·74	161	31	14·6	1	0·44
708	β Centauri... ...	1·2	...	13	54	23·95	149	43	29·9	2	0·35
709	93 Virginis τ ...	4·4	..	13	54	40·67	87	46	21·5	11	0·37
710	Lacaille 5794 ...	6·9	1	13	57	13·97	152	48	5·2	1	0·43
711	94 Virginis	6·5	...	13	59	12·18	98	15	3·1	1	0·40
712	Taylor 6570	8·4	1	14	0	0·00	140	56	35·5	1	0·43
713	R. P. L. 108	7·8	...	14	3	30·57	3	30	3·2	1	0·98
714	Taylor 6616	5·6	...	14	5	35·56	146	27	34·1	1	0·34
715	98 Virginis κ	4·3	...	14	5	46·07	99	33	54·5	2	0·36
716	16 Bootis α (*Arcturus*) ...	0·0	...	14	9	33·01	70	7	8·6	11	0·35
717	... ...	8·1	2	14	14	36·46	130	40	39·0	2	0·43
718	Taylor 6721	7·0	1	14	17	30·09	101	3	32·4	1	0·43
719	Taylor 6740	8·0	1	14	19	12·77	133	43	29·0	1	0·37
720	... ...	8·5	1	14	20	4·88	127	9	26·0	1	0·37
721	25 Bootis ρ	3·6	...	14	26	3·19	59	2	22·4	5	0·39
722	α Centauri ... 1st...	1·0	...	14	30	31·24	150	16	55·3	3	0·41
723	36 Bootis ε (*Mirac.*) ..	2·6	...	14	39	7·99	62	21	35·7	4	0·42
724	9 Libræ α²	3·0	...	14	43	26·09	105	28	39·2	4	0·38
725	O. A. S. 14112	7·9	2	14	50	30·49	109	11	13·3	3	0·43
726	... .	9·0	1	14	51	20·94	30	30	10·0	1	0·43
727	O. A. N. 18004	7·8	1	14	56	30·63	30	21	32·9	1	0·43
728	Taylor 7017	7·0	1	14	57	14·77	150	38	38·0	1	0·44
729	46 Bootis ψ	4·6	...	14	58	42·82	62	31	43·5	5	0·44
730	21 Libræ ν¹	5·4	...	14	59	9·15	105	41	6·8	1	0·38
731	R. P. L. 111.	6·9	..	15	5	29·57	5	31	58·0	1	0·49
732	27 Libræ β	2·7	..	15	9	47·90	98	33	11·3	8	0·46
733	. .	9·0	1	15	14	19·77	123	7	35·6	1	0·54
734	R Serpentis Var. 3. .	10·2	1	15	15	23·30	75	12	12·9	1	0·38
735	R Coronæ Borealis Var 2..	9·5	1	15	18	56·82	38	8	38·5	1	0·40

713.—Groombridge 2099.
722.—The 1st and brightest star.
725.—Comparison star for Iole in 1861.
726.—727.—Comparison stars for Comet 2, of 1861
731.—Groombridge 2218.
734.—R Serpentis Var. 3—Period 361 days.—Range, 6th to 12·5 magnitude.
735.—R Coronæ Borealis Var. 2.—Period 301 days.—Range, 6th to 12th magnitude.

Observed with the Madras Meridian Circle in that Year.

Number.	Star.	In Right Ascension. Annual Precession.	In Right Ascension. Secular Variation.	In Right Ascension. Proper Motion.	In Polar Distance. Annual Precession.	In Polar Distance. Secular Variation.	In Polar Distance. Proper Motion.	Number in B. A. C.
		s	s	s	"	"	"	
701	Taylor 6860	+ 4·0854	+ 0·0909	...	+ 18·294	− 0·268	...	...
702	... .	+ 3·5145	+ 0·0346	...	+ 18·263	− 0·230	...	...
703	... .	+ 3·4323	+ 0·0287	...	+ 18·080	− 0·227	...	...
704	... .	+ 3·4544	+ 0·0298	...	+ 18·044	− 0·238	...	...
705	... .	+ 3·5308	+ 0·0341	...	+ 18·011	− 0·285	...	...
706	8 Bootis η	+ 2·8617	− 0·0006	− 0·004	+ 17·857	− 0·189	+ 0·36	4648
707	... ...	+ 4·2927	+ 0·0960	...	+ 17·602	− 0·300	...	...
708	β Centauri	+ 4·1083	+ 0·0841	− 0·010	+ 17·608	− 0·301	+ 0·05	4669
709	93 Virginis τ ...	+ 3·0470	+ 0·0064	+ 0·001	+ 17·589	− 0·221	+ 0·07	4672
710	Lacaille 5794 ...	+ 4·3457	+ 0·0896	...	+ 17·499	− 0·318	...	...
711	94 Virginis	+ 3·1695	+ 0·0115	− 0·002	+ 17·403	− 0·287	+ 0·01	4685
712	Taylor 6570	+ 4·2898	+ 0·0968	...	+ 17·361	− 0·315	...	...
713	R. P. L. 108 ...	+ 7·8448	+ 2·4959	...	+ 17·204	+ 0·881	...	...
714	Taylor 6616	+ 4·1231	+ 0·0696	...	+ 17·110	− 0·320	...	4709
715	98 Virginis κ ...	+ 3·1907	+ 0·0122	+ 0·001	+ 17·111	− 0·260	− 0·08	4716
716	16 Bootis α (*Arcturus*)	+ 2·8132	+ 0·0004	− 0·079	+ 16·986	− 0·227	+ 1·98	4739
717	... ...	+ 4·3908	+ 0·0900	...	+ 16·896	− 0·362	...	...
718	Taylor 6721	+ 3·2196	+ 0·0132	...	+ 16·853	− 0·272	...	4772
719	Taylor 6740	+ 3·3020	+ 0·0123	...	+ 16·467	− 0·323	...	...
720	... .	+ 3·6886	+ 0·0384	...	+ 16·464	− 0·312	...	...
721	25 Bootis ρ	+ 2·3047	− 0·0015	− 0·008	+ 16·119	− 0·223	− 0·14	4808
722	α Centauri, 1st ..	+ 4·3013	+ 0·0973	− 0·470	+ 15·290	− 0·410	− 0·81	4832
723	36 Bootis ε (*Mirac*)...	+ 2·6240	− 0·0001	− 0·005	+ 15·412	− 0·202	− 0·01	4876
724	9 Libræ α²	+ 3·3143	+ 0·0154	− 0·007	+ 15·166	− 0·294	+ 0·06	4894
725	O. A. S. 14112 ...	+ 3·3879	+ 0·0175	...	+ 14·727	− 0·341	...	...
726	... ...	+ 1·9628	+ 0·0014	...	+ 14·700	− 0·201	...	...
727	O. A. N. 15004 ...	+ 1·9608	+ 0·0017	...	+ 14·560	− 0·202	...	...
728	Taylor 7017	+ 4·7301	+ 0·0960	...	+ 14·350	− 0·408	. .	...
729	45 Bootis ϕ	+ 2·6483	+ 0·0010	− 0·018	+ 14·298	− 0·231	0·00	4960
730	21 Libræ ν¹	+ 3·3375	+ 0·0158	− 0·004	+ 14·233	− 0·349	+ 0·08	4970
731	R. P. L. 111 ..	− 6·9208	+ 1·8443	...	+ 13·836	+ 0·725	. .	5022
732	27 Libræ β	+ 3·2290	+ 0·0117	− 0·009	+ 13·962	− 0·338	+ 0·01	5034
733	... — .	+ 3·7265	+ 0·0281	...	+ 13·267	− 0·414	...	...
734	S Serpentis Var. 3 ..	+ 2·8091	+ 0·0042	...	+ 13·197	− 0·314	. .	...
735	R Coronæ Borealis Var 2	+ 2·4464	+ 0·0014	...	+ 13·161	− 0·275	...	

711.—Proper motions from "*Greenwich Catalogue* 1872."
722.—Proper motions from "*Stone's Cape Catalogue.*"

Mean Positions of Stars for 1866 January 1st.

Number.	Star.	Magnitude.	Estimations.	Mean Right Ascension.			Mean Polar Distance.			Observations.	Fraction of Year.
				h	m	s	°	′	″		
736	Lacaille 6377	8·0	1	15	19	1·84	130	11	20·4	1	0·30
737	... ...	9·3	1	15	20	31·06	130	9	1·0	1	0·30
738	W. B. E. XV. 395 ...	8·0	1	15	22	5·17	101	15	55·8	1	0·39
739	Taylor 7220	8·0	...	15	22	14·08	123	7	0·2	2	0·61
740	38 Libræ γ	4·0	...	15	24	2·03	104	20	34·3	1	0·40
741	5 Cor. Bor. α (*Alphata*) ..	2·4	...	15	29	0·80	62	49	39·0	4	0·45
742	W. B. E. XV. 587 ...	7·8	1	15	31	59·60	103	27	54·9	1	0·48
743	W. B. E. XV. 645 ..	9·1	1	15	34	29·20	102	10	40·2	1	0·39
744	... ...	8·5	1	15	34	58·08	120	1	50·9	1	0·34
745	44 Serpentis α	2·7	...	15	37	40·06	83	9	3·7	2	0·40
746	O. A. S. 14874	8·3	...	15	39	31·98	104	40	3·6	1	0·46
747	O. A. S. 14934	9·4	2	15	42	56·17	107	52	47·7	2	0·54
748	W. B. E. XV. 898 ...	7·7	1	15	44	6·09	104	27	21·6	1	0·48
749	O. A. S. 14963	7·0	1	15	44	25·18	108	1	54·7	1	0·36
750	46 Libræ θ...	4·3	...	15	46	11·82	106	20	1·2	2	0·33
751	Radcliffe 3462	8·0	1	15	46	34·70	47	1	36·2	1	0·39
752	16 Ursæ Minoris ζ ...	4·5	...	15	48	56·09	11	47	40·3	1	0·34
753	49 Libræ	5·6	...	15	52	46·58	106	8	9·1	1	0·41
754	8 Scorpii β¹	3·0	...	15	57	38·37	109	26	10·2	5	0·36
755	O. A. S. 15297	8·2	2	15	59	26·37	106	34	43·6	2	0·47
756	Lalande 29306	8·4	2	15	59	33·13	107	34	20·0	2	0·48
757	R Herculis Var. 2 ...	9·5	1	16	0	12·98	71	16	3·7	1	0·37
758	Lalande 29391	7·0	1	16	1	55·85	102	41	39·8	1	0·41
759	O. A. S. 15342	8·0	3	16	3	33·17	107	46	50·1	3	0·48
760	R. P. L. 116	6·9	...	16	4	10·86	4	19	8·7	2	0·60
761	... .. .	7·7	2	16	4	18·63	107	52	54·0	2	0·40
762	W. B. E. XVI. 98 ..	9·0	1	16	6	9·00	102	41	24·6	1	0·35
763	1 Ophiuchi δ	2·8	...	16	7	19·51	93	20	30·5	6	0·51
764	... ...	10·4	3	16	12	31·34	107	37	2·6	3	0·38
765	... ...	9·6	4	16	16	42·74	107	35	53·8	4	0·46
766	Taylor 8821	9·0	1	16	17	22·31	113	5	37·9	1	0·36
767	7 Ophiuchi χ	5·0		16	19	13·72	108	8	57·0	1	0·40
768	21 Scorpii α (*Antares*) ..	1·1		16	21	11·70	116	7	54·1	4	0·45
769	14 Draconis η ...	2·5	..	16	22	10·92	28	10	31·3	1	0·35
770	9 Ophiuchi ω	4·7	...	16	24	11·84	111	10	38·1	2	0·40

738.—748.—Comparison stars for Sappho in 1864.
742.—746.—749.—758.—Comparison stars for Asia in 1861.
747.—750.—753.—755.—756.—759.—761.—764.—765.—Comparison stars for Sylvia in 1866.
757.—R Herculis Var. 2.—Period 319 days.—Range 8·5 to below 13th magnitude.
752.—Comparison star for Donati's Comet in 1858.
760.—Carrington 2123.

Observed with the Madras Meridian Circle in that Year.

Number.	Star.	In Right Ascension. Annual Precession.	In Right Ascension. Secular Variation.	In Right Ascension. Proper Motion.	In Polar Distance. Annual Precession.	In Polar Distance. Secular Variation.	In Polar Distance. Proper Motion.	Number in B. A. C.
		s	s	s	"	"	"	
736	Lacaille 6677 ..	+ 3·9848	+ 0·0384		+ 12·956	− 0·444	...	
737	.. ..	+ 3·9877	+ 0·0382		+ 12·866	− 0·447	...	
738	W. B. R. XV. 306	+ 3·2777	+ 0·0124		+ 12·750	− 0·374	...	.
739	Taylor 7280 .	+ 3·7466	+ 0·0268	...	+ 12·741	− 0·427	...	..
740	38 Libræ γ	+ 3·3416	+ 0·0130	+ 0·002	+ 12·344	− 0·369	− 0·02	5134
741	5 Cor. Bor. α (*Alphecca*)	+ 2·5295	+ 0·0023	+ 0·009	+ 12·277	− 0·307	+ 0·07	5146
742	W. B. R. XV. 987 ...	+ 3·3270	+ 0·0131	...	+ 12·070	− 0·398	...	...
743	W. B. R. XV. 946	+ 3·3073	+ 0·0135		+ 11·886	− 0·396	...	.
744	... ...	+ 3·9461	+ 0·0302	...	+ 11·860	− 0·471	...	..
745	24 Serpentis α ..	+ 2·9414	+ 0·0062	+ 0·009	+ 11·670	− 0·364	− 0·06	5196
746	O. A. S. 14874 ...	+ 3·3615	+ 0·0138	...	+ 11·588	− 0·405	...	...
747	O. A. S. 14984 ..	+ 3·4966	+ 0·0145	...	+ 11·391	− 0·418	.	...
748	W. B. R. XV. 833 ..	+ 3·3670	+ 0·0130	...	+ 11·207	− 0·411	...	...
749	O. A. S. 14963 ..	+ 3·4636	+ 0·0145	...	+ 11·173	− 0·421	...	...
750	46 Libræ θ	+ 3·3991	+ 0·0136	+ 0·009	+ 11·064	− 0·418	− 0·12	5257
751	Radcliffe 3462	+ 2·0834	+ 0·0035	...	+ 11·099	− 0·363	...	...
752	16 Ursæ Minoris 3 ...	− 2·3116	+ 0·2041	+ 0·020	+ 10·856	+ 0·278	+ 0·05	5265
753	49 Libræ	+ 3·4069	+ 0·0131	− 0·049	+ 10·567	− 0·427	+ 0·37	5294
754	8 Scorpii β¹	+ 3·4781	+ 0·0148	− 0·002	+ 10·306	− 0·441	+ 0·02	5320
755	O. A. S. 15237	+ 3·4163	+ 0·0129	...	+ 10·070	− 0·486	.	. .
756	Lalande 28806 .	+ 3·4384	+ 0·0132		+ 10·061	− 0·438		.
757	R Herculis Var. 2 ...	+ 2·6702	+ 0·0040		+ 10·012	− 0·343		
758	Lalande 28891 ..	+ 3·3342	+ 0·0113		+ 9·851	− 0·427		...
759	O. A. S. 15042 .	+ 3·4460	+ 0·0130		+ 9·757	− 0·448		.
760	B. P. L. 116 ..	− 12·4848	+ 1·7530		+ 9·701	+ 1·532		. .
761	... ...	+ 3·4497	+ 0·0130		+ 9·680	− 0·444	. .	
762	W. B. R. XVI. 80 .	+ 3·3363	+ 0·0110	...	+ 9·557	− 0·431	.	.
763	1 Ophiuchi δ	+ 3·1410	+ 0·0081	− 0·006	+ 9·465	− 0·406	+ 0·13	5414
764	... . ..	+ 3·4603	+ 0·0123	.	+ 9·084	− 0·463	.	...
765	... .	+ 3·4600	+ 0·0119		+ 8·736	− 0·457	.	..
766	Taylor 7581	+ 3·5962	+ 0·0141	...	+ 8·660	− 0·476		5478
767	7 Ophiuchi χ	+ 3·4687	+ 0·0110	− 0·008	+ 8·634	− 0·402	+ 0·01	5480
768	21 Scorpii α (*Antares*)	+ 3·6690	+ 0·0150	− 0·001	+ 8·381	− 0·491	+ 0·08	5485
769	14 Draconis η ..	+ 0·8006	+ 0·0188	+ 0·006	+ 8·302	− 0·111	− 0·07	5612
770	9 Ophiuchi ω . ..	+ 3·5456	+ 0·0126	+ 0·001	+ 8·141	− 0·476	− 0·08	5510

750.—752.—768.—769.—770.—Proper motions from "*Greenwich Catalogue* 1872."

Mean Positions of Stars for 1866 January 1st.

Number.	Star.	Magnitude.	Estimations.	Mean Right Ascension. h.	m.	s.	Mean Polar Distance. °	′	″	Observations.	Fraction of Year.
771	30 Herculis g Var. 5 ...	6·0	2	16	24	14·50	47	40	20·5	2	0·[illegible]
772	Taylor 7728	5·2	...	16	28	49·48	107	26	47·1	2	0·[illegible]
773	α Trianguli Australis ...	2·2	..	16	34	30·39	158	46	28·1	1	0·64
774	40 Herculis ζ	3·1	...	16	36	14·06	58	0	9·7	5	0·64
775	... ...	9·0	1	16	44	32·58	131	1	44·3	1	0·64
776	Taylor 7848	6·8	1	16	48	17·74	106	36	23·3	3	0·40
777	27 Ophiuchi κ	3·4	...	16	51	19·48	80	24	51·9	4	0·[illegible]
778	O. A. S. 16232	9·6	1	16	54	4·72	110	14	52·8	1	0·55
779	... ...	9·0	1	16	56	28·86	109	56	47·3	1	0·64
780	22 Ursæ Minoris ε ..	4·5	...	16	59	46·56	7	44	53·6	3	0·04
781	35 Ophiuchi η	2·6	...	17	2	41·96	105	38	22·1	1	0·41
782	... ...	8·0	...	17	5	52·61	130	50	34·2	1	0·55
783	64 Herculis α Var 1 ...	3·2	...	17	8	32·26	75	27	17·1	2	0·64
784	42 Ophiuchi θ	3·4	...	17	13	46·92	114	51	45·2	5	0·66
785	... ...	7·0	1	17	26	12·21	130	48	36·1	1	0·[illegible]
786	... ...	7·9	1	17	28	23·11	125	14	46·6	1	0·[illegible]
787	55 Ophiuchi α ...	2·2	...	17	28	42·58	77	30	34·5	8	0·60
788	56 Serpentis ζ ...	3·7	...	17	30	54·00	105	18	39·8	2	0·37
789	Lacaille 7406 ...	7·7	1	17	34	50·81	128	44	14·5	2	0·64
790	58 Ophiuchi ...	5·0	...	17	35	24·09	111	36	54·0	1	0·[illegible]
791	... ...	7·0	2	17	37	47·98	126	29	28·5	2	0·[illegible]
792	86 Herculis μ	3·5	...	17	41	12·90	62	11	58·3	4	0·[illegible]
793	Lacaille 7504 ..	7·0	1	17	48	40·45	130	6	52·9	1	0·60
794	Lacaille 7517 ...	7·7	1	17	52	36·31	140	10	26·4	1	0·[illegible]
795	Lacaille 7543	7·8	1	17	52	53·30	140	12	16·4	1	0·62
796	33 Draconis γ	2·4	..	17	53	29·72	38	29	39·9	3	0·62
797	... .	7·9	1	18	3	8·16	89	1	12·8	1	0·[illegible]
798	. .	10·0	1	18	4	1·[illegible]	[illegible]	9	[illegible]	1	0·60
799	13 Sagittarii μ^1	4·1	.	18	5	44·97	111	5	26·9	10	0·64
800	Taylor 8481 ...	5·7	...	18	14	38·17	124	10	22·7	1	0·60
801	23 Ursæ Minoris δ .	4·3	.	18	15	54·30	3	23	40·9	4	0·[illegible]
802	Lalande [illegible]	7·0	1	18	15	46·32	102	4	37·9	2	0·62
803	δ^1 Telescopii	5·7	...	18	22	6·97	135	50	48·5	1	0·63
804	Taylor 8581 . .	7·2	1	18	27	29·74	149	13	38·4	1	0·62
805	.	7·3	1	18	28	26·16	125	31	27·6	1	0·[illegible]

771.—30 Herculis g Var. 5.—Changes irregularly from 5th to 6·8 magnitude.
783.—α Herculis Var 1.—Changes irregularly between 3rd and 4th magnitude.
786.—791.—798.—805.—Comparison stars for Donati's Comet in 1858.
802.—Comparison star for Asia in 1866.

Observed with the Madras Meridian Circle in that Year.

Number.	Star.	In Right Ascension.			In Polar Distance.			Number in B.A.C.
		Annual Precession.	Secular Variation.	Proper Motion.	Annual Precession.	Secular Variation.	Proper Motion.	
		s	s	s	"	"	"	
771	30 Herculis g Var. 5..	+ 1·0619	+ 0·0012	+ 0·006	+ 8·135	− 0·286	− 0·07	5623
772	Taylor 7723	+ 3·4037	+ 0·0105	..	+ 7·366	− 0·478	...	5570
773	α Trianguli Australis.	+ 6·2780	+ 0·0007	0·000	+ 7·309	− 0·858	+ 0·03	5576
774	40 Herculis ζ ..	+ 2·2664	+ 0·0083	− 0·034	+ 7·168	− 0·316	− 0·45	5604
775	...	+ 4·1730	+ 0·0196	...	+ 6·484	− 0·570		...
776	Taylor 7842	+ 3·4611	+ 0·0059	..	+ 6·173	− 0·462	...	5696
777	27 Ophiuchi κ ...	+ 2·8663	+ 0·0044	− 0·023	+ 5·960	− 0·402	− 0·02	5705
778	O. A. S. 16262 ...	+ 3·5461	+ 0·0098	...	+ 5·680	− 0·486	...	..
779	. ... .	+ 3·5382	+ 0·0091	...	+ 5·570	− 0·406	...	...
780	22 Ursæ Minoris ε ...	− 6·4185	+ 0·3044	+ 0·009	+ 5·206	+ 0·901	− 0·01	5780
781	35 Ophiuchi η ..	+ 3·4327	+ 0·0074	+ 0·001	+ 4·062	− 0·467	− 0·12	5781
782	. .. ..	+ 4·1068	+ 0·0146	...	+ 4·692	− 0·567	...	...
783	64 Herculis α Var. 1.	+ 2·7380	+ 0·0085	− 0·008	+ 4·465	− 0·391	− 0·04	5821
784	42 Ophiuchi θ ...	+ 3·6700	+ 0·0080	− 0·003	+ 4·018	− 0·598	− 0·02	5861
785	. ... ..	+ 4·2121	+ 0·0094	...	+ 2·773	− 0·609	..	...
786	... ...	+ 4·0079	+ 0·0079	...	+ 2·742	− 0·580	.	.
787	55 Ophiuchi α ..	+ 2·7745	+ 0·0030	+ 0·004	+ 2·720	− 0·402	+ 0·20	5941
788	55 Serpentis ξ ..	+ 3·4350	+ 0·0047	− 0·004	+ 2·686	− 0·498	+ 0·04	5949
789	Lacaille 7408 ..	+ 4·1353	+ 0·0072	...	+ 2·197	− 0·601	...	...
790	58 Ophiuchi	+ 3·5960	+ 0·0050	− 0·010	+ 2·148	− 0·588	− 0·04	5987
791	.. ..	+ 4·0804	+ 0·0064	...	+ 1·940	− 0·590	.	..
792	86 Herculis μ ...	+ 2·3096	+ 0·0026	− 0·026	+ 1·042	− 0·346	+ 0·74	6021
793	Lacaille 7504 ..	+ 4·1579	+ 0·0042	.	+ 0·990	− 0·608	...	...
794	Lacaille 7517 ...	+ 5·3115	+ 0·0063		+ 0·640	− 0·774	..	.
795	Lacaille 7513 ..	+ 5·3164	+ 0·0062		+ 0·621	− 0·775	..	...
796	33 Draconis γ ...	+ 1·3915	+ 0·0080	0·000	+ 0·580	− 0·206	+ 0·04	6091
797	... .	+ 2·2695	+ 0·0022		− 0·274	− 0·361	...	
798	... .	+ 2·2741	+ 0·0021		− 0·362	− 0·332	.	.
799	13 Sagittarii μ¹ .	+ 3·3575	+ 0·0009	− 0·004	− 0·568	− 0·523	+ 0·01	6108
800	Taylor 8001	+ 4·3684	− 0·0086	..	− 1·272	− 0·696	.	6238
801	23 Ursæ Minoris δ ..	− 19·4001	− 0·4883	+ 0·048	− 1·361	+ 2·985	− 0·08	6291
802	Lalande 33945 ..	+ 3·3575	+ 0·0004		− 1·373	− 0·488	...	
803	δ¹ Telescopii	+ 4·4495	− 0·0056	− 0·008	− 1·383	− 0·643	+ 0·06	6303
804	Taylor 8551 ..	+ 5·3012	− 0·0158		− 2·299	− 0·767	..	.
805	... ...	+ 4·4867	− 0·0073		− 2·481	− 0·640	.	

773, 803.—Proper motions from "*Stone's Cape Catalogue.*"
777.—799.—Proper motions from "*Greenwich Catalogue 1872.*"

Mean Positions of Stars for 1866 January 1st.

Number.	Star.	Magnitude.	Estimations.	Mean Right Ascension.			Mean Polar Distance.			Observations.	Fraction of Year.
				h.	m.	s.	°	′	″		
806	3 Lyræ α (*Vega*)	0·2	...	18	32	24·00	51	20	22·4	8	0·63
807	Lacaille 7832 ...	7·1	1	18	37	46·15	149	5	28·8	1	0·68
808	R Scuti Var. 1 ...	7·1	4	18	40	19·65	95	50	46·9	4	0·64
809	Lacaille 7872 ...	6·5	2	18	42	29·20	136	44	55·4	2	0·66
810	Lacaille 7878 ...	7·1	2	18	43	2·34	136	14	32·4	2	0·65
811	10 Lyræ β Var. 1... ...	3·6	...	18	45	7·89	56	47	26·8	5	0·63
812	... ...	8·2	2	18	47	2·08	137	44	49·4	2	0·69
813	37 Sagittarii ξ²	3·5	...	18	49	44·24	111	16	46·8	1	0·49
814	13 Lyræ Var. 2	4·4	...	18	51	15·85	46	13	44·1	4	0·61
815	O. A. S. 18960	7·6	3	18	53	46·20	121	7	29·7	3	0·66
816	. .. .	8·0	2	18	54	18·40	122	56	5·1	3	0·67
817	39 Sagittarii ο	3·9	...	18	56	30·04	111	53	5·9	2	0·64
818	17 Aquilæ ζ	3·1	...	18	59	14·95	76	20	0·7	5	0·64
819	.. ..	9·4	1	19	0	37·83	92	1	34·2	1	0·61
820	41 Sagittarii π	3·1	...	19	1	47·82	111	14	1·6	2	0·64
821	...	7·0	3	19	3	15·41	109	22	32·4	3	0·68
822	...	7·9	3	19	3	21·27	122	50	57·3	3	0·68
823	...	9·0	1	19	7	14·88	128	47	39·4	1	0·70
824	...	7·1	4	19	8	2·61	122	7	0·6	4	0·59
825	...	9·8	1	19	8	36·72	120	46	50·2	1	0·61
826	..	7·3	2	19	9	30·17	129	46	56·5	2	0·68
827	.	8·0	...	19	10	6·88	107	9	26·5	1	0·70
828	. . ...	7·6	1	19	10	15·20	128	30	47·6	1	0·66
829	26 Aquilæ ω ..	5·1	...	19	11	21·53	79	38	28·6	2	0·64
830	30 Aquilæ δ ...	3·5	...	19	18	44·39	87	9	0·2	9	0·66
831	Taylor 8960	6·8	2	19	22	18·38	143	27	30·7	3	0·60
832	52 Sagittarii h² ..	4·6	.	19	25	33·90	115	10	25·0	5	0·65
833	Lacaille 8173	7·7	1	19	31	46·04	143	18	10·7	1	0·64
834	.	8·4	1	19	34	31·86	127	16	47·6	1	0·60
835	56 Sagittarii *f* ..	5·1	..	19	38	22·31	110	4	51·0	2	0·64
836	50 Aquilæ γ	2·8	...	19	39	53·26	79	43	40·1	5	0·64
837	8 Vulpeculæ Var. 3	8·0	2	19	42	54·05	63	2	49·5	2	0·60
838	53 Aquilæ α (*Altair*) ..	1·0		19	44	14·90	81	28	59·3	4	0·65
839	57 Sagittarii	6·2		19	44	24·75	109	22	57·2	1	0·64
840	60 Aquilæ β .	4·0	. .	19	48	43·60	83	45	28·2	4	0·65

808.—R Scuti Var. 1.—Period 71 days.—Range, 5th to 9th magnitude.
809—810.—Comparison stars for Donati's Comet of 1858.
811.—β Lyræ Var. 1.—Period 12·91 days.—Range, 3·5 to 4·5 magnitude.
814.—13 Lyræ Var. 2.—Period 46 days.—Range, 4·3 to 4·6 magnitude.
815—821.—Comparison stars for Eurydice in 1866.
816.—822.—Comparison stars for Diana in 1864.
828.—Comparison star for Pandora in 1862.
837.—8 Vulpeculæ Var. 3.—Period 67·5 days.—Range, 8·5 to 9·5 magnitude.

Observed with the Madras Meridian Circle in that Year.

Number.	Star.	In Right Ascension. Annual Precession.	Secular Variation.	Proper Motion.	In Polar Distance. Annual Precession.	Secular Variation.	Proper Motion.	Number in B. A. C.
		s	s	s	"	"	"	
806	3 Lyræ α (*Vega*) ...	+ 2·0131	+ 0·0016	+ 0·017	− 2·886	− 0·300	− 0·26	6346
807	Lacaille 7832 ...	+ 5·2760	− 0·0210		− 3·291	− 0·758	..	...
808	R Scuti Var. 1	+ 3·2080	− 0·0011		− 3·510	− 0·468	..	...
809	Lacaille 7872 ...	+ 4·4690	− 0·0122		− 3·696	− 0·686	...	...
810	Lacaille 7878 ...	+ 4·4681	− 0·0124		− 3·744	− 0·686	...	...
811	10 Lyræ β Var. 1 ...	+ 2·2186	+ 0·0015	− 0·002	− 3·924	− 0·315	+ 0·08	6429
812	... ...	+ 4·5130	− 0·0142	...	− 4·088	− 0·643	...	...
813	37 Sagittarii ξ[1] ...	+ 3·5807	− 0·0043	− 0·001	− 4·318	− 0·508	+ 0·08	6461
814	13 Lyræ Var. 2 ...	+ 1·8282	+ 0·0008	− 0·001	− 4·448	− 0·267	0·00	6476
815	O. A. S. 18960 ..	+ 3·8673	− 0·0077	...	− 4·662	− 0·546	...	...
816	... ...	+ 3·9140	− 0·0086	...	− 4·708	− 0·553	...	...
817	39 Sagittarii ο ..	+ 3·5942	− 0·0060	+ 0·001	− 4·907	− 0·508	+ 0·05	6507
818	17 Aquilæ ζ ...	+ 2·7578	+ 0·0003	− 0·006	− 5·127	− 0·397	+ 0·07	6526
819	... .	+ 2·8914	− 0·0004	...	− 5·271	− 0·406	...	...
820	41 Sagittarii π ...	+ 3·5728	− 0·0057	− 0·004	− 5·342	− 0·500	+ 0·08	6548
821	... ...	+ 4·5717	− 0·0208		− 5·465	− 0·640		...
822	... ...	+ 3·9026	− 0·0100		− 5·474	− 0·546		...
823	... ...	+ 4·1381	− 0·0144		− 5·800	− 0·576		...
824	.	+ 3·8747	− 0·0104		− 5·867	− 0·586		...
825	... ...	+ 4·1372	− 0·0146		− 5·901	− 0·574		...
826	... .	+ 4·1346	− 0·0149	...	− 5·969	− 0·578	...	...
827	... ...	+ 3·4867	− 0·0065	...	− 6·040	− 0·479	...	...
828	... ...	+ 3·9162	− 0·0115	...	− 6·052	− 0·542	...	..
829	25 Aquilæ ω	+ 2·8165	− 0·0003	− 0·008	− 6·158	− 0·398	− 0·02	6596
830	30 Aquilæ δ	+ 3·0094	− 0·0018	+ 0·014	− 6·756	− 0·410	− 0·10	6646
831	Taylor 8960	+ 4·7615	− 0·0227	...	− 7·048	− 0·647	...	6689
832	52 Sagittarii h[2] ...	+ 3·6641	− 0·0108	+ 0·002	− 7·558	− 0·490	− 0·02	6708
833	Lacaille 8178 ...	+ 4·7211	− 0·0268	...	− 7·817	− 0·631	...	...
834	... ...	+ 4·0045	− 0·0179	...	− 8·089	− 0·538	...	..
835	54 Sagittarii *f* ...	+ 3·5164	− 0·0091	− 0·013	− 8·380	− 0·482	+ 0·07	6760
836	50 Aquilæ γ	+ 2·8520	− 0·0011	+ 0·001	− 8·466	− 0·373	0·00	6772
837	8 Vulpeculæ Var. 3.	+ 2·4696	+ 0·0011	...	− 8·703	− 0·319	...	.
838	53 Aquilæ α (*Altair*)..	+ 2·8931	− 0·0014	+ 0·036	− 8·811	− 0·374	− 0·38	6802
839	57 Sagittarii ...	+ 3·4045	− 0·0094	− 0·008	− 8·894	− 0·454	+ 0·06	6808
840	60 Aquilæ β	+ 2·9466	− 0·0020	+ 0·002	− 9·162	− 0·378	+ 0·47	[illegible]

814.—835.—Proper motions from "*Greenwich Catalogue* 1872."

Mean Positions of Stars for 1866 January 1st.

Number.	Star.	Magnitude.	Estimations.	Mean Right Ascension. h.	m.	s.	Mean Polar Distance. °	′	″	Observations.	Fraction of Year.
841	... ...	7·9	1	19	56	54·87	130	21	15·2	1	0·69
842	λ Ursæ Minoris	6·5	...	19	58	11·54	1	5	32·7	4	0·14
843	... ...	8·6	3	20	7	47·21	81	22	7·6	3	0·71
844	R Sagittæ Var. 1	9·8	1	20	7	57·46	73	40	38·3	1	0·72
845	O. A. S. 20866	7·7	1	20	8	38·38	110	25	43·6	1	0·69
846	6 Capricorni α²	3·8	...	20	10	36·96	102	57	29·5	6	0·66
847	... ...	7·5	1	20	10	28·87	140	8	41·6	1	0·74
848	... ...	9·7	1	20	12	13·56	85	46	53·2	1	0·69
849	α Pavonis	2·1	...	20	15	1·73	147	9	39·6	1	0·66
850	Lalande 39136	8·9	1	20	15	36·13	106	12	59·2	1	0·71
851	11 Capricorni ρ	5·0	...	20	21	12·80	108	15	16·5	13	0·70
852	... ...	9·0	1	20	26	23·49	121	12	17·8	1	0·61
853	... ...	9·3	1	20	28	50·44	121	6	5·6	1	0·76
854	R. P. L. 148.	6·7	...	20	29	34·84	5	18	3·9	1	0·72
855	... ...	9·0	1	20	31	2·56	149	56	0·4	1	0·72
856	S Capricorni Var. 2 ...	9·3	4	20	34	4·82	109	31	57·1	4	0·70
857	... ...	9·2	1	20	35	54·12	124	49	57·6	1	0·60
858	50 Cygni α (*Deneb*) ...	1·5	...	20	36	51·82	45	11	30·3	4	0·68
859	W. B. E. XX. 988 ...	8·6	2	20	36	58·88	73	22	37·6	2	0·69
860	S Delphini Var. 2. ...	9·2	2	20	38	54·44	73	23	30·8	2	0·72
861	2 Aquarii ε	3·3	...	20	40	25·16	99	50	4·2	2	0·57
862	W. B. E. XX. 1024 ...	9·5	1	20	40	59·74	105	28	30·6	1	0·71
863	... ...	10·4	1	20	41	15·49	106	17	56·5	1	0·78
864	6 Aquarii μ	4·8	...	20	45	26·38	99	39	3·7	1	0·64
865	32 Vulpeculæ .	5·1	..	20	46	50·89	62	27	2·8	8	0·74
866	Lacaille 8620	7·0	2	20	51	31·07	136	28	46·5	2	0·72
867	... .	9·0	1	20	54	6·44	142	54	47·8	1	0·75
868	... ...	10·2	2	20	59	31·18	66	42	20·8	2	0·72
869	Taylor 9773 ... 1st .	7·7	1	21	0	35·11	145	6	30·6	1	0·74
870	Taylor 9773 ... 2nd	7·9	3	21	0	36·20	145	6	30·6	3	0·77
871	61 Cygni . .. 1st ..	5·5	...	21	0	33·90	51	54	39·2	3	0·76
872	18 Aquarii ν ..	4·6	...	21	2	17·63	101	54	45·2	1	0·72
873	...	9·2	1	21	3	2·17	145	16	10·7	1	0·78
874	Lacaille 8712	8·3	1	21	4	39·13	146	48	5·0	1	0·73
875	64 Cygni ζ . .	3·5		21	7	13·91	61	19	17·0	14	0·73

844.—R Sagittæ Var. 1.—Period 70·4 days.—Range, 8·8 to 10·3 magnitude.
845.—Comparison star for Parthenope in 1862.
850.—Comparison star for Nemesis in 1865.
854.—Carrington 3128.
856.—S Capricorni Var. 2.—Supposed to change from 9th to 11th magnitude.
860.—S Delphini Var. 2.—Period 278 days.—Range, 8th to 11th magnitude.

Observed with the Madras Meridian Circle in that Year.

Number.	Star.	In Right Ascension.			In Polar Distance.			Number in B. A. C.
		Annual Precession.	Secular Variation.	Proper Motion.	Annual Precession.	Secular Variation.	Proper Motion.	
		s	s	s	"	"	"	
841	... .	+ 4·0096	− 0·0244	...	− 0·793	− 0·513	...	...
842	λ Ursæ Minoris .	− 57·8942	− 29·8534	− 0·085	− 9·891	+ 7·349	− 0·31	6989
843	... .	+ 2·8899	− 0·0017	...	− 10·011	− 0·354	...	...
844	R Sagittæ Var. 1 .	+ 2·7400	− 0·0080	...	− 10·628	− 0·186	...	...
845	O. A. S. 20966 ...	+ 3·4968	− 0·0116	...	− 10·648	− 0·427	...	...
846	6 Capricorni α²	+ 3·3311	− 0·0084	+ 0·001	− 10·892	− 0·468	0·00	6974
847	... ...	+ 4·0662	− 0·0649	...	− 10·828	− 0·604	...	...
848	... ...	+ 3·0462	− 0·0067	...	− 10·940	− 0·367	...	...
849	α Pavonis	+ 4·7042	− 0·0694	0·000	− 11·144	− 0·674	+ 0·10	7004
850	Lalande 39126 ..	+ 3·3948	− 0·0101	...	− 11·168	− 0·406	...	...
851	11 Capricorni ρ ..	+ 3·4819	− 0·0115	− 0·006	− 11·600	− 0·408	+ 0·01	7042
852	... ...	+ 3·7221	− 0·0200	...	− 11·962	− 0·431	...	...
853	... ...	+ 3·7145	− 0·0201	...	− 12·126	− 0·427	...	...
854	R. P. L. 148 ..	− 6·3817	− 1·2690	...	− 12·168	+ 0·076	...	—
855	... .	+ 4·8907	− 0·0742	...	− 12·281	− 0·560	...	...
856	S Capricorni Var. 2..	+ 3·4482	− 0·0128	...	− 12·459	− 0·395	...	...
857	... .	+ 3·7909	− 0·0242	...	− 12·614	− 0·436	...	...
858	50 Cygni α (*Deneb*) ..	+ 2·0435	+ 0·0021	− 0·002	− 12·680	− 0·236	0·00	7171
859	W. B. E. XX. 983..	+ 2·7080	+ 0·0062	...	− 12·681	− 0·307	...	...
860	S Delphini Var. 2 ...	+ 2·7682	+ 0·0002	...	− 12·683	− 0·307	...	—
861	2 Aquarii ε	+ 3·3621	− 0·0081	− 0·001	− 12·910	− 0·367	+ 0·01	7196
862	W. B. E. XX. 1094..	+ 3·3662	− 0·0169	...	− 12·963	− 0·367	...	...
863	... ...	+ 3·5510	− 0·0169	—	− 12·975	− 0·367	...	...
864	6 Aquarii μ	+ 3·2307	− 0·0058	0·000	− 13·261	− 0·349	+ 0·04	7239
865	32 Vulpeculæ ..	+ 2·5665	+ 0·0026	− 0·003	− 13·476	− 0·270	0·00	7256
866	Lacaille 8680 .	+ 3·8009	+ 0·0272	...	− 13·647	− 0·400	...	..
867	... ...	+ 4·3570	− 0·0653	...	− 13·812	− 0·465	...	...
868	... .	+ 3·6694	+ 0·0092	...	− 14·059	− 0·270	...	...
869	Taylor 9772 ... 1st .	+ 4·4344	− 0·0624	...	− 14·218	− 0·449	...	...
870	Taylor 9772 ... 2nd...	+ 4·4344	− 0·0624	...	− 14·215	− 0·449	...	...
871	61 Cygni ... 1st .	+ 2·3868	+ 0·0044	+ 0·350	− 14·230	− 0·234	− 3·22	7280
872	13 Aquarii ν .	+ 3·2691	− 0·0066	+ 0·001	− 14·328	− 0·328	+ 0·01	7344
873	... .	+ 4·4081	− 0·0590	...	− 14·368	− 0·448	...	...
874	Lacaille 8712 .	+ 4·4698	− 0·0685	..	− 14·448	− 0·449	...	—
875	64 Cygni 3	+ 2·8506	+ 0·0094	− 0·003	− 14·623	− 0·349	+ 0·07	7388

849.—Proper motions from "*Stone's Cape Catalogue.*"

Mean Positions of Stars for 1866 January 1st.

Number.	Star.	Magnitude.	Estimations.	Mean Right Ascension.			Mean Polar Distance.			Observations.	Fraction of Year.
				h.	m.	s.	°	′	″		
876	... ...	9·6	1	21	11	8·83	129	31	29·5	1	0·75
877	Brisbane 7012	8·4	2	21	14	14·11	151	46	31·1	3	0·74
878	T Capricorni Var. 3 ...	9·9	1	21	14	32·36	106	30	41·2	1	0·72
879	5 Cephei α	2·6	...	21	15	22·63	27	58	52·8	2	0·72
880	... ...	7·7	1	21	20	14·71	150	47	21·0	1	0·70
881	22 Aquarii β ...	3·1	...	21	14	30·14	96	9	35·2	9	0·73
882	8 Cephei β ...	3·4	...	21	26	56·24	20	1	37·5	2	0·64
883	... ...	9·0	1	21	28	53·15	134	8	49·6	1	0·70
884	28 Aquarii ξ ...	4·8	...	21	30	36·80	98	27	18·5	2	0·61
885	Taylor 10068 ...	7·5	1	21	34	28·51	134	6	11·8	1	0·74
886	8 Cephei Var. 3	7·0	1	21	36	49·34	11	58	45·3	1	0·66
887	8 Pegasi ε	2·4	...	21	37	30·24	81	44	16·6	7	0·75
888	48 Capricorni λ	5·4	...	21	39	19·02	101	58	57·4	1	0·49
889	μ Cephei Var. 1	3·9	...	21	39	34·48	31	50	3·0	1	0·74
890	W. B. E. XXI. 975 ...	9·7	1	21	41	16·14	97	10	12·1	1	0·78
891	10 Pegasi	5·0	...	21	16	57·90	64	42	16·2	4	0·77
892	... ...	8·0	1	21	47	42·32	153	11	56·7	1	0·70
893	Taylor 10190 ...	6·4	2	21	51	14·14	146	31	28·4	2	0·64
894	... ...	9·5	1	21	52	54·19	136	37	41·6	1	0·78
895	ε Indi	5·2	...	21	53	5·60	147	20	5·5	3	0·71
896	... ...	9·0	1	21	58	53·71	150	48	48·9	1	0·72
897	34 Aquarii α	3·2	...	21	58	53·90	90	58	11·2	9	0·70
898	α Gruis	1·9	...	21	59	46·82	137	36	31·1	1	0·70
899	... ...	8·0	2	22	9	11·77	98	31	30·8	2	0·76
900	.. ..	8·0	1	22	0	16·00	146	36	43·6	1	0·73
901	43 Aquarii θ	4·3	...	22	0	46·64	9	36	58·6	7	0·75
902	.. ..	9·0	1	22	13	17·08	146	28	1·1	1	0·73
903	... ...	9·0	1	22	16	58·44	135	37	49·2	1	0·79
904	55 Aquarii ζ ... 1st ...	4·7	...	22	21	56·66	90	42	15·7	3	0·72
905	55 Aquarii ζ . 2nd ...	4·5	...	22	21	56·67	90	42	19·1	1	0·70
906	... ...	7·9	2	22	21	57·76	100	37	10·8	2	0·66
907	27 Cephei δ Var. 2 ..	4·0	...	22	24	12·00	32	16	13·1	1	0·70
908	... ...	7·9	1	22	26	50·50	141	30	38·7	1	0·79
909	62 Aquarii η . .	4·2	...	22	28	38·14	90	49	37·2	6	0·78
910	Lacaille 9198	7·0	2	22	30	0·61	130	38	7·0	2	0·76

878.—T Capricorni Var. 3.—Period 260 days.—Range, 9th to below 14th magnitude.
886.—T Cephei Var. 3.—Period 425 days.—Range, 6th to 11·5 magnitude.
889.—μ Cephei Var. 1.—Changes irregularly from 4th to 6th magnitude.
890.—Comparison star for Ariadne in 1864.
907.—δ Cephei Var 2.—Period 5·366 days.—Range, 3·7 to 4·8 magnitude.

Observed with the Madras Meridian Circle in that Year.

Number.	Star.	In Right Ascension. Annual Precession.	Secular Variation.	Proper Motion.	In Polar Distance. Annual Precession.	Secular Variation.	Proper Motion.	Number in B. A. C.
		s	s	s	"	"	"	
876	... ...	+ 3·8132	− 0·0880	...	− 14·864	− 0·263	...	...
877	Brisbane 7012 ...	+ 4·7180	− 0·0917	...	− 15·084	− 0·460	...	...
878	T Capricorni Var. 3.	+ 3·3109	− 0·0120	...	− 15·051	− 0·314	...	...
879	5 Cephei α	+ 1·4100	− 0·0071	+ 0·081	− 15·100	− 0·190	− 0·01	7416
880	... ...	+ 4·6072	− 0·0871	...	− 15·376	− 0·486	...	...
881	22 Aquarii β... ...	+ 3·1625	− 0·0071	− 0·001	− 15·614	− 0·262	0·00	7478
882	8 Cephei β	+ 0·8003	− 0·0845	0·000	− 15·746	− 0·066	+ 0·04	7498
883	... ...	+ 3·8045	− 0·0801	...	− 16·856	− 0·385	...	...
884	23 Aquarii ξ	+ 3·1027	− 0·0088	+ 0·004	− 15·948	− 0·276	+ 0·04	7514
885	Taylor 10009 ..	+ 3·8107	− 0·0395	...	− 16·146	− 0·696	...	7538
886	S Cephei Var. 3 ...	− 0·6131	− 0·1622	...	− 16·207	+ 0·059	...	...
887	8 Pegasi ε	+ 2·9451	− 0·0006	+ 0·008	− 16·307	− 0·242	0·00	7561
888	48 Capricorni λ ...	+ 3·2350	− 0·0101	− 0·002	− 16·394	− 0·265	+ 0·01	7577
889	μ Cephei Var. 1 ...	+ 1·8324	+ 0·0080	...	− 16·388	− 0·147	...	7582
890	W. B. K. XXI. 975..	+ 3·1088	− 0·0076	...	− 16·462	− 0·256	...	...
891	16 Pegasi	+ 2·7255	+ 0·0062	+ 0·001	− 16·770	− 0·210	+ 0·01	7627
892	... .	+ 3·7572	− 0·0866	...	− 16·806	− 0·292	...	...
893	Taylor 10190 ...	+ 4·1192	− 0·0696	...	− 16·972	− 0·310	...	7645
894	. ...	+ 3·8173	− 0·0441	...	− 17·040	− 0·296	...	...
895	ι Indi	+ 4·1088	− 0·0724	+ 0·480	− 17·058	− 0·318	+ 2·45	7650
896	. ... .	+ 4·3228	− 0·0872	...	− 17·069	− 0·388	...	...
897	34 Aquarii α	+ 3·0835	− 0·0041	− 0·008	− 17·320	− 0·219	+ 0·02	7689
898	α Gruis	+ 3·8056	− 0·0467	+ 0·011	− 17·360	− 0·270	+ 0·15	7692
899	... .	+ 3·1634	− 0·0077	...	− 17·756	− 0·207	...	...
900	... ...	+ 4·0056	− 0·0681	...	− 17·789	− 0·264	...	...
901	43 Aquarii θ	+ 3·1630	− 0·0075	+ 0·006	− 17·780	− 0·206	+ 0·08	7773
902	... ...	+ 3·9779	− 0·0677	...	− 17·980	− 0·268	...	...
903	... ..	+ 3·6730	− 0·0422	...	− 18·062	− 0·226	...	...
904	55 Aquarii ζ ... 1st..	+ 3·0789	− 0·0083	+ 0·009	− 18·247	− 0·178	− 0·08	7882
905	55 Aquarii ζ .. 2nd ..	+ 3·0789	− 0·0083	+ 0·009	− 18·247	− 0·178	− 0·08	7882
906	... .	+ 3·1761	− 0·0086	...	− 18·268	− 0·189	...	...
907	27 Cephei δ Var. 2 ...	+ 2·2127	+ 0·0165	+ 0·008	− 18·328	− 0·123	+ 0·02	7896
908	... ...	+ 3·7420	− 0·0627	...	− 18·391	− 0·210	..	...
909	62 Aquarii η	+ 3·0791	− 0·0081	+ 0·003	− 18·477	− 0·166	+ 0·06	7908
910	Lacaille 9188... ..	+ 3·3088	− 0·0633	...	− 18·589	− 0·198	..	—

896.—898.—Proper motions from "*Stone's Cape Catalogue.*"

Mean Positions of Stars for 1866 January 1st.

Number.	Star.	Magnitude.	Estimations.	Mean Right Ascension.			Mean Polar Distance.			Observations.	Fraction of Year.
				h.	m.	s.	°	′	″		
911	42 Pegasi ζ	3·6	...	22	34	46·69	79	52	2·7	5	0·74
912	W. B. K. XXII. 844. ...	8·8	1	22	40	40·32	87	48	4·0	1	0·78
913	... ...	9·0	...	22	46	33·73	130	36	2·1	1	0·84
914	O. A. S. 22600	8·0	3	22	49	7·18	110	19	14·2	3	0·77
915	S Aquarii Var. 2.	9·2	1	22	49	55·32	111	3	27·6	1	0·82
916	24 Pis. Aus. α (*Fomalhaut.*)	1·3	...	22	50	14·41	120	19	55·4	6	0·76
917	... ...	7·2	2	22	50	26·82	110	15	27·4	2	0·75
918	... ...	8·5	2	22	51	54·82	86	22	30·3	2	0·88
919	54 Pegasi α (*Markab*) ...	2·0	...	22	58	5·21	75	30	55·4	3	0·78
920	Lacaille 9372	8·0	1	23	0	30·09	150	27	34·8	1	0·77
921	... ...	9·2	1	23	4	25·22	130	46	37·2	1	0·78
922	... ...	8·6	2	23	8	15·22	150	30	39·1	3	0·81
923	Lacaille 9428 ...	7·0	1	23	10	2·88	151	48	57·0	1	0·89
924	6 Piscium γ ...	3·8	...	23	10	13·10	87	26	58·0	9	0·76
925	... ...	8·5	1	23	10	42·06	130	48	56·5	1	0·88
926	96 Aquarii...	5·7	...	23	12	27·07	96	51	38·3	1	0·72
927	... ...	9·0	1	23	15	44·15	130	39	9·4	1	0·84
928	Taylor 10748	6·0	3	23	17	39·84	147	35	4·9	2	0·73
929	... ...	9·5	1	23	19	49·37	151	37	26·8	1	0·84
930	8 Piscium κ	5·0	...	23	20	3·75	89	36	59·8	7	0·77
931	... ...	8·3	2	23	20	51·68	88	51	2·2	2	0·84
932	... ...	9·3	1	23	21	9·37	137	37	5·3	1	0·88
933	10 Piscium θ	4·4	...	23	21	10·39	84	21	33·8	1	0·78
934	... ...	10·0	2	23	26	36·30	129	51	30·0	3	0·84
935	Taylor 10804	6·5	1	23	27	30·19	147	38	57·4	1	0·86
936	R. P. L. 103	5·7	...	23	27	30·09	3	25	54·0	2	0·55
937	... ...	9·7	2	23	30	1·07	137	19	25·8	2	0·84
938	... ...	8·5	...	23	30	31·46	148	54	1·7	1	0·88
939	17 Piscium ι	4·3	...	23	33	3·45	95	5	50·0	6	0·79
940	35 Cephei γ	3·4	...	23	33	32·40	13	6	53·2	2	0·72
941	... ...	9·0	1	23	34	36·96	147	26	40·6	1	0·84
942	.. ..	9·5	2	23	36	53·05	103	1	37·3	3	0·85
943	... ...	9·3	3	23	41	10·77	138	46	0·4	2	0·85
944	δ Sculptoris	4·6	...	23	41	56·46	118	53	16·8	4	0·81
945	Lalande 46280	8·4	1	23	42	12·32	88	18	35·8	1	0·75

914.—Comparison star for Calliope in 1866.
915.—S Aquarii Var. 2.—Period 279 days.—Range, 9th to 13th magnitude.
940.—Groombridge 4101.

Observed with the Madras Meridian Circle in that Year.

Number.	Star.	In Right Ascension. Annual Precession.	In Right Ascension. Secular Variation.	In Right Ascension. Proper Motion.	In Polar Distance. Annual Precession.	In Polar Distance. Secular Variation.	In Polar Distance. Proper Motion.	Number in B. A. C.
		s	s	s	"	"	"	
911	42 Pegasi ζ	+ 2·9852	+ 0·0023	+ 0·001	− 18·885	− 0·149	0·00	7906
912	W. B. E. XXII. 944	+ 3·0547	− 0·0012	...	− 18·885	− 0·148	...	...
913	... ...	+ 3·4473	− 0·0895	...	− 18·950	− 0·137	...	...
914	O. A. S. 22500 ...	+ 3·3006	− 0·0804	...	− 19·108	− 0·139	...	...
915	S Aquarii Var. 2. ...	+ 3·2271	− 0·0140	...	− 19·126	− 0·134	...	..
916	24 Piscis Australis α.	+ 3·3065	− 0·0210	+ 0·022	− 19·134	− 0·135	+ 0·18	7992
917	... ...	+ 3·2256	− 0·0140	...	− 19·135	− 0·138	...	...
918	... ...	+ 3·0404	+ 0·0211	...	− 19·177	− 0·122	...	...
919	54 Pegasi α (*Markab*).	+ 2·9790	+ 0·0066	+ 0·003	− 19·327	− 0·107	+ 0·02	8034
920	Lacaille 9372 ...	+ 3·6777	− 0·0727	...	− 19·393	− 0·130	...	...
921	... ...	+ 3·3493	− 0·0305	...	− 19·468	− 0·109	..	...
922	... ...	+ 3·6014	− 0·0704	...	− 19·540	− 0·110	..	...
923	Lacaille 9428 ...	+ 3·6098	− 0·0748	...	− 19·580	− 0·106	...	8101
924	6 Piscium γ	+ 3·0892	+ 0·0006	+ 0·047	− 19·584	− 0·087	+ 0·01	8106
925	... ...	+ 3·6813	− 0·0704	...	− 19·598	− 0·104	...	...
926	96 Aquarii	+ 3·1004	− 0·0098	+ 0·011	− 19·695	− 0·085	+ 0·01	8119
927	... ...	+ 3·2934	+ 0·0896	...	− 19·698	− 0·094	...	...
928	Taylor 10743 ...	+ 3·4588	− 0·0892	...	− 19·714	− 0·086	...	8137
929	... ...	+ 3·5035	− 0·0708	...	− 19·747	− 0·081	...	...
930	8 Piscium κ	+ 3·0689	0·0000	+ 0·005	− 19·792	− 0·069	+ 0·12	8169
931	... ...	+ 3·0676	+ 0·0008	...	− 19·763	− 0·068	..	...
932	... ...	+ 3·3178	− 0·0875	...	− 19·767	− 0·074	...	...
933	10 Piscium θ	+ 3·0493	+ 0·0086	− 0·011	− 19·768	− 0·067	+ 0·06	8177
934	... ...	+ 3·2890	− 0·0275	...	− 19·830	− 0·063	...	...
935	Taylor 10904 ...	+ 3·3096	− 0·0655	...	− 19·864	− 0·060	...	8208
936	R. P. L. 186 ..	− 0·0466	− 0·5087	+ 0·094	− 19·868	+ 0·010	− 0·01	8213
937	... ...	+ 3·2013	− 0·0380	...	− 19·884	− 0·056	...	...
938	... .	+ 3·3603	− 0·0842	...	− 19·899	− 0·064	...	...
939	17 Piscium ι	+ 3·0886	+ 0·0080	+ 0·096	− 19·917	− 0·048	+ 0·46	8233
940	35 Cephei γ	+ 2·4193	+ 0·0740	− 0·080	− 19·924	− 0·031	− 0·15	8238
941	... ...	+ 3·3051	− 0·0682	...	− 19·960	− 0·043	...	...
942	... ...	+ 3·1108	− 0·0081	...	− 19·964	− 0·037	...	...
943	... .	+, 3·1698	− 0·0946	...	− 19·988	− 0·039	...	...
944	δ Sculptoris	+ 3·1300	− 0·0161	+ 0·009	− 19·968	− 0·036	+ 0·10	8276
945	Lalande 46650 .	+ 3·0820	+ 0·0015	...	− 19·994	− 0·026	...	..

930.—Proper motions from "*Greenwich Catalogue* 1872."
944.—Proper motions from "*Stone's Cape Catalogue.*"

Mean Positions of Stars for 1860 January 1st.

Number.	Star.	Magnitude.	Estimations.	Mean Right Ascension.			Mean Polar Distance.			Observations.	Fraction of Year.
				h.	m.	s.	°	′	″		
946	..	9·7	1	23	42	48·82	150	53	24·0	1	0·85
947	22 Piscium ...	5·0	..	23	46	6·38	87	48	51·9	1	0·73
948	.. ..	9·3	1	23	47	3·10	130	43	2·6	1	0·78
949	Lacaille 9685 ...	7·7	1	23	47	8·04	130	17	10·7	1	[illegible]
950	... .	9·5	...	23	47	37·20	150	46	12·8	1	0·86
951	... .	9·2	2	23	47	40·07	128	50	19·0	2	0·86
952	Lacaille 9611 ...	8·2	2	23	48	8·26	128	6	35·9	2	0·89
953	28 Piscium ω ...	4·2	...	23	52	26·02	88	52	42·6	2	0·76
954	Lacaille 9696 ...	7·0	1	23	53	38·61	143	30	26·2	1	0·82
955	.. ...	8·0	1	23	56	4·61	130	16	22·2	1	0·84
956	Taylor 10990	9·2	1	23	57	0·23	148	34	31·0	1	0·86
957	Lacaille 9721	5·7	..	23	59	22·11	130	40	14·4	1	0·65

Observed with the Madras Meridian Circle in that Year.

Number.	Star.	In Right Ascension.			In Polar Distance.			Number in B. A. C.
		Annual Precession.	Secular Variation.	Proper Motion.	Annual Precession.	Secular Variation.	Proper Motion.	
		s	s	s	"	"	"	
946	. .	+ 3·2821	− 0·0[illegible]	...	− 19·[illegible]	− 0·027	..	.
947	[illegible] Piscium ..	+ 3·0[illegible]	+ 0·0022	− 0·001	− 20·013	− 0·0[illegible]	+ 0·08	[illegible]
948	.. .	+ 3·2067	− 0·0[illegible]	..	− 20·0[illegible]	− 0·018	...	.
949	Lacaille 9[illegible] ..	+ 3·2088	− 0·0557	..	− 20·024	− 0·017	..	..
950	.	+ 3·2010	− 0·0[illegible]	..	− 20·025	− 0·016	...	..
951	. ..	+ 3·1208	− 0·0837	...	− 20·0[illegible]	− 0·015		...
952	Lacaille 9641	+ 3·1261	− 0·0230	...	− 20·027	− 0·015	...	.
953	[illegible] Piscium ω	+ 3·0673	+ 0·0047	+ 0·010	− 20·045	− 0·006	+ 0·13	[illegible]
954	Lacaille 9[illegible]	+ 3·1289	− 0·0407	..	− 20·047	− 0·004	...	.
955	.	+ 3·0915	− 0·0840	.	− 20·052	+ 0·001	.	.
956	Taylor 10[illegible] .	+ 3·1008	− 0·0482		− 20·058	+ 0·008		.
957	Lacaille 9721	+ 3·0765	− 0·0830		− 20·066	+ 0·008		.

SEPARATE RESULTS

OF

OBSERVATIONS

OF THE FIXED STARS,

MADE WITH THE

MADRAS MERIDIAN CIRCLE

IN THE YEAR

1867.

Separate Results of Madras Meridian Circle Observations in 1867.

Number and Date.	Magnitude.	Mean Right Ascension 1867. h.	m.	s.	No. of Wires.	Mean Polar Distance 1867. °	′	″	Observer.
1	*21 Andromedae α Alpherat.*								
Sep. 17	...	0	1	31·06	...	61	38	38·1	R
Oct. 14	...		1	31·05	6		38	40·8	M
20	.		1	31·01	...		38	37·3	R
2	*Anon.*								
Oct. 22	9·2	0	2	1·55		127	29	28·3	R
3	*Lacaille* 9746.								
Nov. 4	8·0	0	2	57·24	.	146	53	56·5	M
4	*Anon.*								
Sep. 24	9·3	0	5	7·85	.	126	17	5·5	R
Nov. 18	9·5		5	7·61	...		17	5·1	R
5	*Anon.*								
Oct. 4	8·0	0	6	20·18	...	131	7	17·9	M
6	88 *Pegasi γ, Algenib.*								
Sep. 27	...	0	6	23·25	...	73	33	23·3	R
Oct. 26			6	23·38	...		33	21·0	R
28	.		6	23·27			33	22·0	R
Nov. 2			6	23·30	...		33	23·1	M
7	*Lalande* 163.								
Nov. 11	7·6	0	7	47·66	5	89	26	38·8	M
13	7·6		7	47·69	...		26	36·1	M
8	*Taylor* 64.								
Nov. 16	7·7	0	14	14·20	...	120	58	38·8	R
18	8·0		14	16·64			58	39·3	R
21	7·5		14	16·70			58	37·5	R
9	*Lalande* 421.								
Oct. 14	7·8	0	15	50·67	...	51	30	1·0	R
25	..		15	50·82	5		30	1·1	R
10	*Lacaille* 61.								
Nov. 4	6·9	0	16	19·26	...	120	40	20·3	M
11	*Lacaille* 68.								
Nov. 28	8·0	0	19	29·87	...	139	38	1·8	R
12	*Anon.*								
Oct. 30	9·0	0	22	51·01	...	94	44	59·0	R
Nov. 12	8·6		22	51·00			45	1·0	M
13	12 *Ceti.*								
Oct. 21		0	23	14·98	5	94	41	34·3	R
Nov. 2	...		23	15·11			41	34·7	M
6	...		23	15·01	...		41	34·4	M
7			23	15·15	.		41	34·5	M
11	...		23	15·07			41	33·9	R
13	...		23	15·04			41	35·0	M
16			23	15·03	5		41	34·8	R
21			23	15·11	.		41	33·7	R
14	*Lalande* 670.								
Nov. 18	6·6	0	23	18·25	.	85	32	33·7	R
15	*Anon.*								
Nov. 28	9·6	0	25	22·08	5	76	8	14·8	R
16	*Anon.*								
Sep. 17	10·5	0	26	13·15	5	76	14	13·7	R
17	*Anon.*								
Oct. 11	8·3	0	28	32·76		144	38	22·3	M
18	*Taylor* 148.								
Nov. 5	6·9	0	29	8·20		143	6	32·0	R
11	6·0		29	7·87	5		6	29·5	M
19	*Anon.*								
Oct. 21	9·2	0	32	5·22		129	43	51·9	R
Nov. 6	9·0		32	5·24			43	51·7	M
13	8·9		32	5·27	.		43	54·0	M

Separate Results of Madras Meridian Circle Observations in 1867.

Number and Date.	Magnitude.	Mean Right Ascension 1867. h.	m.	s.	No. of Wires.	Mean Polar Distance 1867. °	′	″	Observer.
20			*Anon.*						
Nov. 21	9·3	0	36	7·89	...	88	27	53·9	n
23	9·8		36	7·91	5		27	55·5	n
21			16 *Ceti* β						
Oct. 30	...	0	36	51·70	...	108	43	2·2	n
Nov. 1	...		36	51·46	...		43	3·9	n
6	...		36	51·67	...		43	2·9	w
7	...		36	51·80	...		43	3·4	w
16	...		36	51·69	...		43	2·0	n
18	...		36	51·64	...		43	2·7	n
26	...		36	51·65	3		43	2·3	n
22			*W. B. E.* 0·628.						
Nov. 15	9·0	0	37	3·90	...	98	48	19·6	w
23			*W. B. E.* 0·697.						
Oct. 29		0	40	52·74	5	95	13	31·5	n
Nov. 5	8·7		40	52·64	4		13	35·3	w
24			*W. B. E.* 0·705.						
Oct. 26		0	41	17·31	...	94	26	19·6	n
Nov 12	7·9		41	17·29	5		26	22·1	w
25			68 *Piscium* δ						
Nov. 8	..	0	41	46·96	...	80	9	21·3	w
26			*W. B. E.* 0·716.						
Dec 4	8·8	0	44	50·68	...	94	26	49·5	w
27			*Lacaille* 234.						
Nov. 29	7·0	0	46	3·15	4	120	15	7·2	n
31	7·0		46	3·16			15	6·8	n
Dec. 6	7·1		46	3·11			15	9·3	w
28			*Anon.*						
Oct. 30	9·7	0	46	36·11		129	12	54·7	n
Nov 29	9·0		46	36·16	4		12	56·8	n
29			*Anon.*						
Oct. 30	9·5	0	50	29·93	5	129	39	11·2	n
Nov. 19	9·4		50	29·77	3		39	9·8	n
23	...		50	29·98	5		39	11·9	n
30			*Lacaille* 264.						
Oct. 29	7·3	0	50	52·38	5	151	41	8·3	n
Nov. 12	7·7		50	52·10	.		41	9·8	w
15	7·8		50	52·41	6		41	8·6	w
31			*Anon.*						
Nov. 28	9·5	0	51	11·97	...	130	39	56·4	n
Dec. 7	9·0		51	12·01	6		39	57·3	w
32			*Anon.*						
Oct. 18	.	0	51	39·05	5	134	41	3·2	n
33			71 *Piscium* ε						
Oct. 24	..	0	56	2·04		82	49	36·6	w
29			56	2·40			49	36·3	n
Nov. 5			56	2·44			49	37·0	w
6	...		56	2·45			49	36·7	w
7			56	2·39			49	37·0	w
21	.		56	2·51	..		49	37·4	(n
Dec. 10			56	2·52	..		49	36·5	w
34			*Anon.*						
Nov. 18	9·8	1	1	38·80	3	17	36	39·9	n
22	9·4		1	33·98	5		36	38·9	n
35			*Anon.*						
Dec. 9	7·4	1	3	15·84	.	150	15	29·3	w
36			*Anon.*						
Nov. 22	9·0	1	4	1·84	5	[illegible]	38	33·2	n

Separate Results of Madras Meridian Circle Observations in 1867.

Number and Date.	Magnitude.	Mean Right Ascension 1867. h.	m.	s.	No. of Wires.	Mean Polar Distance 1867. °	′	″	Observer.
37			*Anon.*						
Oct. 21	9·2	1	4	55·13	5	127	20	46·1	R
30	10·0		4	55·16	5		20	49·9	R
Nov. 12	9·0		4	55·14	6		20	40·7	M
38			*O. A. N.* 1303.						
Dec. 5	7·0	1	9	2·66	...	18	19	2·6	M
39			1 *Ursae Minoris* α, *Polaris.*						
Sep. 17	...	1	10	16·13	1	1	23	57·4	R
Nov. 11	...		10	17·01	1		23	59·4	M
15	...		10	14·50	2		23	50·1	M
16	...		10	17·51	3		23	58·6	R
21	...		10	17·48	3		23	57·7	R
			1 *Ursae Minoris* α, *Polaris.*—*s.p.*						
Apl. 24	...	1	10	15·00	1	1	24	0·3	R
26	...		10	16·08	3		23	59·6	R
May 4	...		10	16·81	1		24	0·2	M
7	..		10	18·34	1		24	1·3	M
11	..		10	16·38	1		23	58·8	M
June 4	..		10	17·01	1		24	0·0	M
5	...		10	16·39	1		24	0·1	M
40			*Anon.*						
Nov. 18	9·4	1	12	21·17	5	132	16	31·2	R
41			*Brisbane* 203.						
Nov. 2	6·9	1	15	7·20		120	46	34·0	M
42			45 *Ceti* θ						
Oct. 11	...	1	17	22·51	...	98	52	11·4	M
21	...		17	22·54			52	13·9	R
30			17	22·57			52	15·0	R
Nov. 4	...		17	22·47			52	14·6	M
12	...		17	22·43	...		52	16·4	M
22			17	22·60	...		52	14·8	R
25			17	22·36	5		52	11·4	R
26			17	22·46	5		52	14·4	R
43			*Anon.*						
Sep. 17	...	1	19	0·18	5	161	19	27·6	R
44			*Taylor* 465.						
Nov. 1	7·0	1	19	38·91	...	91	5	26·3	M
45			99 *Piscium* η						
Oct. 21	...	1	24	22·15	...	75	20	24·9	R
29	...		24	22·17	...		20	27·6	R
Nov. 16	...		24	22·12	...		20	26·6	R
18	...		24	22·11	4		20	27·1	R
20	...		24	22·13	...		20	26·9	R
21	...		24	22·16	...		20	25·6	R
22	...		24	22·10	...		20	24·9	R
29	..		24	22·07	...		20	29·1	R
Dec. 2	...		24	22·06			20	27·5	M
5			24	22·20	4		20	29·6	M
46			*Taylor* 496.						
Sep. 17	8·0	1	25	49·01	4	140	26	15·8	R
47			*Anon.*						
Nov. 28	8·5	1	25	51·30	5	140	20	42·7	R
48			102 *Piscium* π						
Nov. 20	6·5	1	30	3·03		78	32	22·9	R
49			*Taylor* 525.						
Dec. 4	5·9	1	30	16·64		104	40	13·0	M
50			*Anon.*						
Nov. 16	9·2	1	31	30·80		130	51	19·9	R
Dec. 9	7·9		31	30·79	...		51	22·5	M
51			*Taylor* 530.						
Oct. 26	6·0	1	31	32·23	5	108	45	4·8	R

Separate Results of Madras Meridian Circle Observations in 1867.

Number and Date.	Magnitude.	Mean Right Ascension 1867. h.	m.	s.	No. of Wires.	Mean Polar Distance 1867. °	′	″	Observer.
52				*α Eridani, Achernar.*					
Oct. 19	..	1	32	45·83	5	147	54	50·0	n
29	...		32	45·76	...		54	50·3	n
53				106 *Piscium ν*					
Sep. 16	..	1	34	30·67	5	95	11	11·1	n
17			34	30·68	..		11	11·0	n
Nov. 2	.		34	30·77	...		11	11·0	m
14			34	30·73	3		11	10·8	n
23			34	30·69	...		11	11·4	m
28			34	30·63	...		11	11·7	n
26			34	30·63	5		11	10·9	n
Dec. 10			34	30·68	4		11	12·1	m
54				*Lacaille* 307.					
Dec. 2	5·9	1	37	14·77	4	151	27	35·0	m
55				110 *Piscium ο*					
Sep. 16		1	38	22·22	4	81	30	48·2	n
56				*Anon.*					
Oct. 26	9·2	1	38	39·77	5	102	1	54·3	n
57				*Anon.*					
Dec. 4	8·4	1	39	59·79	...	149	26	28·4	m
58				*Anon.*					
Oct. 19	9·7	1	40	30·46		124	34	9·3	n
Nov. 22	9·6		40	30·40			34	8·1	n
59				*Anon.*					
Dec. 14	9·3	1	43	16·63		168	37	8·6	m
60				*Lacaille* 550.					
Oct. 19	8·3	1	44	08·84	5	160	34	15·2	n
26	9·2		44	08·80	5		34	13·1	n
Dec. 6	7·9		44	08·84	.		34	15·6	m
61				6 *Arietis β.*					
Sep. 17	...	1	47	17·76	...	69	50	37·2	n
Oct. 15	...		47	17·78	...		50	38·0	m
Nov. 1	...		47	17·71	...		50	36·8	m
11	...		47	17·78	...		50	37·7	m
13			47	17·64	...		50	39·3	m
19	...		47	17·70			50	37·0	n
23			47	17·86	4		50	37·9	m
29	...		47	17·73	...		50	37·5	n
Dec. 2	...		47	17·86	...		50	37·0	m
3	...		47	17·78	...		50	37·8	m
62				*Anon.*					
Nov. 22	9·0	1	49	22·47		126	6	37·6	n
63				*Anon.*					
Dec. 10	9·0	1	52	34·32	5	145	47	31·3	m
64				*Anon.*					
Dec. 9	8·8	1	53	3·65	5	151	22	2·7	m
65				*Anon.*					
Oct. 26	9·5	1	56	14·30		109	57	30·8	n
Nov. 21	9·3		56	14·34			57	31·8	n
66				*Anon.*					
Nov. 22	9·3	1	58	14·62	...	120	26	27·5	n
67				13 *Arietis α*					
Sep. 16		1	59	40·90		67	10	6·2	n
Oct. 15	...		59	40·81	6		10	6·7	m
Nov. 11	...		59	40·78	..		10	5·9	m
16	...		59	40·84	.		10	8·8	m
28			59	40·90	—		10	6·6	n
29			59	40·94	..		10	5·5	n
Dec. 2			59	40·86			10	5·4	m
6			59	40·87			10	5·4	m
7			59	40·79	6		10	6·7	m
11			59	40·79			10	5·8	m

Separate Results of Madras Meridian Circle Observations in 1867.

Number and Date.	Magnitude.	Mean Right Ascension 1867. h.	m.	s.	No. of Wires.	Mean Polar Distance 1867. °	′	″	Observer.
68				*Anon.*					
Dec. 10	8·8	2	1	10·63	5	148	40	47·5	M
69				*Anon.*					
Dec. 9	9·3	2	6	11·21	...	131	58	15·8	M
70				21 *Arietis.*					
Dec. 16	6·7	2	8	10·30	...	66	34	51·9	B
71				67 *Ceti.*					
Sep. 16	...	2	10	21·01	...	97	2	11·3	B
Oct. 15	...		10	21·60	...		2	12·5	M
30	...		10	20·98	5		2	11·2	B
Nov. 13	...		10	21·00	...		2	11·3	M
15	...		10	21·01	...		2	12·5	M
Dec. 5	...		10	20·90	5		2	12·7	M
11	...		10	21·19	..		2	12·0	M
15	...		10	21·00	...		2	12·8	M
30	...		10	20·99	...		2	12·7	B
72				68 *Ceti* o *Var.* 1, *Mira.*					
Nov. 14	6·0	2	12	37·80	..	93	36	0·2	M
Dec. 2	6·0		12	37·61	.		36	0·2	M
73				*Anon.*					
Dec. 6	9·0	2	12	46·60	..	93	34	46·2	M
9	9·0		12	46·40			34	47·1	M
74				*Anon.*					
Dec. 18	8·0	2	15	44·70		152	35	1·1	B
75				*Anon.*					
Nov. 15	6·2	2	17	40·81		131	2	20·9	M
Dec. 20	6·8		17	40·29	5		2	19·4	B
76				73 *Ceti* ξ²					
Sep. 16	...	2	21	5·26	..	82	8	15·9	B
17			21	5·31	...		8	15·4	B
Oct. 15			21	5·47	..		8	16·0	M
Nov. 11			21	5·61			8	15·1	M
13	...		21	5·31			8	15·9	M
22	...		21	5·20	...		8	14·8	B
Dec. 2	...		21	5·34	4		8	16·1	M
4	...		21	5·31			8	13·6	M
5			21	5·28	...		8	14·8	M
7			21	5·40	...		8	15·0	M
10			21	5·51			8	13·8	M
11			21	5·36	...		8	16·3	M
14	...		21	5·20	...		8	15·9	M
18			21	5·41	...		8	14·5	B
21	...		21	5·37	...		8	16·4	B
30	.		21	5·28	5		8	15·6	B
77				*Lacaille* 782.					
Dec. 6	7·0	2	26	16·03	...	148	34	8·5	M
78				*Anon.*					
Nov. 21	8·8	2	27	30·81	5	147	11	28·5	B
Dec. 9	7·7		27	30·89	...		11	24·7	M
18	9·6		27	31·19	5		11	26·4	B
79				*Anon.*					
Nov. 23	9·1	2	28	16·95	...	129	45	46·5	B
Dec. 19	9·2		28	16·89	...		45	39·5	B
80				*Anon.*					
Nov. 14	9·7	2	30	51·09	5	147	34	6·1	M
81				*Anon.*					
Dec. 10	9·6	2	31	17·41	5	148	30	22·6	M
82				*Anon.*					
Nov. 22	9·8	2	31	30·63	...	151	29	30·2	B
Dec. 20	9·3		31	30·60			29	30·7	B

Separate Results of Madras Meridian Circle Observations in 1867.

Number and Date	Magnitude	Mean Right Ascension 1867. h.	m.	s.	No. of Wires.	Mean Polar Distance 1867. °	′	″	Observer.
83				*86 Ceti γ*					
Nov. 13		2	36	24·65	...	87	19	34·4	M
Dec. 4			36	24·54	...		19	34·8	M
5	.		36	24·66			19	34·6	M
6			36	24·58	...		19	36·1	M
7			36	24·50			19	31·6	M
14	...		36	24·69	...		19	36·7	M
16	...		36	24·61	...		19	38·4	R
18	...		36	24·66	...		19	34·7	R
20	...		36	24·68	.		19	36·8	R
84				*μ Ceti.*					
Sep. 17	...	2	37	46·27	...	80	36	56·7	R
Nov. 11			37	46·31	...		36	57·2	M
85				*Lacaille* 868.					
Dec. 2	7·9	2	38	38·12		147	12	34·4	M
86				*Taylor* 940.					
Dec. 19	8·0	2	40	22·25	...	129	48	44·9	R
26	7·7		40	22·29			48	46·3	R
87				*Anon.*					
Nov. 22	9·8	2	41	47·02		151	1	53·2	R
Dec. 16	9·5		41	47·46	5		1	57·4	R
19	9·6		41	47·38	2		1	55·8	R
88				*Taylor* 969.					
Dec. 4	7·3	2	45	47·53	4	71	3	41·6	M
7	7·0		45	47·45	6		3	41·1	M
89				*Taylor* 978.					
Dec. 9	6·9	2	47	28·09	6	147	44	26·7	M
11	6·9		47	28·06	5		44	27·7	M
90				*Lacaille* 941.					
Nov. 14	6·9	2	50	31·95	5	144	25	20·6	M
91				*Taylor* 991.					
Dec. 18	7·7	2	50	40·20	...	129	11	25·7	R
21	7·0		50	40·19	.		11	27·5	R
27	7·0		50	40·22	...		11	26·9	R
92				*Lacaille* 969.					
Nov. 21	7·0	2	54	58·74	5	144	13	12·4	R
Dec. 6	7·7		54	58·73	...		13	15·8	M
93				42 *Ceti α, Menkar.*					
Nov. 16	...	2	55	19·66	..	86	26	3·2	R
Dec. 16	.		55	19·72	..		26	3·1	R
27	...		55	19·70	...		26	2·7	R
30	...		55	19·67	...		26	2·6	R
94				25 *Persei ρ Var.* 2.					
Dec. 7	4·7	2	56	59·76	...	51	40	29·7	M
95				26 *Persei β Var.* 1, *Algol.*					
Nov. 5	4·9	2	59	51·25	..	49	28	24·6	M
Dec. 10	2·0		59	51·26	..		28	22·0	R
96				*Taylor* 1047.					
Dec. 21	6·5	2	59	55·47	5	151	10	19·0	R
97				*Anon.*					
Dec. 24	8·9	2	59	57·95		130	37	40·3	R
98				*Taylor* 1062.					
Jan. 3	6·1	3	0	29·39	5	130	13	19·1	R
99				*Taylor* 1067.					
Jan. 10	7·6	3	0	50·23	5	131	21	29·1	M
Dec. 16	8·2		0	50·29	5		21	27·8	R
100				*R. P. L.* 33.					
Dec. 14		3	1	20·75	5	5	34	6·9	R

Separate Results of Madras Meridian Circle Observations in 1867.

Number and Date.	Magnitude.	Mean Right Ascension 1867. h.	m.	s.	No. of Wires.	Mean Polar Distance 1867. °	′	″	Observer.
101				*Anon.*					
Dec. 24	9.3	3	2	10.73	.	130	34	13.4	n
102				57 *Arietis* δ					
Nov. 14		3	4	1.74	...	70	46	41.1	m
20	...		4	1.63	...		46	42.5	n
Dec. 9			4	1.38	...		46	46.5	m
19	...		4	1.70	...		46	48.5	n
20	...		4	1.66	...		46	44.2	n
27	...		4	1.57	...		46	44.1	n
103				*Taylor* 1081.					
Nov. 21	8.0	3	5	10.21	4	151	30	33.2	n
104				*Anon.*					
Dec. 21	9.0	3	6	15.88	...	128	31	22.9	n
105				*Anon.*					
Jan. 9	9.2	3	7	21.56	5	146	39	51.5	m
Dec. 18	9.5		7	21.30	5		39	50.7	n
106				*Taylor* 1112.					
Dec. 27	8.0	3	10	28.03	...	130	30	40.8	n
107				*Taylor* 1113.					
Dec. 19	8.9	3	10	34.30	...	151	43	20.5	n
108				*Taylor* 1127.					
Dec. 16	8.0	3	11	54.24	5	131	45	40.1	n
109				*Anon.*					
Dec. 24	9.3	3	12	29.94	...	130	9	33.0	n
110				*Anon.*					
Dec. 30	9.0	3	12	30.19		130	57	29.2	n
111				*Anon.*					
Dec. 27	8.5	3	12	40.75		129	27	1.9	n
112				*Anon.*					
Dec. 21	7.8	3	13	27.19	...	125	39	12.6	n
113				33 *Persei* α					
Dec. 30		3	14	50.42	...	40	36	54.8	n
114				*Anon.*					
Nov. 2	8.0	3	15	9.63		151	31	37.9	m
14	7.9		15	9.73	3		31	32.8	m
Dec. 18	8.8		16	9.61	5		31	33.6	n
115				*Anon.*					
Dec. 21	9.3	3	16	30.93	...	125	40	48.2	n
116				*Anon.*					
Dec. 19	9.5	3	16	39.65	...	127	6	21.5	n
117				*Anon.*					
Dec. 30	8.8	3	17	36.79	.	130	41	39.6	n
118				*Anon.*					
Nov. 21	9.0	3	20	22.02		140	13	15.3	n
119				*R Persei Var.* 3.					
Dec. 16	9.0	3	21	35.53		54	17	31.8	n
120				*Anon.*					
Dec. 31	9.3	3	22	21.70		130	9	30.0	n
121				5 *Tauri f*					
Sep. 17		3	23	31.86	...	77	31	14.3	n
Nov. 11			23	31.77			31	10.5	m

Separate Results of Madras Meridian Circle Observations in 1867.

Number and Date.	Magnitude.	Mean Right Ascension 1867. h. m. s.	No. of Wires.	Mean Polar Distance 1867. ° ′ ″	Observer.
122 *Anon.*					
Dec 19	8·9	3 23 46·21	5	128 21 34·7	R
27	9·1	23 46·01	5	21 50·0	R
123 *Anon.*					
Dec. 21	8·5	3 24 56·61	...	129 1 6·5	R
124 *Lacaille* 1150.					
Jan. 8	7·0	3 28 30·98		132 27 40·0	M
125 *Lacaille* 1140.					
Jan. 7	7·0	3 28 41·15	5	150 14 33·0	M
126 *Anon.*					
Dec 20	9·1	3 29 52·10		128 24 51·3	R
127 *Anon.*					
Dec. 16	7·0	3 30 17·63		131 30 28·7	R
128 *Anon.*					
Dec. 21	9·6	3 31 36·10	5	129 40 36·8	R
129 *Anon.*					
Dec. 30	8·6	3 32 29·60		131 30 9·3	R
130 *Lacaille* 1166.					
Dec 21	6·6	3 33 8·86	...	129 12 37·6	R
131 *Anon.*					
Dec 27	9·3	3 33 28·25	6	127 42 34·9	R
132 *Anon.*					
Dec. 20	10·0	3 33 48·65	6	128 29 29·2	R
133 *Anon.*					
Jan. 16	8·8	3 35 28·18	...	160 12 42·5	R
134 *Anon.*					
Dec. 21	9·7	3 35 36·51	..	120 16 3·0	R
135 *Anon.*					
Oct. 15	8·0	3 37 2·65	...	148 24 54·2	R
136 25 *Tauri* η, *Alcyone.*					
Jan. 8	..	3 39 34·99	5	66 18 32·6	M
11	...	39 34·81	6	18 31·7	W
Oct. 16	...	39 34·92	...	18 31·7	R
Nov. 12	...	39 35·03	6	18 31·6	M
14	..	39 34·86	...	18 32·8	M
20	..	39 34·89	...	18 31·0	R
Dec. 4		39 34·90	...	18 31·1	M
18		39 34·84	...	18 30·1	R
20	...	39 34·97	...	18 31·1	R
24	...	39 34·93		18 30·5	R
27	..	39 34·94	...	18 31·4	R
137 30 *Tauri* ε					
Sep. 17	..	3 40 59·73	...	79 16 5·7	R
138 *Lacaille* 1242.					
Dec. 2	7·2	3 41 40·31	6	147 4 42·1	W
139 *W. B. E. III.* 900.					
Dec. 21	8·0	3 44 29·11		76 1 4·6	R
27	8·6	44 29·03	5	1 4·8	M
140 *Anon.*					
Nov. 22	9·0	3 49 35·22	5	129 11 57·7	R
141 *Anon.*					
Oct. 15	9·0	3 50 18·94	6	147 24 42·7	W

Separate Results of Madras Meridian Circle Observations in 1867.

Number and Date.	Magnitude.	Mean Right Ascension 1867. h.	m.	s.	No. of Wires.	Mean Polar Distance 1867. °	′	″	Observer.
142				*34 Eridani γ¹*					
Jan. 4		3	51	40·32	..	103	53	20·5	M
8	...		51	40·51	...		53	21·6	M
9	...		51	40·56	...		53	21·0	M
10			51	40·40	...		53	22·0	M
11			51	40·47	5		53	20·6	M
Nov. 14	..		51	40·36	5		53	21·8	M
Dec. 4	.		51	40·47	4		53	20·0	M
6			51	40·39	...		53	20·6	M
20			51	40·30	...		53	20·0	R
21			51	40·51	5		53	21·9	R
24	...		51	40·50	...		53	20·7	M
27			51	40·40	...		53	20·8	R
143				*35 Tauri λ Var.* 1.					
Jan. 16	...	3	53	18·80	...	77	53	17·2	R
Nov. 11	6·0		53	18·64	5		53	15·9	M
144				*Anon.*					
Jan. 15	8·5	3	56	45·06	5	148	7	51·5	R
145				*Anon.*					
Nov. 22	9·1	3	54	48·00	...	129	10	20·6	R
146				*Taylor* 1392.					
Dec. 7	6·9	3	55	51·91	5	147	29	52·5	M
147				*Lalande* 7581.					
Jan 7	8·2	3	56	23·01	..	74	51	30·3	M
148				*Lacaille* 1359.					
Jan. 8	8·3	4	0	4·67	..	147	40	30·8	M
9	8·6		0	4·94	5		40	30·0	M
149				*Lacaille* 1375.					
Jan. 10	7·7	4	2	39·66	..	153	50	18·9	M
14	9·3		2	39·01			50	16·7	R
150				*Lalande* 7764.					
Jan. 12	8·0	4	3	39·08		74	43	23·1	M
151				*Annon.*					
Nov. 21	9·5	4	3	46·18	5	106	56	36·9	R
152				*37 Eridani.*					
Dec. 12	...	4	3	23·37	...	97	16	23·7	M
153				*38 Eridani o¹*					
Nov. 16	...	4	5	22·44	...	97	11	12·6	M
Dec. 24	...		5	22·52			11	13·0	R
154				*Anon.*					
Jan. 4	8·1	4	9	32·66	...	159	16	24·7	M
155				*Taylor* 1460.					
Jan. 8	7·3	4	11	6·32	...	148	21	31·6	M
9	7·3		11	6·38			21	30·6	M
Dec. 13	6·9		11	6·34	5		21	30·7	M
156				*54 Tauri γ*					
Jan. 16	.	4	12	13·73	4	74	41	47·2	R
157				*Lacaille* 1418.					
Jan. 15	7·5	4	12	31·30	...	143	29	16·8	R
Dec. 14	7·1		12	31·02	..		29	14·1	M
158				*Anon.*					
Nov 16	9·5	4	12	35·01		129	10	21·0	R
Dec. 16	8·7		12	35·12	5		10	19·8	R
20	8·0		12	31·08	5		10	20·6	R
159				*Anon.*					
Jan. 5	9·0	4	13	39·03		79	51	3·6	M
7	8·5		13	39·01	.		51	1·0	M

Separate Results of Madras Meridian Circle Observations in 1867.

Number and Date.	Magnitude.	Mean Right Ascension 1867. h.	m.	s.	No. of Wires.	Mean Polar Distance 1867. °	′	″	Observer.
160		*U Tauri Var. 7.*							
Jan. 17	9·9	4	14	4·26	...	70	37	15·3	R
161		*Anon.*							
Nov. 16	10·0	4	15	20·66	5	129	7	25·1	R
162		*Anon.*							
Nov. 14	8·0	4	16	49·23	...	140	3	20·5	M
163		*74 Tauri ε*							
Jan. 4	...	4	20	51·23	...	71	7	3·0	M
8	...		20	51·04	...		7	4·2	M
9	...		20	51·07	...		7	3·9	M
10	...		20	51·10			7	3·8	M
11	...		20	51·20	...		7	3·8	M
12	...		20	51·20	...		7	3·0	M
14	...		20	51·26	...		7	3·8	R
Oct. 15	.		20	50·97	...		7	3·5	M
Dec. 9	..		20	51·19	.		7	3·4	M
16			20	51·07			7	3·1	R
19	...		20	51·16	...		7	2·2	R
164		*Lacaille 1519.*							
Jan. 7	7·0	4	25	27·40	5	153	5	48·2	M
165		*Lacaille 1540.*							
Jan. 10	7·7	4	26	44·46		147	29	20·6	M
166		*Anon.*							
Dec. 13	9·3	4	27	16·77	6	130	31	28·1	M
167		*87 Tauri α, Aldebaran.*							
Jan. 3		4	28	17·48		73	45	39·9	M
5			28	17·37	...		45	39·6	M
6			28	17·37	.		45	41·1	M
9			28	17·46	5		45	40·7	M
12			28	17·48			45	41·3	M
14			28	17·45			45	39·1	R
Jan. 15		4	28	17·40	...	73	45	39·8	R
Oct. 15			28	17·44			45	39·7	M
16	..		28	17·51	.		45	37·3	R
Dec. 9	..		28	17·46			45	40·4	M
12	.		28	17·56	.		45	40·3	M
21			28	17·41	..		45	39·9	R
168		*Anon.*							
Jan. 4	8·3	4	29	32·46	5	110	13	31·9	M
Nov. 14	8·1		29	32·50			13	33·0	M
169		*R Reticuli Var. 1.*							
Jan. 16	10·0	4	32	10·36	6	153	18	22·0	R
17	9·8		32	10·21	...		18	22·2	R
28	9·3		32	10·40	...		18	21·2	R
Feb. 1	8·6		32	10·31	4		18	21·4	M
18	7·7		32	10·40	...		18	19·3	M
Nov. 16	9·3		32	10·36	5		18	20·4	R
Dec. 14	7·8		32	10·14	...		18	20·0	M
170		*Lacaille 1551.*							
Jan. 24	6·0	4	32	11·86	5	160	5	22·9	R
171		*Anon.*							
Dec. 16	9·2	4	34	29·90		153	26	30·7	R
21	9·7		34	29·84	5		26	31·5	R
172		*95 Tauri.*							
Jan. 8	6·9	4	35	10·70	4	66	10	0·0	M
173		*Lacaille 1567.*							
Jan. 18	5·0	4	36	19·10	5	162	20	29·5	R
174		*Lacaille 1566*							
Jan. 10	6·9	4	35	[illegible]		16[illegible]	28	6·2	R

Separate Results of Madras Meridian Circle Observations in 1867.

Number and Date.	Magnitude.	Mean Right Ascension 1867. h.	m.	s.	No. of Wires.	Mean Polar Distance 1867. °	′	″	Observer.
175		*Taylor* 1663.							
Jan. 5	7·7	4	30	54·96	5	138	47	48·4	M
7	7·0		30	55·10	5		47	49·2	M
9	7·7		30	55·26	...		47	48·8	M
176		*Anon.*							
Jan. 24	9·5	4	39	15·81	4	158	15	35·0	R
Dec. 24	9·6		39	15·47	5		15	38·7	R
177		*Anon.*							
Dec. 18	9·0	4	39	27·68	...	158	3	16·9	R
178		*Anon.*							
Jan. 14	9·5	4	48	26·69	...	130	40	58·3	R
17	9·5		48	26·75	5		40	58·7	R
179		*Lacaille* 1629.							
Jan. 4	6·4	4	43	46·66	...	163	55	14·6	M
180		*Anon.*							
Dec. 19	8·9	4	48	57·79	...	127	40	3·2	R
181		*Lacaille* 1625.							
Jan. 8	7·6	4	45	3·79	5	140	1	32·3	M
10	7·6		45	3·72	5		1	31·5	M
182		*Anon.*							
Jan. 9	9·6	4	46	34·96	...	129	24	46·1	M
183		*Lacaille* 1649.							
Dec. 11	6·9	4	46	36·70	5	150	29	19·6	M
16	6·9		46	36·66	...		29	20·2	R
184		3 *Aurigae* ι							
Jan. 3	.	4	48	30·16	5	57	2	52·6	M
5	...		48	30·12	...		2	52·2	M
7			48	30·08			2	53·3	M
11	...		48	30·09			2	53·7	M
12	...		48	30·12			2	51·7	M
15			48	30·02	5		2	53·8	R
Feb. 4			48	30·16	...		2	51·3	M
5	...		48	30·13	4		2	52·8	M
Oct. 16	...		48	30·06	...		2	49·6	R
Dec. 9	...		48	30·18			2	52·7	M
12	...		48	30·19			2	52·5	M
18	...		48	30·19	...		2	53·0	M
24	—		48	30·02	...		2	56·8	R
185		*Taylor* 1764.							
Dec. 10	6·4	4	50	27·56	..	120	50	39·6	M
19	6·2		50	27·69	...		50	37·8	R
186		*Taylor* 1780.							
Jan. 4	7·0	4	53	19·97	5	144	39	32·3	M
15	7·6		53	19·86	5		39	30·4	R
187		*R Leporis Var.* 1.							
Feb. 6	7·5	4	53	32·92	5	104	59	25·6	M
188		*Taylor* 1797.							
Jan. 29	7·9	4	54	54·30	5	104	16	30·5	R
189		*Anon.*							
Jan. 5	8·7	4	56	2·40	..	130	17	24·5	M
190		*Lacaille* 1697.							
Jan. 9	8·9	4	56	57·30		129	6	50·2	M
191		11 *Orionis.*							
Jan. 14		4	56	56·22		74	47	3·6	R
17			56	56·25			47	4·0	R

Separate Results of Madras Meridian Circle Observations in 1867.

Number and Date.	Magnitude.	Mean Right Ascension 1867. h. m. s.	No. of Wires.	Mean Polar Distance 1867. ° ′ ″	Observer.
192	*Taylor* 1611.				
Feb. 9	6·1	4 57 9·12	...	129 54 45·7	m
193	*Taylor* 1614.				
Dec. 11	8·8	4 57 48·40	5	146 41 41·8	m
18	9·6	57 49·33	...	41 38·8	n
194	2 *Leporis* ε				
Jan. 3	...	4 59 49·94	5	112 38 8·6	m
10	.	59 49·84	...	38 8·2	m
16	.	59 49·88	4	38 8·1	n
29		59 49·91	...	38 7·5	n
30	..	59 49·88	...	38 7·7	n
31		59 49·94	...	33 7·6	n
Dec. 13		60 49·73	...	38 7·2	m
19	..	59 49·74	...	38 8·1	n
195	*Anon.*				
Jan. 8	7·9	5 0 8·21	..	136 23 41·6	m
196	*Taylor* 1852.				
Dec. 12	6·9	5 2 13·68	...	144 35 17·2	m
197	*Lacaille* 1739.				
Dec. 20	8·0	5 3 54·90	...	106 37 29·7	m
198	*Anon.*				
Jan. 28	9·6	5 4 30·61	...	135 54 38·2	n
199	*Anon.*				
Jan. 16	9·3	5 6 29·25		86 34 29·7	n
17	9·3	6 29·15	...	34 29·4	n
18	9·0	6 29·34		34 29·5	n
28	9·7	6 29·13	5	34 27·8	n
Dec. 18	9·8	6 29·11	4	34 27·2	n
27	9·8	6 29·09		34 27·6	n

Number and Date.	Magnitude.	Mean Right Ascension 1867. h. m. s.	No. of Wires.	Mean Polar Distance 1867. ° ′ ″	Observer.
200	*R Aurigae Var.* 2.				
Dec. 18	8·7	5 6 38·84	4	36 34 3·8	n
201	13 *Aurigae* α, *Capella.*				
Jan. 7	...	5 6 58·11		44 8 29·4	m
Feb. 2		6 58·18		8 29·1	m
9	...	6 58·08	5	8 29·3	m
202	19 *Orionis* β, *Rigel.*				
Jan. 3		5 8 8·74	...	98 21 29·1	m
15	...	8 8·77	.	21 30·1	n
19	...	8 8·82	...	21 30·1	n
26	...	8 8·80	...	21 29·7	n
203	*Anon.*				
Jan. 9	9·0	5 8 32·01	5	100 36 9·1	m
204	*Anon.*				
Jan. 29	9·2	5 10 57·65		121 19 24·6	n
205	*Anon.*				
Jan. 29	9·7	5 12 3·80	5	120 46 35·9	n
206	*Anon.*				
Jan. 16	9·8	5 13 46·61	.	75 11 35·6	n
17	9·4	13 46·47	5	11 33·3	n
18	9·5	13 46·48	5	11 33·9	n
207	*Anon.*				
Jan. 29	9·3	5 15 1·24	3	121 14 29·6	n
Feb. 1	8·9	16 1·33	...	18 29·2	m
208	*Lacaille* 1828.				
Dec. 18	7·9	5 15 47·04	4	141 08 2·5	m
209	*Anon.*				
Jan. 26	9·2	5 16 12·60	5	101 46 7·6	n

Separate Results of Madras Meridian Circle Observations in 1867.

Number and Date.	Magnitude.	Mean Right Ascension 1867. h.	m.	s.	No. of Wires.	Mean Polar Distance 1867. °	′	″	Observer.
210			*Lacaille* 1824.						
Feb. 13	7·1	5	17	4·40	...	139	37	50·4	M
211			*Anon.*						
Jan. 28	7·0	5	17	30·95	5	153	7	11·2	R
212			112 *Tauri* β						
Jan. 4		5	17	53·16	...	61	30	31·6	M
14	...		17	53·10	...		30	30·6	R
19	...		17	53·07			30	30·3	R
23	...		17	53·11	5		30	31·6	R
30	...		17	53·17	...		30	31·1	R
31	...		17	53·15	..		30	31·4	R
Feb. 2	...		17	53·31	...		30	30·6	M
4	...		17	53·06	...		30	29·7	M
11			17	53·15	5		30	31·1	M
12	.		17	53·26	...		30	31·2	M
14	...		17	53·11	...		30	30·9	M
213			*Anon.*						
Jan. 29	9·0	5	18	25·29	5	121	28	46·4	R
Dec. 18	9·6		18	25·33	...		28	46·2	R
214			*R. P. L.* 40.						
Dec. 19	...	5	19	40·98	3	4	32	51·3	R
			R. P. L. 40.—*s.p.*						
July 5		5	19	40·96	3	4	32	55·1	R
215			*Anon.*						
Jan. 25	9·2	5	22	36·64	5	152	41	55·5	R
216			*Anon.*						
Feb. 6	7·1	5	23	31·90	...	151	13	17·3	M
217			λ *Doradûs.*						
Jan. 24	6·0	5	24	29·14	5	149	1	38·1	R
218			119 *Tauri.*						
Dec. 10	...	5	24	34·97		71	30	39·4	M
11	8·8		24	35·13	5		30	39·2	M
219			34 *Orionis* δ *Var.* 1.						
Jan. 5		5	26	12·75		90	24	2·1	M
14	...		26	12·73	...		24	2·1	R
16	...		26	12·82	...		24	3·1	R
17	...		26	12·82	...		24	2·6	R
23	...		26	12·77	...		24	2·8	R
28			26	12·60	...		24	2·1	R
Feb. 2			26	12·76			24	1·7	M
5	...		26	12·81	5		24	2·1	M
Dec. 13	...		26	12·68	...		24	1·4	M
220			*Anon.*						
Jan. 29	9·5	5	26	12·32		121	24	17·1	R
221			11 *Leporis* α						
Jan. 18	...	5	26	31·91	...	167	55	12·5	R
19	..		26	31·96	...		55	11·7	R
30	...		26	31·80	...		55	12·5	R
31	...		26	31·77			55	12·2	R
Dec. 12	...		26	31·74			55	11·5	M
222			*Taylor* 2037.						
Jan. 25	6·0	5	28	3·29	5	161	14	30·6	R
Feb. 1	7·6		28	3·21	5		14	31·2	M
223			*Lalande* 10532.						
Feb. 4	8·2	5	29	19·84	5	78	15	4·2	M
224			46 *Orionis* ε						
Jan. 16	..	5	29	29·93		91	17	34·4	R
17			29	29·97			17	34·3	R
23			29	29·90			17	33·4	R
26			29	29·91	.		17	30·5	R

Separate Results of Madras Meridian Circle Observations in 1867.

Number and Date.	Magnitude.	Mean Right Ascension 1867. h. m. s.	No. of Wires.	Mean Polar Distance 1867. ° ′ ″	Observer.
225		123 *Tauri* ζ			
Nov. 11		5 29 41·89	.	68 56 31·2	M
Dec. 10	...	29 41·95	4	56 29·6	M
226		*Lalande* 10607.			
Jan. 4	7·0	5 31 27·73	4	68 18 56·2	M
8	7·0	31 27·96	...	19 56·7	M
10	7·1	31 27·85	4	18 56·8	M
227		*Anon.*			
Dec. 20	9·3	5 32 5·08	5	123 55 19·2	R
228		*Anon.*			
Jan. 24	9·0	5 32 14·88	5	123 54 10·9	R
25	8·5	32 14·79	...	54 9·7	R
Dec. 20	8·9	32 14·37	5	54 11·4	R
229		*Lacaille* 1940.			
Jan. 14	6·5	5 32 16·39	5	131 18 29·9	R
230		*Lacaille* 1916.			
Jan. 29	7·0	5 32 29·28	5	121 8 38·9	R
Dec. 18	7·0	32 29·32	.	8 32·5	R
231		α *Columbae.*			
Jan. 5	.	5 34 39·16	...	124 8 49·7	M
7	...	34 39·15	...	8 50·7	M
16	...	34 39·01	...	8 50·5	R
17	..	34 39·06	...	8 48·9	R
19	...	34 39·12	...	8 48·6	R
Feb. 2	...	34 39·98		8 49·2	M
3		34 39·11		8 49·5	M
9		34 39·21		8 49·7	M
232		*Taylor* 2113.			
Feb. 6	7·4	5 35 18·89	5	130 46 27·9	M
233		*Lalande* 10849.			
Jan. 26	8·0	5 37 54·83	...	68 19 40·9	R
26	8·8	37 54·51	...	19 37·5	R
Feb. 1	7·2	37 54·86	..	19 37·9	M
234		*Anon.*			
Feb. 4	8·0	5 38 20·86	...	79 0 9·5	M
Dec. 21	7·5	38 20·71	...	0 9·8	R
235		*Taylor* 2145.			
Jan. 24	6·5	5 39 54·52	5	136 53 46·0	R
236		*Anon.*			
Jan. 29	9·3	5 40 37·23	4	120 57 54·9	R
237		*W. B. E. V.* 1011.			
Nov. 14	8·5	5 44 38·90	4	78 57 37·6	M
238		*Taylor* 2184.			
Jan. 25	9·0	5 43 59·51	5	109 46 29·0	R
239		*Anon.*			
Jan. 28	9·3	5 44 49·68	5	137 10 16·1	R
240		*Lalande* 11096.			
Jan. 5	7·9	5 45 16·73	5	68 29 38·7	M
23	8·6	45 16·76	.	29 37·3	R
24	8·3	45 16·88		29 38·7	R
241		*Lacaille* 2036.			
Jan. 25	8·8	5 45 26·39		129 47 11·3	M
Feb. 8	7·9	45 26·71		47 10·9	M
11	8·0	45 26·61		47 11·0	M
242		54 *Orionis* χ¹			
Feb. 14		5 46 29·31	–	69 46 8·0	M

Separate Results of Madras Meridian Circle Observations in 1867.

Number and Date.	Magnitude.	Mean Right Ascension 1867. h.	m.	s.	No. of Wires.	Mean Polar Distance 1867. °	′	″	Observer.
243				*Anon.*					
Feb. 6	8·0	5	47	6·61	5	126	46	46·5	R
244				*Lalande* 11166.					
Jan. 4	7·8	5	47	11·54	...	78	33	36·9	M
9	8·0		47	11·57	...		33	35·0	M
245				58 *Orionis* = *Var.* 2, *Betelgeux.*					
Jan. 7	.	5	47	58·31	...	82	37	14·2	M
18			47	58·49	...		37	13·6	R
Feb. 5	..		47	58·81	...		37	13·7	M
7	...		47	58·26	5		37	18·7	M
9	..		47	58·25	4		37	15·7	M
12	...		47	58·29	...		37	13·4	M
13	..		47	58·36	...		37	14·4	M
246				*Anon.*					
Jan. 29	9·0	5	49	22·30	5	121	9	50·8	R
247				*Anon.*					
Jan. 26	9·8	5	50	10·06	...	137	10	17·9	R
248				*Lacaille* 2073.					
Jan. 25	7·0	5	50	35·67	5	137	13	34·7	R
249				*Anon.*					
Dec. 13	9·0	5	51	16·48	6	141	51	30·1	R
250				*Lalande* 11208.					
Jan. 24	8·0	5	51	40·65	...	65	34	35·5	R
26	7·3		51	40·62	...		34	35·9	R
Feb. 1	7·1		51	40·85	...		34	36·6	M
Dec. 11	7·1		51	40·42	5		34	34·0	M
251				*Anon.*					
Jan. 29	8·3	5	53	7·61	...	130	34	47·3	R
252				*Anon.*					
Jan. 26	...	5	54	29·70		137	46	14·3	R
253				*Anon.*					
Jan. 29	9·6	5	54	58·96	5	121	30	57·8	R
254				*Lalande* 11455.					
Jan. 5	7·1	5	56	59·40	...	73	19	12·6	M
Feb. 2	7·7		56	59·48	...		19	11·6	M
255				*Anon.*					
Jan. 23	9·2	5	57	4·80	.	126	1	0·7	R
256				*Taylor* 2301.					
Feb. 8	7·0	5	58	32·71	..	149	6	16·3	M
257				*Taylor* 2310.					
Jan. 28	6·7	5	59	39·71	...	150	29	7·4	R
Feb. 13	6·6		59	40·05	...		29	6·0	M
258				67 *Orionis* ν					
Jan. 4	...	5	59	58·65	..	75	13	7·6	M
9	...		59	58·65	5		13	8·1	M
21	.		59	58·61	...		13	7·9	R
25			59	58·66	...		13	7·9	R
Feb. 4			59	58·80			13	6·7	M
7			59	58·67	...		13	6·8	M
11			59	58·57	...		13	8·1	M
12	..		59	58·73			13	8·0	M
15	...		59	58·65			13	7·5	M
259				*Anon.*					
Jan. 29	9·8	6	0	5·40	5	121	30	33·3	R
260				*Taylor* 2321.					
Dec. 13	6·9	6	0	51·73	6	101	5	27·0	M

Separate Results of Madras Meridian Circle Observations in 1867.

Number and Date.	Magnitude.	Mean Right Ascension 1867. h. m. s.	No. of Wires.	Mean Polar Distance 1867. ° ′ ″	Observer.
261		*Anon.*			
Feb. 6	7·9	6 0 35·69	5	137 28 28·1	m
262		*Lalande* 11732.			
Feb. 1	6·8	6 3 32·73	5	77 58 54·9	m
4	7·9	3 32·74	4	58 53·3	m
263		*Anon.*			
Dec. 27	9·3	6 3 44·80	5	139 59 13·0	n
264		*Anon.*			
Jan. 5	8·1	6 5 47·73	...	77 51 30·8	m
265		*Anon.*			
Jan. 29	9·0	6 6 33·10	...	121 29 26·3	n
266		7 *Geminorum* η *Var.* 6.			
Jan. 17	...	6 6 31·00	...	67 27 28·3	n
18	...	6 30·80	5	27 28·1	n
267		*Anon.*			
Jan. 23	9·7	6 7 29·89	...	131 15 39·9	n
268		*Anon.*			
Jan. 24	9·5	6 7 46·69	5	137 6 25·1	n
269		*Anon.*			
Jan. 25	8·3	6 8 33·46	...	131 54 48·5	n
Feb. 6	7·9	8 33·54	6	54 49·3	m
270		*Anon.*			
Jan. 7	9·3	6 9 32·94	6	131 50 56·3	m
11	9·0	9 32·94		50 54·6	m
271		*Anon.*			
Jan. 9	7·6	6 11 4·13	3	139 43 34·8	m
272		*Anon.*			
Jan. 30	8·6	6 11 60·31	..	121 31 31·1	n
273		*Lalande* 12058			
Feb. 2	8·0	6 12 34·70	...	68 41 18·5	m
274		*Lalande* 12094			
Jan. 31	9·3	6 13 45·87	...	68 46 4·8	n
275		*Lalande* 12120			
Feb. 11	7·1	6 14 16·26	...	77 4 67·6	m
13	7·4	14 16·24	6	4 69·7	m
276		13 *Geminorum* μ			
Jan. 17	...	6 14 34·88	...	67 25 16·6	n
18	...	14 34·81	...	25 16·9	n
25	...	14 34·89	...	25 16·7	n
Feb. 1	...	14 34·88	...	25 17·6	m
9	..	14 34·80	6	25 16·5	m
12	...	14 34·76	...	25 16·6	m
Dec. 11	...	14 34·98	...	25 17·4	m
12	...	14 34·87	...	25 19·6	m
277		*Lalande* 12155.			
Feb. 14	6·9	6 16 8·51	5	77 22 3·7	m
Nov. 14	7·3	16 8·73	...	22 3·6	m
278		*Taylor* 2474.			
Jan. 29	6·9	6 18 16·26	...	121 45 25·3	n
Feb. 8	6·5	18 16·01	...	45 27·3	m
279		*Lacaille* 2290.			
Jan. 11	7·9	6 16 49·51	6	133 45 51·0	m
280		*Anon.*			
Jan. 19	8·7	6 19 51·11	5	68 29 39·9	n

Separate Results of Madras Meridian Circle Observations in 1867.

Number and Date.	Magnitude.	Mean Right Ascension 1867. h. m. s.	No. of Wires.	Mean Polar Distance 1867. ° ′ ″	Observer.
281		*α Argûs, Canopus.*			
Feb. 28	...	6 21 0·05	...	142 37 26·1	R
Dec. 16	...	21 0·04	...	37 26·8	R
282		*Anon.*			
Jan. 5	9·0	6 22 2·33	...	129 36 37·1	M
283		*Lacaille* 2321.			
Dec. 11	7·0	6 23 30·68	5	158 20 56·6	M
284		*Taylor* 2524.			
Feb. 4	7·7	6 23 32·98	...	131 8 10·7	M
285		*Taylor* 2529.			
Feb. 7	7·0	6 24 27·02	...	130 59 24·7	M
11	6·9	24 27·12	...	59 23·6	M
286		*Taylor* 2541.			
Feb. 13	6·0	6 24 58·20	...	147 55 6·7	M
287		*Anon.*			
Jan. 21	9·2	6 27 31·43	...	128 46 6·1	R
288		*Anon.*			
Dec. 16	9·5	6 27 51·58	4	131 5 26·5	R
289		*Anon.*			
Jan. 29	6·3	6 28 29·28	...	122 7 37·9	R
Feb. 2	6·9	28 29·16	...	7 37·5	M
290		*Anon.*			
Feb. 8	8·0	6 28 41·14	4	151 10 10·6	M
291		*Taylor* 2589.			
Feb. 12	6·9	6 29 51·58	5	151 44 57·0	M
292		24 *Geminorum γ*			
Jan. 24		6 30 1·61	...	73 29 25·1	R
Feb. 1		30 1·73	...	29 23·7	M
6	...	30 1·65	...	29 25·5	M
14	...	30 1·72	..	29 26·6	M
15	...	30 1·73	.	29 25·3	R
16	...	30 1·81	5	29 24·9	R
293		*Anon.*			
Mar. 2	8·9	6 30 49·10	...	130 55 23·0	M
294		*Anon.*			
Jan. 29	9·5	6 31 7·71	...	122 6 52·2	R
295		*Anon.*			
Jan. 3	8·0	6 31 32·36	...	130 56 32·4	M
296		*R Monocerotis Var.* 1.			
Feb 23	10·3	6 31 54·16	5	81 8 36·5	R
297		*Anon.*			
Jan. 21	8·9	6 32 59·67	...	130 54 35·9	R
298		*Anon.*			
Jan. 5	7·9	6 34 36·48	5	130 29 6·4	M
299		27 *Geminorum ε*			
Feb. 6	..	6 35 45·99	.	64 44 27·3	M
7	.	35 44·98		44 26·6	M
9		35 44·97	...	44 27·4	M
11		35 44·98	.	44 27·1	M
12		35 44·99	4	44 28·3	M
13		35 44·94	5	44 27·8	M
14	...	35 44·94		44 27·6	M
15		35 44·96		44 27·0	M
300		*Anon.*			
Jan. 11	8·0	6 35 53·12	...	130 36 13·7	M

Separate Results of Madras Meridian Circle Observations in 1867.

Number and Date.	Magnitude	Mean Right Ascension 1867. h m s	No. of Wires.	Mean Polar Distance 1867. ° ′ ″	Observer
801		*51 Cephei Hev.*			
Jan 7		6 37 10·19	3	2 45 27·6	M
14	...	37 10·30	3	45 29·0	R
23		37 10·14	3	45 28·2	R
Feb 22		37 10·21	3	45 27·5	R
27		37 9·91	2	45 27·3	R
Mar. 1		37 10·26	3	45 29·0	M
7	...	37 8·69	1	45 27·1	M
		51 Cephei Hev.—s.p.			
July 4		6 37 9·98	3	2 46 29·4	R
Aug. 6		37 10·00	2	45 28·9	M
27		37 8·90	2	46 30·5	R
802		*31 Geminorum* ξ			
Nov. 14		6 37 49·35	...	76 57 30·6	M
803		*Lacaille* 2451.			
Jan. 19	8·2	6 38 11·39		136 57 33·7	R
804		9 *Canis Majoris* α, *Sirius.*			
Jan 12		6 39 17·04		106 32 9·3	M
24		39 17·09	5	32 11·7	R
805		*Anon.*			
Jan. 25	9·0	6 40 30·92	3	151 13 40·2	R
806		*Anon.*			
Jan. 26	8·5	6 42 29·29	4	130 37 8·5	R
Mar. 6	8·6	42 29·47	...	37 9·6	M
807		*Anon.*			
Jan. 26	9·3	6 44 48·10	5	100 39 16·2	R
808		*Anon.*			
Feb. 24	7·7	6 45 47·07	5	104 30 36·6	R
Mar 5	7·7	45 47·01		30 35·3	M
8	7·4	45 46·96		30 36·4	M
809		*Anon.*			
Jan. 24	9·5	6 45 5·93	...	106 23 34·5	R
810		*Lalande* 13279.			
Jan. 16	...	6 46 36·21	...	65 36 21·6	R
19	7·7	46 36·09	4	35 37·2	R
21	8·0	46 36·16	...	35 30·2	R
811		*Lalande* 13313.			
Jan. 3	8·0	6 47 30·90	..	65 40 38·8	M
Feb. 22	8·9	47 31·04	...	40 37·1	R
Dec. 17	8·5	47 31·10	5	40 36·3	R
812		*Anon.*			
Feb. 28	8·2	6 48 34·87	..	125 49 39·5	R
813		30 *Geminorum.*			
Jan. 25		6 50 35·28	..	[illegible] 44 51·4	R
28	...	50 35·54	.	44 52·2	R
29	...	50 35·35	5	44 52·0	R
Feb 1	...	50 35·04	3	44 51·5	M
2	...	50 35·54	..	44 53·6	M
814		*Lacaille* 2588.			
Jan. 11	6·4	6 51 25·22	4	114 47 46·7	M
Mar. 7	7·7	51 25·30	...	47 47·3	M
815		*Anon.*			
Jan. 21	9·5	6 52 49·96	5	100 51 32·7	R
816		21 *Canis Majoris* ε			
Jan. 24	.	6 53 23·94	..	118 47 35·2	R
26		53 23·91		47 35·7	R
Feb. 16	.	53 23·95	..	47 35·3	R
Mar. 8		53 23·92		47 36·0	M
817		*Taylor* 2805.			
Jan 12	8·0	6 54 11·99	6	49 12 22·6	M

Separate Results of Madras Meridian Circle Observations in 1867.

Number and Date.	Magnitude.	Mean Right Ascension 1867. h.	m.	s.	No. of Wires.	Mean Polar Distance 1867. °	′	″	Observer.
818	43 *Geminorum* 3[a] *Var.* 1.								
Jan. 19		6	56	13·10	...	69	14	17·0	R
Nov. 14	...		56	12·77	...		14	15·0	M
819	*Anon.*								
Jan. 23	9·5	6	56	27·29	5	129	17	33·8	R
820	*Taylor* 2825.								
Jan. 15	9·0	6	56	54·09	...	150	54	54·8	M
821	23 *Canis Majoris* γ								
Jan. 25	...	6	57	44·45	...	106	26	21·5	R
26	...		57	44·52	..		26	21·0	R
Feb. 10	...		57	44·40	...		26	21·0	R
18	...		57	44·39	...		26	19·6	R
Dec. 17	...		57	44·44	...		26	21·0	M
822	*Lalande* 13707.								
Jan. 8	8·1	6	59	17·09	4	67	6	54·1	M
Feb. 23	8·0		59	17·08	...		6	53·8	R
26	8·0		59	16·96	5		6	54·5	R
823	*W. B. N. VI.* 1762.								
Jan. 17	9·5	6	59	36·40	...	70	55	5·6	R
21	9·8		59	36·49	...		55	6·0	M
Mar. 1	9·3		59	36·36	...		55	4·8	M
824	*Taylor* 2840.								
Mar. 2	8·2	6	59	5·30	4	150	57	6·2	M
9	8·6		59	5·38	..		57	5·6	M
825	47 *Geminorum*.								
Jan. 19	...	7	3	7·98	...	62	55	41·6	R
21	...		3	8·00	..		55	40·6	R
23	...		3	8·05	..		55	41·1	R
Feb. 4	...		3	7·79	5		55	42·0	M
5			3	7·96	.		55	40·2	M
6			3	8·00			55	41·1	M
7	..		3	8·04	.		55	40·9	M
Feb. 8	..	7	3	8·02	.	62	55	41·9	M
11			3	8·01	...		55	42·2	M
12	...		3	7·95	4		55	41·8	M
13	...		3	8·10	.		55	42·5	M
14	...		3	7·96	...		55	41·9	M
826	*Anon.*								
Feb. 26	9·5	7	5	4·21	5	168	52	24·4	R
827	*Taylor* 2899.								
Feb. 13	8·9	7	5	38·57	...	130	9	5·6	R
Mar. 12	8·6		5	38·60	6		9	7·0	M
828	*Anon.*								
Mar. 2	8·0	7	8	11·34	4	152	5	19·0	M
829	*Anon.*								
Feb. 11	8·9	7	8	59·66	...	160	17	37·4	M
830	55 *Geminorum* δ								
Jan. 16	...	7	12	10·79		67	46	33·7	R
19			12	10·65	...		46	34·3	R
23			12	10·77	.		46	34·9	R
29	...		12	10·80	..		46	33·3	R
Feb. 7			12	10·77	...		46	33·9	M
8			12	10·56	.		46	32·9	M
13	...		12	10·71			46	33·9	M
14	...		12	10·67	...		46	33·6	M
16			12	10·69	.		46	34·3	R
26	.		12	10·61			46	33·2	R
Mar. 6	..		12	10·75			46	31·5	M
13			12	10·85			46	32·6	M
Dec. 12	...		12	10·75	...		46	34·4	M
17			12	10·68	4		46	33·5	M
831	*Anon*								
Mar. 11	8·9	7	12	37·98		134	59	14·7	M

Separate Results of Madras Meridian Circle Observations in 1867.

Number and Date	Magnitude	Mean Right Ascension 1867. h.	m.	s.	No. of Wires.	Mean Polar Distance 1867. °	′	″	Observer.
832		*57 Geminorum* A							
Jan 7	.	7	15	21·91	...	61	41	50·5	M
8	...		15	21·72	5		41	50·1	M
9	...		15	21·91	...		41	50·0	M
10	...		15	21·93	.		41	50·1	M
11			15	21·79	...		41	51·3	M
12			15	21·80	...		41	49·6	M
15			15	21·79	...		41	50·9	R
23			15	21·83			41	49·8	R
24			15	21·80	...		41	50·0	R
25			15	21·66	..		41	51·0	R
28			15	21·92	...		41	51·1	R
29			15	21·70			41	51·8	R
31			15	21·80	5		41	51·1	R
Feb. 1	...		15	21·84	6		41	50·1	M
2	...		15	21·99			41	49·3	M
833		*Taylor* 3005.							
Mar. 1	8·0	7	15	31·95	...	149	1	14·7	M
834		*Lacaille* 2805.							
Feb. 15	7·8	7	17	22·94	...	133	8	25·1	M
835		*Anon.*							
Mar 5	9·0	7	17	48·66	...	129	46	30·4	M
836		*Anon.*							
Mar. 2	9·0	7	18	10·22	6	129	42	51·3	M
837		*Anon.*							
Jan 21	10·0	7	18	57·24	..	60	15	47·5	R
838		*Lacaille* 2807.							
Mar. 7	7·8	7	19	36·96	...	142	15	44·2	M
839		*Anon.*							
Feb. 28	9·0	7	19	47·63	5	128	6	39·2	R
840		*Anon.*							
Mar. 15	9·5	7	19	51·16	6	102	26	56·8	M
841		*Anon.*							
Mar. 8	7·9	7	26	32·04	...	129	14	27·7	M
842		*68 Geminorum.*							
Dec. 12	...	7	26	0·94	...	73	53	27·0	M
18	..		26	0·66	...		53	25·9	M
843		*66 Geminorum* α², *Castor.*							
Jan. 29	.	7	26	6·61	.	57	40	30·0	R
Feb. 11	...		26	6·66	.		40	23·7	M
13			26	6·56	...		40	24·5	M
14	...		26	6·65	..		40	24·6	M
18	...		26	6·65	..		40	29·9	M
26	.		26	6·66	...		40	23·2	R
27			26	6·65	...		40	24·1	R
Mar. 1			26	6·57			40	26·4	M
2			26	6·67	.		40	23·9	M
6	...		26	6·62	..		40	23·4	M
9			26	6·59			40	27·9	M
Dec. 17	.		26	6·66	...		40	22·3	R
844		*69 Geminorum* υ							
Jan. 16	...	7	27	48·35	...	62	46	48·4	R
17	.		27	48·46			46	46·7	R
18	...		27	48·30	.		46	46·9	R
21	...		27	48·46	...		46	49·1	R
845		*Anon.*							
Mar. 11	7·9	7	30	41·50	..	121	51	30·6	M
846		*Taylor* 3133.							
Jan. 4	6·7	7	31	9·87	.	66	23	46·9	M
5	6·7		31	9·77			23	46·9	M
847		*Anon.*							
Mar 7	6·7	7	31	10·92		131	10	39·2	M

Separate Results of Madras Meridian Circle Observations in 1867.

Number and Date.	Magnitude.	Mean Right Ascension 1867. h. m. s.	No. of Wires.	Mean Polar Distance 1867. ° ′ ″	Observer.
848		74 *Geminorum f*			
Mar. 15	..	7 31 47·76	...	72 0 52·8	M
849		10 *Canis Minoris* α, *Procyon.*			
Jan. 21		7 32 20·27	5	84 26 13·6	R
29	...	32 20·25	...	26 13·0	R
Feb. 6	...	32 20·30	...	26 11·9	M
8	...	32 20·33	...	26 11·9	M
11	...	32 20·35	...	26 13·1	M
13	..	32 20·26	5	26 13·6	M
15	...	32 20·26	4	26 13·6	M
18	...	32 20·20	...	26 13·7	R
25	...	32 20·25	...	26 12·4	R
27	...	32 20·20	...	26 12·3	R
Mar. 1	...	32 20·26	...	26 13·9	M
2	...	32 20·29	...	26 12·4	M
5	...	32 20·26	...	26 12·7	M
6	...	32 20·28	...	26 12·7	M
9	...	32 20·31	...	26 11·9	M
Dec. 17	...	32 20·30	...	26 14·2	R
850		*Anon.*			
Feb. 23	9·9	7 34 14·65	...	66 9 16·3	R
26	9·9	34 14·62	..	9 14·9	R
851		*Anon.*			
Feb. 22	10·0	7 34 52·24	...	68 10 18·2	R
26	9·8	34 52·22	3	10 18·4	R
852		*Anon.*			
Jan. 15	9·0	7 35 12·73	5	129 24 16·0	R
853		*Anon.*			
Mar. 12	8·0	7 36 21·04	3	153 0 1·9	M
854		76 *Geminorum c*			
Jan. 10	6·3	7 35 50·26	.	63 54 5·4	M
11	...	35 50·25	3	54 7·6	M
13	...	35 50·30	4	54 7·3	M
855		*Anon.*			
Jan. 15	8·0	7 36 58·90	.	129 57 41·6	R
856		*Anon.*			
Mar. 13	7·6	7 36 58·85		100 51 17·3	M
857		78 *Geminorum* β, *Pollux.*			
Jan. 21		7 37 10·45	...	61 39 21·9	R
Feb. 1	...	37 10·45	...	39 21·5	M
6	...	37 10·47	5	39 22·2	M
8	...	37 10·54	4	39 22·7	M
18	...	37 10·42	...	39 21·4	R
26	...	37 10·39	...	39 19·7	R
27	...	37 10·31	...	39 20·6	R
Mar. 1	...	37 10·40	...	39 21·5	M
2	...	37 10·40	5	39 20·6	M
5	..	37 10·39	5	39 20·6	M
9		37 10·38	...	39 21·3	M
858		*Anon.*			
Jan. 26	10·3	7 37 32·29	6	65 30 8·0	R
859		*Anon.*			
Mar. 7	8·2	7 37 30·00	3	160 25 31·8	M
860		*Anon.*			
Feb. 7	7·2	7 39 11·51		131 9 2·9	M
Mar. 6	7·6	39 11·50		9 2·9	M
861		*Lacaille* 2971.			
Mar. 11	6·9	7 40 22·52	..	148 56 24·4	M
862		*Anon.*			
Jan. 31	9·3	7 40 13·31	5	132 30 22·5	R
863		*Anon.*			
Feb. 25	9·2	7 44 30·12	...	100 34 27·5	R

Separate Results of Madras Meridian Circle Observations in 1867.

Number and Date	Magnitude	Mean Right Ascension 1867. h	m	s	No. of Wires	Mean Polar Distance 1867. °	′	″	Observer
864 *R. P. L. 49.*									
Mar 6		7	44	40·44	3	6	31	7·8	M
865 83 *Geminorum φ*									
Jan 7	...	7	45	21·48		63	53	36·0	M
8	...		45	21·34			53	36·0	M
9			45	21·30			53	35·2	M
866 *Brisbane* 1791.									
Jan 22	8·0	7	46	22·47	..	144	26	7·1	R
867 *Taylor* 3293.									
Mar. 13	7·7	7	46	35·44	5	141	41	33·6	M
14	6·8		46	36·40	...		41	34·5	M
868 *Anon.*									
Mar 12	8·1	7	47	8·14	4	153	21	15·1	M
869 *U Geminorum Var.* 5.									
Dec. 16	10·2	7	46	13·75	5	67	30	0·8	R
870 *Anon.*									
Mar 16	9·3	7	49	40·96	4	120	55	16·4	R
871 1 *Cancri.*									
Feb. 13		7	49	26·23	5	73	51	26·2	M
872 *Taylor* 3323.									
Mar. 11	7·0	7	49	51·28	5	140	16	9·5	M
873 *W. B. N. VII.* 1473.									
Jan 21		7	53	42·30	...	64	32	51·8	R
24	7·2		53	42·34			32	51·6	R
25	7·0		53	42·30			32	51·8	R
874 *Anon.*									
Mar. 5	8·8	7	56	47·19	...	140	56	19·0	M
875 6 *Cancri.*									
Feb. 6	..	7	55	29·77	...	61	50	9·1	M
8	..		55	29·69	...		50	6·2	M
Mar. 4	..		55	29·73	5		50	5·9	R
876 8 *Cancri.*									
Feb. 16	...	7	57	30·06	...	76	30	30·0	R
877 *Anon.*									
Jan. 22	9·0	7	57	41·28	5	166	26	39·6	R
878 *Anon.*									
Jan. 24	10·2	8	0	19·69	4	73	30	53·3	R
879 *Anon.*									
Mar. 11	8·6	8	1	25·98	...	120	31	53·6	M
12	7·9		1	25·69	...		31	53·3	M
880 15 *Argûs.*									
Feb. 4	...	8	1	30·77	...	113	35	28·5	M
22	...		1	32·30	...		35	28·0	R
23	...		1	30·82	...		35	22·6	R
25	...		1	52·94			35	28·3	R
26	...		1	32·98	...		35	28·4	R
27	...		1	32·84	...		35	28·0	R
Mar. 8	...		1	30·90	...		35	28·6	M
9	.		1	32·87	4		35	28·8	M
13	...		1	30·91	...		35	28·7	M
14			1	32·25	.		35	29·3	M
881 *Anon.*									
Jan. 31	9·2	8	2	14·48	5	120	30	30·8	R
882 14 *Cancri ψ*[1]									
Jan 4		8	2	30·26	—	61	5	30·4	M
5			2	25·90			5	31·9	M

Separate Results of Madras Meridian Circle Observations in 1867.

Number and Date.	Magnitude.	Mean Right Ascension 1867. h.	m.	s.	No. of Wires.	Mean Polar Distance 1867. °	′	″	Observer.
383 *Anon.*									
Mar. 10	9·6	8	4	31·68	...	134	41	5·1	R
384 16 *Cancri* 3									
Jan. 19	...	8	4	34·86	...	71	57	13·9	R
385 *Lacaille* 3200.									
Jan. 22	6·5	8	4	49·94	5	153	7	56·9	R
Feb. 11	6·9		4	50·29	...		7	57·1	M
Mar. 2	6·9		4	50·30	5		7	58·3	M
386 *Anon.*									
Mar. 1	9·4	8	5	36·41	3	77	26	31·2	M
387 *Anon.*									
Mar. 18	9·5	8	5	48·67	...	126	30	18·8	R
388 *Anon.*									
Feb. 26	8·0	8	6	10·35	...	128	41	3·6	R
389 *Anon.*									
Jan. 31	9·5	8	6	12·24	3	128	40	14·6	R
390 *Anon.*									
Feb. 25	9·2	8	8	40·56	.	128	38	56·0	R
391 *W. B. E. VIII.* 220.									
Mar. 8	8·0	8	9	37·08		99	22	16·7	M
9	8·1		9	36·95	6		22	16·0	M
11	8·0		9	36·97	...		22	16·0	M
392 *Anon.*									
Jan. 21	9·7	8	10	9·07		150	47	12·9	R
393 *Anon.*									
Mar. 14	9·2	8	10	29·29	5	151	26	51·2	M
394 *Anon.*									
Mar. 12	8·0	8	11	58·90	..	182	1	51·5	M
395 *Anon.*									
Feb. 26	9·5	8	12	1·98	...	131	48	22·6	R
396 *Anon.*									
Mar. 15	8·0	8	12	26·81	5	128	41	13·0	M
397 *Anon.*									
Jan. 26	9·5	8	13	19·98	5	131	48	28·3	R
Mar. 13	9·3		13	19·50	...		48	29·5	M
16	9·0		13	19·76	5		48	30·1	R
19	9·6		13	19·85	5		48	29·1	R
398 *W. B. E. VII.* 383.									
Mar. 4	7·8	8	16	35·46	...	100	19	37·1	R
5	7·6		15	35·46	...		19	35·1	M
6	...		15	35·36	...		19	37·9	M
7	7·8		15	35·55	..		19	35·6	M
399 *Anon.*									
Feb. 25	9·5	8	16	35·[illegible]	...	77	52	30·5	R
400 *Anon.*									
Mar. 2	8·7	8	17	36·62	...	77	40	42·9	M
401 *Anon.*									
Jan. 21	9·3	8	19	1·12	5	134	28	16·6	R
402 *Taylor* 3599.									
Mar. 8	6·9	8	20	21·[illegible]	5	161	53	30·9	W
13	7·0		20	21·83			53	30·7	W
14	6·9		20	21·32			53	31·3	W
403 *Taylor* 3607.									
Mar. 11	6·8	8	21	19·61	5	161	56	37·8	M

Separate Results of Madras Meridian Circle Observations in 1867.

Number and Date.	Magnitude.	Mean Right Ascension 1867. h.	m.	s.	No. of Wires.	Mean Polar Distance 1867. °	′	″	Observer.
404			*Anon.*						
Mar. 9	8·0	8	21	36·53	5	131	42	10·0	M
405			*Anon.*						
Mar. 12	9·3	8	21	41·90	5	138	18	6·4	M
406			*Anon.*						
Mar. 14	7·1	8	23	35·80	...	144	56	42·9	M
407			33 *Cancri* η						
Jan. 19		8	25	0·89	...	69	6	36·7	M
21			25	0·91	.		6	34·1	M
Feb. 22	.		25	0·87	...		6	34·9	M
28	...		25	0·88			6	35·9	M
Mar. 1	...		25	0·91			6	36·2	M
4	.		25	0·80	...		6	33·1	M
7	...		25	0·83	.		6	34·9	M
13	...		25	0·81			6	34·6	M
15	...		25	0·74	3		6	34·7	M
Dec. 13	...		25	0·95			6	34·1	M
408			*Taylor* 3651.						
Feb. 14	6·7	8	23	44·17	5	130	8	50·7	M
409			*Taylor* 3652.						
Feb. 16	7·7	8	25	50·31	5	130	3	15·7	M
410			*Anon.*						
Feb. 8		8	28	11·26	..	165	0	31·9	M
411			*Anon.*						
Mar. 6	9·1	8	28	49·96	5	73	19	48·9	M
412			*Lalande* 16890.						
Mar. 11	7·9	8	29	51·74		78	16	29·5	M
413			*Anon.*						
Mar. 16	9·3	8	29	57·72	—	70	41	16·7	M
414			*W. B. N. VIII.* 690.						
Jan. 22	9·0	8	30	41·42		76	40	16·6	M
415			*Anon.*						
Mar. 19	9·0	8	31	19·17	...	139	46	2·1	M
416			*Taylor* 3710.						
Mar. 2	7·1	8	31	29·56	—	141	21	41·5	M
417			*Anon.*						
Mar. 8	6·4	8	34	11·25	4	134	21	4·6	M
418			*Lacaille* 3475.						
Apl. 4	4·9	8	34	57·13	5	132	28	16·6	M
419			43 *Cancri* γ						
Dec. 18	...	8	35	35·01	...	68	3	29·7	M
420			*Lacaille* 3491.						
Mar. 12	6·6	8	36	4·62	..	136	32	30·5	M
18	7·0		36	4·66	—		32	29·6	M
421			*b Velorum.*						
Mar. 9	6·9	8	36	12·90		136	10	30·4	M
14	6·9		36	12·95	...		10	29·2	M
422			47 *Cancri* δ						
Feb. 16		8	37	7·65	6	71	31	58·5	M
Mar. 18	.		37	7·56			31	54·5	M

Separate Results of Madras Meridian Circle Observations in 1867.

Number and Date.	Magnitude.	Mean Right Ascension 1867. h. m. s.	No. of Wires.	Mean Polar Distance 1867. ° ′ ″	Observer.
423		*11 Hydrae* ε			
Jan. 22	...	8 39 43·84	...	83 5 44·6	R
Feb. 7	...	39 43·88	.	5 42·7	M
22	...	39 43·80	...	5 45·7	R
25	...	39 43·82	...	5 45·5	R
26	...	39 43·91	...	5 43·4	R
Mar. 7	...	39 43·83	.	5 46·7	M
11	...	39 43·91	...	5 44·3	M
19	...	39 43·80	...	5 48·8	R
25	...	39 43·76	...	5 44·0	R
424		*Anon.*			
Mar. 6	7·7	8 40 36·21	..	129 16 13·8	M
425		*Anon.*			
Feb. 10	8·7	8 45 56·15	...	86 27 51·3	R
Mar. 8	8·0	45 56·07	...	27 53·0	M
426		*Anon.*			
Mar. 13	7·8	8 46 50·52	...	132 38 58·3	M
427		*Anon.*			
Mar. 11	8·0	8 46 59·91	4	136 2 36·7	M
20	8·0	46 59·84	5	2 35·0	R
30	7·8	46 59·96	5	2 31·8	R
428		*R. P. L. 60.*			
Feb. 11	.	8 47 3·99	3	5 17 34·5	M
Mar. 12	..	47 4·74	3	17 34·4	M
		R. P. L. 60.—s.p.			
Sep. 6		8 47 4·27	3	5 17 37·6	M
429		*Anon.*			
Mar. 6	7·0	8 48 51·38		133 1 45·0	R
430		*Anon.*			
Jan. 22	8·9	8 49 29·08	..	133 40 40·3	R
431		*Anon.*			
Jan. 22	8·8	8 50 21·97	5	132 57 36·6	R
432		*Anon.*			
Feb. 26	8·5	8 52 22·42	5	77 33 12·6	R
Mar. 15	8·0	52 22·39		33 9·8	M
433		*Anon.*			
Mar. 11	8·0	8 54 8·10	5	132 51 22·9	M
434		*Anon.*			
Mar. 7	8·5	8 54 58·47		130 36 11·6	M
435		*Anon.*			
Mar. 12	7·7	8 55 14·26	5	146 57 23·4	M
Apl. 11	7·9	55 11·16	4	57 22·0	M
436		*Anon.*			
Apl. 2	7·9	8 56 46·86		148 36 23·4	M
437		*Anon.*			
Apl. 3	8·9	8 59 11·40	5	146 19 6·1	M
438		*76 Cancri* κ			
Apl. 13	.	9 0 22·61		78 47 54·5	M
439		*Anon.*			
Mar. 1	8·8	9 4 31·47	5	132 46 11·6	M
440		*Lacaille* 3713.			
Mar. 8	7·0	9 4 40·62		143 49 57·1	M
441		*Anon.*			
Feb. 28	9·5	9 5 14·87	—	124 40 1·9	R
Mar. 29	9·0	5 14·72	...	40 2·9	R
30	9·3	5 14·68		40 2·2	R

Separate Results of Madras Meridian Circle Observations in 1867.

Number and Date	Magnitude	Mean Right Ascension 1867. h	m.	s.	No. of Wires	Mean Polar Distance 1867. °	′	″	Observer.
442			*Taylor* 4026.						
Mar. 11	7·0	9	6	1·63	3	132	48	27·0	M
26	7·3		6	1·69			48	27·4	R
Apl. 10	7·3		6	1·58	...		48	26·0	M
443			*Taylor* 4028.						
Mar. 2	7·8	9	6	11·70	...	132	48	39·8	M
13	7·7		6	11·80	...		48	39·9	M
26	8·5		6	11·78	...		48	40·1	R
444			*Anon.*						
Mar 30	10·0	9	7	30·90	...	134	46	57·5	R
445			*Lacaille* 3747.						
Apl. 1	7·5	9	7	40·98	5	150	34	33·0	M
2	7·8		7	40·99	3		34	34·0	M
4	7·8		7	40·90	5		34	35·1	M
5	8·0		7	40·95	6		34	31·9	M
15	8·6		7	40·90	3		34	32·7	M
446			*Anon.*						
Apl. 3	8·7	9	9	32·46		150	32	31·1	M
447			*Lacaille* 3761.						
Apl. 1	7·5	9	9	34·83	3	150	22	10·1	M
15	7·0		9	34·65	5		22	9·6	M
448			*Anon.*						
Apl. 11	9·6	9	10	16·82	5	150	26	7·1	M
449			*Anon.*						
Apl. 3	7·9	9	10	49·08		150	21	30·4	M
4	7·9		10	49·00	5		21	36·1	M
5	7·9		10	49·10	4		21	32·6	M
450			*Anon.*						
Feb. 22	9·3	9	11	6·85	5	70	41	32·8	R
25	9·4		11	4·88	6		41	31·3	R

Number and Date.	Magnitude	Mean Right Ascension 1867. h.	m.	s.	No. of Wires	Mean Polar Distance 1867. °	′	″	Observer.
451			83 *Cancri*.						
Mar. 11	...	9	11	33·33	...	71	48	56·7	M
12	..		11	38·31			48	50·2	M
14	...		11	39·95			48	56·7	M
25	...		11	39·94	...		48	56·4	R
452			*Anon.*						
Feb. 23	9·2	9	12	27·80	...	134	46	16·0	R
Mar. 15	8·0		12	27·85			46	18·1	M
29	8·9		12	27·60	4		46	16·6	R
453			*Anon.*						
Feb. 20	10·0	9	13	6·16	4	70	32	36·9	R
26	10·2		13	6·10	5		32	36·1	R
454			ι *Argûs*.						
Feb. 27		9	13	52·01	5	148	48	7·1	R
455			*Anon.*						
Apl. 3	8·7	9	15	9·21	..	150	29	29·8	M
6	8·0		15	9·12	..		29	30·6	M
456			*Anon.*						
Apl. 2	8·7	9	15	21·09	5	150	25	5·4	M
4	9·0		15	20·95			25	3·3	M
5	8·1		15	20·87			25	3·3	M
457			*Anon.*						
Mar. 5	7·8	9	15	21·40		141	39	27·4	M
458			*Anon.*						
Feb. 24	8·0	9	16	46·04		134	47	39·6	R
459			*Anon.*						
Feb. 16	9·0	9	16	30·03	5	70	27	24·2	R
28	9·6		16	30·09	5		27	27·2	R

Separate Results of Madras Meridian Circle Observations in 1867.

Number and Date.	Magnitude.	Mean Right Ascension 1867. h.	m.	s.	No. of Wires.	Mean Polar Distance 1867. °	′	″	Observer.
460			*Anon.*						
Mar. 24	9.7	9	16	38.06	...	124	46	18.3	R
461			*Anon.*						
Feb. 11	8.0	9	19	6.07	...	70	22	4.9	M
462			*Anon.*						
Apl. 3	8.5	9	19	25.22	...	150	31	25.8	M
6	8.6		19	25.11	6		31	25.7	M
463			*Anon.*						
Feb. 28	9.0	9	19	51.06	5	128	21	57.8	R
Mar. 6	8.8		19	51.26	...		21	58.3	M
464			*Anon.*						
Feb. 23	8.8	9	20	34.82	5	125	23	42.9	R
465			30 *Hydrae* α, *Var* 2.						
Mar. 7		9	21	3.04	...	98	5	2.0	M
12	...		21	2.94	...		5	3.4	M
13	.		21	3.06	...		5	2.6	M
14			21	3.01	...		5	2.1	M
15			21	3.08	...		5	2.3	M
25			21	3.08	...		5	2.7	R
26	...		21	3.00	...		5	2.5	R
466			*Lalande* 18636.						
Feb. 19	9.0	9	21	51.60	...	68	30	25.3	R
Apl. 16	8.8		21	51.44	6		30	24.5	M
467			*Lalande* 18659.						
Mar. 29	9.3	9	22	47.79	5	67	50	13.6	R
30	9.0		22	47.90	...		50	14.5	R
468			*Lalande* 18683.						
Feb. 24	9.5	9	23	27.92		68	7	36.7	R
Mar. 29	9.2		23	27.75			7	37.9	R
469			6 *Leonis h*						
Jan. 21	..	9	24	49.72		79	41	39.4	R
Mar. 16	..		24	49.67	..		41	38.7	R
470			*Lalande* 18730.						
Feb. 27	9.2	9	25	2.55	...	68	38	37.6	R
Mar. 5	8.7		25	2.64	..		38	37.2	M
Apl. 8	9.3		25	2.61	..		38	37.4	M
471			*Anon.*						
Apl. 5	8.8	9	26	46.02		145	3	14.7	M
472			*Taylor* 4222.						
Feb. 19	7.5	9	27	49.98	...	164	24	7.6	R
23	8.0		27	50.21	5		24	8.3	R
Mar. 6	7.1		27	50.43	...		24	8.9	M
473			*Anon.*						
Feb. 13	8.0	9	28	21.93	4	146	32	46.3	R
Mar 19	9.6		28	22.19	...		32	46.7	R
474			*Anon.*						
Feb. 28	9.0	9	31	23.90	...	126	28	34.3	R
Mar. 14	8.6		31	23.90	...		28	[illegible]	M
30	9.5		31	[illegible]	...		28	[illegible]	R
475			*Anon.*						
Mar. 2	8.4	9	32	19.61	...	120	40	30.3	M
23	9.0		32	19.75	5		40	30.6	R
Apl. 1	8.1		32	19.69	5		40	30.2	M
476			*R. P. L.* 69.						
Mar. 13		9	33	50.63	2	2	47	31.6	R
Apl. 10			33	50.61	1		47	31.5	R
477			14 *Leonis* o						
Mar. 16	...	9	34	3.05		79	30	15.0	R

Separate Results of Madras Meridian Circle Observations in 1867.

Number and Date.	Magnitude.	Mean Right Ascension 1867. h. m. s.	No. of Wires.	Mean Polar Distance 1867. ° ′ ″	Observer.
478		*Lacaille* 3980.			
Apl. 6	7·0	9 34 36·71		148 31 26·0	M
479		*Anon.*			
Apl. 2	9·0	9 35 36·60	...	149 30 24·2	M
480		*Anon.*			
Feb. 27	8·0	9 35 55·04	...	149 40 57·6	R
Apl. 3	7·8	36 54·36	...	40 59·6	M
11	7·4	35 54·86	...	40 58·6	M
481		*Anon.*			
Apl. 13	8·6	9 36 53·69	5	151 51 28·9	M
482		*Anon.*			
Apl. 5	6·9	9 37 0·22	...	148 26 5·0	M
483		*Anon.*			
Apl. 4	7·9	9 37 10·56	5	149 32 5·8	M
6	8·0	37 10·43	5	32 6·3	M
484		*R Leonis Minoris Var.* 1.			
Feb. 13	7·0	9 37 36·29	...	54 52 44·9	R
Mar. 1	7·7	37 36·61	...	52 44·1	M
6	7·3	37 36·43	...	52 42·6	M
485		17 *Leonis* ε			
Feb. 16		9 38 17·62	...	65 36 55·4	R
22	...	38 17·70	...	36 54·7	R
24	.	38 17·73	...	36 55·8	R
Mar. 5	...	38 17·87	5	36 51·7	M
7	.	38 17·66	.	36 56·2	M
8		38 17·80	...	36 55·4	M
19		38 17·74	...	36 56·9	R
26		38 17·88	5	36 54·9	R
28		38 17·92		36 54·8	R
29	...	38 17·66		36 55·0	R
30		38 17·53		36 55·8	R
Apl. 12	...	38 17·96	5	36 56·9	M
486		*Anon.*			
Apl. 6	7·9	9 39 44·70	5	149 34 33·2	M
487		*l Carinae Var.* 1.			
Apl. 1	5·7	9 41 26·60	5	151 58 45·8	M
488		*Anon.*			
Feb. 19	8·0	9 41 36·77	5	130 46 40·3	R
489		*Taylor* 4337.			
Apl. 5	7·0	9 41 56·04	...	149 26 5·4	M
490		*Anon.*			
Apl. 10	8·0	9 42 49·48	5	130 48 29·4	M
491		*Anon.*			
Apl. 8	9·5	9 44 15·40	...	149 30 49·0	M
492		*Anon.*			
Feb. 28	9·2	9 46 4·65	5	129 48 10·8	R
Mar. 29	9·5	46 4·87	...	48 11·6	R
493		*R. P. L.* 70.			
Feb. 23	...	9 46 49·77	3	5 26 38·9	R
27	...	46 51·86	3	26 38·1	R
Mar. 26	...	46 50·81	3	26 37·3	R
494		*Taylor* 4381.			
Feb. 26	6·5	9 47 11·40	5	138 7 24·1	R
Mar. 8	6·7	47 11·29	5	7 23·8	M
13	6·8	47 11·28		7 24·2	M
495		*Anon.*			
Feb. 28	10·3	9 47 16·53	5	73 31 28·4	R
Mar. 29	..	47 16·60		31 20·4	R
496		*Anon.*			
Feb. 28	9·0	9 52 40·79		129 41 39·1	R
Apl. 4	8·6	52 40·57	...	41 32·0	M
15	9·3	52 49·80		41 34·4	M

Separate Results of Madras Meridian Circle Observations in 1867.

Number and Date.	Magnitude.	Mean Right Ascension 1867. h.	m.	s.	No. of Wires.	Mean Polar Distance 1867. °	'	"	Observer.
497			*29 Leonis π*						
Feb. 18	...	9	53	11·09	...	81	19	9·1	R
19			53	10·90	...		19	9·5	R
Mar. 2			53	10·95	...		19	10·1	M
5	...		53	10·96	5		19	8·9	M
6			53	10·98	...		19	8·5	M
11			53	11·02	...		19	8·1	M
12	...		53	10·96	...		19	9·8	M
28	...		53	11·02	...		19	8·2	R
29			53	10·97	...		19	9·2	R
Apl. 2	...		53	11·03	...		19	8·9	M
3	...		53	10·88	5		19	9·1	M
8			53	11·01	...		19	8·6	M
11	...		53	11·01	5		19	9·3	M
13	..		53	11·05	...		19	7·8	M
498			*Anon.*						
Feb 28	9·6	9	54	54·39	5	126	18	14·0	R
499			*Anon.*						
Apl. 1	8·0	9	55	57·49	5	147	25	7·3	M
500			*Anon.*						
Mar. 16	9·0	9	57	54·23	...	146	38	51·6	R
Apl. 6	8·6		57	54·12	...		38	55·1	M
501			*Taylor 4476.*						
Feb. 28	8·0	9	57	57·80	5	146	36	55·7	R
502			*Anon.*						
Mar. 14	7·9	9	59	31·74	...	160	39	51·6	M
503			*Taylor 4484.*						
Apl. 12	7·1	9	58	46·06	...	151	30	53·2	M
504			*Anon.*						
Mar. 13	9·4	10	0	8·73	5	98	23	11·4	M
Apl. 6	9·5		0	8·66	.		23	7·7	M
16	9·7		0	8·68			23	8·6	R
505			*32 Leonis α, Regulus.*						
Jan. 21	...	10	1	17·14	4	77	23	3·0	R
22	...		1	17·16	...		23	4·8	R
Feb. 28	...		1	17·17	...		23	3·2	R
Mar. 8	...		1	17·14	...		23	2·6	M
11	...		1	17·07	..		23	3·1	M
12	...		1	17·14	...		23	3·0	M
15	...		1	17·23	..		23	3·1	M
18	...		1	17·16	...		23	2·9	R
26			1	17·14			23	3·0	R
28	.		1	17·13	..		23	2·8	R
Apl. 11	...		1	17·07	...		23	1·9	M
13	..		1	17·10	.		23	1·4	M
506			*Anon.*						
Apl. 2	8·0	10	1	28·32		130	2	31·7	M
507			*Anon.*						
Apl. 4	8·1	10	2	9·03	6	129	57	19·4	M
508			*Anon.*						
Feb. 28	9·0	10	2	27·90		128	28	16·1	R
Mar. 20	9·6		2	27·89	5		28	16·4	R
509			*Anon.*						
Feb. 28	8·5	10	4	38·60	..	122	54	30·9	R
510			*Anon.*						
Feb. 28	9·0	10	4	41·59	5	123	29	42·4	R
Mar. 19	9·5		4	41·58			29	42·3	R
29	9·3		4	41·65			29	40·3	R
511			*Taylor 4532.*						
Apl. 1	7·0	10	7	8·66	5	167	34	17·9	M
512			*Anon.*						
Apl. 10	7·9	10	8	0·94	5	169	45	11·3	M
13	8·0		8	[illegible]			45	11·1	M

Separate Results of Madras Meridian Circle Observations in 1867.

Number and Date.	Magnitude.	Mean Right Ascension 1867. h. m. s.	No. of Wires.	Mean Polar Distance 1867. ° ′ ″	Observer.
513		*Anon.*			
Apl. 3	8·7	10 0 8·80	5	139 52 35·6	M
514		*R. P. L.* 72.			
Mar. 23		10 9 50·41	2	5 4 31·0	R
		R. P. L. 72—*s.p.*			
Sep. 24	...	10 9 50·78	3	5 4 33·2	R
Oct. 8	...	9 50·83	1	4 32·3	M
515		*Anon.*			
Feb. 26	8·0	10 9 58·15	...	145 26 16·2	R
516		*Anon.*			
Mar. 6	8·0	10 10 23·27	5	139 52 6·3	M
16	9·5	10 23·22	5	52 5·9	R
Apl. 5	8·9	10 23·26	6	52 5·2	M
517		41 *Leonis* γ¹			
Mar. 13	...	10 12 38·14	5	69 29 13·9	M
14	...	12 38·07	...	29 13·7	M
Apl. 2	...	12 38·13	...	29 14·4	M
4	...	12 38·16	...	29 14·1	M
8	..	12 38·12	...	29 13·7	M
11	...	12 38·22	...	29 13·4	M
13	..	12 38·12	5	29 14·3	M
518		*Anon.*			
Apl. 16	9·5	10 12 50·06	5	128 57 51·6	R
519		*Anon.*			
Feb. 26	8·0	10 16 6·31	5	125 33 18·1	R
Mar. 13	7·9	16 6·76	...	33 15·0	M
29	8·2	16 6·72	..	33 13·5	R
520		*Lalande* 20139.			
Feb. 16	8·8	10 16 17·85	...	76 25 24·9	R
521		*Anon.*			
Feb. 25	9·7	10 19 52·01	5	146 0 21·0	R
522		*Anon.*			
Mar. 16	9·7	10 19 30·33	...	146 9 38·9	R
523		*Anon.*			
Feb. 27	9·5	10 21 25·02	5	126 34 7·7	R
Mar. 19	9·5	21 25·00	5	34 11·6	R
524		*Anon.*			
Feb. 20	7·2	10 22 17·71	...	125 32 29·5	R
Mar. 19	7·0	22 17·60	5	32 29·6	R
29	7·9	22 17·68	...	32 29·4	R
525		*Anon.*			
Apl. 2	8·0	10 24 12·36	5	146 56 1·5	M
526		*Anon.*			
Apl. 3	8·4	10 24 32·14	...	140 50 50·6	M
527		*Anon.*			
Feb. 27	8·0	10 24 58·47	..	125 34 48·3	R
Apl. 13	8·3	24 53·40	..	34 46·7	M
528		*Lalande* 20402.			
Apl. 4	7·3	10 25 8·24	...	79 58 41·6	M
6	7·6	25 8·14	...	58 46·2	M
17	8·0	25 8·16	...	58 40·9	R
529		47 *Leonis* ρ			
Jan. 21	...	10 25 48·27	...	80 0 37·2	R
22	...	25 49·41	...	0 37·8	R
Feb. 18	...	25 48·41	...	0 37·0	R
19	...	25 48·41	...	0 38·6	R
26	..	25 49·47	...	0 37·3	R
Mar. 16	...	25 49·37	...	0 36·2	R
18	...	25 49·40	...	0 31·7	R
20	..	25 48·34	...	0 37·8	R
Apl. 5	...	25 48·34	...	0 36·3	M
8	...	25 49·06	.	0 36·0	M
11	..	25 49·33	...	0 35·9	R
12	...	25 49·42	...	0 38·6	M

Separate Results of Madras Meridian Circle Observations in 1867.

Number and Date.	Magnitude.	Mean Right Ascension 1867. h.	m.	s.	No. of Wires.	Mean Polar Distance 1867. °	′	″	Observer.
530	*Taylor* 4769.								
Apl. 15	5·7	10	30	29·30	...	140	52	12·9	R
531	*Anon.*								
Feb. 20	9·0	10	31	10·48	...	151	10	36·9	R
Apl. 1	8·6		31	10·40	5		10	34·5	M
532	*R Ursae Majoris Var* 1.								
Apl. 4	7·7	10	35	11·60	3	20	31	37·7	M
10	7·3		35	11·75	5		31	38·2	M
11	7·6		35	11·36	4		31	38·2	R
533	*Anon.*								
Mar. 16	0·3	10	36	29·86	5	137	29	28·7	R
534	*Anon.*								
Apl. 2	8·7	10	36	48·04	...	140	54	26·7	M
13	8·0		36	47·87	5		54	25·6	M
15	8·0		36	47·89	5		54	26·2	R
535	*Taylor* 4850—1*st.*								
Apl. 26	9·0	10	38	40·16	3	146	51	7·8	R
536	*Taylor* 4850—2*nd.*								
Apl. 26	7·7	10	38	40·27	4	146	51	3·0	R
537	*Taylor* 4852—1*st.*								
Mar. 15	8·6	10	38	59·00	...	148	53	42·6	M
538	*Taylor* 4852—2*nd.*								
Apl. 3	8·9	10	39	0·28	...	148	52	46·0	M
539	*Brisbane* 3194.								
Apl. 5	8·3	10	39	24·40	...	140	2	41·0	M
540	*Anon.*								
Apl. 25	9·0	10	39	31·78	5	139	3	7·0	R
541	*η Argûs Var.* 1.								
Mar. 11		10	39	51·47	...	148	59	11·0	R
18	...		39	51·42	...		59	11·3	M
542	*Taylor* 4872.								
Apl. 17	7·7	10	41	11·48	5	151	14	31·8	R
543	53 *Leonis l.*								
Feb. 18	...	10	42	15·87	5	79	43	8·3	R
19	...		42	15·91	...		43	7·6	R
20	...		42	15·86	...		43	9·1	R
Mar. 16	...		42	15·89	..		43	7·9	R
19	..		42	15·91	...		43	8·4	R
26	.		42	15·82	...		43	7·6	R
Apl. 2	...		42	15·84	...		43	7·5	M
6			42	15·92	5		43	7·5	M
27	...		42	15·82	..		43	6·9	R
544	*Anon.*								
Mar. 15	9·5	10	42	31·99	5	142	52	17·0	R
545	*Anon.*								
Feb. 21	9·7	10	43	33·13	...	76	5	49·6	R
546	*Taylor* 4896.								
Mar. 9	6·9	10	43	56·34	...	137	2	57·7	R
547	*Lacaille* 4602.								
Jan. 22	7·0	10	43	57·75		111	3	31·8	R
548	*Anon.*								
Apl. 23	9·2	10	44	8·88	—	147	48	1·6	R
549	*Anon.*								
Apl. 12	8·2	10	44	23·98	6	141	45	49·4	M
17	9·0		44	23·35	5		45	47·9	R

Separate Results of Madras Meridian Circle Observations in 1867.

Number and Date.	Magnitude.	Mean Right Ascension 1867. h.	m.	s.	No. of Wires.	Mean Polar Distance 1867. °	′	″	Observer.
550			*Anon.*						
Mar. 13	8.6	10	43	38.91	6	148	52	33.5	M
Apl. 5	8.7		43	34.03	5		52	33.0	M
551			*Anon.*						
Mar. 16	...	10	50	29.34	5	148	40	2.6	R
552			7 *Crateris* α						
Feb. 21		10	53	17.79	...	107	35	29.5	R
553			58 *Leonis* d						
Apl. 16	...	10	58	41.44	...	86	40	0.7	M
554			*R Crateris Var.* 1.						
Feb. 20	8.0	10	54	0.94	5	107	36	44.5	R
Mar. 30	9.0		54	1.16	...		36	43.7	R
Apl. 13	9.0		54	0.90	...		36	42.7	M
26	9.0		54	1.00	...		36	48.2	R
May 1	9.0		54	0.98	...		36	43.6	M
555			*Anon.*						
Apl. 11	9.1	10	54	11.17	6	107	30	14.7	M
May 6	9.1		54	11.13	3		30	14.9	M
556			50 *Ursae Majoris* α, *Dubhe.*						
May. 3	—	10	55	29.92	...	27	31	54.8	M
557			*Anon.*						
Mar. 14	8.9	10	57	9.71	...	146	34	40.6	M
558			53 *Leonis* χ						
Mar. 16	...	10	58	9.81	...	81	56	46.5	R
18	...		58	9.27	...		56	45.4	R
19	...		58	9.36	...		56	44.8	R
21	...		58	9.30	.		56	44.5	R
29	...		58	9.31	...		56	45.1	R
Apl. 1	..		58	9.33	4		56	45.8	M
Apl. 6	...	10	58	9.17	...	81	56	44.9	M
15	...		58	9.28	3		56	44.8	M
16	...		58	9.32	..		56	46.1	R
17	...		58	9.27	...		56	46.6	R
26	...		58	9.30	...		56	44.0	R
29	...		58	9.39	—		56	43.1	R
30	...		58	9.23	...		56	44.6	R
May 3	...		58	9.35	3		56	41.8	M
11	...		58	9.21	4		56	46.2	M
559			*Lacaille* 4612.						
Apl. 26	8.6	11	1	1.88	5	154	47	31.8	R
560			67 *Leonis.*						
Apl. 27	6.2	11	1	40.60	...	64	37	39.0	R
561			*Lalande* 21367.						
Mar. 30	8.0	11	3	28.30	...	78	6	46.3	R
May 8	7.7		3	28.25	...		6	45.7	M
13	7.5		3	29.32	5		6	45.4	M
562			*Lalande* 21366.						
Feb. 20	9.0	11	3	30.17	...	64	8	44.1	R
Apl. 17	9.0		3	30.27	5		8	44.2	R
563			*S Leonis Var.* 2.						
Feb. 21	10.5	11	3	36.37	3	98	49	7.8	R
564			*Anon.*						
May. 1	8.8	11	4	27.69	5	150	13	32.9	M
565			*Anon.*						
Apl. 16	9.7	11	7	9.36	..	65	26	3.3	R

Separate Results of Madras Meridian Circle Observations in 1867.

Number and Date.	Magnitude.	Mean Right Ascension 1867. h.	m.	s.	No. of Wires.	Mean Polar Distance 1867. °	′	″	Observer.
566				*68 Leonis δ*					
Mar. 16	...	11	7	1·84	...	68	44	56·2	R
21	...		7	1·84	...		44	54·9	R
Apl. 1	...		7	1·98	...		44	54·8	M
3	...		7	1·91	...		44	54·2	M
10	...		7	1·90	...		44	54·0	M
12	...		7	1·89	..		44	54·2	M
13	...		7	1·78	..		44	53·6	M
15	...		7	1·87	...		44	54·3	R
26	...		7	1·88	...		44	53·9	R
29	...		7	1·84	...		44	53·6	R
May 2	...		7	1·83	...		44	55·1	M
11	...		7	1·07	...		44	54·6	M
567				*Anon.*					
Feb. 20	8·3	11	10	33·08	5	148	56	47·7	R
Mar. 30	9·2		10	33·45	5		56	49·0	R
568				*Anon.*					
May 6	8·0	11	11	16·77	...	127	30	22·7	M
569				*12 Crateris δ*					
Mar. 21	...	11	12	41·56	...	104	3	33·2	R
Apl 1	...		12	41·48	...		3	34·0	M
3	...		12	41·71	...		3	33·2	M
4	...		12	41·46	...		3	34·2	M
10	...		12	41·59	...		3	33·0	M
13	...		12	41·68	...		3	32·5	M
17	...		12	41·33	...		3	36·5	R
25	...		12	41·60	...		3	33·3	R
26	...		12	41·50	..		3	32·7	R
29	..		12	41·40	...		3	32·2	R
30	...		12	41·50	...		3	33·5	R
May 2	...		12	41·57	...		3	33·6	M
4	...		12	41·53	...		3	33·3	M
570				*77 Leonis σ*					
Jan 23	...	11	14	16·68	...	83	14	33·6	R
Mar. 18	–		14	16·60	...		14	32·8	R
19	...		14	16·60			14	32·7	R
571				*Anon.*					
Feb. 20	9·0	11	16	35·30	...	125	40	0·0	R
Mar. 30	8·5		16	35·07	...		40	0·4	R
May 7	8·6		16	35·37	.		39	59·4	M
572				*Anon.*					
Feb. 20	9·3	11	18	7·61	3	125	40	3·6	R
21	9·5		18	7·85	5		40	2·6	R
Mar. 30	9·0		18	7·81	5		40	3·7	R
May 1	9·2		18	7·70	6		40	3·2	M
573				*O. A. N.* 11612.					
May 3	8·0	11	23	30·34	...	22	58	48·9	M
574				*Anon.*					
Apl. 17	9·0	11	24	40·74	5	22	06	58·5	R
575				*Anon.*					
Apl. 16	9·7	11	24	48·24	5	146	9	57·8	R
576				*Anon.*					
Mar. 30	9·5	11	24	50·12	5	84	43	20·0	R
577				*Anon.*					
May 8	9·2	11	25	17·40	5	151	33	29·7	R
578				*Anon.*					
Feb. 21	8·9	11	26	29·85	5	126	27	47·0	R
579				*Anon.*					
Apl. 3	9·0	11	26	32·76		98	54	29·6	M
26	9·6		26	32·65	5		54	29·7	R

Separate Results of Madras Meridian Circle Observations in 1867.

Number and Date	Magnitude.	Mean Right Ascension 1867. h.	m.	s.	No. of Wires.	Mean Polar Distance 1867. °	′	″	Observer.
580			*Anon.*						
Apl. 25	10·0	11	26	33·73	5	151	39	55·0	n
581			*Anon.*						
Apl. 30	9·7	11	26	50·68	...	151	31	58·4	n
582			91 *Leonis* ν						
Feb. 19	...	11	30	8·31	...	90	5	23·8	n
20	...		30	8·30	...		5	26·5	n
Mar. 20	...		30	8·30	...		5	23·1	n
Apl. 1			30	8·27	...		5	24·4	m
3	...		30	8·27	...		5	24·3	m
4	..		30	8·20	5		5	24·6	m
5	...		30	8·14	...		5	23·5	m
6	...		30	8·31	...		5	24·0	m
15	...		30	8·30	..		5	25·5	m
16	...		30	8·32	5		5	24·7	n
27			30	8·36	...		5	23·6	n
May 1	...		30	8·33	...		5	23·3	m
2	...		30	8·33	...		5	24·2	m
4	...		30	8·28	...		5	23·8	m
6	...		30	8·33	...		5	24·8	m
7			30	8·25	5		5	24·1	m
9	...		30	8·19	...		5	23·4	m
11	...		30	8·38	...		5	23·6	m
18	...		30	8·41	...		5	23·9	m
583			*Anon.*						
Feb. 21	8·6	11	34	6·01		127	50	17·3	n
584			*W. B. E. XI.* 579.						
Mar. 30	9·3	11	34	52·94	.	93	16	5·0	n
May 3	9·1		34	52·89	5		16	5·3	m
585			*Anon.*						
Apl. 30	8·8	11	36	12·08	5	139	41	16·9	n
586			*Anon.*						
Apl. 25	7·9	11	39	17·98		109	39	49·3	n
587			*Anon.*						
May 8	7·9	11	41	5·56	5	149	53	8·0	m
11	7·8		41	5·40	...		53	4·6	m
588			*Anon.*						
May 6	8·5	11	41	17·07	...	126	31	26·7	m
7	8·7		41	17·20	6		31	25·7	m
589			94 *Leonis* β, *Deneb.*						
Feb. 21	...	11	42	16·52	...	74	41	4·7	n
Mar. 20	...		42	16·46	...		41	5·2	n
30	...		42	16·48	...		41	5·6	n
Apl. 4	...		42	16·50	...		41	6·2	m
5	...		42	16·39	...		41	6·4	m
6	...		42	16·44	...		41	4·8	m
10	...		42	16·28	...		41	5·0	m
17	...		42	16·37	...		41	5·2	n
26	...		42	16·40	...		41	4·7	n
27	...		42	16·30	...		41	5·3	n
30	...		42	16·40	...		41	5·0	n
May 1	...		42	16·46	...		41	6·3	m
2	...		42	16·48	...		41	6·0	m
4	...		42	16·36	...		41	6·0	m
9	...		42	16·41	...		41	5·3	m
14	...		42	16·44	5		41	5·0	.
15	...		42	16·41	..		41	4·8	m
590			5 *Virginis* β						
Feb. 19	...	11	43	49·19	...	87	29	10·5	n
20	..		43	48·97	..		29	11·6	n
Apl. 15	...		43	48·97	...		29	7·8	m
16	..		43	49·01	..		29	8·1	n
591			*Anon.*						
Mar. 29	9·0	11	49	34·20	5	124	55	47·3	n
Apl. 25	8·6		49	34·15			55	49·4	n
May 6	8·8		49	34·21	5		55	47·6	m
8	8·7		49	33·79	4		55	46·9	m
592			*Lacaille* 4956.						
Feb. 21	8·6	11	51	18·37	...	154	34	58·1	n

Separate Results of Madras Meridian Circle Observations in 1867.

Number and Date.	Magnitude.	Mean Right Ascension 1867. h.	m.	s.	No. of Wires.	Mean Polar Distance 1867. °	′	″	Observer.
593				*Anon.*					
Apl. 27	9·5	11	51	45·88	5	144	13	56·1	R
30	8·7		51	46·98	5		13	58·4	R
594				*R. P. L.* 87.					
Apl. 1	..	11	52	28·32	2	2	15	53·9	M
May 3	...		52	28·47	1		15	53·6	M
4	..		52	27·87	1		15	58·5	M
7	...		52	29·07	1		15	56·7	M
11	...		52	27·28	1		15	53·8	M
				R. P. L. 87—*s.p.*					
Nov. 2	...	11	52	27·70	2	2	15	53·6	M
595				*Anon.*					
Feb. 21	9·4	11	52	29·64	...	154	33	31·4	R
596				8 *Virginis* π					
Mar. 19	...	11	54	3·44	...	82	38	38·9	R
20	...		54	3·88	..		38	39·7	R
597				*Taylor* 5535.					
Apl. 24	8·0	11	57	12·75	...	70	26	30·5	R
598				*Anon.*					
Apl. 26	9·0	11	59	45·65	...	144	19	40·4	R
27	8·8		59	45·94	5		19	40·0	R
May 7	8·1		59	46·18	...		19	40·7	M
11	8·6		59	46·28	5		19	40·8	M
599				*Anon.*					
May 4	8·9	12	1	46·16	...	130	2	36·1	M
6	8·9		1	46·29	..		2	37·9	M
600				2 *Corvi* ε					
Apl. 2	...	12	3	17·22	5	111	52	40·3	M
5	...		3	17·27	...		52	40·8	M
10	...		3	17·29	...		52	40·8	M
May 1	..		3	17·22	...		52	42·5	M
3			3	17·27	5		52	40·6	M
14	...		3	17·28	.		52	40·9	M

Number and Date.	Magnitude.	Mean Right Ascension 1867. h.	m.	s.	No. of Wires.	Mean Polar Distance 1867. °	′	″	Observer.
601				*Anon.*					
May 8	8·2	12	6	21·96	...	135	24	32·7	M
602				*R. P. L.* 90.					
Apl. 13	...	12	7	3·67	1	2	19	39·7	M
603				*Anon.*					
Mar. 30	9·5	12	7	44·63	..	90	15	19·1	R
Apl. 18	..		7	46·74	...		15	17·5	R
604				15 *Virginis* η					
Feb. 20	...	12	13	6·06	...	89	55	40·0	R
21	...		13	6·09	...		55	40·5	R
Apl. 16	...		13	6·07	...		55	39·7	R
19	...		13	5·97	...		55	39·1	R
26	.		13	6·07	...		55	40·3	R
May 1	...		13	6·06	...		55	40·0	M
3	...		13	6·16	...		55	40·7	M
4	...		13	6·04	...		55	39·7	M
7	...		13	5·98	...		55	39·6	M
13	...		13	5·98	...		55	39·6	M
14	...		13	6·10	...		55	39·7	M
22	...		13	6·02	...		55	39·9	R
605				16 *Virginis* c.					
Mar. 19	...	12	13	28·77	..	86	54	47·9	R
606				*R. P. L.* 93—*s.p.*					
Sep. 17	...	12	14	16·46	1	1	33	46·6	R
607				*Anon.*					
Mar. 31		12	17	32·75	5	94	44	7·9	R
29	9·0		17	32·76	5		44	7·4	R
Apl. 18	8·4		17	32·71	4		44	6·2	R
608				*Anon.*					
May 18	7·8	12	20	36·90	5	141	20	17·5	M

Separate Results of Madras Meridian Circle Observations in 1867.

Number and Date.	Magnitude.	Mean Right Ascension 1867. h.	m.	s.	No. of Wires.	Mean Polar Distance 1867. °	′	″	Observer.
609		*W. B. E. XII. 446.*							
Apl. 11	8·9	12	27	23·09	...	92	20	7·7	m
May 1	8·9		27	23·07	6		20	9·0	m
6	8·4		27	22·92	...		20	9·1	m
610		*D Corvi*							
May 9	...	12	27	34·25	...	112	39	39·9	m
16	...		27	34·19	5		39	39·9	n
18	...		27	34·15	...		39	40·5	n
22	.		27	34·22	...		39	40·3	n
611		*Anon.*							
Apl. 16	10·0	12	27	38·36	4	99	38	40·8	n
27	10·0		27	38·38	3		38	40·6	n
30	9·8		27	38·31	...		38	38·7	n
612		*Lalande 23532.*							
Apl. 13	7·9	12	28	37·88	...	92	48	47·7	m
May 7	7·9		28	37·76	...		48	45·9	m
8	7·9		28	37·88	...		48	46·7	m
613		*Anon.*							
May 11	8·6	12	31	0·69	...	112	29	42·1	m
614		*R Virginis Var. 2.*							
May 3	7·9	12	31	46·10	...	82	16	47·3	m
615		*Anon.*							
Apl. 18	9·8	12	32	46·25	...	100	5	42·2	n
24	9·7		32	46·28	.		5	43·1	n
May 8	9·3		32	46·06	...		5	46·5	m
616		*Anon.*							
May 28	7·0	12	32	59·98		98	14	23·3	n
617		*29 Virginis γ¹*							
Feb. 20	...	12	34	55·16	...	90	46	9·4	n
21	...		34	55·17	...		43	9·9	n
Apl. 16	...		34	56·27	6		46	8·8	n
17	...		34	55·96	...		43	7·8	n
May 13	...		34	56·19	...		43	8·9	m
15	...		34	56·30	6		46	7·1	m
18	...		34	55·19	..		46	9·6	n
24	...		34	56·19	..		46	7·7	n
618		*Anon.*							
Apl. 19	9·5	12	39	38·00	...	141	26	13·0	n
619		*Anon.*							
May 9	9·6	12	39	46·51	...	94	2	52·5	m
620		*Anon.*							
Apl. 18	9·7	12	42	13·99	...	147	17	27·9	n
621		*Anon.*							
Apl. 24	9·0	12	43	1·92	...	139	36	16·2	n
622		*Anon.*							
Apl. 30	9·9	12	46	18·41	5	89	42	20·1	n
May 7	9·9		46	18·36	...		42	19·5	m
623		*Anon.*							
May 27	10·5	12	46	40·26	6	89	41	39·9	n
31	10·4		46	40·14	8		41	36·3	n
624		*38 Virginis.*							
Apl. 16	..	12	46	22·80	5	92	49	46·9	n
17	...		46	22·69	4		49	45·8	n
625		*40 Virginis ψ*							
Mar. 21	...	12	47	26·26	..	98	49	32·6	n

Separate Results of Madras Meridian Circle Observations in 1867.

Number and Date.	Magnitude.	Mean Right Ascension 1867. h. m. s.	No. of Wires.	Mean Polar Distance 1867. ° ′ ″	Observer.
626		*W. B. E. XII.* 799.			
Apl. 11	8·6	12 47 46·77		80 46 26·2	M
12	8·4	47 46·77	...	46 28·5	M
15	8·6	47 46·67	...	46 27·2	M
627		*R. P. L.* 99.			
May 30	...	12 48 11·26	3	5 51 48·1	n
628		*Anon.*			
Apl. 19	9·7	12 48 38·64	...	125 26 18·8	n
May 11	8·3	48 38·74	...	26 18·5	M
629		*Anon.*			
Apl. 4	7·9	12 49 3·64	...	140 26 19·9	M
May 13	7·9	49 3·35	...	26 17·3	M
630		12 *Canum Venaticorum.*			
Apl. 18	...	12 49 46·08	...	50 57 45·4	n
24	...	49 46·00	...	57 46·0	n
May 6	...	49 46·07	...	57 46·2	M
8	...	49 46·14	5	57 46·7	M
18	...	49 46·00	...	57 45·9	n
22	...	49 46·04	—	57 46·4	n
28	...	49 46·07	5	57 46·3	n
631		*Anon.*			
May 25	9·0	12 53 18·89	...	143 25 3·3	n
632		*Anon.*			
May 9	8·0	12 53 35·72		126 45 30·4	M
633		*Anon.*			
May 7	9·1	12 54 49·34	...	130 19 21·0	M
634		*Anon.*			
May 30	9·2	12 57 9·26		123 24 10·9	n

Number and Date.	Magnitude.	Mean Right Ascension 1867. h. m. s.	No. of Wires.	Mean Polar Distance 1867. ° ′ ″	Observer.
635		*Lacaille* 5381.			
Apl. 25	8·0	12 57 17·88		109 58 5·3	n
May 13	7·3	57 18·00	4	58 5·9	M
636		*Anon.*			
May 11	8·3	12 58 15·98	..	124 30 41·2	M
637		51 *Virginis θ*			
Feb. 21	...	13 3 3·84	.	94 40 41·3	n
Mar. 21		3 3·94	...	40 43·0	n
22		3 3·91		40 42·6	n
Apl. 15	...	3 3·85	...	40 42·4	n
18	..	3 3·92	4	40 41·3	n
19	...	3 3·90	...	40 42·1	n
24	...	3 3·80	—	40 42·6	n
May 14	...	3 3·84	..	40 42·9	M
16	...	3 3·84	..	40 41·8	n
18	...	3 3·88	...	40 41·5	n
22	...	3 3·82	..	40 43·1	n
23		3 3·92	...	40 42·1	n
25	...	3 3·85	...	40 43·1	n
27	...	3 3·88	...	40 42·1	n
638		*Taylor* 6057.			
May 8	5·6	13 4 0·73	..	140 12 44·9	M
639		*Anon.*			
May 9	9·0	13 4 42·50	...	135 11 32·4	M
640		*Anon.*			
Apl. 30	9·3	13 4 46·30	5	143 13 19·7	n
641		*Anon.*			
May 30	9·5	13 5 47·38	...	134 17 39·9	n
642		*Lacaille* 5184.			
Apl. 25	7·8	13 6 9·27	5	162 52 09·6	n

Separate Results of Madras Meridian Circle Observations in 1867.

Number and Date.	Magnitude.	Mean Right Ascension 1867. h.	m.	s.	No. of Wires.	Mean Polar Distance 1867. °	′	″	Observer.
643				*R. P. L.* 101—*s.p.*					
Nov. 16	...	13	9	56·84	3	1	26	15·8	R
21	...		9	53·71	3		25	15·1	R
644				*Taylor* 6129.					
May 13	7·0	13	12	22·30	5	130	20	20·3	M
645				*Taylor* 6148.					
May 26	7·5	13	14	21·75	...	123	0	17·3	R
646				*Taylor* 6160.					
May 27	8·0	13	15	46·26	5	123	55	56·0	R
30	8·3		15	45·92	...		56	54·4	R
647				67 *Virginis* α, *Spica.*					
Feb. 21	...	13	18	11·16	...	100	28	0·7	R
Mar. 22	...		18	11·23	...		27	58·5	R
Apl. 15	...		18	11·22	...		27	56·7	M
17	...		18	11·26	...		27	58·4	R
18	...		18	11·30	...		27	58·5	R
May 3	...		18	11·13	5		27	58·7	M
14	..		18	11·28	...		27	58·8	M
16	..		18	11·34	...		27	57·9	R
20	...		18	11·21	...		27	58·4	R
June 4	...		18	11·27	...		27	58·6	M
648				79 *Ursae Majoris* ζ—2nd.					
June 7	6·5	13	19	34·75	...	34	22	57·4	M
649				*O. A. S.* 12872.					
Apl. 29	9·7	13	19	30·85	.	116	37	28·7	R
May 31	9·9		19	30·80	5		37	21·7	R
650				*Lacaille* 5546.					
May 23	9·0	13	19	58·27	3	148	28	26·2	R
651				*R. P. L.* 103.					
Apl. 24	...	13	20	6·71	3	4	32	10·8	R
25			20	6·61	3		32	9·9	R
				R. P. L. 103—*s.p.*					
Dec. 6	...	13	20	5·77	3	4	33	2·1	M
652				*Anon.*					
May 30	10·6	13	23	26·90	5	88	30	10·4	R
653				*Anon.*					
May 24	...	13	24	37·20	6	128	9	28·7	R
654				*Anon.*					
Apl. 25	8·2	13	25	41·80	4	108	11	18·6	R
655				*Taylor* 6257.					
May 25	9·2	13	25	47·38	...	148	40	20·5	R
June 5	8·2		25	47·76	...		40	20·9	M
656				76 *Virginis* h					
Apl. 17	...	13	25	57·88	5	99	25	46·9	R
657				*Anon.*					
May 27	8·3	13	26	45·45	...	131	36	7·5	R
658				79 *Virginis* ζ					
Mar. 22	..	13	27	56·00	...	89	54	55·3	R
23			27	55·96	...		54	54·3	R
Apl. 19			27	56·04	...		54	55·9	R
24	...		27	55·98	...		54	56·2	R
May 23	..		27	55·93	...		54	54·9	R
29	...		27	56·06	...		54	53·8	R
31	.		27	56·07	...		54	55·6	R
June 1			27	54·90			54	54·2	M
4	...		27	55·12	..		54	56·2	M
6	...		27	55·07	.		54	50·1	M
10	...		27	54·95	..		54	54·6	M

Separate Results of Madras Meridian Circle Observations in 1867.

Number and Date.	Magnitude.	Mean Right Ascension 1867. h. m. s.	No. of Wires.	Mean Polar Distance 1867. ° ′ ″	Observer.
659 *Lacaille 5614.*					
Apl. 29	8·0	13 30 7·20		123 13 3·1	R
660 *Lacaille 5639.*					
May 31	7·0	13 33 28·82	5	123 46 59·6	R
June 1	7·1	33 28·73	5	47 0·6	M
661 *Anon.*					
May 30	9·5	13 34 13·30	5	129 10 53·7	R
662 *Anon.*					
May 27	8·3	13 35 43·62	...	128 4 23·1	R
663 *O. A. S.* 13079.					
June 5	9·3	13 36 1·71	6	116 40 26·0	M
7	9·2	36 1·74	...	40 23·0	M
664 *Anon.*					
Apl. 16	9·2	13 36 26·17	5	129 6 16·7	R
665 *Anon.*					
Apl. 26	9·5	13 36 42·66	...	141 39 14·4	R
666 *Taylor* 6366.					
Apl. 30	7·0	13 37 5·16	5	151 47 0·3	R
667 *Lacaille* 5669.					
Apl. 26	7·8	13 37 30·42	..	128 14 39·7	R
668 *O. A. S.* 13100.					
June 4	5·0	13 37 28·78	...	116 53 37·0	M
669 *Anon.*					
May 31	9·5	13 37 49·65	...	128 44 56·7	R
670 *Anon.*					
May 30	8·9	13 40 40·97	5	120 24 55·9	R
671 *O. A. S.* 13186.					
May 1	8·4	13 43 55·82	...	116 59 34·9	M
3	7·8	43 55·29	...	59 34·2	M
4	8·4	43 56·12	...	59 33·3	M
672 *O. A. S.* 13198.					
Apl. 30	9·2	13 44 57·78	...	117 11 26·2	R
May 6	9·0	44 57·84	5	11 26·7	R
7	9·1	44 57·94	4	11 26·8	M
673 *Anon.*					
Apl. 26	7·9	13 45 36·92	...	123 23 39·4	R
May 24	6·0	45 34·14	4	23 39·6	R
674 *Anon.*					
Apl. 27	10·0	13 46 41·61	5	129 34 5·0	R
May 8	9·8	46 41·49	3	34 6·1	M
27	10·0	46 41·44	...	34 5·3	R
675 *X Virginis Var.* 5.					
Apl. 26	9·0	13 47 39·90	...	79 16 49·5	R
May 18	8·0	47 39·81	4	16 47·5	M
30	9·2	47 39·92	.	16 45·2	R
June 1	9·2	47 38·88	5	16 48·3	M
5	9·9	47 38·98	4	16 47·8	M
7	9·2	46 34·09	5	16 49·7	M
676 8 *Boötis* η					
Mar. 28	—	13 48 21·11	...	70 36 4·9	R
Apl. 22	...	48 21·17	...	36 4·3	M
May 9	...	48 21·10	.	36 4·6	M
15	...	48 21·12		36 4·7	M
31	...	48 21·04	—	36 5·1	R
June 6	—	48 21·08		36 5·0	M
10	...	48 21·09		36 5·7	M
14	...	48 21·08	...	36 5·9	M
26	..	48 21·09	—	36 4·8	M

Separate Results of Madras Meridian Circle Observations in 1867.

Number and Date.	Magnitude.	Mean Right Ascension 1867. h. m. s.	No. of Wires.	Mean Polar Distance 1867. ° ′ ″	Observer.
677		*Anon.*			
May 25	9·2	13 50 11·81	5	140 55 8·5	R
678		*Anon.*			
Apl. 19	9·5	13 52 36·56	—	129 2 13·0	R
679		*Anon.*			
May 30	9·7	13 53 16·96	5	135 41 44·5	R
680		*β Centauri.*			
May 1		13 54 [illegible]·01	...	149 48 47·9	M
681		93 *Virginis* τ			
Mar. 23	...	13 54 52·71	...	67 46 36·6	R
Apl. 18	...	54 52·76	...	46 36·6	R
May 7	...	54 52·70	...	46 35·0	M
8	...	54 52·69	...	46 35·1	M
11	...	54 52·69	...	46 36·3	M
24	...	54 52·73	...	46 37·1	R
27	...	54 52·77	...	46 37·4	R
June 4	...	54 52·73	...	46 36·6	M
6	...	54 52·75	5	46 36·3	M
10	...	54 52·77	...	46 35·5	M
14	...	54 52·75	...	46 35·7	M
15	...	54 52·65	...	46 35·3	M
682		*Lacaille* 5794.			
Apl. 25	7·0	13 57 17·99	...	152 46 35·6	R
683		*Anon.*			
May 26	8·7	13 58 25·16	...	140 20 57·3	R
30	8·0	58 25·22	5	20 57·2	R
684		94 *Virginis.*			
Mar. 21		13 59 15·37	6	98 15 50·6	R
685		95 *Virginis.*			
May 15	...	13 59 40·88	4	98 40 39·3	R
16	...	59 40·82	...	40 39·9	R
686		*Lalande* 25896.			
Apl. 22	7·0	14 0 2·61	...	67 11 45·5	W
687		*Anon.*			
May 27	10·2	14 0 34·41	...	150 52 1·7	R
688		*Taylor* 6585.			
May 13	7·8	14 1 38·07	...	124 14 56·5	W
689		*Anon.*			
May 31	9·2	14 2 12·87	5	124 17 6·0	R
690		*Anon.*			
May 25	8·0	14 5 14·78	...	129 21 11·0	R
30	8·0	5 14·96	...	21 10·2	R
691		*Lacaille* 5844			
Apl. 26	7·8	14 5 17·51	5	151 4 57·1	R
692		98 *Virginis* κ			
Mar. 21	...	14 5 48·22	5	99 39 11·3	R
22	...	5 48·18	...	39 11·2	R
Apl. 18	...	5 48·07	...	39 11·6	R
19	...	5 48·19	...	39 11·4	R
693		*Anon.*			
May 11	8·5	14 6 30·16	...	136 2 10·1	M
June 5	8·3	6 30·25	6	2 10·9	M
7	8·6	6 30·20	...	2 10·5	M
694		90 *Virginis* ι			
May 15	...	14 9 2·25	6	95 21 32·4	M
16	...	9 2·67	4	21 33·1	R

Separate Results of Madras Meridian Circle Observations in 1867.

Number and Date.	Magnitude.	Mean Right Ascension 1867. h. m. s.	No. of Wires.	Mean Polar Distance 1867. ° ′ ″	Observer.
695		16 *Bootis* α, *Arcturus*.			
May 6	...	14 9 36·71	...	70 7 27·8	M
7	...	9 35·85	...	7 26·6	M
8	...	9 36·06	...	7 26·9	M
13	...	9 36·78	...	7 27·2	M
25	...	9 35·73	...	7 27·6	R
27	...	9 35·77	...	7 27·0	R
28	...	9 35·72	...	7 26·3	R
30	...	9 35·77	...	7 26·5	R
June 1	...	9 35·77	...	7 27·2	M
4	...	9 35·67	...	7 27·0	M
6	...	9 36·06	5	7 28·2	M
10	...	9 35·76	...	7 27·3	M
14	...	9 35·75	...	7 28·0	M
15		9 35·74	...	7 27·2	M
696		*Anon.*			
May 31	9·5	14 11 46·83	...	121 25 34·4	R
697		100 *Virginis* λ			
Apl. 19	...	14 11 54·98	...	102 46 26·7	R
698		*W. B. E. XIV.* 192.			
Apl. 18	7·5	14 12 1·26	...	106 47 33·0	R
699		*Anon.*			
Apl. 28	9·8	14 12 42·34		136 30 42·5	R
700		*Anon.*			
May 30	9·6	14 15 30·26	...	122 12 29·3	R
701		*Taylor* 6740.			
June 7	7·4	14 19 14·84	...	133 43 47·4	M
702		*Anon.*			
May 31	8·8	14 19 44·67		134 39 10·8	R
June 1	8·8	19 44·68	...	39 12·0	R
703		*Anon.*			
June 5	8·6	14 20 8·82	..	127 9 44·7	M
704		*Lacaille* 5962.			
Apl. 30	7·0	14 22 53·23	4	129 47 34·6	R
705		*Anon.*			
Apl. 30	9·0	14 23 7·11	6	129 44 47·5	R
706		25 *Bootis* ρ			
Apl. 22	...	14 26 5·79	...	59 2 37·4	M
May 6	...	26 5·83	...	2 37·7	M
8	..	26 5·81	...	2 37·9	M
13	...	26 5·79	...	2 37·7	M
15	...	26 5·80	...	2 38·6	M
23	...	26 5·90	...	2 36·9	R
30	...	26 5·77	...	2 37·3	R
June 14	..	26 5·87	...	2 37·9	M
15	...	26 5·80	...	2 37·5	M
22	...	26 6·01	4	2 37·8	M
24	...	26 5·86	...	2 38·0	R
707		*Anon.*			
June 6	9·3	14 26 54·52	4	123 30 51·2	R
708		*Anon.*			
June 5	7·7	14 29 37·88	...	104 56 30·6	R
7	7·0	29 37·64	6	55 29·3	M
10	7·9	29 37·57	...	56 19·2	M
709		*Lacaille* 6027.			
June 4	7·9	14 31 18·11	.	152 49 5·5	M
710		*Taylor* 6848.			
June 1	7·3	14 33 0·34	5	136 42 7·3	M
711		α *Lupi*.			
May 16		14 33 5·80		136 53 54·9	R
June 28	...	33 5·72	5	53 53·6	M

Separate Results of Madras Meridian Circle Observations in 1867.

Number and Date.	Magnitude.	Mean Right Ascension 1867. h. m. s.	No. of Wires.	Mean Polar Distance 1867. ° ′ ″	Observer.
712		*Anon.*			
May 23	..	14 33 6·25	5	120 18 35·1	R
713		*Anon.*			
June 24	8·9	14 36 44·48	...	150 18 0·5	M
714		5 *Librae.*			
Mar. 22	...	14 38 38·60	...	104 53 40·7	R
23	...	38 37·90	...	53 40·6	R
715		36 *Bootis* ε, *Mirac.*			
Apl. 28	...	14 39 10·66	...	62 21 40·9	M
May 26	...	39 10·65	...	21 40·4	R
28	...	39 10·73	...	21 40·7	R
June 5	...	39 10·60	..	21 40·7	M
7	...	39 10·57	..	21 40·3	M
26	...	39 10·72	..	21 40·6	M
716		*Anon.*			
May 28	...	14 40 26·91	5	127 4 32·7	R
717		*Brisbane* 5069.			
June 10	8·0	14 41 35·40	...	131 17 31·8	M
718		*Lalande* 27022.			
June 25	7·9	14 46 22·06	6	78 57 10·4	M
719		9 *Librae* α¹			
Mar. 22	...	14 43 31·51	..	105 20 15·6	R
May 16	...	43 31·43	..	20 12·9	R
27	...	43 31·37	...	20 14·9	R
30	—	43 31·47	..	20 14·1	R
June 5	...	43 31·42	5	20 14·8	M
22	...	43 31·39	...	20 14·7	M
720		*Anon.*			
June 21	9·5	14 47 44·57	4	150 41 48·7	M
721		*Anon.*			
June 6	8·9	14 50 34·91	6	130 33 05·5	M
722		*O. A. S.* 14112.			
June 4	7·9	14 51 3·30	...	109 11 27·5	M
28	8·3	51 3·36	...	11 27·8	M
723		7 *Ursae Minoris* β *Var.* 1, *Kochab.*			
June 27	...	14 51 7·65	...	15 18 2·8	M
724		*Radcliffe* 3306.			
May 30	7·5	14 56 6·46	5	48 11 46·3	R
725		*Taylor* 7086.			
May 29	...	14 58 5·86	...	62 28 40·4	R
726		43 *Bootis* ψ			
May 24	...	14 58 44·79	...	62 31 55·5	R
June 7	...	58 44·70	...	31 55·5	M
July 2	...	58 44·71	..	31 56·5	R
727		21 *Librae* ν¹			
May 16	...	14 59 12·62	...	105 44 19·6	R
728		*W. B. E. XV.* 7.			
June 14	8·5	15 2 30·34	...	97 31 42·2	M
729		*Taylor* 7079.			
June 19	6·6	15 3 30·09	5	128 7 39·2	M
730		*W. B. E. XV.* 32.			
June 24	8·9	15 3 34·41	6	97 2 16·4	M

Separate Results of Madras Meridian Circle Observations in 1867.

Number and Date.	Magnitude.	Mean Right Ascension 1867. h.	m.	s.	No. of Wires.	Mean Polar Distance 1867. °	′	″	Observer.
731				*R. P. L.* 111.					
June 1	...	15	3	28·56	3	5	32	6·1	M
5	...		3	22·49	1		32	4·5	M
732				*W. B. E. XV.* 86.					
June 28	9·3	15	6	40·88	...	98	2	33·8	M
July 4	9·2		6	41·23	5		2	34·6	R
733				27 *Librae β*					
Apl. 10	...	15	9	51·08	...	98	53	24·7	R
22	...		9	51·13	...		53	24·8	M
May 24	...		9	51·15	...		53	24·4	R
June 22	...		9	50·98	4		53	25·4	M
27	...		9	51·03	...		53	24·2	M
734				*Anon.*					
June 10	7·7	15	12	2·88	6	130	24	41·7	M
735				*Anon.*					
May 27	9·0	15	17	16·87	5	130	4	21·4	R
736				32 *Librae ζ*¹					
Mar. 28	...	15	20	46·64	...	106	13	3·2	R
Apl. 19	...		20	46·47	5		13	1·7	R
June 14	...		20	46·69	...		13	4·2	M
737				*R. P. L.* 114—*s.p.*					
Dec. 14	...	15	21	30·41	1	2	15	42·4	M
738				*Anon.*					
June 1	9·0	15	22	42·73	5	131	37	41·1	R
July 4	...		22	42·91			37	39·9	R
739				*Anon.*					
May 21	9·7	15	24	48·15	5	130	9	31·3	R
740				*Anon.*					
July 5	8·3	15	26	11·94	5	122	44	11·9	R
741				*Lacaille* 6421.					
June 28	7·9	15	26	50·19	...	122	43	11·1	M
July 5	7·7		26	50·31	4		43	10·5	R
742				38 *Librae γ*					
Mar. 28	...	15	28	5·86	...	104	20	39·3	R
May 18	...		28	5·46	4		20	39·0	R
June 14	...		28	5·42	...		20	39·1	M
743				5 *Coronae Borealis α, Alphecta.*					
June 7	...	15	29	3·58	...	62	30	10·1	M
24	...		29	3·37	4		30	10·7	R
27	...		29	3·41	...		30	10·0	R
July 4	...		29	3·33	5		30	10·6	R
744				*Anon.*					
July 6	10·0	15	29	10·22	4	110	35	28·6	R
745				*Anon.*					
July 6	9·7	15	29	28·71	4	110	41	3·9	R
746				*Lalande* 28530.					
May 16	6·0	15	31	36·40	...	47	25	53·4	R
747				*Taylor* 7300.					
May 27	7·7	15	32	24·24	...	103	37	10·0	R
29	7·7		32	24·50			37	10·8	R
748				*Anon.*					
May 21	9·9	15	33	47·51		108	36	49·3	R

Separate Results of Madras Meridian Circle Observations in 1867.

Number and Date.	Magnitude.	Mean Right Ascension 1867. h.	m.	s.	No. of Wires.	Mean Polar Distance 1867. °	′	″	Observer.
749	*24 Serpentis α*								
June 7	...	15	37	43·02	...	88	9	14·3	M
22	...		37	43·03	...		9	14·4	M
24	...		37	43·00	...		9	14·1	M
27	...		37	43·07	..		9	13·6	M
28	...		37	43·08	...		9	13·9	M
July 2	...		37	43·19	4		9	14·0	R
4	...		37	43·01	...		9	13·4	R
5	..		37	43·03	5		9	13·0	R
11	...		37	42·97	...		9	13·0	R
12			37	43·00	..		9	13·4	R
750	*O. A. S.* 14874.								
May 27	8·9	15	39	33·29	5	104	49	12·1	R
751	*Lalande* 28787.								
July 10	8·8	15	42	15·43	..	92	49	33·0	R
752	*Anon.*								
May 24	9·3	15	42	19·98	...	104	24	56·6	R
753	*Anon.*								
May 29	10·5	15	42	39·97	3	61	47	14·2	R
754	*O. A. S.* 14934.								
July 5	9·7	15	43	39·76	...	107	52	58·2	R
755	*Anon.*								
May 24	9·0	15	44	7·97	5	104	22	9·7	R
756	*W. B. E. XV.* 833.								
June 5	8·9	15	44	9·90	5	104	27	34·3	M
757	36 *Serpentis b*								
July 11		15	44	30·20	.	92	41	9·3	R
758	46 *Libræ θ*								
May 18		15	46	15·24	...	106	30	13·6	R
759	*O. A. S.* 14996.								
May 27	8·9	15	46	41·14		106	16	1·1	R
760	*R. P. L.* 115—*s.p.*								
Jan. 7	...	13	48	8·01	3	4	41	27·1	M
Dec. 19	...		48	9·01	3		41	31·5	R
761	*O. A. S.* 15063.								
June 28	8·0	15	49	12·38	...	105	26	30·7	M
July 6	8·0		49	12·32	5		26	28·6	R
762	*W. B. E.* 923.								
July 5	9·8	15	49	33·16	...	104	56	37·5	R
763	*O. A. S.* 15146.								
May 24	9·5	15	54	57·22	...	107	29	20·3	R
764	*O. A. S.* 15146.								
June 26	8·9	15	54	59·07	...	107	47	38·6	M
765	*W. B. E. XV.* 1044.								
May 27	7·7	15	55	30·58	...	96	27	42·1	R
July 6	7·6		55	30·49	...		27	44·6	R
766	8 *Scorpii β¹*								
June 5	...	15	57	42·26	5	109	26	20·8	M
27	...		57	42·20	5		26	19·6	M
July 2	...		57	42·37	...		26	20·9	R
4	...		57	42·43	...		26	19·7	R
10	...		57	42·33	...		26	19·4	R
11	...		57	42·40	4		26	21·5	R
12	.		57	42·42	...		26	19·3	R
767	*Lalande* 29306.								
June 28	8·0	15	59	36·06	.	107	31	29·5	M

Separate Results of Madras Meridian Circle Observations in 1867.

Number and Date.	Magnitude.	Mean Right Ascension 1867. h. m. s.	No. of Wires.	Mean Polar Distance 1867. ° ′ ″	Observer.
768		14 *Scorpii* ν			
June 14	...	16 4 16·02	...	109 6 47·4	M
15	...	4 16·05	...	6 46·1	M
July 11	...	4 15·98	5	6 41·3	R
12	...	4 16·06	5	6 45·1	R
769		1 *Ophiuchi* δ			
June 24		16 7 22·61	...	98 20 50·0	M
July 11	...	7 22·60	...	20 50·1	R
770		*Lalande* 29610.			
July 10	...	16 8 30·45	...	106 32 58·9	R
771		*O. A. S.* 15470.			
May 30	8·0	16 9 19·13	5	112 35 20·5	R
772		*R Scorpii Var.* 1.			
July 5	9·5	16 9 48·67	4	112 36 58·6	R
6	9·5	9 48·61	...	36 48·6	R
773		*S Scorpii Var.* 2.			
June 1	9·9	16 9 44·94	6	112 33 42·2	M
4	10·0	9 44·74	6	33 45·2	M
5	9·9	9 44·97	5	33 46·3	M
7	10·0	9 44·98	4	33 44·1	M
774		*Anon.*			
July 5	8·7	16 9 56·99	3	112 34 4·3	R
775		*Anon.*			
July 4	8·0	16 14 27·86	...	144 11 32·2	R
776		4 *Ophiuchi* ψ			
June 14	.	16 16 19·30	5	109 46 25·4	M
15		16 19·22	...	46 24·9	M
777		*O. A. S.* 15613.			
June 28	7·9	16 17 36·91	...	118 9 2·5	M
778		21 *Scorpii* α, *Antares.*			
July 10	...	16 21 16·38	...	116 8 1·4	R
779		40 *Herculis* ζ			
July 5	...	16 36 16·26	...	58 9 16·7	R
6		36 16·29	...	9 18·2	R
10	...	36 16·30	...	9 17·2	R
780		*Lacaille* 6984.			
July 4	8·7	16 38 54·74	5	129 37 46·6	R
781		*Anon.*			
July 5	9·3	16 44 45·50	...	130 14 39·6	R
782		*Anon.*			
July 6	10·3	16 46 1·63	4	76 17 8·9	R
783		*Taylor* 7815.			
July 5	7·5	16 46 42·39	4	130 16 13·2	R
784		*S Herculis Var.* 3.			
Aug. 6	6·9	16 45 50·81	4	74 40 37·9	M
785		49 *Herculis.*			
July 4	7·0	16 46 1·33	5	74 45 3·2	R
786		*Taylor* 7832.			
July 5	8·0	16 47 41·28	5	130 17 49·5	R
787		*Taylor* 7842.			
July 12		16 48 31·25	5	108 35 39·0	R

Separate Results of Madras Meridian Circle Observations in 1867.

Number and Date.	Magnitude.	Mean Right Ascension 1867. h. m. s.	No. of Wires.	Mean Polar Distance 1867. ° ′ ″	Observer.
788		*Anon.*			
July 23	...	16 49 5·74	3	126 31 35·6	R
789		27 *Ophiuchi* κ			
June 1	.	16 51 22·36	6	80 34 58·4	M
23		51 22·29	5	34 57·7	M
July 20	...	51 22·41	...	34 56·8	R
790		29 *Ophiuchi.*			
Apl. 22	.	16 54 4·56	3	108 41 13·9	M
791		*O. A. S.* 16233.			
July 30	8·6	16 54 9·45	4	110 23 50·8	R
792		*Anon.*			
July 23	9·3	16 55 27·48	6	109 56 49·5	R
793		*O. A. S.* 16288.			
Aug. 10	7·7	16 56 38·63	4	119 50 34·5	M
794		22 *Ursae Minoris* ε—*s.p.*			
Jan. 14	...	16 59 48·79	3	7 46 0·0	R
Feb. 11	...	59 42·02	3	44 50·6	M
795		*R Ophiuchi Var.* 2.			
June 4	7·8	17 0 7·81	...	105 54 45·5	M
7	7·9	0 7·83	...	54 46·6	M
10	8·0	0 7·98	...	54 47·1	M
796		35 *Ophiuchi* η			
June 15	..	17 2 44·98		105 33 20·6	M
July 13	...	2 44·67	...	33 26·3	R
797		*Anon.*			
Aug. 10	8·0	17 5 58·72	...	130 50 30·0	M
798		*Anon.*			
July 31	9·0	17 6 19·26	...	130 34 14·7	R
799		64 *Herculis* α *Var.* 1.			
June 1	...	17 8 35·08	4	75 27 26·1	M
July 28	...	8 34·91	...	27 21·0	R
30	...	8 34·94	...	27 22·3	R
800		*Taylor* 8017.			
June 4	6·9	17 13 32·80	4	114 46 8·3	M
801		42 *Ophiuchi* θ			
June 17	...	17 13 50·50	...	114 51 50·1	M
July 6	...	13 50·66	...	51 49·9	R
23	...	13 50·36	...	51 49·2	R
30	..	13 50·54	...	51 49·4	R
802		*Anon.*			
July 28	9·0	17 21 16·76	...	130 48 44·7	R
Aug. 6	8·6	21 16·65	5	48 43·0	M
803		*Anon.*			
July 30	9·3	17 21 25·74	.	130 45 30·2	R
804		*Anon.*			
July 31	9·7	17 27 14·81	...	130 35 40·1	R
805		23 *Draconis* β			
Aug. 8	...	17 27 25·61	...	37 36 37·1	M
806		55 *Ophiuchi* α			
June 5	..	17 28 45·80	...	77 30 28·1	M
July 3	.	28 45·86	...	30 26·8	R
6	...	28 45·89	...	30 26·6	R
20		28 45·64	...	30 26·2	R
22	.	28 45·55	...	30 25·7	R
30	...	28 45·50	...	30 28·4	R
Aug. 7	...	28 45·36	...	30 26·5	M
23	..	28 45·67	...	30 27·5	R

Separate Results of Madras Meridian Circle Observations in 1867

Number and Date.	Magnitude.	Mean Right Ascension 1867. h. m. s.	No. of Wires.	Mean Polar Distance 1867. ° ′ ″	Observer.
807		*Taylor* 8141.			
Apl. 22	6·0	17 30 43·50	...	111 49 52·0	M
808		*Anon.*			
July 31	9·8	17 34 46·30	4	128 35 28·6	R
Aug 21	10·0	34 46·36	3	35 29·4	R
809		*Anon.*			
July 23	9·9	17 34 48·51	5	126 15 7·4	R
810		*Anon.*			
July 6	9·2	17 36 16·98		150 36 11·6	R
811		58 *Ophiuchi*.			
Apl. 22	...	17 36 27·75	2	111 36 55·7	M
812		*Anon.*			
July 6	9·4	17 36 30·44	5	160 37 13·4	R
813		*Anon.*			
July 30	9·9	17 37 52·05	..	128 20 24·4	R
814		*Anon.*			
July 8	8·0	17 39 46·58		127 21 48·8	R
815		*Anon.*			
July 23	8·6	17 39 56·07	4	127 14 41·1	R
816		*Anon.*			
Aug 7	8·3	17 40 7·98		126 28 30·9	M
817		86 *Herculis* μ			
June 17	..	17 41 16·21	...	62 11 50·2	M
25	...	41 15·27	4	11 50·6	M
July 13	...	41 15·15	...	11 58·9	R
20		41 16·21	5	11 59·3	R
Aug. 6		41 15·19		11 50·9	M
21	.	41 15·30		11 59·2	R
818		*Radcliffe* 3765.			
July 31	8·3	17 43 41·47	4	17 32 1·3	R
819		*Anon.*			
July 8	8·3	17 45 13·70	5	128 35 29·4	R
10		45 13·85	5	35 21·5	R
23	8·9	45 13·71	4	35 22·5	R
820		*Lacaille* 7499.			
Aug. 7	7·2	17 48 27·88	4	120 4 44·8	M
8	7·1	48 27·84	...	4 44·4	M
10	7·0	48 27·82	...	4 44·6	M
821		*Taylor* 8288.			
June 17	6·0	17 49 49·02	...	105 47 10·7	M
822		*Lacaille* 7504.			
Aug 6	6·9	17 49 14·46	5	120 6 53·1	M
823		4 *Sagitarii* b			
Aug. 21		17 51 49·31	...	113 17 44·9	R
824		*Anon.*			
July 28	9·0	17 50 38·81	..	150 26 10·6	R
31	9·0	50 38·73	5	26 9·7	R
825		13 *Sagitarii* μ¹			
June 17		18 5 49·50		111 5 27·4	M
28	..	5 49·54		5 28·9	M
July 8		5 48·51	.	5 27·8	R
Aug. 3	...	5 49·22	.	5 28·9	R
6	..	5 49·45		5 28·9	M
7		5 49·64	.	5 27·1	M
8		5 48·36	...	5 26·6	M
10		5 49·60		5 27·7	W
14		5 48·47		5 28·4	W

Separate Results of Madras Meridian Circle Observations in 1867.

Number and Date.	Magnitude.	Mean Right Ascension 1867. h.	m.	s.	No. of Wires.	Mean Polar Distance 1867. °	′	″	Observer.
826 *Lacaille* 7644.									
July 10	6·7	18	9	9·68	...	132	19	30·6	R
827 *Anon.*									
July 31	8·8	18	13	7·90		127	40	58·4	R
828 23 *Ursae Minoris* δ									
July 4		18	15	14·65	3	3	23	41·7	R
5			15	15·11	3		23	40·7	R
Aug. 6			15	15·60	3		23	42·1	M
23 *Ursae Minoris* δ—*s.p.*									
Jan. 23	...	18	15	15·10	2	3	23	40·9	R
829 21 *Sagittarii.*									
July 8	...	18	17	24·61	...	110	36	36·7	R
830 *Taylor* 8509.									
July 8		18	21	36·90		104	38	52·5	R
Aug. 10	...		21	36·97	...		38	51·8	M
831 *Anon.*									
July 31	.	18	23	14·79	5	136	15	42·2	R
832 θ *Coronae Australis.*									
Aug. 6	6·6	18	25	30·97	...	132	24	17·6	M
833 *Anon.*									
July 6	10·2	18	29	30·48	3	136	55	20·1	R
834 *Anon.*									
July 8	8·9	18	29	30·20	5	135	51	44·4	R
23	9·0		29	30·28			51	47·9	R
835 3 *Lyrae* α, *Vega.*									
July 18		18	32	26·09	5	51	20	19·9	R
31	...		32	26·04	3		20	18·7	R
Aug. 8			32	26·05	3		20	16·7	M
14	...		32	26·06			20	20·2	M
20	...		32	26·05	.		20	16·8	R
836 *Anon.*									
July 10	9·3	18	36	12·89	3	137	16	7·4	R
837 *Anon.*									
July 20	9·3	18	36	2·46	...	137	10	02·1	R
838 *Anon.*									
Aug. 28	9·0	18	40	37·19	...	127	27	26·8	R
839 10 *Lyrae* β *Var.* 1.									
June 17	...	18	45	10·08	...	56	47	25·0	M
19	...		45	10·07	...		47	25·0	M
July 6			45	9·97	...		47	23·8	R
25	...		45	10·05	5		47	26·0	R
31			45	10·04	4		47	24·6	R
Aug. 3	.		45	10·00	3		47	24·2	M
8	...		45	10·18	3		47	23·4	M
10	...		45	10·10	.		47	23·9	M
20	..		45	10·07	5		47	25·5	R
21	...		45	9·95	3		47	26·6	R
24	...		45	10·09	..		47	26·1	R
27	...		45	10·12	5		47	24·6	R
Sep. 6	...		45	10·16	...		47	23·5	M
840 *Anon.*									
July 6	8·2	18	46	9·04	...	126	40	44·3	R
841 *Lacaille* 7919.									
Aug. 29	8·9	18	47	25·24	...	129	4	16·6	R
842 *O. A. S.* 18960.									
Aug. 31	7·9	18	53	49·95	...	121	7	25·2	R

Separate Results of Madras Meridian Circle Observations in 1867.

Number and Date.	Magnitude.	Mean Right Ascension 1867. h.	m.	s.	No. of Wires.	Mean Polar Distance 1867. °	′	″	Observer.
843		*39 Sagittarii ο*							
June 17	...	18	56	48·68	...	111	56	6·9	M
844		*17 Aquilae ζ*							
June 19	...	18	59	17·59	...	76	19	56·3	M
July 8	...		59	17·72	...		19	56·4	R
18	...		59	17·77	...		19	56·7	R
23	...		59	17·71	...		19	55·4	R
25			59	17·72	...		19	56·5	R
Aug. 3			59	17·66	...		19	57·3	M
8			59	17·69	...		19	56·5	M
10			59	17·65	...		19	56·5	M
20			59	17·72	...		19	55·3	R
21			59	17·72	...		19	55·6	R
24			59	17·77	...		19	55·2	R
Sep. 19	...		59	17·86	...		19	55·2	M
845		*41 Sagittarii π*							
June 17	...	19	1	51·16	...	111	13	56·5	M
846		*Anon.*							
Aug. 27	9·3	19	5	21·17	...	136	23	6·1	R
29	9·2		5	21·07	...		23	6·7	R
847		*Anon.*							
July 25	...	19	7	18·69	4	120	47	23·3	R
848		*Anon.*							
July 25	...	19	8	30·65	4	120	48	46·5	R
849		*Anon.*							
Aug. 29	8·8	19	10	19·96	5	146	12	40·1	R
850		*23 Aquilae ω*							
June 19	...	19	11	34·32	...	75	35	32·9	M
July 16	...		11	34·30			35	32·7	R
Aug. 16	...		11	34·31	...		35	31·4	M
20	...		11	34·30	...		35	31·5	R
24	...		11	34·36	...		36	31·3	R
Sep. 11	...		11	34·47	6		36	32·1	M
851		*30 Aquilae δ*							
June 18	..	19	18	47·46	...	87	8	53·4	M
19	...		18	47·46	...		8	52·7	M
July 18	..		18	47·47	...		8	53·9	R
25	.		18	47·42	...		8	54·2	R
Aug. 24	...		18	47·47	...		8	52·6	R
27	...		18	47·49	—		8	52·5	R
Sep. 6	...		18	47·41	...		8	56·4	M
852		*Anon.*							
Aug. 28	9·3	19	19	21·55	...	128	36	7·2	R
853		*Anon.*							
Aug. 28	9·0	19	26	22·14	...	127	48	58·6	R
854		*52 Sagittarii h^2*							
Aug. 21	...	19	28	36·57	...	115	10	29·4	R
27	...		28	36·56	—		10	27·6	R
855		*Anon.*							
Aug. 28	8·0	19	31	11·27	...	127	42	5·7	R
856		*Anon.*							
Aug. 28	8·0	19	34	2·07	...	127	41	58·3	R
857		*55 Sagittarii e^2*							
June 19	.	19	34	54·46	...	106	25	29·4	M
858		*50 Aquilae γ*							
July 27	.	19	39	56·09		79	42	31·2	R
31			39	56·11	...		42	31·3	R
Aug. 6	.		39	56·13			42	31·7	M
26	...		39	56·04			42	32·2	R
Sep. 6	...		39	56·07	...		42	31·8	M
10			39	56·01	...		42	31·9	M
859		*O. A. S. 1[illegible]6.*							
July 16	9·5	19	42	27·29		108	11	37·3	R

Separate Results of Madras Meridian Circle Observations in 1867.

Number and Date.	Magnitude.	Mean Right Ascension 1867. h.	m.	s.	No. of Wires.	Mean Polar Distance 1867. °	′	″	Observer.
860		33 *Aquilae* α, *Altair.*							
July 27	...	19	44	17·51	...	81	29	49·5	R
31	...		44	17·55	...		28	50·8	R
Aug. 26	...		44	17·49	5		29	49·9	R
28	.		44	17·50	...		29	50·2	R
861		55 *Aquilae* η *Var.* 1.							
Sep. 10	...	19	45	41·71	5	89	20	1·2	M
862		60 *Aquilae* β							
June 20	...	19	48	46·61	...	83	55	24·0	M
July 19	...		48	46·74	...		55	24·6	R
27	...		48	46·76	5		55	24·0	R
31	...		48	46·73	...		55	23·6	R
Aug. 6	...		48	46·67	...		55	24·5	M
27	...		48	46·70	...		55	23·9	R
Sep. 6	..		48	46·70	...		55	23·4	M
863		λ *Ursae Minoris.*							
Aug. 27	..	19	57	13·06	1	1	5	23·3	R
Sep. 6	.		57	13·88	1		5	26·0	M
		λ *Ursae Minoris—s.p.*							
Feb. 27	..	19	57	15·32	7	1	5	25·1	R
Mar. 6	...		57	13·08	2		5	24·5	M
7	..		57	12·90	3		5	24·7	M
15	..		57	13·91	1		5	21·9	M
864		*Anon.*							
Aug. 28	9·3	19	59	27·77	5	120	10	56·1	R
865		*S Cygni Var.* 4.							
July 19	9·5	20	2	49·80	...	32	23	1·4	R
866		*S Aquilae Var.* 4.							
Aug. 28	.	20	5	31·08	...	74	47	53·3	R

Number and Date.	Magnitude.	Mean Right Ascension 1867. h.	m.	s.	No. of Wires.	Mean Polar Distance 1867. °	′	″	Observer.
867		*Lacaille* 8370.							
Oct. 1	7·6	20	7	18·93	...	108	13	42·3	M
868		*O. A. S.* 20356.							
Aug. 16	9·6	20	8	31·97	...	110	26	36·6	R
869		6 *Capricorni* α²							
Aug. 7	...	20	10	49·31	...	102	57	19·5	M
14	...		10	49·31	...		57	19·0	M
23	—		10	49·30	5		57	18·1	R
Sep. 10	...		10	49·28	..		57	17·2	M
11	...		10	49·30	—		57	17·8	M
870		*Anon.*							
Oct. 2	7·6	20	11	26·51	5	100	16	10·7	M
871		*Lalande* 39045.							
Oct. 5	6·6	20	12	11·88	5	90	2	29·0	M
872		34 *Cygni Var.* 1.							
Sep. 6	6·0	20	12	53·06	6	52	22	48·4	M
873		9 *Capricorni* β							
June 19	...	20	13	32·09	...	105	11	26·3	M
20	...		13	32·08	...		11	24·6	M
874		*Lalande* 39125.							
July 19	8·0	20	15	41·23	5	108	12	49·0	R
875		*Anon.*							
Oct. 3	9·6	20	15	51·04	5	106	16	51·2	M
876		*Anon.*							
Aug. 27	9·8	20	17	1·92	5	121	11	11·7	R

Separate Results of Madras Meridian Circle Observations in 1867.

Number and Date.	Magnitude.	Mean Right Ascension 1867. h.	m.	s.	No. of Wires.	Mean Polar Distance 1867. °	′	″	Observer.
877				*11 Capricorni ρ*					
June 18	...	20	21	16·20	5	108	15	4·0	M
19	...		21	16·23	...		15	4·3	M
20	...		21	16·19	...		15	3·7	M
Aug. 15	...		21	16·14	5		15	3·9	M
26	...		21	16·35	4		16	6·5	R
28	...		21	16·24	...		15	4·8	R
Sep. 10	...		21	16·46	...		15	4·4	M
Oct. 1	...		21	16·20	4		15	3·2	M
4	...		21	16·24	...		15	4·5	M
878				*Anon.*					
Oct. 2	8·6	20	23	10·14	6	125	57	53·8	M
5	8·6		23	10·12	5		57	53·4	M
879				*Anon.*					
July 19	8·7	20	23	27·08	...	124	56	27·3	R
880				*Lalande 39525.*					
Oct. 3	7·0	20	25	5·04	6	80	1	54·2	M
881				*Anon.*					
July 18	...	20	26	32·30	5	121	12	4·1	R
882				*Anon.*					
Aug. 16	9·0	20	27	34·47	...	121	5	21·7	R
883				*Anon.*					
Aug. 16	9·5	20	28	54·11	5	121	5	51·3	R
884				*Anon.*					
Sep. 23	9·0	20	29	50·60		148	51	36·6	R
885				*Anon.*					
July 19	9·3	20	31	42·60		124	40	12·9	R
886				*Taylor 9518.*					
Aug. 14	7·0	20	32	6·85	.	105	36	27·3	M
Sep. 6	7·1		32	6·80	...		36	26·3	M
887				*Anon.*					
Sep. 30	9·2	20	34	52·15	...	129	13	8·2	R
Oct. 2	8·7		34	51·86	...		13	8·3	M
888				*Anon.*					
Oct. 3	8·6	20	35	58·17	...	134	50	41·1	M
889				*Anon.*					
Sep. 26	..	20	36	46·49	4	148	22	57·0	R
28	8·5		36	46·42	..		22	57·6	R
890				*50 Cygni α, Deneb.*					
June 18	...	20	36	53·61	...	45	11	37·9	M
Aug. 15	...		36	53·39	...		11	37·8	M
Sep. 11	...		36	53·71	...		11	39·6	M
19	...		36	53·83	...		11	39·7	R
891				*Anon.*					
Sep. 26	10·4	20	39	22·25	4	74	4	40·4	R
892				*O. A. S. 20841.*					
Oct. 5	7·9	20	39	28·79	.	116	53	55·8	M
7	7·9		39	28·60	...		53	55·9	M
8	7·9		39	28·83	...		53	54·9	M
893				*2 Aquarii ε*					
Aug. 24	...	20	40	28·45	5	99	54	50·5	R
894				*Anon.*					
Sep. 28	9·3	20	44	44·16		124	57	31·1	R
895				*6 Aquarii μ*					
Sep. 19	.	20	45	[illegible]	..	99	[illegible]	[illegible]	R

Separate Results of Madras Meridian Circle Observations in 1867.

Number and Date.	Magnitude.	Mean Right Ascension 1867. h. m. s.	No. of Wires.	Mean Polar Distance 1867. ° ′ ″	Observer.
896		32 *Vulpeculae.*			
Aug. 15	...	20 48 53·40	...	62 26 49·2	M
16	...	48 53·51	...	26 49·3	M
Sep. 19	...	48 53·48	5	26 49·1	R
20	...	48 53·46	...	26 49·1	R
23	...	48 53·47	5	26 51·0	R
24	...	48 53·46		26 48·9	R
26	...	48 53·47	5	26 49·9	R
30	...	48 53·49	5	26 49·6	R
Oct. 10	...	48 53·50	...	26 49·6	M
897		*Anon.*			
Oct. 1	9·0	20 50 56·00	5	148 15 10·8	M
898		*Lacaille* 8630.			
Sep. 27	7·0	20 51 35·61	5	126 38 31·4	R
899		*Anon.*			
Sep. 26	9·6	20 57 56·15	...	120 24 36·3	R
27	9·7	57 56·26	...	24 38·1	R
900		ι *Microscopii.*			
Sep. 24	...	20 57 50·14	...	120 30 8·7	R
Oct. 7	6·8	57 50·01	...	30 4·8	M
9	6·7	57 55·98	...	30 4·0	M
901		23 *Capricorni* θ			
Oct. 8	...	20 58 39·27	...	107 46 34·0	M
902		*Anon.*			
Aug. 16	9·7	20 59 58·99	6	129 0 45·7	R
21	..	59 58·78	4	0 46·6	R
903		*Anon.*			
Oct. 9	9·0	21 0 46·11	4	120 4 8·2	M
10	8·9	0 46·01		4 8·5	M
904		61 *Cygni—1st*			
Sep. 19	..	21 0 56·22	5	51 54 12·2	R
23	...	0 56·05	...	54 13·3	R
30	...	0 56·12	5	54 12·6	R
905		61 *Cygni—2nd*			
Sep. 20	...	21 0 57·40	.	51 54 13·6	R
906		*Anon.*			
Oct. 2	9·0	21 1 22·14	...	119 40 42·8	M
3	9·2	1 22·25	...	40 41·0	M
907		13 *Aquarii* ν			
July 18	..	21 2 20·68	..	101 54 20·8	R
Oct. 8	...	2 20·35	5	54 20·3	M
908		64 *Cygni* ζ			
June 20	...	21 7 16·51	...	60 19 3·1	M
July 19	...	7 16·44	...	19 4·1	R
Sep. 19	. .	7 16·51	...	19 3·9	R
20	...	7 16·47	...	19 2·9	R
23	...	7 16·46	...	19 4·0	R
24	...	7 16·56	..	19 2·7	R
26	...	7 16·49	...	19 2·4	R
Oct. 1	...	7 16·55	..	19 2·5	M
909		*Anon.*			
Aug. 16	9·3	21 13 50·48	...	128 59 31·2	R
Sep. 26	9·5	13 50·57	..	59 35·6	R
910		*Anon.*			
Aug. 27	7·8	21 14 30·35	5	125 28 57·7	R
Sep. 30	...	14 30·73	...	28 58·6	R
911		32 *Capricorni* ι			
June 30	...	21 15 49·09	3	107 28 57·4	M
Sep. 10	.	15 50·12	...	28 57·6	M
11		15 50·36	4	28 57·6	M

Separate Results of Madras Meridian Circle Observations in 1867.

Number and Date.	Magnitude.	Mean Right Ascension 1867. h. m. s.	No. of Wires.	Mean Polar Distance 1867. ° ′ ″	Observer.
912		*Anon.*			
July 19	...	21 18 48·39	5	153 51 51·7	B
Oct. 3	9·1	18 48·49	.	51 47·5	M
5	9·1	18 48·46	...	51 47·4	M
913		*Anon.*			
Oct. 1	7·0	21 20 10·60	5	150 47 4·0	M
914		*Anon.*			
Aug. 16	8·0	21 21 21·65	...	128 55 33·4	B
Oct. 2	8·0	21 21·78	...	55 32·0	M
915		*Lacaille* 8829.			
Aug. 27	7·0	21 22 46·40	...	127 7 46·5	B
Sep. 6	7·4	22 46·51	5	7 47·0	M
916		*Anon.*			
Oct. 7	9·9	21 23 7·93	...	110 6 41·8	M
917		22 *Aquarii* β			
July 19	..	21 24 33·31	...	96 9 17·3	B
Sep. 20		24 33·34	6	9 16·9	B
23	...	24 33·31	...	9 19·3	B
27	...	24 33·13	...	9 17·7	B
30	..	24 33·35	4	9 17·3	B
Oct. 4		24 33·38	.	9 17·6	M
8	.	24 33·30	...	9 18·1	B
918		*Anon.*			
Sep. 24	..	21 27 16·97	...	132 37 30·9	B
919		*Anon.*			
Aug. 27	7·7	21 30 54·06	...	127 46 25·1	B
920		40 *Capricorni* γ			
June 20		21 32 38·94	3	107 15 42·7	M
Sep. 10	...	32 38·90	...	15 41·7	M
921		*Taylor* 10068.			
Sep. 26	7·2	21 31 32·53		131 5 54·5	B
Oct. 2	7·1	31 32·38	.	5 56·0	M
922		*Taylor* 10065.			
Sep. 6	6·6	21 31 40·08	5	145 6 17·8	M
923		*Anon.*			
Oct. 3	8·3	21 34 34·23		133 50 40·1	M
924		8 *Pegasi* ε			
Aug. 16	...	21 37 39·17	...	80 44 1·6	B
Sep. 13	...	37 39·00	..	43 59·9	M
24	...	37 39·13	..	43 59·9	B
27	...	37 39·13	...	44 0·1	B
Oct. 1	...	37 39·06	–	43 59·8	M
5	...	37 39·09	..	44 0·9	M
7	...	37 39·25	...	44 1·2	M
10	...	37 38·92	.	44 1·0	M
925		*Anon.*			
Aug. 27	9·2	21 37 51·08	...	127 47 32·8	B
926		48 *Capricorni* λ			
Oct. 8	...	21 39 22·09		101 50 40·9	M
9	.	39 22·33	...	50 39·9	M
927		49 *Capricorni* δ			
Aug. 14		21 39 41·60		106 44 45·5	M
15		39 41·66	4	43 46·5	M
928		*Anon.*			
Aug. 27	9·3	21 41 6·10	.	127 48 39·7	B
Sep. 25	9·7	41 6·30		48 40·1	B
929		*Taylor* 10126.			
Oct. 4	7·0	21 41 11·11	4	132 13 34·7	M

Separate Results of Madras Meridian Circle Observations in 1867.

Number and Date.	Magnitude.	Mean Right Ascension 1867. h	m	s	No. of Wires.	Mean Polar Distance 1867. °	'	"	Observer.
930				*Anon.*					
Oct. 2	8·8	21	43	4·19	4	132	30	57·8	M
931				*Anon.*					
July 18	7·7	21	46	13·06	...	127	31	7·3	R
Sep. 6	7·3		45	12·99	6		31	6·5	M
932				51 *Capricorni* μ					
Aug. 14		21	46	2·46	..	104	10	35·7	M
Oct. 8			46	2·30	..		10	36·3	M
9			46	2·02	...		10	35·5	M
933				16 *Pegasi.*					
Aug. 16		21	47	0·58	...	64	41	58·5	R
26	...		47	0·53	5		41	58·9	R
Sep. 26	...		47	0·61	...		41	59·9	R
27	...		47	0·65	5		42	0·2	R
934				*Anon.*					
Sep. 24	9·5	21	50	52·91		127	35	35·8	R
Oct. 7	9·3		50	52·83	5		35	36·3	M
935				*Anon.*					
July 18	9·0	21	52	56·64	...	127	29	47·0	R
Aug. 27	9·2		52	56·56			29	46·2	R
Sep. 24	9·2		52	56·64			29	47·8	R
936				*Anon.*					
Sep. 26	9·0	21	52	59·44		129	31	50·6	R
27	9·0		52	59·32			31	49·5	R
Oct. 3	8·9		52	59·52			31	50·9	M
937				ι *Indi.*					
Oct. 1	5·8	21	53	10·19	6	147	19	50·7	M
10	5·9		53	10·05	..		19	51·1	M
938				*Lacaille* 9006.					
Sep. 27	7·6	21	56	20·25	.	129	31	1·1	R
Oct. 5	7·1		56	20·24			31	2·7	M

Number and Date.	Magnitude.	Mean Right Ascension 1867. h	m	s	No. of Wires.	Mean Polar Distance 1867. °	'	"	Observer.
939				34 *Aquarii* α					
Sep. 13		21	58	56·96	5	90	57	55·9	M
26			58	57·14	.		57	54·0	R
Oct. 2			58	57·07	...		57	54·8	M
8			58	57·01			57	54·1	M
940				33 *Aquarii* ι					
July 18		21	59	16·08	...	104	30	59·1	R
19	...		59	16·01	...		30	58·3	R
941				*W. B. E. XXI.* 1418.					
Sep. 24	9·0	22	1	58·24		78	8	36·6	R
942				*Anon.*					
Sep. 27	9·7	22	3	21·97	...	129	4	37·0	R
943				43 *Aquarii* θ					
July 16		22	9	48·77	...	98	26	41·0	R
Oct. 2	...		9	48·74			26	40·9	M
3			9	48·68	...		26	40·7	M
4	..		9	48·73			26	40·7	M
5	...		9	48·84			26	39·8	M
7	...		9	48·78	...		26	40·6	M
10			9	48·70	...		26	41·1	M
19			9	48·73	...		26	41·3	R
944				*Anon.*					
Sep. 20	8·2	22	12	14·21	..	129	26	49·2	R
945				*Anon.*					
Sep. 27	7·6	22	12	20·64	5	150	34	46·6	R
946				*Anon.*					
Sep. 26	9·0	22	14	54·62		129	28	28·3	R
Oct. 23	9·2		14	54·70	6		28	27·9	R
947				*Anon.*					
Oct. 5	9·2	22	19	2·14	6	100	44	49·6	M
7	9·2		19	2·13	...		44	52·5	M

Separate Results of Madras Meridian Circle Observations in 1867.

Number and Date.	Magnitude.	Mean Right Ascension 1867. h.	m.	s.	No. of Wires.	Mean Polar Distance 1867. °	′	″	Observer.
948		*R. P. L.* 150.							
Sep. 24	..	22	23	27·55	3	4	33	46·6	R
Oct. 1			23	27·41	3		33	45·9	M
8	...		23	27·56	2		33	47·0	M
		R. P. L. 150.—*s.p.*							
Mar. 1		22	23	28·13	2	4	33	46·3	M
12			23	27·76	2		33	48·6	M
Apl. 1			23	27·48	3		33	48·3	M
10	...		23	27·04	1		33	48·8	M
15	..		23	27·18	3		33	46·3	M
949		57 *Aquarii* σ							
Aug. 15	...	22	23	36·23	...	101	21	27·8	M
16	...		23	36·51	...		21	26·4	R
Oct. 9	...		23	36·34	...		21	27·9	M
10	...		23	36·39	...		21	26·9	M
950		*Anon.*							
Oct. 22	8·2	22	26	52·92	5	130	39	34·2	R
951		27 *Cephei* δ *Var.* 1.							
Oct. 2	...	22	24	14·17	6	32	15	54·7	M
3	...		24	14·26	...		15	54·8	M
952		*Anon.*							
Oct. 19	9·3	22	24	47·95	5	129	46	46·9	R
953		62 *Aquarii* η							
Oct. 2	...	22	28	31·27	4	90	49	8·8	M
9	...		28	31·17	4		49	6·5	M
16	...		28	31·17	...		49	7·8	R
Nov. 1	...		28	31·30	..		49	8·2	M
954		*R. P. L.* 153.							
Nov. 2	.	22	29	8·04	1	7	30	36·7	M
955		63 *Aquarii* κ							
Aug. 15		22	30	51·89	...	94	54	49·1	M
16	.		30	52·02	...		54	49·2	R
956		*Anon.*							
Oct. 7	8·9	22	34	24·87	4	135	39	34·8	M
957		42 *Pegasi* ζ							
Oct. 1	...	22	34	49·64	..	79	51	46·3	M
2	..		34	49·59	.		51	43·8	M
3			34	49·65	5		51	46·6	M
5			34	49·70	..		51	43·3	M
9	...		34	49·68	...		51	44·1	M
18	...		34	49·70	..		51	44·6	R
19	...		34	49·74			51	44·2	R
21	...		34	49·79	...		51	46·5	R
23	...		34	49·73	..		51	43·4	R
Nov. 4	...		34	49·74			51	43·5	M
5	...		34	49·78	..		51	44·0	M
958		67 *Aquarii.*							
July 19	—	22	36	17·83	...	97	30	32·0	R
959		*Anon.*							
Oct. 8	8·2	22	38	40·70	5	130	25	2·3	M
960		*Anon.*							
Oct. 21	9·5	22	48	37·26	.	130	26	42·7	R
961		*Anon.*							
Nov. 5	8·0	22	44	52·01	...	105	40	20·0	M
962		73 *Aquarii* λ							
July 19	.	22	45	40·11		98	17	13·2	R
Oct. 9			45	40·06			17	12·2	M
10	.		45	40·25			17	12·2	M

Separate Results of Madras Meridian Circle Observations in 1867.

Number and Date	Magnitude	Mean Right Ascension 1867. h.	m.	s.	No. of Wires	Mean Polar Distance 1867. °	′	″	Observer
963			*Anon.*						
Oct. 22	9.5	23	47	33.88	5	128	56	54.4	B
964			*Anon.*						
Sep. 24	9.5	22	48	13.84	5	132	34	14.8	B
Oct. 3	9.0		48	16.84	5		34	14.3	M
965			*O. A. S. 22500.*						
Oct. 5	8.0	22	49	10.48	4	119	18	54.4	M
7	8.5		49	10.39	6		18	56.6	M
966			*S Aquarii Var. 2.*						
Sep. 23	10.2	22	49	58.98	5	111	3	9.8	B
967			*2 Piscis Australis α, Fomalhaut.*						
Oct. 4	...	22	50	17.64	...	120	19	36.4	M
18	...		50	17.48	5		19	36.5	B
Nov. 6	...		50	17.58	...		19	36.4	M
8	...		50	17.56	5		19	35.7	B
968			*Anon.*						
Sep. 30	8.0	22	53	34.88	...	128	4	20.4	B
Oct. 8	7.9		53	34.60	6		4	21.2	M
969			*Anon.*						
Sep. 26		22	54	49.79	3	101	42	6.6	B
970			*58 Pegasi β Var. 1, Scheat.*						
Nov. 11	5.9	22	57	19.29	6	62	39	17.4	M
971			*Anon.*						
Oct. 31	9.5	22	57	39.98	...	87	10	48.7	B
972			*Anon.*						
Oct. 21	9.0	22	57	23.47	4	149	37	3.2	B
973			*54 Pegasi α, Markab.*						
Oct. 7	...	22	58	8.18	3	75	30	36.0	M
9	...		58	8.20			30	36.8	M
22			58	8.22			30	34.8	B
24	.		58	8.13	.		30	35.4	B
26			58	8.16	...		30	35.8	B
974			*R Pegasi Var. 2.*						
Oct. 29	8.5	22	59	56.90	5	80	10	29.2	B
975			*Anon.*						
Oct. 30	9.2	23	3	16.60	...	137	29	6.2	B
976			*90 Aquarii φ*						
Aug. 16		23	7	26.03	...	96	45	56.7	B
Sep. 18	..		7	26.13	2		45	56.9	M
977			*Anon.*						
Oct. 21	8.7	23	7	35.78	..	129	54	6.8	B
29	8.0		7	35.94	5		54	6.1	B
Nov. 2	7.9		7	35.86	...		54	6.6	M
978			*Anon.*						
Oct. 2	9.0	23	10	0.79	5	151	48	30.3	M
8	8.9		10	0.91	6		48	30.6	M
979			*6 Piscium γ*						
Oct. 10	...	23	10	16.31	...	87	26	39.5	M
19	...		10	16.17	...		26	38.9	B
22	...		10	16.15	...		26	39.3	B
24	...		10	16.38	..		26	39.1	B
25	...		10	16.19	...		26	38.4	B
26	...		10	16.20	...		26	39.0	B
29	...		10	16.21	...		26	38.6	B
31	...		10	16.29			26	38.1	B
Nov. 1	...		10	16.28			26	39.1	M
4			10	16.18			26	37.6	M

Separate Results of Madras Meridian Circle Observations in 1867.

Number and Date	Magnitude	Mean Right Ascension 1867. h.	m.	s.	No. of Wires.	Mean Polar Distance 1867. °	′	″	Observer.
980			*93 Aquarii ψ³*						
Nov. 6	...	23	10	59·37	...	99	54	30·1	M
981			*Anon.*						
Sep. 26	9·0	23	11	41·27	5	131	7	16·7	R
Oct. 20	9·3		11	44·23	...		7	12·0	R
982			*Anon.*						
Oct. 19	9·5	23	17	4·35	...	127	26	29·6	R
983			*Anon.*						
Sep. 26	9·7	23	18	24·68	...	131	7	20·3	R
Oct. 29	9·7		18	24·78	5		7	27·0	R
984			*Anon.*						
Sep. 30	8·8	23	18	30·61	...	127	16	46·0	M
Oct. 21	8·3		18	30·78	5		16	46·5	R
Nov. 2	9·1		18	30·61	...		16	48·3	M
985			*8 Piscium κ*						
Aug. 16	...	23	20	6·96	5	89	28	19·6	R
Oct. 10	...		20	6·80	5		28	21·2	M
21	...		20	6·87	...		28	20·6	R
25	...		20	6·90	...		28	20·7	R
26	...		20	6·84	...		28	20·0	R
31	...		20	6·85	6		28	20·2	R
Nov. 1	...		20	6·81	...		28	20·9	M
5	...		20	6·99	...		28	20·1	M
15	...		20	6·94	...		28	20·2	M
986			*Anon.*						
Nov. 4	7·5	23	20	55·78	...	86	50	30·5	M
987			*Lacaille 9496.*						
Oct. 19	8·0	23	23	26·66	.	127	41	29·1	R
30	8·0		23	25·61	5		41	26·4	R
988			*Anon.*						
Sep. 23	7·4	23	24	27·79	.	127	1	16·6	R
26	8·0		24	27·80	5		1	17·5	R
989			*Lacaille 9514.*						
Oct. 20	9·0	23	26	2·05	.	131	51	46·9	R
990			*Lacaille 9517.*						
Sep. 20	7·0	23	26	36·47		136	30	58·3	R
26	7·2		26	36·68	...		30	58·1	R
Oct. 24	7·0		26	36·61	.		30	56·8	R
991			*Anon.*						
Oct. 19	9·3	23	30	50·26		130	6	8·5	R
21	9·5		30	50·30			6	7·4	R
992			*Anon.*						
Oct. 30	7·8	23	30	51·91	...	127	32	8·2	R
Nov. 2	7·9		30	51·80	...		32	10·3	M
4	7·8		29	51·81			32	8·6	M
993			*17 Piscium ι*						
Oct. 3	...	23	33	6·65	5	84	5	39·8	M
5	...		33	6·80	...		5	39·8	M
7	...		33	6·81			5	40·2	M
9	...		33	6·68			5	39·5	M
22	...		33	6·67			5	39·8	R
31			33	6·65			5	39·1	R
Nov. 1			33	6·59			5	39·4	M
5			33	6·80			5	39·7	M
7			33	6·57			5	40·3	M
994			*18 Piscium λ*						
Oct. 10		23	35	15·40	4	88	57	6·1	M
995			*Anon.*						
Oct. 29	6·7	23	35	21·42	5	109	41	25·7	R
Nov. 11	6·0		35	21·27	5		41	26·3	M
996			*Anon.*						
Sep. 26	9·3	23	36	5·36		128	5	41·1	R
Oct. 30	9·5		36	5·28	5		5	40·6	R

Separate Results of Madras Meridian Circle Observations in 1867.

Number and Date.	Magnitude.	Mean Right Ascension. 1867. h.	m.	s.	No. of Wires.	Mean Polar Distance. 1867. °	′	″	Observer.
997		*R Aquarii Var. 1.*							
Oct 19	9·5	23	36	50·47	5	106	1	19·8	R
21	9·9		36	50·37			1	16·7	R
998		*Lacaille* 9583.							
Nov. 2	8·0	23	39	0·65	...	128	42	54·5	M
4	7·9		39	0·63	...		42	53·4	M
999		*Lacaille* 9597.							
Sep. 27	8·2	23	40	33·31	5	128	53	43·5	R
Nov. 5	7·8		40	32·99	4		53	41·0	M
12	7·9		40	32·40	...		53	45·7	M
1000		δ *Sculptoris.*							
Oct 3	...	23	41	50·67		118	51	57·1	M
8	...		41	50·58			51	57·3	M
14	...		41	50·48	4		51	58·2	M
28			41	50·61			51	57·1	R
1001		*Anon.*							
Sep 30	7·5	23	42	8·16		112	3	27·1	R
1002		21 *Piscium.*							
Nov 7		23	42	39·86	...	90	30	45·5	M
1003		*Anon.*							
Oct 22	9·3	23	42	56·63	...	120	42	48·5	R
30	9·2		42	56·48			42	48·2	R
1004		*Anon.*							
Sep. 27	9·0	23	47	1·44	-	128	8	51·4	R
Oct. 19	9·0		47	1·30	5		8	52·5	R
Nov 11	8·0		47	1·28	5		8	51·2	M
1005		*Lacaille* 9641.							
Nov 5	7·8	23	49	11·66	4	129	6	17·1	M
12	7·8		49	11·80			6	17·7	M
1006		*Lacaille* 9650.							
Oct. 30	8·2	23	49	15·60	...	130	47	10·3	R
1007		*Anon.*							
Sep. 24	9·0	23	51	54·98	5	132	19	49·5	R
1008		28 *Piscium*							
Oct. 8		23	52	29·00	.	88	32	23·4	M
20			52	28·80	6		32	22·6	R
Nov. 15	...		52	28·90	...		32	23·3	M
1009		29 *Piscium.*							
Nov. 7	...	23	55	0·82	6	93	46	5·7	M
1010		*Anon.*							
Oct 30	9·5	23	56	7·75		100	16	1·3	R
Nov. 16	9·5		56	7·90	...		16	2·1	R
1011		*Anon.*							
Oct 21	9·3	23	56	35·79	6	124	42	22·8	R
1012		*Taylor* 10990.							
Nov 12	8·3	23	57	3·48	...	148	24	13·3	M
1013		*Taylor* 10997.							
Sep. 27	9·0	23	58	12·65	.	120	46	30·6	R
1014		*Anon.*							
Sep. 24	9·2	23	59	44·60	.	125	52	14·1	R
Nov. 5	8·7		59	44·41	...		52	16·6	M
1015		*Lacaille* 9721.							
Nov. 11	6·0	23	59	25·27		129	45	51·8	M
1016		*Lacaille* 9723.							
Nov. 6	7·7	23	59	51·62	6	126	49	27·8	M

MEAN POSITIONS OF STARS

OBSERVED WITH THE

MADRAS MERIDIAN CIRCLE

IN THE YEAR

1867

REDUCED TO JANUARY 1, OF THAT YEAR.

Mean Positions of Stars for 1867 January 1st.

Number.	Star.	Magnitude.	Estimations.	Mean Right Ascension.			Mean Polar Distance.			Observations.	Fraction of Year.
				h	m	s	°	′	″		
1	21 Androm. α (*Alpherat*) ..	2·0	...	0	1	31·04	61	38	38·7	3	0·77
2	... ...	9·2	1	0	2	1·86	127	29	28·8	1	0·81
3	Lacaille 9746	8·0	1	0	2	37·24	146	55	55·5	1	0·84
4	... ...	9·4	2	0	5	7·73	126	17	5·8	2	0·80
5	... ...	8·0	1	0	6	30·18	131	7	17·9	1	0·70
6	88 Pegasi γ (*Algenib*) ...	3·0	...	0	6	23·30	75	38	22·4	4	0·80
7	Lalande 163	7·6	2	0	7	47·98	89	30	35·0	2	0·86
8	Taylor 64	7·7	3	0	14	16·77	129	38	39·2	3	0·88
9	Lalande 421	7·4	1	0	15	56·60	81	39	1·1	2	0·90
10	Lacaille 61	6·0	1	0	16	10·26	120	39	39·3	1	0·84
11	Lacaille 86 ...	8·0	1	0	19	30·67	130	28	1·8	1	0·89
12	... ...	8·8	2	0	22	51·01	94	46	0·5	2	0·84
13	12 Ceti	6·2	...	0	23	15·06	94	41	34·8	6	0·86
14	Lalande 670 ...	6·6	1	0	23	18·35	86	32	38·7	1	0·88
15	... ...	9·6	1	0	26	32·08	76	8	14·8	1	0·69
16	... .	10·5	1	0	26	13·15	76	14	13·7	1	0·71
17	.. ..	8·3	1	0	28	32·76	141	58	22·3	1	0·78
18	Taylor 148... ...	6·5	2	0	29	8·09	143	6	30·8	2	0·86
19	... ...	9·0	3	0	32	5·31	128	43	52·5	3	0·84
20	... ...	9·1	2	0	35	7·90	88	27	58·2	2	0·80
21	16 Ceti β	2·1	...	0	36	54·65	108	40	2·6	7	0·96
22	W. B. E. 0.686	9·0	1	0	37	3·30	98	48	19·6	1	0·97
23	W. B. E. 0.697	8·7	1	0	40	52·60	96	13	38·4	2	0·98
24	W. B. E. 0.705	7·9	1	0	41	17·36	98	36	30·9	2	0·98
25	63 Piscium δ	4·6	...	0	41	46·96	83	8	21·3	1	0·96
26	W. B. E. 0.716	8·8	1	0	41	50·63	91	35	49·5	1	0·92
27	Lacaille 294	7·0	3	0	46	3·15	129	15	8·1	3	0·90
28	.. ..	9·4	2	0	46	19·13	129	12	53·3	2	0·86
29	... ...	9·5	2	0	50	28·91	120	39	11·0	3	0·87
30	Lacaille 304	7·6	3	0	50	52·38	151	41	8·7	3	0·86
31	... ...	9·3	2	0	51	11·90	120	39	36·9	2	0·91
32	.. ..	8·8		0	51	59·06	120	41	8·2	1	0·79
33	71 Piscium ε .	4·5		0	56	2·46	82	49	36·6	5	0·85
34	... ...	9·8	2	1	1	33·72	17	36	22·3	2	0·89
35	... .	7·4	1	1	3	15·94	150	15	30·3	1	0·94

7.—Star occulted by the moon, when totally eclipsed, on 1866 Sep. 24.
22.—23.—24.—26.—Comparison stars for Europa in 1881.

Observed with the Madras Meridian Circle in that Year.

Number.	Star.	In Right Ascension. Annual Precession.	Secular Variation.	Proper Motion.	In Polar Distance. Annual Precession.	Secular Variation.	Proper Motion.	Number in B. A. C.
		s	s	s	"	"	"	
1	21 Androm. α *Alpheratz*	+ 3·0768	+ 0·0182	+ 0·009	− 20·086	+ 0·013	+ 0·15	4
2	. . .	+ 3·0681	− 0·0807	.	− 20·084	+ 0·013	...	...
3	Lacaille 9746 ..	+ 3·0456	− 0·0429	...	− 20·052	+ 0·015	.	...
4	..	+ 3·0501	− 0·0194	...	− 20·051	+ 0·019	...	...
5	..	+ 3·0398	− 0·0288	...	− 20·046	+ 0·021	...	...
6	88 Pegasi γ (*Algenib*)	+ 3·0816	+ 0·0100	0·000	− 20·048	+ 0·022	+ 0·02	26
7	Lalande 168 . ..	+ 3·0724	+ 0·0026	...	− 20·044	+ 0·024	...	...
8	Taylor 64	+ 3·0023	− 0·0211		− 20·016	+ 0·037	. .	60
9	Lalande 421	+ 3·1447	+ 0·0271		− 20·007	+ 0·041	...	...
10	Lacaille 61 ..	+ 2·9930	− 0·0309		− 20·005	+ 0·040	...	...
11	Lacaille 88	+ 2·9747	− 0·0207	...	− 19·961	+ 0·046	...	...
12	... .	+ 3·0611	+ 0·0007	..	− 19·960	+ 0·054	...	...
13	12 Ceti	+ 3·0609	+ 0·0008	− 0·002	− 19·952	+ 0·055	+ 0·01	112
14	Lalande 670	+ 3·0819	+ 0·0054	...	− 19·951	+ 0·054	...	118
15	..	+ 3·1058	+ 0·0103	..	− 19·931	+ 0·060	...	—
16	... .	+ 3·1000	+ 0·0100		− 19·934	+ 0·061	...	...
17	.. ..	+ 2·8490	− 0·0326		− 19·917	+ 0·066	...	—
18	Taylor 148	+ 2·8540	− 0·0308		− 19·904	+ 0·060	...	148
19	.	+ 2·9226	− 0·0176		− 19·859	+ 0·069	...	...
20	... ..	+ 3·0965	+ 0·0073		− 19·819	+ 0·078	...	...
21	16 Ceti β .	+ 2·9991	− 0·0065	+ 0·013	− 19·796	+ 0·080	− 0·08	196
22	W. B. E. 0.690	+ 3·0670	+ 0·0020	...	− 19·794	+ 0·080	...	...
23	W. B. E. 0.697 ..	+ 3·0503	+ 0·0015	...	− 19·737	+ 0·087	...	...
24	W. B. E. 0.706 ..	+ 3·0535	+ 0·0010	...	− 19·731	+ 0·087	...	...
25	63 Piscium δ ..	+ 3·1013	+ 0·0077	+ 0·008	− 19·738	+ 0·090	+ 0·05	220
26	W. B. E. 0.716 ..	+ 3·0586	+ 0·0018		− 19·721	+ 0·088	...	...
27	Lacaille 284 ..	+ 2·8687	− 0·0161		− 19·668	+ 0·080	...	...
28	... . ..	+ 2·8578	− 0·0160		− 19·664	+ 0·090	...	...
29	... .. .	+ 2·6300	− 0·0155		− 19·570	+ 0·096	...	...
30	Lacaille 304 ...	+ 2·4409	− 0·0318		− 19·563	+ 0·097	...	...
31	... ..	+ 2·8076	− 0·0160	...	− 19·555	+ 0·099		...
32	.. .. ..	+ 2·8136	− 0·0160	...	− 19·548	+ 0·100	...	...
33	71 Piscium ε	+ 3·1189	+ 0·0067	− 0·002	− 19·460	+ 0·110	0·00	296
34	... . .	+ 4·1014	+ 0·1622	...	− 19·384	+ 0·171	...	...
35	... .	+ 2·4846	− 0·0287	...	− 19·296	+ 0·166	...	. .

13.—Proper motions from "*Greenwich Catalogue* 1872."

Mean Positions of Stars for 1867 January 1st.

Number.	Star.	Magnitude.	Estimations.	Mean Right Ascension.			Mean Polar Distance.			Observations.	Fraction of Year.
				h.	m.	s.	°	′	″		
36	... ...	9·0	1	1	4	1·81	18	36	56·2	1	0·90
37	... ...	9·4	3	1	4	56·14	127	20	48·2	3	0·83
38	O. A. N. 1308	7·0	1	1	9	2·56	18	18	2·6	1	0·93
39	1 Urs. Min. α (*Polaris*) ...	2·2	...	1	10	16·60	1	28	59·4	12	0·56
40	... ...	9·4	1	1	12	21·17	132	16	34·2	1	0·88
41	Brisbane 208 ...	6·9	1	1	15	7·20	150	46	24·0	1	0·84
42	45 Ceti θ¹	3·8	...	1	17	22·49	98	52	14·7	6	0·86
43	... ...	8·0	...	1	19	0·18	151	19	27·8	1	0·71
44	Taylor 486 ...	7·0	1	1	19	38·01	91	5	28·3	1	0·83
45	99 Piscium η ...	4·5	...	1	24	22·13	75	30	27·2	10	0·95
46	Taylor 496	8·0	1	1	25	40·04	140	36	15·8	1	0·71
47	... ...	8·5	1	1	25	51·30	160	20	42·7	1	0·89
48	102 Piscium π	5·6	...	1	30	3·03	78	32	22·0	1	0·89
49	Taylor 525	6·0	...	1	30	15·64	148	49	13·0	1	0·92
50	... ...	8·6	2	1	31	30·84	130	51	30·7	2	0·91
51	Taylor 539	6·0	1	1	31	52·28	148	57	0·8	1	0·52
52	α Eridani (*Achernar*) ...	1·0	...	1	32	46·69	147	54	50·2	2	0·61
53	106 Piscium ν	4·7	...	1	34	30·68	85	11	11·5	8	0·94
54	Lacaille 507	6·0	...	1	37	14·77	151	27	35·6	1	0·92
55	110 Piscium ο	4·4	...	1	38	28·22	81	30	46·2	1	0·71
56	... ...	9·2	1	1	38	30·77	152	1	56·3	1	0·92
57	... ...	8·4	1	1	39	50·79	140	26	28·4	1	0·92
58	... ...	9·7	2	1	40	30·43	128	36	6·7	2	0·84
59	... .	9·3	1	1	40	16·03	145	57	3·6	1	0·96
60	Lacaille 550 ...	8·5	3	1	46	43·48	129	24	14·6	3	0·86
61	6 Arietis β	2·8	...	1	47	17·76	69	50	37·6	10	0·46
62	... .	9·0	1	1	49	22·47	156	6	37·6	1	0·89
63	... ..	9·0	1	1	52	24·32	145	47	31·3	1	0·94
64	... ..	8·8	1	1	58	3·63	151	22	2·7	1	0·94
65	... ...	9·4	2	1	53	14·85	129	57	30·7	2	0·96
66	... .	9·3	1	1	56	11·52	129	26	22·6	1	0·89
67	13 Arietis α ..	2·0	..	1	59	40·81	67	10	6·2	10	0·47
68	.. ..	8·8	1	2	1	10·63	144	46	47·6	1	0·94
69	... ...	9·3	1	2	6	11·21	151	38	16·8	1	0·94
70	21 Arietis	5·6		2	8	14·30	65	34	31·9	1	0·90

36.—38.—Stars observed for map of S Cassiopeæ Var. 4

Observed with the Madras Meridian Circle in that Year.

Number.	Star.	In Right Ascension. Annual Precession.	In Right Ascension. Secular Variation.	In Right Ascension. Proper Motion.	In Polar Distance. Annual Precession.	In Polar Distance. Secular Variation.	In Polar Distance. Proper Motion.	Number in B. A. C.
		s	s	s	"	"	"	
36	... .	+ 4·1608	+ 0·1517	...	− 19·277	+ 0·177	...	
37	... .	+ 2·7800	− 0·0128	...	− 19·256	+ 0·122	...	..
38	O. A. N. 1303 ..	+ 4·2718	+ 0·1510	...	− 19·152	+ 0·194	...	...
39	1 Urs. Min. α (*Polaris*)	+ 10·5920	+ 13·6992	+ 0·066	− 19·119	+ 0·871	0·00	300
40	... .	+ 2·3898	− 0·0208	...	− 19·064	+ 0·111		...
41	Brisbane 308	+ 2·3098	− 0·0190	...	− 18·987	+ 0·116	...	...
42	45 Ceti θ¹	+ 3·0080	+ 0·0018	− 0·007	− 18·928	+ 0·154	+ 0·22	430
43	... ...	+ 2·3462	− 0·0173	...	− 18·875	+ 0·110	...	...
44	Taylor 465	+ 3·0634	+ 0·0066	...	− 18·856	+ 0·160	...	455
45	99 Piscium η	+ 3·1080	+ 0·0142	0·000	− 18·711	+ 0·176	0·00	463
46	Taylor 496	+ 2·4769	− 0·0141	...	− 18·665	+ 0·140	...	462
47	... .	+ 2·2129	− 0·0148	...	− 18·604	+ 0·126	...	...
48	108 Piscium ε ..	+ 3·1760	+ 0·0125	− 0·007	− 18·526	+ 0·186	− 0·08	498
49	Taylor 585	+ 2·2948	− 0·0185	...	− 18·590	+ 0·183	...	408
50	... .	+ 2·6226	− 0·0101	...	− 18·477	+ 0·157	...	...
51	Taylor 589	+ 2·2066	− 0·0129	...	− 18·465	+ 0·183	...	497
52	α Eridani (*Achernar*)	+ 2·2234	− 0·0128	+ 0·008	− 18·436	+ 0·187	+ 0·07	507
53	106 Piscium ν ...	+ 3·1172	+ 0·0091	− 0·004	− 18·374	+ 0·191	− 0·04	518
54	Lacaille 507	+ 2·0600	− 0·0090	...	− 18·276	+ 0·182	...	601
55	110 Piscium ο ...	+ 3·1561	+ 0·0111	+ 0·006	− 18·236	+ 0·109	− 0·01	507
56	... .	+ 2·0213	− 0·0090		− 18·226	+ 0·181		...
57	... ...	+ 2·1161	− 0·0100		− 18·176	+ 0·138		...
58	... .	+ 2·6188	− 0·0081		− 18·156	+ 0·171		...
59	... .	+ 2·0790	− 0·0082		− 17·938	+ 0·144		...
60	Lacaille 550	+ 2·5788	− 0·0076		− 17·910	+ 0·177		...
61	6 Arietis β	+ 3·2935	+ 0·0186	+ 0·002	− 17·907	+ 0·236	+ 0·11	577
62	... ...	+ 2·6341	− 0·0068	...	− 17·815	+ 0·184	...	...
63	.. .. ..	+ 2·1436	− 0·0077	...	− 17·601	+ 0·156	...	..
64	... ...	+ 1·9124	− 0·0040	...	− 17·604	+ 0·140	...	...
65	... .	+ 2·5879	− 0·0066	...	− 17·561	+ 0·187	...	...
66	... ...	+ 2·5379	− 0·0064	...	− 17·580	+ 0·188	...	–
67	13 Arietis α	+ 3·3530	+ 0·0203	+ 0·012	− 17·362	+ 0·242	+ 0·16	669
68	... .	+ 1·9893	− 0·0036	...	− 17·317	+ 0·151	...	...
69	– ...	+ 1·7997	+ 0·0006	...	− 17·092	+ 0·144	...	
70	21 Arietis ...	+ 3·3942	+ 0·0215	− 0·006	− 16·999	+ 0·269	+ 0·07	698

52 — Proper motions from "*Stone's Cape Catalogue.*"
70 — Proper motions from "*Greenwich Catalogue* 1861."

Mean Positions of Stars for 1867 January 1st.

Number.	Star.	Magnitude.	Estimations.	Mean Right Ascension.			Mean Polar Distance.			Observations.	Fraction of Year.
				h.	m.	s.	°	′	″		
71	67 Ceti	5·5	...	2	10	21·03	97	2	12·1	9	0·87
72	68 Ceti o Var. 1 (*Mira*) ...	6·0	2	2	12	37·71	93	36	0·2	3	0·90
73	... ...	9·0	2	2	12	46·40	98	34	46·7	3	0·98
74	... ...	8·0	1	2	15	44·70	152	38	21·1	1	0·96
75	.. ..	8·5	2	2	17	40·30	151	3	20·2	2	0·92
76	73 Ceti ξ²	4·4	...	2	21	5·36	82	8	15·4	16	0·89
77	Lacaille 782 ...	7·0	1	2	26	15·38	148	24	8·5	1	0·93
78	... ...	8·5	3	2	27	30·97	147	11	24·2	3	0·93
79	... ...	9·2	2	2	28	15·92	120	45	40·0	2	0·93
80	... ...	9·7	1	2	30	51·00	147	34	6·1	1	0·87
81	... ...	9·0	1	2	31	17·41	148	30	32·8	1	0·94
82	... ...	9·6	2	2	31	20·61	161	28	36·5	2	0·93
83	86 Ceti γ	3·6	...	2	36	24·61	87	10	31·8	9	0·88
84	87 Ceti μ	4·4	...	2	37	45·20	80	26	57·0	2	0·78
85	Lacaille 868 ...	7·0	1	2	38	38·12	147	12	24·4	1	0·92
86	Taylor 940 ...	7·0	2	2	40	22·32	128	43	45·6	2	0·97
87	... ...	9·6	3	2	41	47·20	161	1	56·1	3	0·94
88	Taylor 960 ...	7·2	2	2	46	47·40	74	3	41·5	2	0·98
89	Taylor 978 ...	6·9	2	2	47	36·42	147	44	27·2	2	0·94
90	Lacaille 941 ...	6·9	1	2	50	31·96	146	26	30·8	1	0·87
91	Taylor 991	7·2	3	2	50	40·34	129	11	26·7	3	0·97
92	Lacaille 969	7·8	2	2	54	58·75	144	18	14·1	2	0·91
93	42 Ceti α (*Menkar*) ...	2·7	...	2	55	18·71	86	26	2·0	4	0·96
94	25 Persei ρ Var. 2 ..	3·7	...	2	56	30·76	51	40	29·7	1	0·93
95	26 Persei β Var. 1 (*Algol*).	2·3	...	2	59	31·30	49	33	33·2	3	0·90
96	Taylor 1047	6·5	1	2	59	56·67	151	10	10·0	1	0·97
97	... ...	8·0	1	2	59	57·96	130	37	40·8	1	0·98
98	Taylor 1052 ...	5·3	...	3	0	29·38	160	15	16·1	1	0·92
99	Taylor 1057	8·0	2	3	0	50·41	161	21	38·5	2	0·89
100	R. P. L. 33	5·0	..	3	1	20·76	5	34	6·0	1	0·96
101	... .	9·3	1	3	2	16·73	130	38	13·4	1	0·98
102	57 Arietis δ	4·5	.	3	4	1·66	70	40	48·7	6	0·94
103	Taylor 1081	8·0	1	3	5	16·31	151	30	33·2	1	0·90
104	—	9·0	1	3	6	15·83	128	31	22·9	1	0·97
105	... ...	9·4	2	3	7	21·46	146	30	51·1	2	0·89

72.—o Ceti Var. 1 (*Mira*)—Period 331 days.—Range, 2nd to 10th magnitude.
94.—ρ Persei Var. 2.—Changes irregularly from 3·5 to 4·3 magnitude.
95.—β Persei Var. 1 (*Algol*).—Period 2·867 days.—Range, 2·3 to 4th magnitude
100.—Groombridge 566.

Observed with the Madras Meridian Circle in that Year.

Number.	Star.	In Right Ascension. Annual Precession.	In Right Ascension. Secular Variation.	In Right Ascension. Proper Motion.	In Polar Distance. Annual Precession.	In Polar Distance. Secular Variation.	In Polar Distance. Proper Motion.	Number in B. A. C.
		s	s	s	"	"	"	
71	67 Ceti	+ 2·9832	+ 0·0040	+ 0·003	− 16·897	+ 0·242	+ 0·14	704
72	68 Ceti o Var. 1 (*Mira.*)	+ 3·0263	+ 0·0064	− 0·001	− 16·780	+ 0·248	+ 0·23	720
73	. .. ..	+ 3·0263	+ 0·0064	...	− 16·758	+ 0·249	...	...
74	... .	+ 1·6340	+ 0·0065	...	− 16·638	+ 0·140	.	...
75	. ..	+ 1·7502	+ 0·0037	...	− 16·536	+ 0·148		...
76	73 Ceti 1[1]	+ 3·1787	+ 0·0117	+ 0·001	− 16·874	+ 0·270	+ 0·02	760
77	Lacaille 782 .. .	+ 1·7770	+ 0·0086	...	− 16·105	+ 0·161	...	...
78	... .	+ 1·8275	+ 0·0016	...	− 16·042	+ 0·167	.	...
79	... .	+ 2·4016	− 0·0027	...	− 16·003	+ 0·218	.	...
80	... .	+ 1·7840	+ 0·0027	...	− 15·885	+ 0·160	..	..
81	... ...	+ 1·7337	+ 0·0039	...	− 15·841	+ 0·162	.	. .
82	... .	+ 1·5527	+ 0·0084	.	− 15·839	+ 0·146	..	. .
83	86 Ceti γ	+ 3·1114	+ 0·0094	− 0·011	− 15·564	+ 0·294	+ 0·19	837
84	87 Ceti μ	+ 3·2150	+ 0·0120	+ 0·017	− 15·480	+ 0·306	+ 0·07	846
85	Lacaille 848	+ 1·7478	+ 0·0040	...	− 15·442	+ 0·170	..	...
86	Taylor 840	+ 2·3814	+ 0·0015		− 15·348	+ 0·201		...
87	... .	+ 1·5063	+ 0·0090		− 15·263	+ 0·140		..
88	Taylor 940	+ 3·3249	+ 0·0157		− 15·082	+ 0·326		[illegible]
89	Taylor 978	+ 1·6065	+ 0·0062		− 14·986	+ 0·168		[illegible]
90	Lacaille 941	+ 1·7080	+ 0·0063		− 14·734	+ 0·176		...
91	Taylor 991	+ 2·3382	− 0·0006	...	− 14·740	+ 0·237	. .	917
92	Lacaille 969	+ 1·7694	+ 0·0040	...	− 14·488	+ 0·158	.	..
93	92 Ceti α (*Menkar*) ...	+ 3·1297	+ 0·0098	− 0·008	− 14·467	+ 0·323	+ 0·11	940
94	25 Persei ρ Var. 2 ...	+ 3·6084	+ 0·0382	+ 0·010	− 14·266	+ 0·366	+ 0·11	963
95	26 Persei β Var. 1 (*Algol*)	+ 3·8762	+ 0·0366	− 0·002	− 14·210	+ 0·406	− 0·01	963
96	Taylor 1047	+ 1·3446	+ 0·0139		− 14·186	+ 0·145	..	968
97	... .	+ 2·2611	+ 0·0002		− 14·183	+ 0·230	. .	...
98	Taylor 1052	+ 1·4142	+ 0·0120		− 14·150	+ 0·132	...	972
99	Taylor 1057	+ 1·3346	+ 0·0142		− 14·129	+ 0·144	.	973
100	R. P. L. 83	+ 12·8260	+ 1·8870		− 14·097	+ 1·336	+ 0·12	960
101	. ...	+ 2·2627	+ 0·0004	...	− 14·089	+ 0·241	..	.
102	57 Arietis δ	+ 3·4074	+ 0·0171	+ 0·010	− 13·929	+ 0·364	0·00	988
103	Taylor 1081 .	+ 1·2704	+ 0·0186	. .	− 13·861	+ 0·141	. .	992
104	... ...	+ 2·9991	+ 0·0006	...	− 13·780	+ 0·230	...	...
105	... .	+ 1·6444	+ 0·0080	...	− 13·719	+ 0·151	...	..

76.—83.—100.—Proper motions from "Greenwich Catalogue 1872."

Mean Positions of Stars for 1867 January 1st.

Number.	Star.	Magnitude.	Estimations.	Mean Right Ascension.			Mean Polar Distance.			Observations.	Fraction of Year.
				h.	m.	s.	°	′	″		
106	Taylor 1112 ...	8·0	1	3	10	28·68	129	29	40·8	1	0·96
107	Taylor 1113 ...	8·9	1	3	10	34·00	131	46	29·5	1	0·96
108	Taylor 1127 ...	8·0	1	3	11	54·24	131	45	40·1	1	0·96
109	... ...	9·8	1	3	12	20·94	130	0	53·0	1	0·98
110	... ...	9·0	1	3	12	30·19	130	57	20·2	1	0·97
111	... ...	8·5	1	3	12	40·75	130	27	1·9	1	0·99
112	... ...	7·8	1	3	13	27·13	125	30	12·0	1	0·97
113	33 Persei α ...	1·9	...	3	14	50·42	40	36	54·8	1	0·99
114	... ...	8·2	3	3	15	9·62	161	31	33·4	3	0·80
115	... ...	9·3	1	3	15	30·03	125	40	46·2	1	0·97
116	... ...	9·5	1	3	16	55·05	127	6	21·8	1	0·95
117	... ...	8·8	1	3	17	26·70	160	44	30·6	1	0·97
118	... ...	9·0	1	3	20	32·62	140	16	16·3	1	0·99
119	R Persei Var. 3	9·0	1	3	21	36·58	54	47	24·8	1	0·96
120	... ...	9·3	1	3	23	21·70	130	9	39·0	1	0·98
121	5 Tauri *f*	4·3	2	3	23	31·82	77	31	16·4	2	0·73
122	... ...	9·0	2	3	23	46·11	126	21	35·9	2	0·95
123	... ...	8·5	1	3	24	50·51	130	1	6·5	1	0·97
124	Lacaille 1150 ...	7·0	1	3	26	30·95	152	27	40·0	1	0·92
125	Lacaille 1149 ...	7·0	1	3	28	41·15	150	18	32·0	1	0·92
126	... ...	9·2	1	3	28	62·10	129	26	54·8	1	0·97
127	... ...	7·0	1	3	30	17·68	151	50	28·7	1	0·96
128	... .	8·0	1	3	31	52·10	130	40	30·0	1	0·95
129	... ...	8·0	1	3	32	30·00	131	30	0·8	1	0·98
130	Lacaille 1166 ...	8·8	1	3	33	8·36	130	12	37·6	1	0·97
131	... ...	6·2	1	3	33	25·56	127	42	33·0	1	0·90
132	... ...	10·0	1	3	33	46·33	128	29	27·2	1	0·97
133	... ...	8·6	1	3	35	33·18	160	12	42·5	1	0·68
134	... .	9·7	1	3	36	58·51	129	10	3·0	1	0·97
135	.. ..	8·0	1	3	37	2·66	146	35	55·2	1	0·79
136	25 Tauri η (*Alcyone*) ...	3·0	..	3	39	34·91	66	18	31·6	11	0·75
137	30 Tauri *e*	5·1		3	40	38·73	79	16	5·7	1	0·71
138	Lacaille 1262	7·2	1	3	41	49·31	117	4	42·1	1	0·92
139	W. B. K. III. 880 .	5·3	2	3	46	20·06	76	1	4·7	2	0·86
140	... .	9·0	1	3	49	35·32	109	11	37·7	1	0·99

119.—R Persei Var. 3.—Period 209 days.—Range, 9·5 to 12·5 magnitude.
139.—Comparison star for Asia in 1862.

Observed with the Madras Meridian Circle in that Year.

Number.	Star.	In Right Ascension.			In Polar Distance.			Number in B. A. C.
		Annual Precession.	Secular Variation.	Proper Motion.	Annual Precession.	Secular Variation.	Proper Motion.	
		s	s	s	"	"	"	
106	Taylor 1112	+ 2·2684	+ 0·0009	...	− 13·623	+ 0·249	...	...
107	Taylor 1113 . .	+ 2·1916	+ 0·0011	...	− 13·623	+ 0·242	...	...
108	Taylor 1127	+ 2·1853	+ 0·0012	...	− 13·466	+ 0·246	...	...
109	... ..	+ 2·2818	+ 0·0012	...	− 13·387	+ 0·249	...	...
110	... ...	+ 2·2080	+ 0·0011	...	− 13·387	+ 0·246	...	...
111	.. ..	+ 2·2623	+ 0·0011		− 13·370	+ 0·251	...	...
112	... ...	+ 2·3553	+ 0·0011	...	− 13·285	+ 0·268	...	...
113	83 Persei a	+ 4·2484	+ 0·0483	+ 0·002	− 13·284	+ 0·472	+ 0·06	1043
114	... .	+ 1·2178	+ 0·0166	...	− 13·212	+ 0·140	...	...
115	.. ...	+ 2·3469	+ 0·0012	...	− 13·190	+ 0·264	...	...
116	... ...	+ 2·3060	+ 0·0013	...	− 13·094	+ 0·261	...	...
117	... .	+ 2·1979	+ 0·0015	...	− 13·062	+ 0·249	...	...
118	... .. .	+ 1·3446	+ 0·0161	...	− 12·860	+ 0·156	...	...
119	B Persei Var. 3 ..	+ 3·7991	+ 0·0273	...	− 12·784	+ 0·482	...	...
120	.. ... .	+ 2·1072	+ 0·0018	...	− 12·665	+ 0·264	...	...
121	5 Tauri f	+ 3·3017	+ 0·0130	+ 0·002	− 12·652	+ 0·370	+ 0·08	1097
122	... ...	+ 2·3076	+ 0·0016	...	− 12·637	+ 0·266	...	...
123	... ...	+ 2·2272	+ 0·0018	...	− 12·567	+ 0·259	...	...
124	Lacaille 1130 ...	+ 1·0490	+ 0·0808	...	− 12·311	+ 0·126	...	...
125	Lacaille 1149 ...	+ 1·7201	+ 0·0166	...	− 12·300	+ 0·146	...	...
126	.. ..	+ 2·2820	+ 0·0020		− 12·267	+ 0·262	...	...
127	.. ..	+ 1·0987	+ 0·0190		− 12·186	+ 0·131	...	...
128	. .. ...	+ 2·1820	+ 0·0023		− 12·079	+ 0·259	...	...
129	... .	+ 2·1318	+ 0·0026		− 12·047	+ 0·260	...	...
130	Lacaille 1166 ...	+ 2·1076	+ 0·0025		− 11·989	+ 0·262	...	...
131	.. ...	+ 2·2420	+ 0·0021		− 11·970	+ 0·268	...	...
132	.. ...	+ 2·2162	+ 0·0022		− 11·946	+ 0·266	...	...
133	... ...	+ 1·1863	+ 0·0180		− 11·882	+ 0·146	...	...
134	.. ...	+ 2·1012	+ 0·0084		− 11·790	+ 0·264	...	...
135	... ...	+ 1·3046	+ 0·0181		− 11·714	+ 0·160	...	...
136	25 Tauri η (*Alcyone*) ..	+ 3·5620	+ 0·0477	− 0·001	− 11·585	+ 0·460	+ 0·06	1166
137	30 Tauri e	+ 3·2608	+ 0·0115	0·000	− 11·466	+ 0·398	+ 0·06	1174
138	Lacaille 1248 .	+ 1·3712	+ 0·0114	...	− 11·373	+ 0·170	...	...
139	W. B. E. III. 860 .	+ 3·3492	+ 0·0127	...	− 11·104	+ 0·412	...	...
140	.. .. .	+ 2·1829	+ 0·0029	..	− 10·781	+ 0·270	...	...

121.—Proper motions from "*Greenwich Catalogue* 1872."

Mean Positions of Stars for 1867 January 1st.

Number.	Star.	Magnitude.	Estimations.	Mean Right Ascension.			Mean Polar Distance.			Observations.	Fraction of Year.
				h.	m.	s.	°	′	″		
141	... ...	8·0	1	3	50	53·94	147	28	42·7	1	0·79
142	34 Eridani γ^1	3·0	...	3	51	40·43	103	53	20·9	12	0·56
143	35 Tauri λ Var. 1	3·6	...	3	53	18·77	77	53	16·6	2	0·45
144	... ...	8·5	1	3	53	46·06	143	7	54·5	1	0·04
145	... ...	9·1	1	3	54	46·00	129	10	20·6	1	0·59
146	Taylor 1392 ..	6·9	1	3	56	31·04	147	28	52·5	1	0·03
147	Lalande 7581 ...	8·2	1	3	58	23·01	74	51	50·8	1	0·02
148	Lacaille 1359 ...	6·5	2	4	0	4·86	147	49	36·4	2	0·02
149	Lacaille 1375 ...	6·1	2	4	2	38·08	143	50	17·8	2	0·03
150	Lalande 7764 ...	8·0	1	4	3	33·46	74	48	23·1	1	0·03
151	... ...	9·5	1	4	3	46·13	146	55	35·9	1	0·89
152	37 Eridani ...	5·8	...	4	3	33·37	97	16	26·7	1	0·95
153	38 Eridani o^1 ...	4·1	...	4	5	22·48	97	11	12·6	2	0·92
154	... ...	6·1	1	4	9	32·86	129	13	24·7	1	0·01
155	Taylor 1480 ..	7·2	3	4	11	6·48	148	21	34·6	3	0·83
156	54 Tauri γ	3·9	...	4	12	13·78	74	41	47·2	1	0·04
157	Lacaille 1418	7·5	2	4	12	31·61	143	39	17·3	2	0·49
158	... ...	8·7	3	4	12	36·05	139	10	30·5	3	0·04
159	... ...	8·3	2	4	13	58·02	70	51	2·8	2	0·01
160	U Tauri Var. 7	9·9	1	4	14	4·25	70	30	15·3	1	0·04
161	... ...	10·0	1	4	15	20·86	129	7	36·1	1	0·67
162	... ...	8·0	1	4	16	49·28	140	8	50·5	1	0·37
163	74 Tauri ϵ	3·7	...	4	20	51·13	71	7	3·8	11	0·86
164	Lacaille 1519 ...	7·0	1	4	25	37·48	138	5	48·2	1	0·02
165	Lacaille 1520 ...	7·7	1	4	26	44·46	147	38	30·0	1	0·03
166	... ...	9·3	1	4	27	36·77	130	33	35·1	1	0·96
167	87 Tauri α (*Aldebaran*) ...	1·0	...	4	28	17·47	73	46	40·0	12	0·88
168	... ...	8·2	2	4	28	52·48	140	13	54·0	2	0·41
169	R Reticuli Var. 1	9·0	7	4	32	10·28	153	13	20·9	7	0·31
170	Lacaille 1551 .. .	6·0	1	4	32	11·35	153	5	32·9	1	0·96
171	... .	9·5	2	4	34	39·82	153	20	31·1	2	0·36
172	96 Tauri	6·9	1	4	36	10·70	80	10	0·0	1	0·02
173	Lacaille 1567	8·0	1	4	35	13·10	132	30	26·5	1	0·01
174	Lacaille 1568 . ..	6·9	1	4	35	46·70	148	38	6·2	1	0·03
175	Taylor 1683	7·6	3	4	36	36·13	128	47	49·8	3	0·09

143.—λ Tauri Var. 1.—Period 3·96 days.—Range, 3·5 to 4·3 magnitude.
147.—150.—Comparison stars for Asia in 1862.
160.—U Tauri Var. 7.—Period unknown—Range, 9th to 10·5 magnitude.
169.—R Reticuli Var. 1.—Period 281 days.—Range, 7th to below 13th magnitude.
170.—171.—173.—Stars for map of R Reticuli Var. 1.

Observed with the Madras Meridian Circle in that Year.

Number.	Star.	In Right Ascension. Annual Precession.	Secular Variation.	Proper Motion.	In Polar Distance. Annual Precession.	Secular Variation.	Proper Motion.	Number in B. A. C.
		s	s	s	"	"	"	
141	... .	+ 1·2901	+ 0·0134	...	− 10·709	+ 0·165	...	...
142	34 Eridani γ¹	+ 2·7918	+ 0·0047	+ 0·002	− 10·611	+ 0·361	+ 0·12	1284
143	35 Tauri λ Var. 1 ...	+ 3·3164	+ 0·0115	− 0·002	− 10·389	+ 0·416	+ 0·02	1241
144	... .	+ 1·5331	+ 0·0082	...	− 10·497	+ 0·198	...	...
145	... ..	+ 2·1413	+ 0·0031	...	− 10·410	+ 0·271	...	...
146	Taylor 1392	+ 1·2731	+ 0·0126		− 10·340	+ 0·164		1265
147	Lalande 7581 .	+ 3·2810	+ 0·0124		− 10·148	+ 0·430		...
148	Lacaille 1350 ...	+ 1·2314	+ 0·0131		− 10·082	+ 0·160		...
149	Lacaille 1375 ..	+ 1·1433	+ 0·0144		− 9·902	+ 0·140		...
150	Lalande 7704	+ 3·3013	+ 0·0121		− 9·750	+ 0·486		...
151	... .	+ 1·2778	+ 0·0200	...	− 9·740	+ 0·167	...	...
152	37 Eridani	+ 2·9230	+ 0·0058	− 0·002	− 9·732	+ 0·377	+ 0·04	1294
153	38 Eridani ο¹ .	+ 2·9241	+ 0·0058	− 0·008	− 9·617	+ 0·379	− 0·07	1290
154	... ...	+ 2·1016	+ 0·0085	...	− 9·270	+ 0·276	...	...
155	Taylor 1469	+ 1·1127	+ 0·0187	..	− 9·175	+ 0·132	...	1326
156	54 Tauri γ	+ 3·3962	+ 0·0115	+ 0·009	− 9·087	+ 0·446	+ 0·03	1329
157	Lacaille 1418 ..	+ 1·4613	+ 0·0088	..	− 9·064	+ 0·193	...	...
158	... ...	+ 2·1002	+ 0·0086	...	− 9·080	+ 0·277	...	...
159	... ...	+ 3·4873	+ 0·0128	...	− 8·961	+ 0·469	...	...
160	U Tauri Var. 7 ..	+ 3·4030	+ 0·0129	...	− 8·943	+ 0·460	...	...
161	... .	+ 2·0960	+ 0·0085	...	− 8·842	+ 0·279	...	...
162	... ...	+ 1·0694	+ 0·0146	...	− 8·727	+ 0·144	...	...
163	74 Tauri ε	+ 3·4872	+ 0·0120	+ 0·006	− 8·408	+ 0·466	+ 0·08	1376
164	Lacaille 1519 ..	+ 0·6570	+ 0·0212	...	− 8·027	+ 0·091	...	...
165	Lacaille 1520 ...	+ 1·1460	+ 0·0122	...	− 7·907	+ 0·157	...	...
166	... ...	+ 0·8945	+ 0·0162	...	− 7·886	+ 0·123	...	...
167	87 Tauri α (*Aldebaran*)	+ 3·4307	+ 0·0166	+ 0·004	− 7·813	+ 0·464	+ 0·17	1420
168	... .	+ 1·6018	+ 0·0067	...	− 7·792	+ 0·217	...	...
169	R Reticuli Var. 1 ..	+ 0·6061	+ 0·0210	...	− 7·499	+ 0·085	...	...
170	Lacaille 1551 ..	+ 0·6860	+ 0·0205	...	− 7·407	+ 0·089	...	...
171	... .	+ 0·8802	+ 0·0209	...	− 7·307	+ 0·083	...	...
172	96 Tauri	+ 3·6288	+ 0·0125	+ 0·004	− 7·284	+ 0·496	0·00	1452
173	Lacaille 1567	+ 0·6937	+ 0·0186	...	− 7·261	+ 0·097	..	...
174	Lacaille 1580	+ 1·0886	+ 0·0158	...	− 7·202	+ 0·144	...	..
175	Taylor 1663 ..	+ 1·6448	+ 0·0060	..	− 7·112	+ 0·227	...	

143.—Proper motions from "*Greenwich Catalogue* 1872."
172.—Proper motions from "*Greenwich Catalogue* 1864."

Mean Positions of Stars for 1867 January 1st.

Number.	Star.	Magnitude.	Estimations.	Mean Right Ascension. h.	m.	s.	Mean Polar Distance. °	′	″	Observations.	Fraction of Year.
176	... ...	0·6	2	4	30	15·49	163	15	34·4	2	0·52
177	... ...	9·0	1	4	30	27·09	163	3	16·9	1	0·96
178	... ...	9·5	2	4	43	26·68	130	40	53·5	2	0·04
179	Lacaille 1629 ..	6·4	1	4	46	44·61	168	28	14·6	1	0·01
180	... ...	8·0	1	4	48	57·70	127	40	3·2	1	0·96
181	Lacaille 1626 ...	7·3	2	4	46	3·76	140	1	31·9	2	0·02
182	... ...	9·0	1	4	46	34·95	120	24	45·1	1	0·02
183	Lacaille 1640 ...	6·9	2	4	46	56·69	130	20	19·9	2	0·06
184	3 Aurigæ ι ..	2·7	...	4	48	20·12	57	2	52·4	13	0·38
185	Taylor 1704 ...	6·3	2	4	50	27·02	130	50	35·8	2	0·06
186	Taylor 1730	7·0	2	4	52	10·23	144	38	31·4	2	0·03
187	R Leporis Var. 1	7·5	1	4	53	32·02	104	59	26·6	1	0·10
188	Taylor 1797	7·0	1	4	54	54·20	148	16	30·6	1	0·07
189	... ...	8·7	1	4	56	2·40	130	17	24·5	1	0·01
190	Lacaille 1697	8·0	1	4	56	57·20	129	6	36·3	1	0·02
191	11 Orionis .. .	4·7	...	4	56	58·20	74	47	3·0	2	0·01
192	Taylor 1811 ..	6·1	1	4	57	9·12	120	54	47·7	1	0·11
193	Taylor 1814 ...	9·2	2	4	57	48·47	148	41	40·3	2	0·96
194	2 Leporis ε ..	3·3	...	4	59	49·95	112	33	7·3	8	0·36
195	... ...	7·9	1	5	0	5·21	185	28	41·6	1	0·02
196	Taylor 1852	6·9	1	5	2	13·68	144	36	17·2	1	0·96
197	Lacaille 1739	8·0	1	5	2	54·81	146	57	36·7	1	0·94
198	... ...	9·5	1	5	4	39·04	136	34	26·2	1	0·07
199	... ...	9·3	6	5	6	29·17	36	31	39·0	6	0·06
200	R Aurigæ Var. 2 ..	8·7	1	5	6	33·94	36	31	3·8	1	0·96
201	13 Aurigæ α (*Capella*) ...	0·2	...	5	6	32·11	41	8	28·6	3	0·07
202	19 Orionis β (*Rigel*)	0·3	...	5	8	8·81	98	21	39·6	4	0·04
203	... ...	9·0	1	5	8	32·01	150	30	9·1	1	0·92
204	.. ..	9·2	1	5	10	57·05	121	19	24·4	1	0·98
205	. .. .	9·7	1	5	12	3·39	129	40	35·9	1	0·97
206	.. ...	9·6	3	5	13	46·60	75	11	34·1	3	0·04
207	... .	9·1	2	5	15	1·34	121	13	29·4	2	0·03
208	Lacaille 1822	7·9	1	5	15	47·84	141	43	2·5	1	0·96
209	.. ..	9·2	1	5	16	12·49	131	46	7·8	1	0·07
210	Lacaille 1824	7·4	1	5	17	4·44	129	37	30·4	1	0·12

187.—R Leporis Var. 1.—Period [illegible] days.—Range, 6th to 9th magnitude.
200.—R Aurigæ Var. 2.—Period 465 days.—Range, 7th to 12·7 magnitude.
205.—Comparison star for Ann in 1866.

Observed with the Madras Meridian Circle in that Year.

Number.	Star.	In Right Ascension. Annual Precession.	In Right Ascension. Secular Variation.	In Right Ascension. Proper Motion.	In Polar Distance. Annual Precession.	In Polar Distance. Secular Variation.	In Polar Distance. Proper Motion.	Number in B.A.C.
		s	s	s	"	"	"	
176	... .	+ 0·3815	+ 0·0190	...	− 6·720	+ 0·068		.
177	... .	+ 0·6086	+ 0·0196	...	− 6·704	+ 0·085		...
178	... ...	+ 1·9803	+ 0·0040	...	− 6·676	+ 0·277		...
179	Lacaille 1689 ..	+ 0·5409	+ 0·0197	...	− 6·550	+ 0·077		...
180	... .	+ 2·0962	+ 0·0086	...	− 6·533	+ 0·202		...
181	Lacaille 1695 ..	+ 1·9610	+ 0·0068	...	− 6·442	+ 0·219	...	...
182	... ...	+ 2·0800	+ 0·0087	...	− 6·399	+ 0·284	...	...
183	Lacaille 1699 ..	+ 0·8291	+ 0·0145	...	− 6·267	+ 0·118	...	...
184	3 Aurigæ ι	+ 3·8066	+ 0·0144	− 0·003	− 6·170	+ 0·844	+ 0·08	1580
185	Taylor 1764	+ 2·0074	+ 0·0088	...	− 5·992	+ 0·292	...	1583
186	Taylor 1780	+ 1·2695	+ 0·0084		− 5·886	+ 0·150	...	...
187	R Leporis Var. 1 ...	+ 2·7287	+ 0·0088		− 5·784	+ 0·382	...	...
188	Taylor 1797	+ 0·0980	+ 0·0111		− 5·620	+ 0·141	...	...
189	... .	+ 1·9925	+ 0·0088		− 5·524	+ 0·280	...	...
190	Lacaille 1697 ..	+ 2·0280	+ 0·0086		− 5·447	+ 0·266	...	...
191	11 Orionis	+ 3·4221	+ 0·0070	0·000	− 5·446	+ 0·462	+ 0·03	1587
192	Taylor 1811	+ 1·9966	+ 0·0083	...	− 5·431	+ 0·282	...	1591
193	Taylor 1814	+ 0·0541	+ 0·0111	...	− 5·376	+ 0·156	...	...
194	2 Leporis ε	+ 2·5869	+ 0·0082	+ 0·001	− 5·205	+ 0·390	+ 0·08	1576
195	... ...	+ 1·7627	+ 0·0046	...	− 5·158	+ 0·280	...	...
196	Taylor 1852	+ 1·9510	+ 0·0077	...	− 5·002	+ 0·179		1596
197	Lacaille 1780 ...	+ 1·0798	+ 0·0093	...	− 4·944	+ 0·186		...
198	... ...	+ 1·7477	+ 0·0045	...	− 4·700	+ 0·280		...
199	... .	+ 4·8251	+ 0·0280	...	− 4·641	+ 0·687		...
200	R Aurigæ Var. 2 ..	+ 4·6256	+ 0·0280	...	− 4·634	+ 0·697		...
201	13 Aurigæ α (*Capella*)	+ 4·4129	+ 0·0173	+ 0·008	− 4·605	+ 0·620	+ 0·48	1613
202	19 Orionis β (*Rigel*) ..	+ 2·8806	+ 0·0040	− 0·001	− 4·490	+ 0·412	+ 0·02	1623
203	... ...	+ 0·7587	+ 0·0117	...	− 4·467	+ 0·110	...	..
204	... .	+ 3·2772	+ 0·0080	...	− 4·287	+ 0·386	...	...
205	... ...	+ 1·9863	+ 0·0086	...	− 4·165	+ 0·295	...	...
206	... .	+ 3·4164	+ 0·0068	...	− 4·018	+ 0·490		..
207	... .	+ 2·8741	+ 0·0080	...	− 3·911	+ 0·398		..
208	Lacaille 1822 ..	+ 1·1096	+ 0·0067	...	− 3·846	+ 0·204		...
209	... .	+ 1·8881	+ 0·0037	...	− 3·810	+ 0·274		
210	Lacaille 1824 ..	+ 1·0818	+ 0·0034	...	− 3·786	+ 0·380		

Mean Positions of Stars for 1867 January 1st.

Number.	Star.	Magnitude.	Estimations.	Mean Right Ascension.			Mean Polar Distance.			Observations.	Fraction of Year.
				h.	m.	s.	°	′	″		
211	... ...	7·0	1	5	17	39·32	153	7	11·2	1	0·07
212	112 Tauri β ..	1·9	...	5	17	58·15	61	30	30·9	12	0·08
213	... .	9·3	2	5	18	28·31	121	38	46·8	2	0·62
214	R. P. L. 40 ...	6·2	...	5	19	40·91	4	32	58·2	2	0·74
215	... ...	9·2	1	5	22	36·04	152	41	55·6	1	0·07
216	... ...	7·1	1	5	23	31·99	151	13	17·3	1	0·10
217	λ Doradûs... . ..	6·0	1	5	24	23·14	149	1	33·1	1	0·06
218	119 Tauri	4·6	...	5	24	26·05	71	30	28·9	2	0·94
219	34 Orionis δ Var. 1 ...	2·4	..	5	25	12·75	90	21	9·2	9	0·16
220	... ...	9·5	1	5	26	12·32	121	21	17·1	1	0·08
221	11 Leporis α ...	2·7	...	5	26	51·82	107	55	12·1	5	0·24
222	Taylor 2057 ...	7·8	2	6	28	3·42	151	55	30·9	2	0·08
223	Lalande 10582 ...	8·2	1	5	29	19·51	73	15	4·2	1	0·09
224	46 Orionis ε ..	1·8	...	5	29	27·93	91	17	23·9	4	0·05
225	123 Tauri ζ ...	3·0	...	5	29	41·92	68	56	30·4	2	0·90
226	Lalande 10607 ...	7·0	3	5	31	27·86	69	18	36·6	3	0·02
227	... ...	9·3	1	5	32	5·08	120	55	10·2	1	0·97
228	... ...	8·8	3	6	32	14·73	128	54	10·7	3	0·37
229	Lacaille 1949 ...	6·5	1	5	32	16·36	154	18	59·9	1	0·04
230	Lacaille 1916 ...	7·0	2	5	32	20·28	121	8	38·2	2	0·62
231	α Columbæ	2·7	...	5	34	50·10	124	8	40·4	8	0·05
232	Taylor 2118	7·4	1	5	36	18·88	130	46	27·9	1	0·10
233	Lalande 10840	8·0	3	5	37	54·54	68	19	38·6	3	0·07
234	... ...	7·8	2	5	39	20·79	79	0	9·7	2	0·63
235	Taylor 2145	9·5	1	5	39	54·82	136	53	40·0	1	0·07
236	... ...	9·3	1	5	40	37·23	120	57	54·9	1	0·08
237	W. B. R. V. 1011	8·5	1	5	40	38·90	78	57	37·0	1	0·97
238	Taylor 2194	9·0	1	5	43	36·51	160	46	32·0	1	0·07
239	... ...	9·8	1	5	44	49·08	137	10	16·1	1	0·07
240	Lalande 11088 .	8·4	3	5	46	16·70	68	29	36·9	3	0·05
241	Lacaille 2036 ..	8·2	3	5	46	28·64	129	47	11·1	3	0·10
242	54 Orionis χ^1	4·6	..	5	46	30·31	69	45	8·0	1	0·12
243	.. ...	8·0	1	5	47	6·61	135	46	49·6	1	0·10
244	Lalande 11166	7·9	2	5	47	11·56	78	38	35·0	2	0·02
245	58 Orionis α Var 2	0·9	.	5	47	54·31	82	37	14·1	7	0·09

211.—Groombridge 944.
219.—δ Orionis Var. 1.—Supposed to vary irregularly from 2·3 to 2·7 magnitude.
223.—234.—237.—244.—Comparison stars for Sappho in 1866.
226.—233.—240.—Comparison stars for Harmonia in 1861.
245.—α Orionis Var. 2. (*Betelgeuse*).—Irregularly variable from 0·9 to 1·5 magnitude.

Observed with the Madras Meridian Circle in that Year.

Number.	Star.	In Right Ascension. Annual Precession.	Secular Variation.	Proper Motion.	In Polar Distance. Annual Precession.	Secular Variation.	Proper Motion.	Number in B. A. C.
		s	s	s	"	"	"	
211	... .	+ 0·4794	+ 0·0189	...	− 3·684	+ 0·070	.	.
212	112 Tauri β .	+ 3·7655	+ 0·0082	+ 0·003	− 3·606	+ 0·545	+ 0·20	1681
213	... ..	+ 2·2668	+ 0·0080	...	− 3·619	+ 0·327	..	...
214	B. P. L. 40 .	+ 18·4663	+ 0·6730	...	− 3·540	+ 2·657		1692
215	... ...	+ 0·5163	+ 0·0113	...	− 3·257	+ 0·075		.
216	... ...	+ 0·6688	+ 0·0098		− 3·175	+ 0·097	.	.
217	λ Doradûs	+ 0·8716	+ 0·0081		− 3·104	+ 0·127	..	1729
218	119 Tauri .	+ 3·5138	+ 0·0057	0·000	− 3·101	+ 0·307	+ 0·01	1726
219	34 Orionis δ Var 1	+ 3·0088	+ 0·0038	+ 0·001	− 3·082	+ 0·443	+ 0·04	1730
220	... .	+ 2·2693	+ 0·0028	...	− 2·917	+ 0·384	.	.
221	11 Leporis α .	+ 2·6412	+ 0·0029	+ 0·001	− 2·890	+ 0·369	0·00	1741
222	Taylor 2087 ..	+ 0·5898	+ 0·0098	..	− 2·790	+ 0·086	..	
223	Lalande 10532 ..	+ 3·3170	+ 0·0045	...	− 2·670	+ 0·456	.	.
224	46 Orionis ε .	+ 3·0422	+ 0·0035	− 0·002	− 2·604	+ 0·411	+ 0·01	1765
225	123 Tauri ζ ...	+ 3·5821	+ 0·0055	0·000	− 2·644	+ 0·519	+ 0·06	1767
226	Lalande 10607	+ 2·3980	+ 0·0053		− 2·540	+ 0·522		..
227	... .	+ 2·1798	+ 0·0090		− 2·437	+ 0·317		.
228	... . ..	+ 2·1902	+ 0·0029		− 2·494	+ 0·317		.
229	Lacaille 1949...	+ 0·3184	+ 0·0106		− 2·410	+ 0·046		1780
230	Lacaille 1016	+ 2·2680	+ 0·0027		− 2·162	+ 0·380		.
231	α Columbæ	+ 2·1707	+ 0·0027	+ 0·008	− 2·190	+ 0·316	0·00	1802
232	Taylor 2118	+ 1·0205	+ 0·0031	...	− 2·103	+ 0·290	..	...
233	Lalande 10949	+ 3·6010	+ 0·0046	.	− 1·930	+ 0·394	.	.
234	.. . ...	+ 3·3300	+ 0·0036		− 1·805	+ 0·496	.	.
235	Taylor 2145	+ 1·0080	+ 0·0063		− 1·723	+ 0·348		1836
236	... ...	+ 2·2727	+ 0·0026		− 1·604	+ 0·331		..
237	W. B. E. V. 1011	+ 3·3820	+ 0·0034		− 1·601	+ 0·466		...
238	Taylor 2184 .	+ 0·6996	+ 0·0080		− 1·404	+ 0·101		.
239		+ 1·0830	+ 0·0038		− 1·327	+ 0·230		
240	Lalande 11098	+ 3·5978	+ 0·0030		− 1·287	+ 0·528		
241	Lacaille 2030	+ 1·0606	+ 0·0039		− 1·193	+ 0·260	.	.
242	51 Orionis χ[1]	+ 3·5015	+ 0·0034	− 0·016	− 1·190	+ 0·520	+ 0·10	1876
243		+ 1·7004	+ 0·0030	..	− 1·127	+ 0·248	..	
244	Lalande 11160	+ 3·3496	+ 0·0030		− 1·120	+ 0·487		.
245	58 Orionis α Var. 2 .	+ 3·2430	+ 0·0027	+ 0·001	− 1·062	+ 0·173	0·01	1883

212.—242.—Proper motions from "Greenwich Catalogue 1872."

Mean Positions of Stars for 1867 January 1st.

Number.	Star.	Magnitude.	Estimations.	Mean Right Ascension.			Mean Polar Distance.			Observations.	Fraction of Year.
				h.	m.	s.	°	′	″		
246	... ...	9·0	1	5	49	22·30	121	9	50·3	1	0·08
247	... ...	9·8	1	5	50	19·08	137	10	17·9	1	0·07
248	Lacaille 2073 ...	7·0	1	5	50	35·57	137	12	36·7	1	0·06
249	... .	9·0	1	5	51	15·48	141	51	39·1	1	0·06
250	Lalande 11293 ...	7·4	4	5	51	40·38	68	24	34·8	4	0·29
251	... ...	8·3	1	5	53	7·51	130	24	57·2	1	0·07
252	... .	9·5	...	5	54	39·70	137	46	14·2	1	0·07
253	... ...	9·6	1	5	54	50·66	121	30	57·8	1	0·08
254	Lalande 11456	7·4	2	5	56	59·12	78	19	12·1	2	0·06
255	... ...	9·2	1	5	57	4·30	136	1	0·7	1	0·08
256	Taylor 2301 ...	7·0	1	5	58	32·71	148	6	19·3	1	0·10
257	Taylor 2310 ...	6·7	2	5	59	39·88	150	29	6·7	2	0·10
258	67 Orionis ν ...	4·4	...	5	59	58·69	75	13	7·6	9	0·08
259	... .	9·8	1	6	0	5·40	121	30	58·8	1	0·08
260	Taylor 2321 ...	6·9	1	6	0	61·72	141	5	27·0	1	0·06
261	... ...	7·0	1	6	0	56·69	137	28	38·1	1	0·10
262	Lalande 11732	8·0	2	6	3	32·74	77	38	61·1	2	0·09
263	... ...	9·2	1	6	3	44·89	129	08	12·0	1	0·09
264	... ...	8·1	1	6	5	47·78	77	51	30·3	1	0·01
265	... .	9·0	1	6	6	33·10	121	29	20·3	1	0·08
266	7 Geminorum η Var. 6 ...	3·5	..	6	6	50·90	67	27	29·2	2	0·05
267	... ...	9·7	1	6	7	29·80	151	18	36·9	1	0·07
268	.. . .	9·5	1	6	7	45·08	137	6	28·1	1	0·06
269	... ...	8·1	2	6	8	55·50	131	54	40·1	2	0·09
270	.. .. . .	9·2	2	6	9	32·40	131	30	55·0	2	0·02
271	...	7·6	1	6	11	4·98	140	58	54·6	1	0·02
272	.. .	8·9	1	6	11	50·21	121	31	31·1	1	0·08
273	Lalande 12063 .	8·0	1	6	12	34·70	86	51	16·8	1	0·09
274	Lalande 12094	9·3	1	6	13	45·87	86	42	4·3	1	0·05
275	Lalande 12120	7·3	2	6	14	15·32	77	4	49·2	2	0·12
276	13 Geminorum μ	3·2	...	6	14	54·32	67	26	14·2	8	0·30
277	Lalande 12156	7·1	2	6	15	8·62	77	22	3·8	2	0·09
278	Taylor 2474	6·3	2	6	18	15·63	121	40	20·4	2	0·09
279	Lacaille 2268	7·0	1	6	18	49·51	153	45	51·0	1	0·08
280		8·7	1	6	19	51·11	86	39	59·0	1	0·06

250 – 254.—262.—264.—273 – 277.—Comparison stars for Sappho in 1866.
266.—η Geminorum Var. 6.—Period 229 days. Range, 3rd to 4th magnitude.
278. – 279.—Comparison stars for Ariadne in 1866.

Observed with the Madras Meridian Circle in that Year.

Number.	Star.	In Right Ascension. Annual Precession.	Secular Variation.	Proper Motion.	In Polar Distance. Annual Precession.	Secular Variation.	Proper Motion.	Number in B. A. C.
		s	s	s	″	″	″	
246	..	+ 2·3618	+ 0·0024	...	− 0·080	+ 0·380	...	..
247	.. ..	+ 1·6311	+ 0·0030	...	− 0·848	+ 0·298	..	...
248	Lacaille 2073 ..	+ 1·0291	+ 0·0030	...	− 0·883	+ 0·237	...	...
249	... .	+ 1·3701	+ 0·0033	...	− 0·765	+ 0·200	...	..
250	Lalande 11393	+ 3·6008	+ 0·0028	...	− 0·728	+ 0·526	...	...
251	. . ..	+ 1·8842	+ 0·0026		− 0·602	+ 0·282	...	..
252	... ...	+ 1·6005	+ 0·0029		− 0·467	+ 0·238	...	..
253	.. ..	+ 2·2523	+ 0·0024		− 0·447	+ 0·320	..	...
254	Lalande 11455 ..	+ 3·3464	+ 0·0021		− 0·360	+ 0·488	...	..
255	... ...	+ 1·6870	+ 0·0027		− 0·236	+ 0·216	...	...
256	Taylor 2301 ...	+ 0·1236	+ 0·0030	...	− 0·127	+ 0·136	...	1964
257	Taylor 2310	+ 0·7105	+ 0·0030	...	− 0·030	+ 0·104	...	...
258	67 Orionis ν	+ 3·4940	+ 0·0017	+ 0·001	− 0·002	+ 0·500	+ 0·02	1986
259	. ...	+ 2·2522	+ 0·0028	...	+ 0·008	+ 0·320	...	...
260	Taylor 2321	+ 1·4156	+ 0·0026	...	+ 0·076	+ 0·307	...	...
261	... ...	+ 1·6143	+ 0·0024		+ 0·081	+ 0·286		...
262	Lalande 11732 ...	+ 3·3667	+ 0·0015		+ 0·310	+ 0·400		...
263	.. ..	+ 1·0516	+ 0·0028		+ 0·320	+ 0·295		..
264	... .	+ 3·3907	+ 0·0016		+ 0·306	+ 0·460		...
265	... ...	+ 2·3633	+ 0·0022		+ 0·573	+ 0·329		...
266	7 Geminorum η Var. 6.	+ 3·6267	+ 0·0007	− 0·007	+ 0·600	+ 0·389	+ 0·02	2002
267	... .	+ 0·6805	+ 0·0016	...	+ 0·680	+ 0·092	..	...
268	... .	+ 1·6298	+ 0·0028	..	+ 0·679	+ 0·285	...	...
269	. .. ...	+ 1·8729	+ 0·0021	...	+ 0·780	+ 0·273	...	...
270	... . ..	+ 1·6757	+ 0·0021	...	+ 0·800	+ 0·273	..	.
271	. ... ..	+ 0·7636	+ 0·0010		+ 0·960	+ 0·112		.
272	... . ..	+ 2·3680	+ 0·0021		+ 1·036	+ 0·398		..
273	Lalande 12068 ..	+ 3·6934	+ 0·0002		+ 1·100	+ 0·388		..
274	Lalande 12094	+ 3·3924	0·0000		+ 1·107	+ 0·822		.
275	Lalande 12120 ..	+ 3·3782	+ 0·0005		+ 1·217	+ 0·667		...
276	13 Geminorum μ ..	+ 3·6269	− 0·0003	+ 0·006	+ 1·204	+ 0·627	+ 0·11	2017
277	Lalande 12153 ..	+ 3·3711	+ 0·0004	...	+ 1·234	+ 0·460	.	.
278	Taylor 2174	+ 2·3461	+ 0·0030	.	+ 1·290	+ 0·336	...	2073
279	Lacaille 2283	+ 0·3096	− 0·0017	...	+ 1·046	+ 0·083	..	2078
280	... ..	+ 3·6716	− 0·0011	...	+ 1·736	+ 0·508	...	.

258.—Proper motions from "*Greenwich Catalogue 1872.*"

Mean Positions of Stars for 1867 January 1st.

Number.	Star.	Magnitude.	Estimations.	Mean Right Ascension.			Mean Polar Distance.			Observations.	Fraction of Year.
				h	m	s	°	′	″		
281	α Argûs (*Canopus*) ...	0·4	...	6	21	0·07	142	37	20·5	2	0·36
282	... ...	9·0	1	6	22	2·33	120	36	37·1	1	0·01
283	Lacaille 2321	7·0	1	6	23	30·63	133	20	56·6	1	0·04
284	Taylor 2534	7·7	1	6	23	32·08	131	8	10·7	1	0·00
285	Taylor 2560	7·0	2	6	24	27·07	130	59	21·2	2	0·11
286	Taylor 2541 ...	6·0	1	6	24	55·20	117	56	6·7	1	0·12
287	... ...	9·2	1	6	27	24·48	128	46	6·1	1	0·08
288	... ...	9·5	1	6	27	51·56	131	5	25·5	1	0·28
289	... ...	6·6	2	6	28	39·22	122	7	37·9	2	0·01
290	... ...	8·0	1	6	28	41·14	151	10	10·6	1	0·10
291	Taylor 2589	6·9	1	6	29	51·53	151	46	57·0	1	0·12
292	24 Geminorum γ	2·0	...	6	30	1·71	73	20	26·2	6	0·10
293	... ...	8·9	1	6	30	48·10	180	56	28·0	1	0·17
294	... ...	9·5	1	6	31	7·71	122	6	32·2	1	0·08
295	... ...	8·0	1	6	31	32·56	130	56	52·6	1	0·01
296	R Monocerotis Var. 1 ...	10·3	1	6	31	54·16	81	8	38·5	*1	0·15
297	... ...	8·9	1	6	33	39·67	130	54	36·9	1	0·08
298	... ...	7·9	1	6	34	36·46	139	28	6·4	1	0·01
299	27 Geminorum ε	3·2	...	6	35	41·93	64	44	27·2	8	0·11
300	... ...	8·0	1	6	36	53·12	130	38	13·7	1	0·08
301	51 Cephei (*Hev.*)	5·3	..	6	37	9·00	2	45	29·1	10	0·26
302	31 Geminorum ξ	3·4	...	6	37	40·36	76	57	50·5	1	0·67
303	Lacaille 2461	8·2	1	6	38	11·29	135	57	38·7	1	0·06
304	9 Canis Majoris α (*Sirius*)	−1·4	...	6	39	17·08	106	32	10·5	2	0·06
305	... .	9·0	1	6	40	30·82	154	13	40·2	1	0·07
306	... . .	8·6	2	6	42	19·35	130	57	9·1	2	0·12
307	. ... .	9·8	1	6	42	48·10	130	53	16·2	1	0·07
308	... ...	7·6	3	6	43	44·05	128	30	34·8	3	0·17
309	... –	9·5	1	6	45	5·62	106	32	56·5	1	0·06
310	Lalande 13279	7·9	2	6	46	36·16	66	35	20·7	2	0·05
311	Lalande 13313 . .	8·5	3	6	47	31·01	65	40	37·9	3	0·37
312	. .	8·2	1	6	48	34·57	125	48	39·3	1	0·15
313	39 Geminorum	6·7	.	6	50	36·02	68	44	22·1	5	0·08
314	Lacaille 2538 .	7·1	2	6	51	28·31	114	47	47·0	2	0·10
315	... . .	9·5	1	6	53	48·00	132	54	32·7	1	0·08

296.—R Monocerotis Var. 1.—Irregularly variable from 9·8 to 11·5 magnitude.
310.–311.—Comparison stars for [illegible] in 1861.

Observed with the Madras Meridian Circle in that Year.

Number.	Star.	In Right Ascension. Annual Precession.	Secular Variation.	Proper Motion.	In Polar Distance. Annual Precession.	Secular Variation.	Proper Motion.	Number in B. A. C.
		s	s	s	"	"	"	
281	α Argûs *(Canopus)*	+ 1·3392	+ 0·0010	0·000	+ 1·886	+ 0·102	0·00	2086
282	. .	+ 1·0708	+ 0·0018	...	+ 1·986	+ 0·266	...	..
283	Lacaille 2321	+ 0·4222	− 0·0028	...	+ 2·064	+ 0·080		..
284	Taylor 2324 .	+ 1·0139	+ 0·0018		+ 2·069	+ 0·277		...
285	Taylor 2520	+ 1·0108	+ 0·0017		+ 2·130	+ 0·277		2121
286	Taylor 2341 ...	+ 0·0590	− 0·0006		+ 2·180	+ 0·137		2134
287	...	+ 2·0061	+ 0·0018		+ 2·407	+ 0·280		..
288		+ 1·0148	+ 0·0010		+ 2·432	+ 0·276		..
289		+ 2·2390	+ 0·0018		+ 2·480	+ 0·323		...
290		+ 0·6026	− 0·0026		+ 2·504	+ 0·096		...
291	Taylor 2540 .	+ 0·0016	− 0·0031	...	+ 2·600	+ 0·086	.	2166
292	24 Geminorum γ ...	+ 3·4640	− 0·0015	+ 0·001	+ 2·680	+ 0·560	+ 0·04	2168
293		+ 1·9280	+ 0·0016	. .	+ 2·687	+ 0·277	.	...
294		+ 2·2406	+ 0·0018	...	+ 2·717	+ 0·336	. .	...
295	..	+ 1·0280	+ 0·0015	...	+ 2·752	+ 0·277	.	..
296	R Monocerotis Var. 1	+ 3·2783	− 0·0007		+ 2·780	+ 0·473		...
297	. .	+ 1·9264	+ 0·0016		+ 2·964	+ 0·277		...
298	. ... ..	+ 1·9446	+ 0·0015	...	+ 3·017	+ 0·270	...	...
299	27 Geminorum ε ...	+ 3·6952	− 0·0026	0·000	+ 3·110	+ 0·531	+ 0·02	2194
300	. . .	+ 1·9967	+ 0·0014	.	+ 3·136	+ 0·278	...	...
301	51 Cephei *(Hev.)*	+ 30·4036	− 1·3075	− 0·027	+ 3·333	+ 4·394	+ 0·03	2157
302	31 Geminorum ξ	+ 3·3774	− 0·0017	− 0·007	+ 3·386	+ 0·486	+ 0·22	2206
303	Lacaille 2451 ...	+ 0·1187	− 0·0002	...	+ 3·386	+ 0·016	...	...
304	9 Canis Maj. α *(Sirius)*	+ 2·6458	+ 0·0010	− 0·035	+ 3·480	+ 0·384	+ 1·24	2213
305	.. . .	+ 0·8461	− 0·0076	...	+ 3·686	+ 0·046	...	...
306	. .	+ 1·9317	+ 0·0013		+ 3·697	+ 0·273		...
307	... ...	+ 1·9312	+ 0·0013		+ 3·717	+ 0·275		..
308	. .. ...	+ 2·0277	+ 0·0013		+ 3·303	+ 0·286		...
309	... ...	+ 2·6936	+ 0·0009		+ 3·921	+ 0·382		...
310	Lalande 13279	+ 3·6643	− 0·0049		+ 4·050	+ 0·532		..
311	Lalande 13313	+ 3·6996	− 0·0048	...	+ 4·129	+ 0·521		...
312	...	+ 2·0208	+ 0·0018	...	+ 4·280	+ 0·286		.
313	30 Geminorum	+ 3·7156	− 0·0058	− 0·009	+ 4·291	+ 0·527	− 0·10	2276
314	Lacaille 2694 .	+ 2·4690	+ 0·0013	...	+ 4·462	+ 0·349	.	[illegible]
315	.	+ 0·5276	− 0·0086	.	+ 4·381	+ 0·073	. .	...

281.—Proper motions from "*Stone's Cape Catalogue.*"
301.—302.—313.—Proper motions from "*Greenwich Catalogue 1872.*"

Mean Positions of Stars for 1867 January 1st.

Number.	Star.	Magnitude.	Estimations.	Mean Right Ascension.			Mean Polar Distance.			Observations.	Fraction of Year.
				h.	m.	s.	°	′	″		
316	21 Canis Majoris ε ...	1·5	..	6	58	28·89	118	47	35·9	4	0·11
317	Taylor 2806	8·0	1	6	56	11·09	69	12	42·3	1	0·08
318	48 Geminorum ζ Var. 1 ..	4·0	.	6	56	12·04	60	14	16·0	2	0·46
319	... ...	9·5	1	6	56	27·29	139	17	33·8	1	0·06
320	Taylor 2826	9·0	1	6	56	54·09	150	54	51·8	1	0·04
321	23 Canis Majoris γ ...	4·1	...	6	57	44·44	105	26	20·8	5	0·27
322	Lalande 13707	8·0	3	6	58	17·04	67	6	54·1	3	0·10
323	W. B. N. VI. 1762 ...	9·4	3	6	58	36·42	70	55	5·3	3	0·09
324	Taylor 2840	8·1	2	6	59	5·82	150	57	6·9	2	0·17
325	47 Geminorum	5·5	...	7	3	7·09	62	55	41·5	12	0·09
326	... ...	9·5	1	7	5	4·24	158	52	24·4	1	0·18
327	Taylor 2899	8·5	2	7	5	58·62	130	9	6·8	2	0·16
328	... ...	8·0	1	7	8	11·34	152	5	19·0	1	0·17
329	... ...	8·9	1	7	8	30·63	130	17	37·4	1	0·11
330	55 Geminorum δ	3·6	...	7	12	10·07	67	46	33·6	14	0·28
331	... ...	8·0	1	7	12	37·06	152	46	14·7	1	0·30
332	57 Geminorum A	5·0	...	7	15	21·84	64	41	50·4	16	0·06
333	Taylor 3005	8·0	1	7	15	31·98	149	1	14·7	1	0·16
334	Lacaille 2905	7·8	1	7	17	23·94	158	8	28·1	1	0·12
335	... ...	9·0	1	7	17	48·06	129	46	30·4	1	0·17
336	... ...	9·0	1	7	18	10·22	139	42	54·8	1	0·17
337	... ...	10·0	2	7	18	57·24	80	15	46·5	2	0·06
338	Lacaille 2807 ...	7·8	1	7	19	36·98	142	15	41·2	1	0·18
339	... ...	9·0	1	7	19	42·63	138	8	20·2	1	0·15
340	... .	9·5	1	7	19	51·16	158	36	56·8	1	0·20
341	... ...	7·9	1	7	25	32·94	130	13	27·7	1	0·13
342	65 Geminorum	5·0	...	7	26	0·57	73	53	26·9	2	0·06
343	66 Geminorum α¹ (*Castor*).	1·6	...	7	26	6·63	57	49	28·9	12	0·21
344	69 Geminorum υ	4·2	.	7	27	48·42	62	46	42·5	4	0·05
345	... ...	7·9	1	7	30	41·60	121	54	50·5	1	0·19
346	Taylor 3128	6·7	2	7	31	9·67	65	38	44·4	2	0·01
347	... ...	8·7	1	7	31	10·92	131	10	55·2	1	0·13
348	74 Geminorum *f*	5·2	..	7	31	47·76	72	0	52·8	1	0·20
349	10 Canis Min. α (*Procyon*).	0·5		7	32	30·29	84	36	13·6	16	0·18
350	... ...	9·9	2	7	34	14·66	88	9	14·1	3	0·15

318.—ζ Geminorum Var. 1.—Period 10·16 days.—Range, 3·7 to 4·5 magnitude.
328.—Comparison star for Hestia in 1867.
350.—Observed by mistake for Thetis.

Observed with the Madras Meridian Circle in that Year.

Number.	Star.	In Right Ascension.			In Polar Distance.			Number in B.A.C.
		Annual Precession.	Secular Variation.	Proper Motion.	Annual Precession.	Secular Variation.	Proper Motion.	
		s	s	s	"	"	"	
316	21 Canis Majoris ε ..	+ 2·3571	+ 0·0013	0·000	+ 4·680	+ 0·382	+ 0·02	2293
317	Taylor 2805	+ 3·5045	− 0·0050	...	+ 4·869	+ 0·508	...	.
318	43 Gem. ζ² Var. 1 ..	+ 3·5689	− 0·0060	− 0·001	+ 4·570	+ 0·503	+ 0·01	2305
319	. . .	+ 2·0111	+ 0·0012	...	+ 4·891	+ 0·292	...	.
320	Taylor 2823	+ 0·7424	− 0·0070	..	+ 4·928	+ 0·108	...	.
321	23 Canis Majoris γ ..	+ 2·7145	+ 0·0005	+ 0·002	+ 5·000	+ 0·381	+ 0·01	2319
322	Lalande 13707 ..	+ 3·6153	− 0·0067	...	+ 5·045	+ 0·509	...	..
323	W. B. N. VI. 1762...	+ 3·5190	− 0·0049	...	+ 5·073	+ 0·495	...	...
324	Taylor 2840	+ *0·7446	− 0·0074	...	+ 5·113	+ 0·103	...	.
325	47 Geminorum .	+ 3·7297	− 0·0077	− 0·003	+ 5·455	+ 0·522	+ 0·03	2348
326	. .. ..	+ 0·4532	− 0·0125	...	+ 5·617	+ 0·062	...	..
327	Taylor 2899	+ 1·9905	+ 0·0010	...	+ 5·687	+ 0·277	...	..
328	.. ..	+ 0·6591	− 0·0102	...	+ 5·870	+ 0·090	...	...
329	. ... ..	+ 1·9896	+ 0·0010	...	+ 5·940	+ 0·274	...	...
330	Geminorum δ ...	+ 3·5915	− 0·0072	0·000	+ 0·212	+ 0·496	+ 0·02	2410
331	.. . ...	+ 0·5993	− 0·0119	...	+ 0·250	+ 0·050	...	...
332	57 Geminorum A ...	+ 3·6708	− 0·0057	0·000	+ 0·477	+ 0·504	+ 0·02	2431
333	Taylor 3065	+ 0·9651	− 0·0071	...	+ 0·431	+ 0·130	..	...
334	Lacaille 2805 ...	+ 0·5812	− 0·0132	...	+ 0·044	+ 0·077	...	..
335	... ...	+ 2·0296	+ 0·0009	...	+ 6·071	+ 0·275	...	..
336	. ...	+ 2·0256	+ 0·0010		+ 0·709	+ 0·275		...
337	.. ..	+ 3·5460	− 0·0074		+ 6·773	+ 0·494		...
338	Lacaille 2807 ..	+ 1·4479	− 0·0023		+ 6·827	+ 0·196		...
339	.. ..	+ 2·2615	+ 0·0018		+ 0·586	+ 0·300		...
340	... .	+ 0·6473	− 0·0127		+ 0·946	+ 0·086		...
341	... .	+ 2·0629	+ 0·0009	...	+ 7·314	+ 0·276	...	...
342	65 Geminorum ...	+ 3·4314	− 0·0066	− 0·004	+ 7·352	+ 0·463	0·00	2455
343	66 Gem. α² Castor ..	+ 3·8540	− 0·0138	− 0·013	+ 7·389	+ 0·519	+ 0·05	2465
344	69 Geminorum υ ..	+ 3·7092	− 0·0110	− 0·001	+ 7·490	+ 0·490	+ 0·11	2478
345	... .	+ 2·2038	+ 0·0012	...	+ 7·731	+ 0·307	...	..
346	Taylor 3133	+ 3·6344	− 0·0102	...	+ 7·763	+ 0·497	..	2514
347	... .	+ 1·9907	+ 0·0008	...	+ 7·770	+ 0·265	...	.
348	74 Geminorum f .	+ 3·4719	− 0·0078	0·000	+ 7·990	+ 0·463	− 0·01	2519
349	10 Can. Min. α (Procyon)	+ 3·1918	− 0·0041	− 0·048	+ 7·898	+ 0·436	+ 1·03	2532
350	. ...	+ 3·6684	− 0·0094	...	+ 8·010	+ 0·478	...	...

332.—Proper motions from "*Greenwich Catalogue* 1872."

Mean Positions of Stars for 1867 January 1st.

Number.	Star.	Magnitude.	Estimations.	Mean Right Ascension.			Mean Polar Distance.			Observations.	Fraction of Year.
				h.	m.	s.	°	′	″		
361	... ...	0·9	2	7	34	52·38	68	10	13·3	2	0·15
362	... .	9·0	1	7	36	12·72	160	58	16·0	1	0·04
363	... ...	8·0	1	7	36	31·04	163	0	1·9	1	0·19
364	76 Geminorum c	5·3	..	7	36	69·86	63	54	7·8	3	0·09
365	... .	8·0	1	7	36	50·30	129	57	41·8	1	0·04
366	... .	7·6	1	7	36	58·86	180	51	17·3	1	0·20
367	78 Geminorum β (*Pollux*).	1·1	...	7	37	10·41	61	39	20·0	11	0·13
368	.. ..	10·3	1	7	37	32·30	68	30	8·0	1	0·07
369	... ...	8·2	1	7	37	30·50	180	58	31·3	1	0·18
370	... ...	7·1	2	7	39	11·57	131	0	2·0	2	0·14
361	Lacaille 2971	6·9	1	7	40	22·59	143	55	34·4	1	0·10
362	... ...	9·5	1	7	42	12·51	162	88	22·8	1	0·06
363	... ...	9·2	1	7	44	16·12	180	86	27·5	1	0·15
364	B. P. L. 40	6·5	...	7	44	40·44	5	34	7·8	1	0·18
365	83 Geminorum φ	4·9	...	7	45	21·37	62	50	35·5	3	0·02
366	Brisbane 1791	8·0	1	7	46	32·47	144	36	7·1	1	0·06
367	Taylor 3399	7·8	2	7	46	35·47	144	44	34·1	2	0·20
368	... ...	8·1	1	7	47	6·14	168	21	15·1	1	0·19
369	U Geminorum Var. 5 ...	10·2	1	7	47	12·75	67	39	6·8	1	0·06
370	... ...	9·5	1	7	48	49·96	120	56	19·4	1	0·00
371	1 Cancri	5·9	...	7	49	30·33	73	61	36·2	1	0·12
372	Taylor 3323	7·0	1	7	49	51·35	149	16	9·5	1	0·19
373	W. B. N. VII. 1473 ..	7·5	2	7	53	42·27	64	32	51·7	3	0·06
374	... ...	8·3	1	7	53	47·19	140	53	19·0	1	0·17
375	6 Cancri	5·0	...	7	56	30·73	61	60	8·7	3	0·13
376	8 Cancri . ..	5·1	..	7	57	30·85	76	30	20·0	1	0·13
377	.. .. ..	9·0	1	7	57	41·32	156	34	35·3	1	0·06
378	.. ..	10·2	1	8	0	15·60	78	34	23·3	1	0·06
379	... .. .	8·0	2	8	1	26·91	160	31	32·0	2	0·10
380	15 Argûs .. .	2·9	..	8	1	52·15	113	54	22·8	10	0·16
381	. .	9·2	1	8	2	19·65	129	30	36·3	1	0·09
382	14 Cancri ψ[1] ...	5·3		8	2	36·11	64	5	31·2	2	0·01
383	.	9·6	1	8	4	31·[illegible]	154	41	5·1	1	0·21
384	16 Cancri ζ .	5·4	...	8	4	31·[illegible]	71	57	13·9	1	0·06
385	Lacaille 3200	6·8	3	8	4	50·21	163	7	57·4	3	0·11

381.—Groombridge 1369.
369.—U Geminorum Var. 5.—Period 90 days but irregular.—Range, 9th to below 14th magnitude.

Observed with the Madras Meridian Circle in that Year.

Number.	Star.	In Right Ascension. Annual Precession.	In Right Ascension. Secular Variation.	In Right Ascension. Proper Motion.	In Polar Distance. Annual Precession.	In Polar Distance. Secular Variation.	In Polar Distance. Proper Motion.	Number in B. A. C.
		s	s	s	"	"	"	
351		+ 3·5684	− 0·0095	.	+ 8·067	+ 0·472	...	
352	...	+ 2·0467	+ 0·0008	. .	+ 8·094	+ 0·270		...
353	.. . ..	+ 0·0721	− 0·0152	...	+ 8·106	+ 0·097	...	...
354	76 Geminorum c	+ 3·6704	− 0·0134	0·000	+ 8·157	+ 0·486	+ 0·03	2640
355	... ...	+ 2·0508	+ 0·0008	...	+ 8·224	+ 0·269	...	...
356	... .	+ 2·0170	+ 0·0000		+ 8·228	+ 0·264	...	...
357	78 Gem. β (*Pollux*) .	+ 3·7294	− 0·0128	+ 0·049	+ 8·251	+ 0·491	+ 0·06	2656
358	... ...	+ 3·5618	− 0·0096	...	+ 8·279	+ 0·469	..	...
359	... ...	+ 2·0148	+ 0·0008	...	+ 8·289	+ 0·264	...	..
360	. ..	+ 2·0114	+ 0·0009	...	+ 8·411	+ 0·262	...	...
361	Lacaille 2971 ...	+ 1·4104	− 0·0088	...	+ 8·506	+ 0·182	...	..
362	... .	+ 0·7076	− 0·0158	...	+ 8·650	+ 0·089	...	...
363	... ...	+ 2·0300	+ 0·0008	...	+ 8·812	+ 0·262	...	...
364	R. P. L. 49	+ 15·3790	− 1·2263	...	+ 8·845	+ 2·009	...	2695
365	83 Geminorum φ ..	+ 3·6844	− 0·0130	− 0·004	+ 8·898	+ 0·478	+ 0·06	2617
366	Brisbane 1791 .	+ 1·4010	− 0·0043	...	+ 8·978	+ 0·179	...	...
367	Taylor 3898	+ 1·3813	− 0·0045	.	+ 8·994	+ 0·176	...	...
368	... .	+ 0·6935	− 0·0170	...	+ 9·098	+ 0·096	...	...
369	U Geminorum Var. 5.	+ 3·5627	− 0·0168	...	+ 9·044	+ 0·460	...	...
370	... .	+ 2·0771	+ 0·0010	...	+ 9·168	+ 0·266	...	...
371	1 Cancri	+ 3·4159	− 0·0084	− 0·001	+ 9·217	+ 0·469	+ 0·04	2680
372	Taylor 3328	+ 1·0765	− 0·0096	...	+ 9·249	+ 0·126	...	...
373	W. B. N. VII. 1473..	+ 3·0817	− 0·0130	...	+ 9·547	+ 0·462	...	...
374	. .. ..	+ 1·0451	− 0·0104	...	+ 9·558	+ 0·130	...	...
375	6 Cancri	+ 3·0901	− 0·0148	− 0·005	+ 9·672	+ 0·468	+ 0·07	2672
376	8 Cancri .	+ 3·3516	− 0·0079	− 0·009	+ 9·850	+ 0·462	+ 0·05	2690
377	. ...	+ 0·4063	− 0·0273	...	+ 9·862	+ 0·048	...	...
378	. ...	+ 3·3077	− 0·0074	...	+ 10·062	+ 0·413	...	...
379	.. .	+ 1·0906	− 0·0116	...	+ 10·126	+ 0·126	...	...
380	15 Argûs ..	+ 2·6808	+ 0·0009	− 0·007	+ 10·170	+ 0·318	− 0·06	[illegible]
381	... .	+ 2·1510	+ 0·0016	...	+ 10·202	+ 0·280	...	...
382	14 Cancri ψ²	+ 3·6311	− 0·0140	− 0·006	+ 10·212	+ 0·462	+ 0·35	2730
383	... ..	+ 0·0831	− 0·0217	...	+ 10·373	+ 0·078	...	...
384	10 Cancri 3	+ 3·4460	− 0·0103	+ 0·004	+ 10·373	+ 0·466	+ 0·11	2744
385	Lacaille 3200 ..	+ 0·8151	− 0·0172	...	+ 10·391	+ 0·068	...	...

371.—Proper motions from "*Greenwich Catalogue* 1872."
376.—Proper motions from "*Greenwich Catalogue* 1864."

Mean Positions of Stars for 1867 January 1st.

Number.	Star.	Magnitude.	Estimations.	Mean Right Ascension.			Mean Polar Distance.			Observations.	Fraction of Year.
				h.	m.	s.	°	′	″		
386	...	9·4	1	8	5	36·44	77	26	31·2	1	0·16
387	...	9·5	1	8	5	43·07	125	39	18·3	1	0·21
388	...	8·0	1	8	6	10·35	128	41	8·0	1	0·15
389	...	9·5	1	8	6	12·34	123	40	14·6	1	0·08
390	...	9·2	1	8	8	40·56	129	38	56·0	1	0·15
391	W. B. E. VIII. 220 ...	8·0	3	8	9	36·88	99	32	10·2	3	0·19
392	... ...	9·7	1	8	10	9·07	150	47	12·9	1	0·06
393	... ...	9·2	1	8	10	29·29	151	26	51·2	1	0·20
394	... .	8·0	1	8	11	53·80	152	1	51·5	1	0·19
395	.. ..	9·5	1	8	12	1·98	131	43	22·6	1	0·15
396	... .	8·0	1	8	12	28·81	186	44	13·0	1	0·20
397	... ...	9·4	4	8	13	10·74	131	46	29·3	4	0·17
398	W. B. E. VIII. 383 ...	7·8	3	8	15	35·46	100	10	37·0	4	0·17
399	... ...	9·5	1	8	16	38·00	77	32	20·5	1	0·15
400	... ...	8·7	1	8	17	32·62	77	40	42·9	1	0·17
401	... ...	9·5	1	8	18	1·12	154	38	16·6	1	0·08
402	Taylor 3599 ...	6·9	3	8	20	21·35	144	58	21·0	3	0·19
403	Taylor 3607	6·8	1	8	21	19·61	144	55	57·8	1	0·10
404	... ...	8·0	1	8	21	36·58	131	42	10·0	1	0·15
405	... ...	9·8	1	8	21	41·90	168	13	6·4	1	0·21
406	... .	7·1	1	8	23	35·39	144	53	42·9	1	0·20
407	33 Cancri η ...	6·5	...	8	25	0·86	69	6	36·0	10	0·28
408	Taylor 3651 ...	6·7	1	8	25	46·17	130	3	56·7	1	0·13
409	Taylor 3652 ...	7·7	1	8	25	50·31	130	3	15·7	1	0·13
410	... .	8·0	...	8	26	11·35	145	0	34·0	1	0·09
411	. ...	9·1	1	8	28	45·06	76	19	48·9	1	0·18
412	Lalande 16890 . ..	7·9	1	8	28	51·74	73	13	30·5	1	0·19
413	... ...	9·8	1	8	28	57·72	70	41	10·7	1	0·20
414	W. B. N. VIII. 690 ...	9·0	1	8	29	41·42	70	40	16·6	1	0·06
415	.. .	9·0	1	8	31	19·17	128	46	2·1	1	0·21
416	Taylor 3710	7·1	1	8	31	30·53	141	21	41·3	1	0·17
417	.	8·4	1	8	34	11·36	154	21	4·6	1	0·15
418	Lacaille 3475 ...	6·0		8	34	37·13	132	38	13·3	1	0·26
419	43 Cancri γ ...	4·3	...	8	35	30·01	68	3	30·7	1	0·06
420	Lacaille 3491	6·9	2	8	36	4·44	132	22	30·1	2	0·19

391.—398.—Comparison star for Phocea in 1867.
412.—Comparison star for Freia in 1864.

Observed with the Madras Meridian Circle in that Year.

Number.	Star.	In Right Ascension. Annual Precession.	Secular Variation.	Proper Motion.	In Polar Distance. Annual Precession.	Secular Variation.	Proper Motion.	Number in B.A.C.
		s	s	s	"	"	"	
386	..	+ 3·3967	− 0·0081		+ 10·[illegible]	+ 0·410	...	...
387	...	+ 2·1596	+ 0·0016		+ 10·[illegible]	+ 0·265	...	...
388	..	+ 2·1597	+ 0·0016		+ 10·491	+ 0·254	...	...
389	..	+ 2·1600	+ 0·0010		+ 10·[illegible]	+ 0·[illegible]	...	...
390	..	+ 2·1671	+ 0·0017		+ 10·678	+ 0·[illegible]	...	...
391	W. B. E. VIII. 290..	+ 2·8865	− 0·0015		+ 10·747	+ 0·[illegible]		...
392	... .	+ 1·0664	− 0·0120		+ 10·737	+ 0·126		...
393	... ..	+ 1·0027	− 0·0133		+ 10·811	+ 0·118		...
394	... ...	+ 0·9590	− 0·0146		+ 10·915	+ 0·112		...
395	..	+ 2·0723	+ 0·0014		+ 10·925	+ 0·248		...
396	... ..	+ 2·1736	+ 0·0018		+ 10·982	+ 0·261		...
397	...	+ 2·0760	+ 0·0015		+ 11·020	+ 0·248		...
398	W. B. E. VIII. 393..	+ 2·8690	− 0·0014		+ 11·185	+ 0·243		...
399	.. ... .	+ 3·3166	− 0·0085		+ 11·[illegible]	+ 0·[illegible]		...
400	...	+ 3·3101	− 0·0084		+ 11·[illegible]	+ 0·[illegible]		...
401	.. ..	+ 0·7795	− 0·0207		+ 11·361	+ 0·058		...
402	Taylor 3599	+ 1·5161	− 0·00[illegible]		+ 11·[illegible]	+ 0·176		...
403	Taylor 3607	+ 1·5162	− 0·00[illegible]		+ 11·[illegible]	+ 0·176		...
404	... .	+ 2·1010	+ 0·0018		+ 11·617	+ 0·246		..
405	... .	+ 0·9086	− 0·0174		+ 11·[illegible]	+ 0·103		...
406	.. .	+ 1·2814	− 0·0037	..	+ 11·780	+ 0·176	.	...
407	38 Cancri η ..	+ 3·[illegible]	− 0·0129	− 0·006	+ 11·[illegible]	+ 0·401	+ 0·05	[illegible]
408	Taylor 3651 ..	+ 2·1675	+ 0·0022		+ 11·913	+ 0·249	...	...
409	Taylor 3652	+ 2·1651	+ 0·00[illegible]		+ 11·917	+ 0·[illegible]		...
410	... .	+ 1·5376	− 0·00[illegible]		+ 11·[illegible]	+ 0·175		...
411	. .. ..	+ 3·[illegible]	− 0·0092		+ 12·127	+ 0·[illegible]		...
412	Lalande 16[illegible] .	+ 3·[illegible]	− 0·0110		+ 12·130	+ 0·[illegible]		...
413	.. ..	+ 3·4131	− 0·0124		+ 12·137	+ 0·[illegible]		..
414	W. B. N. VIII. [illegible]	+ 3·4446	− 0·0124		+ 12·187	+ 0·[illegible]		...
415	...	+ 2·10[illegible]	+ 0·0014		+ 12·[illegible]	+ 0·232		...
416	Taylor 3710 .	+ 1·7619	− 0·0006	...	+ 12·812	+ 0·197	...	..
417	..	+ 0·[illegible]	− 0·0197	...	+ 12·407	+ 0·097	...	..
418	Lacaille 3473 ..	+ 1·0784	− 0·0142	...	+ 12·[illegible]	+ 0·119	...	[illegible]
419	43 Cancri γ .	+ 3·4012	− 0·0143	− 0·011	+ 12·[illegible]	+ 0·[illegible]	− 0·01	[illegible]
420	Lacaille 3491 ...	+ 1·0573	− 0·0141	...	+ 12·[illegible]	+ 0·118	..	[illegible]

Mean Positions of Stars for 1867 January 1st.

Number.	Star.	Magnitude.	Estimations.	Mean Right Ascension. h.	m.	s.	Mean Polar Distance. °	′	″	Observations.	Fraction of Year.
421	b Velorum ...	4·1	...	8	36	12·94	136	10	37·3	2	0·19
422	47 Cancri δ ...	4·3	...	8	37	7·41	71	21	51·0	2	0·16
423	11 Hydræ ε ...	3·6	...	8	39	48·85	83	5	46·7	9	0·16
424	... ...	7·7	1	8	40	36·21	129	16	18·3	1	0·13
425	... ...	8·4	2	8	45	56·11	95	27	52·2	2	0·15
426	... ...	7·8	1	8	46	50·52	132	53	55·8	1	0·30
427	... ...	7·0	8	8	46	59·73	130	2	36·5	3	0·28
428	R. P. L. 60 ...	6·3	...	8	47	4·33	5	17	35·5	3	0·38
429	... ...	7·0	1	8	48	51·36	128	1	46·0	1	0·18
430	... .	7·5	...	8	49	26·63	132	50	40·3	1	0·06
431	... ...	8·8	1	8	50	21·07	132	57	36·6	1	0·06
432	... ...	8·3	2	8	52	22·85	77	58	11·2	2	0·16
433	... ...	8·0	1	8	54	8·10	132	54	22·9	1	0·19
434	... ...	8·5	1	8	54	58·87	130	56	11·6	1	0·19
435	... ...	7·8	2	8	55	14·21	146	37	22·7	2	0·23
436	... ...	7·9	1	8	56	44·38	145	56	26·4	1	0·36
437	... .	8·0	1	8	58	14·40	146	19	6·1	1	0·35
438	76 Cancri κ ...	5·0	...	9	0	32·01	78	47	54·5	1	0·38
439	... ...	8·3	1	9	4	31·47	132	46	11·8	1	0·10
440	Lacaille 3713 ...	7·0	1	9	4	40·88	148	49	57·1	1	0·16
441	... .	9·3	3	9	5	14·64	134	40	2·4	3	0·21
442	Taylor 4026	7·3	3	9	6	1·72	132	48	26·8	3	0·23
443	Taylor 4028	8·0	3	9	6	11·60	132	43	39·9	3	0·30
444	... ...	10·0	1	9	7	39·90	134	47	57·5	1	0·34
445	Lacaille 3747 ...	7·8	5	9	7	49·94	150	34	28·5	5	0·25
446	... .	8·7	1	9	9	22·46	150	32	31·1	1	0·35
447	Lacaille 3761 ...	7·3	2	9	9	34·74	150	22	0·9	2	0·27
448	...	9·6	1	9	10	15·52	150	36	7·1	1	0·37
449	...	7·9	3	9	10	49·69	150	21	34·7	3	0·36
450	...	9·5	2	9	11	4·42	70	41	32·1	2	0·15
451	83 Cancri	6·0	..	9	11	38·26	71	43	59·0	4	0·30
452	.. ..	8·7	3	9	12	27·46	124	48	17·5	3	0·30
453	.. ..	10·1	2	9	13	6·17	70	52	59·5	2	0·15
454	ι Argûs ...	2·5		9	13	32·04	148	48	7·1	1	0·16
455	. . .	8·4	2	9	15	9·17	150	50	20·3	2	0·28

425.—Star for map of S Hydræ Var. 3.
428.—Carrington 1324.
430.—Observed by mistake for Diana.

Observed with the Madras Meridian Circle in that Year.

Number	Star.	In Right Ascension. Annual Precession.	In Right Ascension. Secular Variation.	In Right Ascension. Proper Motion.	In Polar Distance. Annual Precession.	In Polar Distance. Secular Variation.	In Polar Distance. Proper Motion.	Number in B. A. C.
		s	s	s	"	"	"	
421	b Velorum ..	+ 1·9906	+ 0·0018	.	+ 12·686	+ 0·221	.	2947
422	47 Cancri δ	+ 3·4212	− 0·0126	− 0·002	+ 12·697	+ 0·382	+ 0·24	2963
423	11 Hydræ ε	+ 3·1062	− 0·0071	− 0·013	+ 12·878	+ 0·351	+ 0·04	2971
424	.. ...	+ 2·2893	+ 0·0081	...	+ 12·981	+ 0·244		...
425	...	+ 3·1340	− 0·0065		+ 13·264	+ 0·277		...
426		+ 2·1446	+ 0·0082		+ 13·344	+ 0·228		...
427	. ...	+ 2·0376	+ 0·0096		+ 13·354	+ 0·216		...
428	R. P. L. 60 ..	+ 13·8358	− 1·7200		+ 13·380	+ 1·490		...
429	.. ..	+ 2·1478	+ 0·0083		+ 13·475	+ 0·226		...
430	... ...	+ 2·1611	+ 0·0083		+ 13·513	+ 0·226		...
431	..	+ 2·1500	+ 0·0084		+ 13·578	+ 0·226		...
432	.	+ 3·2870	− 0·0098		+ 13·701	+ 0·344		...
433	...	+ 2·1712	+ 0·0087		+ 13·818	+ 0·283		...
434	...	+ 2·2489	+ 0·0069		+ 13·862	+ 0·281		...
435		+ 1·5989	− 0·0030		+ 13·883	+ 0·161		...
436	... ...	+ 1·3091	− 0·0027	...	+ 13·977	+ 0·161	...	...
437	... .	+ 1·6426	− 0·0018	...	+ 14·071	+ 0·166	...	...
438	76 Cancri κ . ..	+ 3·2966	− 0·0098	− 0·002	+ 14·215	+ 0·329	0·00	3111
439	... ...	+ 2·2150	+ 0·0046	...	+ 14·463	+ 0·218	...	...
440	Lacaille 3713	+ 1·8065	+ 0·0010	...	+ 14·467	+ 0·170	...	...
441	... .	+ 2·4896	+ 0·0045		+ 14·502	+ 0·260		...
442	Taylor 4086 . ..	+ 2·2223	+ 0·0048		+ 14·549	+ 0·217		...
443	Taylor 4088	+ 2·2226	+ 0·0048		+ 14·560	+ 0·217		...
444	..	+ 2·4874	+ 0·0046		+ 14·647	+ 0·237		...
445	Lacaille 3747 ...	+ 1·4061	− 0·0057		+ 14·667	+ 0·140		...
446	... .	+ 1·4691	− 0·0067		+ 14·740	+ 0·186	...	...
447	Lacaille 3761...	+ 1·4605	− 0·0062		+ 14·761	+ 0·140	...	...
448	. ... ..	+ 1·4622	− 0·0062		+ 14·804	+ 0·140	...	...
449	. . .	+ 1·4909	− 0·0060		+ 14·894	+ 0·140	...	...
450	. . ..	+ 3·3309	− 0·0140		+ 14·849	+ 0·336		...
451	88 Cancri . .	+ 3·3480	− 0·0134	− 0·012	+ 14·877	+ 0·328	+ 0·16	3171
452	.. .	+ 2·4516	+ 0·0060	..	+ 14·980	+ 0·238	..	...
453	... . ..	+ 3·3068	− 0·0142	.	+ 14·988	+ 0·329	...	...
454	ι Argûs .. .	+ 1·6106	− 0·0022	..	+ 14·998	+ 0·150	...	3180
455	.. .. ..	+ 1·5167	− 0·0044	.	+ 15·037	+ 0·140	...	...

Mean Positions of Stars for 1867 January 1st.

Number.	Star.	Magnitude.	Estimations.	Mean Right Ascension.			Mean Polar Distance.			Observations.	Fraction of Year.
				h.	m.	s.	°	′	″		
456	...	8·3	3	9	15	30·07	150	36	4·7	3	0·25
457	...	7·5	1	9	15	21·49	142	49	27·4	1	0·17
458	...	8·0	1	9	15	48·04	124	47	39·8	1	0·16
459	...	9·3	2	9	16	30·58	70	37	26·7	2	0·14
460	...	9·7	1	9	16	38·98	124	46	19·3	1	0·24
461	... ...	8·0	1	9	19	6·07	70	22	4·9	1	0·12
462	... ...	8·5	2	9	19	26·17	130	31	25·8	2	0·26
463	... ...	8·7	2	9	19	51·17	136	21	57·8	2	0·17
464	... ...	8·3	1	9	20	34·62	135	23	42·9	1	0·16
465	30 Hydræ α Var. 2 ...	2·0	...	9	21	3·03	98	5	2·5	7	0·20
466	Lalande 18636	8·7	2	9	21	51·62	68	30	24·0	2	0·21
467	Lalande 18660	9·2	2	9	22	47·20	67	50	14·1	2	0·24
468	Lalande 18683	8·4	2	9	23	27·84	68	7	37·3	2	0·20
469	6 Leonis h	5·4	...	9	24	49·70	79	41	50·1	2	0·13
470	Lalande 18730	9·1	3	9	25	2·60	68	38	37·4	3	0·20
471	... ...	8·8	1	9	26	46·02	145	3	14·7	1	0·25
472	Taylor 4228 ...	7·5	3	9	27	50·21	146	24	8·3	3	0·15
473	... ...	8·5	2	9	28	22·06	146	32	46·0	2	0·17
474	... ...	9·0	3	9	31	26·96	136	26	34·8	3	0·20
475	... ...	8·5	3	9	32	19·69	129	40	39·4	3	0·22
476	R. P. L. 60 ...	7·9	...	9	33	30·84	2	47	34·6	2	0·24
477	14 Leonis o ...	3·8	...	9	34	3·08	79	30	15·0	1	0·20
478	Lacaille 3960 ...	7·0	1	9	34	36·71	145	34	36·0	1	0·26
479	.. ..	9·0	1	9	35	36·60	145	30	24·2	1	0·25
480	... ...	7·7	3	9	36	54·83	146	40	53·7	3	0·23
481	... .	6·6	1	9	36	53·89	131	54	39·9	1	0·26
482	. ...	6·0	1	9	37	0·22	146	25	5·9	1	0·26
483	... ...	8·0	2	9	37	10·60	145	32	0·1	2	0·26
484	R Leonis Minoris Var. 1...	7·3	3	9	37	36·44	54	32	43·9	3	0·16
485	17 Leonis ε	3·1	...	9	38	17·91	65	36	55·4	12	0·20
486	. . .	7·9	1	9	39	44·76	148	34	33·2	1	0·26
487	l Carinæ Var. 1 .	.		9	41	36·09	151	53	40·6	1	0·25
488	... .	8·0	1	9	41	36·77	130	44	40·3	1	0·14
489	Taylor 4637 .	7·0	1	9	41	56·91	165	36	5·4	1	0·26
490	... ...	8·0	1	9	42	49·48	130	48	39·4	1	0·27

459.—461.—Observed by mistake for Diana.
465.—α Hydræ Var. 2.—Supposed to vary irregularly from 2·0 to 2·5 magnitude.
466.—467.—468.—470.—Comparison stars for Metis in 1851.
476.—Carrington 1418.
484.—R Leonis Minoris Var. 1.—Period 376 days.—Range, 6·5 to below 11th magnitude.
487.—l Carinæ Var. 1.—Period 31 days.—Range, 3·7 to 5·2 magnitude.

Observed with the Madras Meridian Circle in that Year.

Number.	Star.	In Right Ascension. Annual Precession.	In Right Ascension. Secular Variation.	In Right Ascension. Proper Motion.	In Polar Distance. Annual Precession.	In Polar Distance. Secular Variation.	In Polar Distance. Proper Motion.	Number in B. A. C.
		s	s	s	"	"	"	
456	..	+ 1·5220	− 0·0042	...	+ 15·096	+ 0·140	...	...
457	...	+ 1·8657	+ 0·0090	...	+ 15·098	+ 0·174	...	...
458	...	+ 2·4017	+ 0·0053	...	+ 15·119	+ 0·289	...	...
459	...	+ 3·3827	− 0·0142	...	+ 15·165	+ 0·316	...	...
460	...	+ 2·4048	+ 0·0058	...	+ 15·178	+ 0·288	...	...
461	... ...	+ 3·3801	− 0·0114	...	+ 15·312	+ 0·311	...	...
462	...	+ 1·5460	− 0·0065	...	+ 15·380	+ 0·138	...	...
463	... ...	+ 2·4616	+ 0·0060	...	+ 15·366	+ 0·294	...	...
464	... .	+ 2·4638	+ 0·0057	...	+ 15·396	+ 0·288	...	...
465	30 Hydræ α Var. 2 ...	+ 2·9800	− 0·0013	− 0·004	+ 15·423	+ 0·266	− 0·08	[illegible]
466	Lalande 18686 ...	+ 3·4072	− 0·0158	...	+ 15·467	+ 0·310	...	...
467	Lalande 18699 .	+ 3·4170	− 0·0161	...	+ 15·519	+ 0·309	...	...
468	Lalande 18680 ..	+ 3·4109	− 0·0150	...	+ 15·556	+ 0·308	...	...
469	6 Leonis *h* .	+ 3·2844	− 0·0092	− 0·002	+ 15·631	+ 0·288	+ 0·02	3361
470	Lalande 18730 ...	+ 3·3992	− 0·0156	...	+ 15·643	+ 0·304	...	...
471	... .	+ 1·8863	+ 0·0037		+ 15·738	+ 0·164	...	...
472	Taylor 4222	+ 1·8820	+ 0·0023		+ 15·796	+ 0·158	...	...
473	. .. ...	+ 1·8958	+ 0·0029		+ 15·866	+ 0·157	...	...
474	... ...	+ 2·4762	+ 0·0067		+ 15·980	+ 0·211	...	...
475	... .	+ 2·4059	+ 0·0072		+ 16·034	+ 0·304	...	...
476	R. P. L. 69	+ 19·3986	− 5·7291	...	+ 16·113	+ 1·677	...	...
477	14 Leonis *o* . .	+ 3·2198	− 0·0098	− 0·013	+ 16·194	+ 0·272	+ 0·04	3312
478	Lacaille 3980	+ 1·7750	+ 0·0024	...	+ 16·152	+ 0·147	..	...
479	... ..	+ 1·7866	+ 0·0027	..	+ 16·205	+ 0·147	...	...
480	. ..	+ 1·7792	+ 0·0026	..	+ 16·230	+ 0·146	...	...
481	.	+ 1·6076	− 0·0016		+ 16·271	+ 0·130	...	...
482		+ 1·8006	+ 0·0031		+ 16·270	+ 0·146	...	...
483	... .	+ 1·7970	+ 0·0030	...	+ 16·278	+ 0·140	...	...
484	R Leonis Min. Var. 1.	+ 3·6196	− 0·0276	...	+ 16·306	+ 0·301	...	...
485	17 Leonis *ε* ...	+ 3·4283	− 0·0180	− 0·004	+ 16·342	+ 0·288	+ 0·02	3381
486	... .	+ 1·8180	+ 0·0036	...	+ 16·415	+ 0·146	...	...
487	*l* Carinæ Var. 1	+ 1·6604	− 0·0001	− 0·008	+ 16·307	+ 0·130	− 0·02	[illegible]
488	... ..	+ 2·4174	+ 0·0083	...	+ 16·809	+ 0·198	...	...
489	Taylor 4337 .	+ 1·8870	+ 0·0042	...	+ 16·586	+ 0·146	..	..
490	. .	+ 2·4217	+ 0·0084	..	+ 16·588	+ 0·192	..	...

457.—Proper motions from "*Stone's Cape Catalogue.*"

Mean Positions of Stars for 1867 January 1st.

Number.	Star.	Magnitude.	Estimations.	Mean Right Ascension.			Mean Polar Distance.			Observations.	Fraction of Year.
				h	m	s	°	′	″		
491	... ...	9·5	1	9	44	15·40	148	30	48·0	1	0·36
492	... ...	9·4	2	9	45	4·61	109	48	11·3	2	0·18
493	R. P. L. 70 ...	6·5	...	9	46	50·38	5	26	35·1	8	0·18
494	Taylor 4091 ...	6·8	3	9	47	11·38	162	7	31·0	3	0·19
495	... .	10·3	1	9	47	16·57	73	34	21·0	2	0·20
496	... ...	8·3	3	9	52	40·72	130	41	32·2	3	0·28
497	29 Leonis π ...	5·0	...	9	53	10·90	81	19	6·9	14	0·21
498	... .	9·6	1	9	54	54·30	135	13	14·0	1	0·16
499	... .	8·0	1	9	56	57·99	147	25	7·3	1	0·26
500	... .	8·8	2	9	57	54·18	145	38	54·9	2	0·28
501	Taylor 4470	8·0	1	9	57	57·80	146	30	56·7	1	0·15
502	... ...	7·9	1	9	58	31·74	130	39	51·6	1	0·20
503	Taylor 4494	7·1	1	9	58	46·06	151	30	53·2	1	0·28
504	... ...	9·5	3	10	0	3·81	80	28	6·0	3	0·26
505	32 Leonis α (*Regulus*) ...	1·4	...	10	1	17·10	77	38	2·9	12	0·19
506	... ...	8·0	1	10	1	38·83	130	2	51·7	1	0·26
507	... ...	8·1	1	10	2	9·03	130	57	19·8	1	0·28
508	... ...	9·3	2	10	2	27·30	108	38	17·3	2	0·19
509	... ...	8·5	1	10	4	33·50	122	54	50·9	1	0·14
510	... ...	9·6	3	10	4	41·51	123	39	41·7	3	0·20
511	Taylor 4652 ...	7·0	1	10	7	8·60	147	34	17·9	1	0·26
512	... ...	7·5	2	10	8	6·86	130	37	11·2	2	0·28
513	... .	8·7	1	10	9	8·80	100	52	35·8	1	0·26
514	R. P. L. 72 ..	5·9	...	10	9	30·67	5	4	33·2	3	0·28
515	... .	8·0	1	10	9	38·15	145	26	16·2	1	0·16
516	... .	9·1	3	10	10	38·25	130	32	5·8	3	0·21
517	41 Leonis γ¹	2·2	...	10	12	28·14	69	30	13·0	7	0·26
518	... ...	9·5	1	10	12	30·00	128	37	51·6	1	0·29
519	... ...	8·0	3	10	16	6·78	126	38	14·5	3	0·20
520	Lalande 20130	8·8	1	10	16	17·83	75	26	34·9	1	0·13
521	.. .. ..	9·7	1	10	18	52·64	140	9	21·0	1	0·15
522	... .	9·7	1	10	19	30·83	140	9	36·9	1	0·20
523	. . ..	9·5	2	10	21	26·04	125	34	9·7	2	0·18
524	... .	7·4	3	10	22	17·68	106	32	30·2	3	0·20
525	... ...	8·0	1	10	24	12·38	146	55	1·6	1	0·28

493.—Carrington 1451.
514.—Groombridge 1600.

Observed with the Madras Meridian Circle in that Year.

Number.	Star.	In Right Ascension. Annual Precession.	In Right Ascension. Secular Variation.	In Right Ascension. Proper Motion.	In Polar Distance. Annual Precession.	In Polar Distance. Secular Variation.	In Polar Distance. Proper Motion.	Number in B. A. C.
		s	s	s	"	"	"	
491		+ 1·8836	+ 0·0047	...	+ 16·639	+ 0·144	..	
492	..	+ 2·4535	+ 0·0036	...	+ 16·679	+ 0·192	...	...
493	R. P. L. 70	+ 10·7719	− 1·5839	...	+ 16·764	+ 0·866		...
494	Taylor 4381	+ 1·6840	+ 0·0013	..	+ 16·781	+ 0·138		3380
495	..	+ 3·2603	− 0·0113		+ 16·786	+ 0·266		...
496	... ...	+ 2·4475	+ 0·0005	..	+ 17·046	+ 0·184	...	..
497	29 Leonis π	+ 3·1798	− 0·0090	− 0·008	+ 17·062	+ 0·286	+ 0·03	3415
498	.	+ 2·5806	+ 0·0066	...	+ 17·141	+ 0·198	...	..
499	.	+ 1·9944	+ 0·0098	...	+ 17·169	+ 0·148	...	...
500	..	+ 2·0817	+ 0·0099	...	+ 17·275	+ 0·147	...	...
501	Taylor 4476	+ 2·0802	+ 0·0100	...	+ 17·273	+ 0·147	..	...
502	... .	+ 1·8696	+ 0·0067	...	+ 17·308	+ 0·131	..	...
503	Taylor 4494	+ 1·8987	+ 0·0058	.	+ 17·314	+ 0·128	...	...
504	.. ..	+ 3·1143	− 0·0085	...	+ 17·371	+ 0·280	...	...
505	32 Leonis α (*Regulus*)	+ 3·2202	− 0·0102	− 0·010	+ 17·494	+ 0·285	− 0·01	3450
506	...	+ 2·5165	+ 0·0105		+ 17·488	+ 0·175	..	--
507	...	+ 2·8212	+ 0·0105		+ 17·461	+ 0·178	...	..
508	...	+ 2·6898	+ 0·0094		+ 17·475	+ 0·198	...	...
509	..	+ 2·6544	+ 0·0089		+ 17·564	+ 0·179	...	...
510	..	+ 2·6465	+ 0·0091		+ 17·570	+ 0·178	...	...
511	Taylor 4562 .. .	+ 2·0836	+ 0·0116		+ 17·672	+ 0·136	...	3491
512	. .	+ 2·3260	+ 0·0130	...	+ 17·712	+ 0·151	...	...
513	. .	+ 2·3844	+ 0·0131	...	+ 17·754	+ 0·150	...	...
514	R. P. L. 72	+ 10·0320	− 1·6899	− 0·070	+ 17·788	+ 0·680	+ 0·05	3496
515	.. .	+ 2·1707	+ 0·0129	...	+ 17·788	+ 0·136	...	...
516	.. ..	+ 2·3422	+ 0·0134	...	+ 17·801	+ 0·140	...	...
517	41 Leonis γ¹ . ..	+ 3·2870	− 0·0148	+ 0·010	+ 17·894	+ 0·308	+ 0·15	[illegible]
518	.. ..	+ 2·8912	+ 0·0115	...	+ 17·908	+ 0·162	...	...
519	.. ..	+ 2·8580	+ 0·0110	...	+ 18·089	+ 0·161	...	...
520	Lalande 20139	+ 3·2942	− 0·0110	...	+ 18·086	+ 0·196	...	...
521	.. . .	+ 2·3806	+ 0·0152		+ 18·186	+ 0·131	...	...
522	.. ...	+ 2·3864	+ 0·0155		+ 18·197	+ 0·130	..	...
523	.. ...	+ 2·6795	+ 0·0116		+ 18·288	+ 0·154	..	...
524	.. ..	+ 2·6771	+ 0·0116		+ 18·260	+ 0·158	...	...
525	... ...	+ 2·2896	+ 0·0166	..	+ 18·398	+ 0·125	...	...

Mean Positions of Stars for 1867 *January* 1*st*.

Number.	Star.	Magnitude.	Estimations.	Mean Right Ascension.			Mean Polar Distance.			Observations.	Fraction of Year.
				h.	m.	s.	°	′	″		
526	... ...	8·4	1	10	24	32·14	146	30	50·6	1	0·25
527	... ...	8·2	2	10	24	53·44	135	34	47·0	2	0·22
528	Lalande 20402	7·7	3	10	25	8·18	79	53	41·9	3	0·27
529	47 Leonis ρ	4·0	...	10	25	46·39	80	0	36·7	12	0·19
530	Taylor 4769 ...	5·7	1	10	30	29·30	146	52	12·9	1	0·29
531	... ...	8·8	2	10	31	10·46	151	10	33·7	2	0·19
532	R Ursæ Majoris Var. 1 ...	7·5	3	10	35	11·65	20	31	35·0	3	0·27
533	... ...	9·3	1	10	35	20·86	137	20	26·7	1	0·20
534	... ...	8·2	3	10	36	47·98	146	54	26·2	3	0·27
535	Taylor 4850 ... 1st...	9·0	1	10	38	46·16	148	51	8·2	1	0·32
536	Taylor 4850 ... 2nd...	7·7	1	10	38	46·27	148	51	2·6	1	0·32
537	Taylor 4862 ... 1st...	8·0	1	10	38	59·00	146	52	42·6	1	0·20
538	Taylor 4862 ... 2nd...	8·9	1	10	39	0·58	146	52	46·0	1	0·25
539	Brisbane 3194	8·8	1	10	39	24·40	149	2	41·0	1	0·26
540	... ...	9·0	1	10	39	31·73	159	3	7·9	1	0·31
541	η Argûs Var. 1	6·0	...	10	39	54·45	148	59	11·2	2	0·19
542	Taylor 4872	7·7	1	10	41	11·48	151	14	31·8	1	0·29
543	53 Leonis *l*	5·3	...	10	42	15·87	78	46	7·8	9	0·21
544	... ...	9·5	1	10	42	31·99	148	52	17·9	1	0·20
545	... ...	9·7	1	10	42	32·13	75	5	46·0	1	0·14
546	Taylor 4886	6·9	1	10	42	55·34	137	2	37·7	1	0·13
547	Lacaille 4602	7·0	1	10	46	37·75	141	5	34·5	1	0·46
548	... ...	9·2	1	10	48	3·29	147	48	1·4	1	0·31
549	... ...	8·6	2	10	48	28·73	141	45	46·7	2	0·28
550	... ...	8·7	2	10	48	38·98	146	38	38·3	2	0·28
551	... ...	9·1	...	10	50	30·34	148	49	2·0	1	0·20
552	7 Crateris α	4·1	...	10	53	17·78	107	36	29·5	1	0·14
553	58 Leonis *d*	5·0	...	10	53	41·44	86	40	9·7	1	0·29
554	R Crateris Var. 1 ...	8·8	5	10	54	0·99	107	36	43·5	5	0·26
555	... .	9·1	2	10	54	11·15	107	39	14·8	2	0·31
556	50 Ursæ Majoris α (*Dubhe*).	2·0	...	10	55	29·92	27	31	54·8	1	0·33
557	... ...	8·9	1	10	57	9·71	146	36	40·6	1	0·30
558	63 Leonis χ	4·7	...	10	58	9·30	81	36	44·7	15	0·27
559	Lacaille 4612	8·8	1	11	1	1·88	154	47	31·8	1	0·32
560	67 Leonis	5·6	...	11	1	40·68	64	37	30·0	1	0·32

532.—R Ursæ Majoris Var. 1.—Period 302 days.—Range, 6th to 13th magnitude.
541.—η Argûs Var. 1.—Irregularly variable from 1st to 5th magnitude.
554.—R Crateris Var. 1.—Changes between 8th and 9th magnitude.
560.—Comparison star for Thalia in 1866.

Observed with the Madras Meridian Circle in that Year.

Number.	Star.	In Right Ascension. Annual Precession.	In Right Ascension. Secular Variation.	In Right Ascension. Proper Motion.	In Polar Distance. Annual Precession.	In Polar Distance. Secular Variation.	In Polar Distance. Proper Motion.	Number in B. A. C.
		s	s	s	"	"	"	
526	.. ..	+ 2·2393	+ 0·0163	...	+ 18·340	+ 0·124	...	...
527	... .	+ 2·0365	+ 0·0120	...	+ 18·353	+ 0·150	...	...
528	Lalande 20402 ..	+ 3·1670	− 0·0084	..	+ 18·361	+ 0·176	...	...
529	47 Leonis ρ	+ 3·1661	− 0·0080	0·000	+ 18·386	+ 0·170	+ 0·08	3600
530	Taylor 4769	+ 2·3932	+ 0·0186	...	+ 18·544	+ 0·119	...	3626
531	... .	+ 2·1589	+ 0·0170		+ 18·567	+ 0·111		...
532	R Ursæ Maj. Var. 1...	+ 4·3685	− 0·1402		+ 18·697	+ 0·288		...
533	... ...	+ 2·5402	+ 0·0177		+ 18·707	+ 0·126		...
534	... ...	+ 2·3332	+ 0·0202		+ 18·717	+ 0·114		...
535	Taylor 4860 .. 1st...	+ 2·3043	+ 0·0212		+ 18·808	+ 0·109		...
536	Taylor 4860 ... 2nd ..	+ 2·3048	+ 0·0211		+ 18·809	+ 0·109	...	...
537	Taylor 4882 ... 1st...	+ 2·3065	+ 0·0213		+ 18·815	+ 0·108	...	...
538	Taylor 4882 .. 2nd ..	+ 2·3057	+ 0·0212		+ 18·816	+ 0·108	...	...
539	Brisbane 3104 ...	+ 2·3043	+ 0·0214		+ 18·828	+ 0·108	...	...
540	... ...	+ 2·5421	+ 0·0191		+ 18·831	+ 0·119	...	...
541	η Argûs Var. 1 ...	+ 2·3106	+ 0·0216	...	+ 18·842	+ 0·107	...	3686
542	Taylor 4872	+ 2·2605	+ 0·0220	...	+ 18·881	+ 0·102	...	...
543	53 Leonis l	+ 3·1605	− 0·0080	− 0·003	+ 18·912	+ 0·146	+ 0·02	3703
544	... ...	+ 2·3379	+ 0·0236	...	+ 18·920	+ 0·105	...	...
545	... ...	+ 3·1901	− 0·0104	...	+ 18·920	+ 0·147	...	...
546	Taylor 4886	+ 2·3962	+ 0·0180		+ 18·931	+ 0·117		...
547	Lacaille 4602 ...	+ 2·5508	+ 0·0215		+ 19·086	+ 0·169		...
548	... ...	+ 2·4187	+ 0·0242		+ 19·075	+ 0·101		...
549	... .	+ 2·8506	+ 0·0222		+ 19·086	+ 0·106		...
550	...	+ 2·3961	+ 0·0247		+ 19·088	+ 0·099		...
551	... ...	+ 2·4139	+ 0·0254	...	+ 19·144	+ 0·097	...	...
552	7 Crateris α	+ 2·9605	+ 0·0072	− 0·088	+ 19·211	+ 0·115	− 0·14	3766
553	63 Leonis d	+ 3·1009	− 0·0039	− 0·002	+ 19·221	+ 0·120	+ 0·09	3755
554	R Crateris Var. 1 ...	+ 2·9617	+ 0·0068	...	+ 19·280	+ 0·114	...	...
555	.. ..	+ 2·9616	+ 0·0069	..	+ 19·294	+ 0·114	...	...
556	50 Urs. Maj. α (*Dubhe*).	+ 3·7846	− 0·0821	− 0·017	+ 19·305	+ 0·141	+ 0·00	3777
557	... ...	+ 2·5432	+ 0·0268	...	+ 19·306	+ 0·099	...	...
558	68 Leonis χ .. .	+ 3·1294	− 0·0066	− 0·094	+ 19·329	+ 0·113	+ 0·08	3780
559	Lacaille 4612 ..	+ 2·3494	+ 0·0315	...	+ 19·395	+ 0·079	...	...
560	67 Leonis	+ 3·2866	− 0·0164	+ 0·008	+ 19·398	+ 0·111	+ 0·04	3809

560.—Proper motions from "*Greenwich Catalogue* 1872."

Mean Positions of Stars for 1867 January 1st.

Number.	Star.	Magnitude.	Estimations.	Mean Right Ascension. h.	m.	s.	Mean Polar Distance. °	′	″	Observations.	Fraction of Year.
561	Lalande 21867 ...	7·7	3	11	3	28·25	78	0	46·8	3	0·31
562	Lalande 21866	9·0	2	11	3	30·27	64	8	41·2	2	0·21
563	S Leonis Var. 2	10·5	1	11	3	58·27	83	49	7·8	1	0·14
564	... ...	8·8	1	11	4	27·40	150	15	32·0	1	0·23
565	... ...	9·7	1	11	7	0·30	66	20	3·3	1	0·20
566	68 Leonis δ ...	2·8	...	11	7	1·88	68	44	54·4	12	0·28
567	... ...	8·8	2	11	10	33·27	148	50	46·4	2	0·19
568	... ...	8·0	1	11	11	16·77	127	30	22·7	1	0·34
569	12 Crateris δ ...	3·9	...	11	12	41·55	104	3	38·3	13	0·29
570	77 Leonis σ ...	4·1	...	11	14	16·64	83	14	33·0	3	0·16
571	... ...	8·7	3	11	16	36·51	135	00	59·9	3	0·24
572	... ...	9·3	4	11	18	7·72	135	40	8·3	4	0·21
573	O. A. N. 11812	8·0	1	11	23	20·34	23	58	49·9	1	0·23
574	... ...	9·0	1	11	24	40·74	22	56	58·5	1	0·20
575	... ...	9·7	1	11	24	48·94	146	9	57·8	1	0·29
576	*	9·5	1	11	24	50·12	84	43	30·0	1	0·24
577	... ...	9·2	1	11	25	17·49	151	32	39·7	1	0·36
578	... ...	8·9	1	11	25	29·33	133	27	47·0	1	0·14
579	... ...	9·3	2	11	25	32·71	22	56	59·2	2	0·28
580	... ...	10·0	1	11	26	38·76	151	38	56·0	1	0·31
581	... ...	9·7	1	11	26	59·68	151	31	55·4	1	0·36
582	91 Leonis υ	4·5	...	11	30	8·31	90	5	24·1	19	0·28
583	.. ..	8·0	1	11	34	6·01	127	50	17·3	1	0·14
584	W. B. E. XI. 697	9·2	2	11	34	52·32	88	10	5·2	2	0·29
585	.. ..	8·0	1	11	36	12·08	139	41	16·9	1	0·33
586	... .	7·9	1	11	38	17·80	140	30	46·3	1	0·31
587	... ...	7·9	2	11	41	5·52	140	53	4·8	2	0·35
588	.. ..	8·6	2	11	41	17·94	126	31	36·2	2	0·34
589	94 Leonis β (Drach)	2·2	..	11	42	16·48	74	41	5·4	17	0·29
590	5 Virginis β	3·7	...	11	43	46·04	87	29	9·8	1	0·21
591	.. ... .	8·7	4	11	46	34·09	134	35	47·1	4	0·31
592	Lacaille 4866	8·0	1	11	51	18·37	154	34	28·1	1	0·14
593	.. .. .	9·1	2	11	51	[illegible]	144	13	[illegible]	2	0·32
594	R. P. L. 87	8·0	..	11	52	38·07	2	15	38·4	6	0·41
595	... .	9·4	1	11	52	39·64	154	38	31·4	1	0·14

562.—595.—Comparison stars for Thalia in 1862.
563.—S Leonis Var. 2.—Period [illegible] days.—Range, 9th to below 13th magnitude.
573.—574.—579.—Comparison stars for Comet 2, 1861.
576.—Observed by mistake for Amphitrite in 18[illegible].
591.—Comparison star for Amphitrite in 18[illegible].
594.—Carrington 1778

Observed with the Madras Meridian Circle in that Year.

Number.	Star.	In Right Ascension. Annual Precession.	In Right Ascension. Secular Variation.	In Right Ascension. Proper Motion.	In Polar Distance. Annual Precession.	In Polar Distance. Secular Variation.	In Polar Distance. Proper Motion.	Number in B. A. C.
		s	s	s	"	"	"	
561	Lalande 21207 ...	+ 3·1408	− 0·0075	...	+ 19·448	+ 0·104	...	...
562	Lalande 21360 ..	+ 3·2302	− 0·0168	...	+ 19·460	+ 0·107	...	...
563	δ Leonis Var. 2 ..	+ 3·1071	− 0·0044	...	+ 19·469	+ 0·101	...	...
564	.. ..	+ 2·5406	+ 0·0313	...	+ 19·489	+ 0·080	...	...
565	... .	+ 3·2132	− 0·0156	...	+ 19·521	+ 0·100	...	...
566	68 Leonis δ ...	+ 3·1912	− 0·0132	+ 0·011	+ 19·582	+ 0·098	+ 0·14	3884
567	... ...	+ 2·5969	+ 0·0398	...	+ 19·560	+ 0·078	...	...
568	... ...	+ 2·8545	+ 0·0166	...	+ 19·604	+ 0·080	...	...
569	12 Crateris δ ..	+ 3·0084	+ 0·0064	− 0·009	+ 19·690	+ 0·081	− 0·13	3890
570	77 Leonis σ	+ 3·1086	− 0·0042	− 0·009	+ 19·657	+ 0·081	+ 0·03	3892
571	... ...	+ 2·8914	+ 0·0132		+ 19·697	+ 0·071	...	...
572	... .	+ 2·8973	+ 0·0164		+ 19·721	+ 0·069	...	...
573	O. A. N. 11612 ...	+ 3·5744	− 0·0980		+ 19·799	+ 0·075	...	...
574	... ...	+ 3·5309	− 0·0988		+ 19·817	+ 0·071	...	...
575	... ...	+ 2·7868	+ 0·0882		+ 19·818	+ 0·058	...	...
576	...	+ 3·0910	− 0·0085		+ 19·819	+ 0·080	...	...
577	...	+ 2·7000	+ 0·0415		+ 19·886	+ 0·080	...	...
578	...	+ 2·9128	+ 0·0210		+ 19·888	+ 0·066	...	...
579	...	+ 3·5450	− 0·0913		+ 19·880	+ 0·068	...	...
580	...	+ 2·7127	+ 0·0424		+ 19·843	+ 0·046	...	...
581	... ..	+ 2·7188	+ 0·0428	...	+ 19·847	+ 0·045	...	...
582	91 Leonis υ	+ 3·0715	+ 0·0008	− 0·006	+ 19·886	+ 0·040	− 0·08	3946
583	... ...	+ 2·9580	+ 0·0210	...	+ 19·906	+ 0·040	...	...
584	W. B. E. XI. 507 ..	+ 3·0766	− 0·0004	...	+ 19·904	+ 0·041	...	...
585	... ...	+ 2·9087	+ 0·0820	...	+ 19·946	+ 0·086	...	...
586	... .	+ 2·8861	+ 0·0444	...	+ 19·966	+ 0·030	...	...
587	... ...	+ 2·8822	+ 0·0466	...	+ 19·967	+ 0·030	...	...
588	... .	+ 2·9914	+ 0·0218	...	+ 19·988	+ 0·027	...	...
589	94 Leonis β (Deneb)...	+ 3·1004	− 0·0074	− 0·036	+ 19·986	+ 0·036	+ 0·10	3996
590	5 Virginis β	+ 3·0768	− 0·0008	+ 0·048	+ 20·004	+ 0·023	+ 0·36	4002
591	... .	+ 3·0286	+ 0·0215	...	+ 20·030	+ 0·018	...	...
592	Lacaille 4856 ..	+ 2·9986	+ 0·0699	...	+ 20·040	+ 0·008	...	...
593	... .	+ 3·0064	+ 0·0410	...	+ 20·042	+ 0·007	...	...
594	R. P. L. 87	+ 4·1888	− 1·2996	...	+ 20·044	+ 0·011	...	...
595	... ...	+ 2·9801	+ 0·0804	...	+ 20·044	+ 0·006	...	...

Mean Positions of Stars for 1867 January 1st.

Number.	Star.	Magnitude.	Estimations.	Mean Right Ascension.			Mean Polar Distance.			Observations.	Fraction of Year.
				h.	m.	s.	°	′	″		
596	8 Virginis ω ..	4·4	...	11	54	3·41	82	38	33·8	2	0·21
597	Taylor 5585 ...	8·0	1	11	57	12·75	70	36	30·5	1	0·31
598	... ...	8·8	4	11	59	46·13	144	19	49·5	4	0·38
599	... ...	8·9	2	12	1	46·28	130	2	36·0	2	0·34
600	2 Corvi ε	3·1	...	12	3	17·26	111	52	46·0	6	0·30
601	... ...	8·2	1	12	6	21·96	133	28	32·7	1	0·35
602	R. P. L. 90 ...	7·7	...	12	7	3·67	2	19	38·7	1	0·29
603	... ...	9·5	1	12	7	40·79	90	15	17·8	2	0·27
604	15 Virginis η ...	4·0	...	12	13	6·06	90	56	39·9	12	0·30
605	16 Virginis c ...	5·2	...	12	13	36·77	86	56	47·9	1	0·31
606	R. P. L. 93	6·5	...	12	14	18·46	1	38	46·6	1	0·71
607	... ...	8·7	2	12	17	32·74	94	44	6·9	3	0·36
608	... ...	7·8	1	12	20	56·80	141	30	17·5	1	0·36
609	W. B. E. XII. 446 ...	8·4	8	12	27	38·00	92	59	8·6	3	0·32
610	9 Corvi β	2·8	...	12	27	24·20	112	39	40·2	4	0·37
611	... ...	9·0	3	12	27	38·36	90	53	40·0	3	0·31
612	Lalande 23382	7·9	3	12	28	37·77	92	48	48·3	3	0·32
613	... ...	8·6	1	12	31	0·60	142	30	42·1	1	0·36
614	R Virginis Var. 2... ...	7·9	1	12	31	46·10	82	16	47·3	1	0·33
615	... ...	9·6	3	12	32	46·04	100	5	42·6	3	0·32
616	... ...	7·0	1	12	32	50·58	38	14	33·3	1	0·39
617	29 Virginis γ ... 1st...	3·1	...	12	34	56·32	90	48	8·5	8	0·29
618	... ...	9·5	1	12	39	36·07	141	36	13·0	1	0·30
619	... ...	9·6	1	12	39	46·51	94	2	32·5	1	0·36
620	... ...	9·7	1	12	42	13·60	147	17	27·9	1	0·39
621	... ...	9·0	1	12	43	1·22	130	26	16·2	1	0·31
622	... ...	9·0	2	12	43	14·50	80	42	19·5	2	0·34
623	... ...	10·6	2	12	43	40·20	80	41	38·6	2	0·41
624	33 Virginis	6·8	...	12	46	22·65	98	49	47·4	2	0·20
625	40 Virginis ψ	5·0	.	12	47	38·33	98	45	58·6	1	0·32
626	W. B. E. XII. 790 ...	8·5	3	12	47	46·74	90	46	26·6	3	0·35
627	R. P. L. 99	5·6	...	12	48	11·26	5	51	46·1	1	0·41
628	... ...	9·0	2	12	48	38·71	136	38	16·7	2	0·36
629	... ...	7·0	2	12	49	3·60	140	36	16·1	2	0·31
630	12 Canum Venaticorum ...	3·0	...	12	49	48·06	50	57	46·0	7	0·36

602.—Carrington 1816.
606.—Groombridge 1864.
607.—Comparison star for Comet 2, 1861.
609.—612.—Comparison stars for Polyhymnia in 1867.
611.—615.—Comparison stars for Sappho in 1867.
614.—R Virginis Var. 2.—Period 146 days.—Range, 6·6 to 11th magnitude.
619.—Comparison star for Hestia in 1864
622.—623.—625.—Comparison stars for Isis in 1867.
627.—Groombridge 1840.

Observed with the Madras Meridian Circle in that Year.

Number.	Star.	In Right Ascension. Annual Precession.	In Right Ascension. Secular Variation.	In Right Ascension. Proper Motion.	In Polar Distance. Annual Precession.	In Polar Distance. Secular Variation.	In Polar Distance. Proper Motion.	Number in B. A. C.
		s	s	s	"	"	"	
596	8 Virginis π	+ 3·0766	− 0·0022	0·000	+ 20·048	+ 0·002	+ 0·01	4082
597	Taylor 5586	+ 3·0770	− 0·0080	...	+ 20·064	− 0·006	...	...
598	... .	+ 3·0701	+ 0·0496	...	+ 20·066	− 0·000	...	...
599	... ...	+ 3·0808	+ 0·0273	...	+ 20·064	− 0·012	...	...
600	2 Corvi ε	+ 3·0708	+ 0·0142	− 0·005	+ 20·058	− 0·016	− 0·01	4097
601	... ...	+ 3·1140	+ 0·0869	...	+ 20·047	− 0·022	...	...
602	R. P. L. 90	+ 2·0680	− 0·2471	...	+ 20·045	− 0·018	...	...
603	... ...	+ 3·0728	+ 0·0024	...	+ 20·048	− 0·084	...	..
604	15 Virginis η ...	+ 3·0720	+ 0·0027	− 0·007	+ 20·022	− 0·085	+ 0·03	4146
605	16 Virginis c ...	+ 3·0865	+ 0·0006	− 0·010	+ 20·019	− 0·086	+ 0·09	4151
606	R. P. L. 98	+ 0·0126	+ 1·0876	− 0·152	+ 20·016	− 0·009	− 0·07	4165
607	... ...	+ 2·8502	− 0·0623	...	+ 19·986	− 0·041	...	...
608	... .	+ 3·2246	+ 0·0438	...	+ 19·971	− 0·061	...	...
609	W. B. K. XII. 446 ...	+ 3·0804	+ 0·0051	...	+ 19·912	− 0·083	..	...
610	9 Corvi β	+ 3·1386	+ 0·0164	− 0·008	+ 19·912	− 0·064	+ 0·07	4201
611	... ...	+ 3·0894	+ 0·0086		+ 19·900	− 0·064	..	...
612	Lalande 23632 ...	+ 3·0808	+ 0·0060		+ 19·890	− 0·065	...	...
613	... ...	+ 3·3068	+ 0·0476		+ 19·871	− 0·074	..	...
614	R Virginis Var. 2 ...	+ 3·0471	− 0·0098		+ 19·868	− 0·070	...	...
615	... ...	+ 3·1060	+ 0·0091		+ 19·850	− 0·074	...	..
616	... ...	+ 2·7140	− 0·0804	...	+ 19·848	− 0·086	...	...
617	29 Virginis γ ... 1st ...	+ 3·0748	+ 0·0048	− 0·037	+ 19·808	− 0·078	+ 0·05	4265
618	... ...	+ 3·3886	+ 0·0487	...	+ 19·766	− 0·092	...	...
619	.. ..	+ 3·0826	+ 0·0062	...	+ 19·763	− 0·080	...	...
620	... ...	+ 3·4586	+ 0·0611	...	+ 19·715	− 0·100	...	...
621	... ...	+ 3·3686	+ 0·0440	...	+ 19·702	− 0·090	..	...
622	... ...	+ 3·0810	− 0·0006	...	+ 19·606	− 0·091	...	...
623	... ...	+ 3·0800	− 0·0008	...	+ 19·692	− 0·092	...	...
624	38 Virginis	+ 3·0864	+ 0·0060	− 0·010	+ 19·646	− 0·099	+ 0·08	4288
625	40 Virginis ψ ...	+ 3·1148	+ 0·0092	− 0·002	+ 19·627	− 0·101	+ 0·04	4290
626	W. B. K. XII. 799	+ 3·0272	− 0·0001	...	+ 19·621	− 0·100	...	..
627	R. P. L. 99	+ 0·2661	+ 0·2944	− 0·017	+ 19·613	− 0·010	− 0·04	4302
628	. ... ..	+ 3·2726	+ 0·0270	...	+ 19·606	− 0·108	...	..
629	.	+ 3·5528	+ 0·0692	...	+ 19·597	− 0·118	...	..
630	12 Can. Venaticorum.	+ 2·8368	− 0·0152	− 0·023	+ 19·594	− 0·094	− 0·06	4340

617.—Proper motions from "*Greenwich Catalogue* 1872."

Mean Positions of Stars for 1867 January 1st.

Number.	Star.	Magnitude.	Estimations.	Mean Right Ascension.			Mean Polar Distance.			Observations.	Fraction of Year.
				h.	m.	s.	°	′	″		
631	... ...	9·0	1	12	53	18·29	142	25	8·3	1	0·40
632	... ...	8·0	1	12	56	35·72	135	46	20·4	1	0·85
633	... .	9·1	1	12	56	46·28	139	19	21·0	1	0·86
634	... .	9·2	1	12	57	0·36	123	26	10·8	1	0·41
635	Lacaille 5361 ...	7·7	2	12	57	17·02	129	53	5·6	2	0·34
636	. ...	8·3	1	12	58	15·93	121	30	41·2	1	0·36
637	51 Virginia θ	4·4	...	13	3	3·88	94	40	42·5	14	0·32
638	Taylor 6057	5·6	1	13	4	0·78	119	12	41·9	1	0·85
639	... ...	9·0	1	13	4	42·80	138	11	32·4	1	0·26
640	... ...	9·3	1	13	4	46·30	140	18	19·7	1	0·33
641	... ...	9·5	1	13	6	47·26	134	17	29·9	1	0·41
642	Lacaille 5434 ...	7·5	1	13	6	9·27	152	52	31·6	1	0·31
643	B. P. L. 101	7·5	...	13	9	53·73	1	08	16·5	2	0·69
644	Taylor 6129	7·0	1	13	12	28·80	130	20	29·8	1	0·30
645	Taylor 6148	7·5	1	13	14	21·75	128	9	17·2	1	0·40
646	Taylor 6160	8·1	2	13	15	46·80	128	55	31·7	2	0·40
647	67 Virginis α (*Spica*) ...	1·2	...	13	18	11·23	100	27	39·5	10	0·31
648	70 Ursæ Majoris ζ ...2nd	4·2	...	13	18	34·73	34	32	57·4	1	0·48
649	O. A. S. 12872	9·8	2	13	19	30·87	110	57	32·7	2	0·37
650	Lacaille 5540	9·0	1	13	19	58·27	143	53	26·2	1	0·30
651	B. P. L. 103 ...	7·3	...	13	20	6·83	4	28	0·6	3	0·82
652	... ...	10·0	1	13	23	26·80	88	30	10·4	1	0·44
653	... ...	8·0	...	13	24	57·90	129	9	35·7	1	0·30
654	... ...	8·2	1	13	26	41·80	138	11	13·8	1	0·81
655	Taylor 6307 ..	8·7	2	13	25	47·87	146	40	30·7	2	0·44
656	76 Virginis λ ...	6·6	...	13	26	57·68	99	35	43·9	1	0·89
657	. ...	8·2	1	13	26	46·46	131	36	7·5	1	0·40
658	79 Virginis ζ ..	3·6	...	13	27	56·01	90	51	54·3	11	0·86
659	Lacaille 5614 ...	8·0	1	13	30	7·29	124	13	3·1	1	0·83
660	Lacaille 5680 ..	7·1	2	13	33	38·73	128	47	0·1	2	0·41
661	... .	9·5	1	13	34	18·30	120	10	33·7	1	0·41
662	... ...	8·3	1	13	35	58·63	124	4	20·1	1	0·80
663	O. A. S. 13079	9·3	2	13	36	1·73	116	49	34·8	2	0·48
664	... ...	9·2	1	13	36	28·17	128	6	16·7	1	0·29
665	. ...	9·5	1	13	36	42·64	144	39	14·4	1	0·32

648.—Groombridge 2006.
649.—Comparison star for Eunomia in 1863
651.—Groombridge 2017
652.—Comparison star for Europa in 1861

Observed with the Madras Meridian Circle in that Year.

Number.	Star.	In Right Ascension. Annual Precession.	In Right Ascension. Secular Variation.	In Right Ascension. Proper Motion.	In Polar Distance. Annual Precession.	In Polar Distance. Secular Variation.	In Polar Distance. Proper Motion.	Number in B. A. C.
		s	s	s	"	"	"	
631	...	+ 3·4736	+ 0·0682	...	+ 19·515	− 0·125	.	.
632	...	+ 3·3903	+ 0·0407		+ 19·509	− 0·128		...
633	...	+ 3·4400	+ 0·0466		+ 19·484	− 0·127		.
634	... ...	+ 3·2900	+ 0·0368		+ 19·486	− 0·126		..
635	Lacaille 5381 ...	+ 3·3494	+ 0·0395		+ 19·482	− 0·128		..
636	... .	+ 3·3032	+ 0·0276	...	+ 19·409	− 0·139	..	.
637	51 Virginis θ ...	+ 3·1096	+ 0·0078	− 0·004	+ 19·301	− 0·152	+ 0·04	4401
638	Taylor 6057	+ 3·6903	+ 0·0729	...	+ 19·276	− 0·153		4412
639	... ...	+ 3·4537	+ 0·0459	...	+ 19·260	− 0·150		...
640	.. ..	+ 3·5710	+ 0·0560	...	+ 19·260	− 0·153		...
641	... .. .	+ 3·3908	+ 0·0292		+ 19·236	− 0·145		...
642	Lacaille 5484... ...	+ 3·8151	+ 0·0850		+ 19·226	− 0·166		..
643	R. P. L. 101	− 10·9688	+ 8·1284		+ 19·129	+ 0·472		...
644	Taylor 6129	+ 3·4963	+ 0·0368		+ 19·068	− 0·169		...
645	Taylor 6146	+ 3·4070	+ 0·0388		+ 19·009	− 0·166		...
646	Taylor 6160	+ 3·4226	+ 0·0398	...	+ 18·969	− 0·170	...	...
647	67 Virginis α (*Spica*).	+ 3·1547	+ 0·0116	− 0·005	+ 18·899	− 0·168	+ 0·04	4480
648	70 Urs. Maj. 3 2nd...	+ 2·4151	− 0·0172	+ 0·014	+ 18·836	− 0·127	+ 0·06	4486
649	O. A. S. 12672 ...	+ 3·3098	+ 0·0284	...	+ 18·880	− 0·172	.	...
650	Lacaille 5646 ...	+ 3·6866	+ 0·0669	...	+ 18·849	− 0·192	...	...
651	R. P. L. 108 ...	− 2·6818	+ 0·9967	...	+ 18·842	+ 0·186		4498
652	... ...	+ 3·0609	+ 0·0055	...	+ 18·740	− 0·167		...
653	... ...	+ 3·4628	+ 0·0384	...	+ 18·692	− 0·190		...
654	... ...	+ 3·4568	+ 0·0384	...	+ 18·689	− 0·192		...
655	Taylor 6267	+ 3·8801	+ 0·0761	...	+ 18·686	− 0·215		..
656	76 Virginis λ ...	+ 3·1889	+ 0·0113	− 0·004	+ 18·660	− 0·176	+ 0·02	4521
657	... ...	+ 3·5111	+ 0·0870	...	+ 18·626	− 0·197	...	...
658	70 Virginis 3... ...	+ 3·0718	+ 0·0004	− 0·019	+ 18·596	− 0·176	− 0·06	4562
659	Lacaille 5614 ...	+ 3·4756	+ 0·0367	...	+ 18·504	− 0·202	..	...
660	Lacaille 5689 ...	+ 3·4966	+ 0·0391	...	+ 18·412	− 0·207	...	...
661	... ...	+ 3·5076	+ 0·0351	...	+ 18·394	− 0·212	...	...
662	... ...	+ 3·4977	+ 0·0383	..	+ 18·336	− 0·215	...	...
663	O. A. S. 13079 ...	+ 3·3472	+ 0·0229	...	+ 18·330	− 0·207	...	...
664	... ...	+ 3·6004	+ 0·0359	...	+ 18·306	− 0·216	...	...
665	... ...	+ 3·8443	+ 0·0642	...	+ 18·296	− 0·237	..	...

Mean Positions of Stars for 1867 *January* 1*st.*

Number.	Star.	Magnitude.	Estimations.	Mean Right Ascension.			Mean Polar Distance.			Observations.	Fraction of Year.
				h.	m.	s.	°	′	″		
666	Taylor 6366	7·0	1	13	37	5·10	161	47	0·3	1	0·33
667	Lacaille 5689	7·8	1	13	37	20·42	158	14	25·7	1	0·31
668	O. A. S. 13100	8·0	1	13	37	03·78	110	36	37·0	1	0·42
669	... ...	9·5	1	13	37	40·85	123	46	56·7	1	0·41
670	... ...	8·0	1	13	40	40·07	120	24	55·9	1	0·41
671	O. A. S. 13186	8·2	3	13	43	35·24	116	58	24·3	3	0·33
672	O. A. S. 13193	9·1	3	13	44	57·85	117	11	26·6	3	0·31
673	... ...	7·9	2	13	46	34·03	128	23	50·0	2	0·35
674	... ...	9·7	3	13	46	41·51	128	24	5·5	3	0·36
675	X Virginis Var. 5... ...	8·9	6	13	47	28·92	78	16	47·7	6	0·30
676	8 Bootis η	2·0	...	13	48	21·09	70	56	5·0	9	0·34
677	... ...	9·2	1	13	50	11·81	149	55	3·6	1	0·40
678	.. ..	9·5	1	13	52	36·60	128	2	13·0	1	0·30
679	... ...	9·7	1	13	53	13·96	136	41	44·5	1	0·41
680	β Centauri ..	1·2	...	13	54	28·01	149	48	47·9	1	0·33
681	93 Virginis τ ...	4·4	...	13	54	53·73	87	46	38·3	12	0·35
682	Lacaille 5794 ...	7·0	1	13	57	17·90	132	48	38·3	1	0·31
683	... ...	8·4	2	13	58	26·19	169	30	57·3	2	0·40
684	94 Virginis ...	7·0	...	13	59	16·37	98	15	30·6	1	0·32
685	95 Virginis ...	5·7	...	13	59	40·85	98	40	39·5	2	0·37
686	Lalande 25596 ...	7·8	1	14	0	2·58	67	11	45·5	1	0·30
687	... ...	10·2	1	14	0	34·41	150	52	1·7	1	0·40
688	Taylor 6385 ...	7·8	1	14	1	33·07	134	14	50·5	1	0·36
689	... ...	9·2	1	14	2	13·57	134	17	6·0	1	0·41
690	... ...	8·0	2	14	5	14·86	139	21	10·6	2	0·40
691	Lacaille 5846	7·3	1	14	5	17·51	161	4	57·1	1	0·32
692	98 Virginis κ	4·3	...	14	5	48·17	99	39	11·4	4	0·36
693	... ...	5·5	3	14	6	30·24	135	2	10·5	3	0·30
694	99 Virginis ι	4·3	...	14	9	2·51	95	21	52·8	2	0·37
695	16 Bootis α (*Arcturus*) ...	0·0	...	14	9	38·74	70	7	27·3	14	0·30
696	... ...	9·5	1	14	11	40·98	124	23	24·4	1	0·41
697	100 Virginis λ	4·6	...	14	11	54·98	102	46	38·7	1	0·30
698	W. B. B. XIV. 192 .	7·5	1	14	13	1·36	108	47	38·0	1	0·30
699	... ...	9·5	1	14	12	46·81	136	50	46·5	1	0·32
700	... ...	9·5	1	14	15	30·96	132	12	35·3	1	0·41

671.—672.—Comparison stars for Atalanta in 1867.
675.—X Virginis Var. 5.—Range, 8·9 to below 12th magnitude.

Observed with the Madras Meridian Circle in that Year.

Number.	Star.	In Right Ascension.			In Polar Distance.			Number in B. A. C.
		Annual Precession.	Secular Variation.	Proper Motion.	Annual Precession.	Secular Variation.	Proper Motion.	
		s	s	s	"	"	"	
666	Taylor 6360 ...	+ 4·0963	+ 0·0909	...	+ 18·398	− 0·263	...	...
667	Lacaille 5689	+ 4·1186	+ 0·0983	...	+ 18·278	− 0·256	...	...
668	O. A. S. 13100	+ 3·3527	+ 0·0231	...	+ 18·271	− 0·210	...	...
669	... .	+ 3·4184	+ 0·0292	...	+ 18·261	− 0·215	...	...
670	.. .	+ 3·6395	+ 0·0356	...	+ 18·151	− 0·233	..	...
671	O. A. S. 13168	+ 3·3702	+ 0·0283	...	+ 18·086	− 0·221		.
672	O. A. S. 13198 .	+ 3·3783	+ 0·0284	...	+ 17·966	− 0·226		.
673	.. ..	+ 3·5131	+ 0·0346	...	+ 17·965	− 0·238		...
674	... . ..	+ 3·5437	+ 0·0346	...	+ 17·960	− 0·288		...
675	X Virginis Var. 5	+ 2·9468	+ 0·0022	...	+ 17·898	− 0·202		...
676	8 Bootis η .	+ 2·8616	− 0·0006	− 0·001	+ 17·886	− 0·199	+ 0·36	4685
677	... ..	+ 4·1372	+ 0·0942	...	+ 17·781	− 0·287	...	...
678	. ... ..	+ 3·3637	+ 0·0348	...	+ 17·692	− 0·253	..	...
679	... . ..	+ 3·7222	+ 0·0463	...	+ 17·658	− 0·265	...	...
680	β Centauri .. .	+ 4·1602	+ 0·0641	− 0·310	+ 17·606	− 0·301	+ 0·05	4660
681	93 Virginis τ	+ 3·0477	+ 0·0064	+ 0·001	+ 17·587	− 0·221	+ 0·07	4672
682	Lacaille 5794 ..	+ 4·3467	+ 0·0990	...	+ 17·495	− 0·316	...	...
683	. ...	+ 3·6137	+ 0·0361	...	+ 17·487	− 0·266	..	..
684	94 Virginis	+ 3·1636	+ 0·0116	− 0·002	+ 17·400	− 0·287	+ 0·01	4696
685	96 Virginis	+ 3·1739	+ 0·0113	− 0·010	+ 17·382	− 0·233	− 0·01	4690
686	Lalande 25696 .	+ 2·7910	− 0·0023		+ 17·366	− 0·210		..
687	... ...	+ 4·2707	+ 0·0897		+ 17·348	− 0·330		
688	Taylor 6935	+ 3·5396	+ 0·0302		+ 17·309	− 0·265		...
689	... .	+ 3·5635	+ 0·0302		+ 17·271	− 0·269		...
690	... ...	+ 3·6119	+ 0·0362		+ 17·134	− 0·294		...
691	Lacaille 5944	+ 4·3908	+ 0·0912	...	+ 17·181	− 0·336	.	...
692	98 Virginis κ	+ 3·1908	+ 0·0122	+ 0·001	+ 17·109	− 0·250	− 0·02	4716
693	... .	+ 3·7738	+ 0·0446	..	+ 17·084	− 0·296	..	.
694	99 Virginis ι	+ 3·1391	+ 0·0102	+ 0·001	+ 16·969	− 0·262	+ 0·41	4727
695	16 Bootis α (*Arcturus*)	+ 2·8132	+ 0·0001	− 0·079	+ 16·984	− 0·227	+ 1·93	4729
696	... .	+ 3·3710	+ 0·0304	...	+ 16·885	− 0·260	...	.
697	100 Virginis λ .	+ 3·2949	+ 0·0140	− 0·002	+ 16·884	− 0·264	− 0·02	4733
698	W. B. E. XIV. 192	+ 3·2809	+ 0·0146	...	+ 16·846	− 0·266	..	..
699	... .	+ 3·3984	+ 0·0477	..	+ 16·786	− 0·314	...	..
700	... ...	+ 3·5446	+ 0·0381	...	+ 16·650	− 0·294	.	..

680.—Proper motions from "*Stone's Cape Catalogue.*"
694.—Proper motions from "*Greenwich Catalogue 1872.*"

Mean Positions of Stars for 1867 January 1st.

Number.	Star.	Magnitude.	Estimations.	Mean Right Ascension.			Mean Polar Distance.			Observations.	Fraction of Year.
				h	m	s	°	′	″		
701	Taylor 6740 ...	7·4	1	14	19	16·54	133	48	47·4	1	0·45
702	... ...	8·7	2	14	19	44·67	134	30	11·4	2	0·41
703	... ...	6·0	1	14	20	8·62	127	9	44·7	1	0·43
704	Lacaille 5962 ...	7·0	1	14	24	53·23	129	47	33·6	1	0·33
705	... ...	9·0	1	14	28	7·11	129	46	47·5	1	0·38
706	26 Bootis ρ ...	3·6	...	14	26	5·84	59	2	37·4	11	0·40
707	... ...	8·3	1	14	26	54·52	123	20	51·2	1	0·45
708	... ...	7·5	3	14	29	37·08	124	56	10·3	3	0·46
709	Lacaille 6027 ...	7·9	1	14	31	15·11	122	46	5·5	1	0·42
710	Taylor 6848 ...	7·3	1	14	33	0·84	136	42	7·2	1	0·41
711	α Lupi	2·6	...	14	33	5·77	136	46	54·3	2	0·43
712	... ...	9·0	...	14	33	6·45	136	18	25·1	1	0·40
713	... ...	8·9	1	14	36	44·43	150	13	9·5	1	0·45
714	5 Libræ	6·6	...	14	38	37·08	101	53	40·7	2	0·32
715	36 Bootis ε (*Mirac*) ...	2·6	...	14	39	10·06	62	21	40·4	8	0·41
716	... ...	8·3	...	14	40	35·61	127	4	32·7	1	0·39
717	Brisbane 5069	8·0	1	14	41	56·46	131	17	31·3	1	0·44
718	Lalande 27082	7·8	1	14	43	22·06	75	57	10·4	1	0·45
719	9 Libræ α²	3·0	...	14	43	31·48	105	30	14·6	6	0·38
720	... ...	9·5	1	14	47	46·96	150	41	43·7	1	0·48
721	... ...	8·9	1	14	50	34·91	130	32	53·5	1	0·48
722	O. A. S. 14112	8·1	2	14	51	3·32	109	11	27·7	2	0·46
723	7 Urs. Min. β Var 1 (*Kochab*)	2·1	...	14	51	7·65	15	13	2·3	1	0·40
724	Radcliffe 3306	7·5	1	14	53	6·46	42	11	46·3	1	0·41
725	Taylor 7086	6·7	...	14	53	5·56	62	23	49·4	1	0·41
726	43 Bootis ψ	4·5	...	14	54	44·76	62	31	56·3	6	0·44
727	21 Libræ ν¹	5·4	...	14	59	12·62	105	44	19·6	1	0·37
728	W. B. E. XV. 7	8·5	1	15	2	30·34	97	21	42·2	1	0·45
729	Taylor 7070	6·6	1	15	3	30·99	128	7	50·1	1	0·44
730	W. B. E. XV. 32 ...	8·2	1	15	3	36·41	97	2	16·4	1	0·48
731	R. P. L. 111	9·0	...	15	5	38·02	6	32	5·3	2	0·42
732	W. B. E. XV 56... ...	9·3	2	15	6	41·06	98	2	34·2	3	0·50
733	27 Libræ β	2·7	...	15	9	51·11	98	55	24·7	5	0·39
734	... ...	7·7	1	15	12	2·38	130	34	41·7	1	0·44
735	... ...	9·0	1	15	17	15·87	130	4	21·4	1	0·40

702—Comparison star for Isis in 1861.
723.—β Ursæ Minoris Var. 1.—(*Kochab*)—Supposed to vary irregularly from 3·0 to 2·5 magnitude
728.—730.—732.—Comparison stars for Comet 2, 1867.
731.—Groombridge 2218.

Observed with the Madras Meridian Circle in that Year.

Number.	Star.	In Right Ascension. Annual Precession.	In Right Ascension. Secular Variation.	In Right Ascension. Proper Motion.	In Polar Distance. Annual Precession.	In Polar Distance. Secular Variation.	In Polar Distance. Proper Motion.	Number in B. A. C.
		s	s	s	"	"	"	
701	Taylor 6740 ...	+ 3·5034	+ 0·0423	...	+ 16·464	− 0·283	.	.
702		+ 3·6010	+ 0·0806	...	+ 16·441	− 0·306	...	.
703	. ..	+ 3·6940	+ 0·0884	...	+ 16·421	− 0·313		. .
704	Lacaille 5962	+ 3·7224	+ 0·0565	.	+ 16·282	− 0·394		...
705	.. .. ..	+ 3·7230	+ 0·0364		+ 16·270	− 0·334		...
706	26 Bootis ρ .	+ 2·5947	− 0·0015	− 0·008	+ 16·117	− 0·283	− 0·14	4808
707	... .	+ 3·5982	+ 0·0201		+ 16·074	− 0·321	...	...
708	.. .	+ 3·6305	+ 0·0306		+ 15·981	− 0·329	...	...
709	Lacaille 6027	+ 3·6004	+ 0·0281		+ 15·844	− 0·329	...	...
710	Taylor 6848 ..	+ 3·7600	+ 0·0469		+ 15·740	− 0·364	...	...
711	α Lupi ...	+ 3·9546	+ 0·0472	...	+ 15·744	− 0·304	...	[illegible]
712	... ..	+ 3·6807	+ 0·0619	...	+ 15·743	− 0·380	...	...
713	... ..	+ 4·5582	+ 0·0874	. .	+ 15·544	− 0·496	...	...
714	5 Libræ .	+ 3·3992	+ 0·0152	− 0·008	+ 15·469	− 0·314	+ 0·01	[illegible]
715	36 Bootis ε (*Mirac*)	+ 2·6240	− 0·0001	− 0·006	+ 15·409	− 0·282	− 0·01	4876
716	.	+ 3·7236	+ 0·0826	. .	+ 15·329	− 0·367	...	...
717	Brisbane 5069 .	+ 3·8362	+ 0·0679	...	+ 15·273	− 0·369	...	...
718	Lalande 27022 .	+ 2·0014	+ 0·0045	...	+ 15·171	− 0·286	...	...
719	9 Libræ α¹ .	+ 3·3146	+ 0·0154	− 0·007	+ 15·163	− 0·294	+ 0·06	[illegible]
720	. ..	+ 4·6039	+ 0·0888	...	+ 14·920	− 0·460	...	...
721	. . .	+ 3·8471	+ 0·0363	...	+ 14·751	− 0·386	...	...
722	O. A. S. 14112 .	+ 3·3381	+ 0·0175	...	+ 14·724	− 0·341	...	...
723	7 Urs. Min. β Var. 1	− 0·2470	+ 0·1022	− 0·005	+ 14·720	+ 0·013	+ 0·08	4936
724	Radcliffe 3306 .	+ 2·0174	+ 0·0009	...	+ 14·416	− 0·213	...	4940
725	Taylor 7006	+ 2·8610	+ 0·0011	...	+ 14·396	− 0·270	...	4962
726	48 Bootis ψ .	+ 2·5688	+ 0·0010	− 0·013	+ 14·289	− 0·271	0·00	[illegible]
727	21 Libræ ν¹	+ 3·3377	+ 0·0153	− 0·004	+ 14·280	− 0·349	+ 0·08	4970
728	W. B. E. XV. 7	+ 3·1966	+ 0·0112	...	+ 14·026	− 0·289	...	...
729	Taylor 7079 . .	+ 3·6936	+ 0·0273	...	+ 13·962	− 0·398	...	...
730	W. B. E. XV. 32	+ 3·1906	+ 0·0112	...	+ 13·956	− 0·340	...	. .
731	R. P. L. 111 ...	− 6·9100	+ 1·1927	. .	+ 13·944	+ 0·728	...	5022
732	W. B. E. XV. 86	+ 3·2696	+ 0·0124	..	+ 13·702	− 0·347	...	...
733	27 Libræ β ..	+ 3·2262	+ 0·0117	− 0·009	+ 13·866	− 0·366	+ 0·01	5054
734	... . .	+ 3·9162	+ 0·0846	...	+ 13·416	− 0·461	.	.
735	.. .. ..	+ 3·9061	+ 0·0894	..	+ 13·073	− 0·440	. .	. .

Mean Positions of Stars for 1867 January 1st.

Number.	Star.	Magnitude.	Estimations.	Mean Right Ascension.			Mean Polar Distance.			Observations.	Fraction of Year.
				h.	m.	s.	°	′	″		
736	32 Libræ ξ[1] ...	6·2	...	15	20	46·60	106	15	8·0	3	0·32
737	R. P. L. 114 ...	6·9	...	15	21	20·41	2	15	42·4	1	0·96
738	... ...	9·0	1	15	22	42·68	151	37	40·5	2	0·46
739	... ...	9·7	1	15	24	46·15	130	9	34·5	1	0·39
740	... ...	8·3	1	15	26	11·94	122	44	14·0	1	0·51
741	Lacaille 6421	7·3	2	15	26	50·26	122	43	10·8	2	0·50
742	38 Libræ γ	4·0	...	15	28	5·41	104	20	38·8	3	0·35
743	5 Cor. Bor. α (*Alphecca*) ...	2·4	...	15	29	3·41	62	50	10·4	4	0·47
744	... ...	10·0	1	15	29	10·22	119	28	26·6	1	0·51
745	... ...	9·7	1	15	29	23·71	119	41	3·0	1	0·51
746	Lalande 28530	8·0	1	15	31	56·46	47	25	55·4	1	0·37
747	Taylor 7800	7·7	2	15	32	24·92	103	37	10·4	2	0·40
748	... ...	9·0	1	15	33	47·51	126	26	48·5	1	0·39
749	24 Serpentis α	2·7	...	15	37	43·04	83	0	13·7	10	0·49
750	O. A. S. 14874	8·0	1	15	39	38·20	104	40	12·1	1	0·40
751	Lalande 28787	6·8	1	15	42	15·48	92	40	25·0	1	0·38
752	... ...	9·5	1	15	42	19·68	104	24	56·6	1	0·39
753	... ...	10·5	1	15	42	30·07	61	47	14·2	1	0·41
754	O. A. S. 14934	9·7	1	15	43	50·76	107	52	58·2	1	0·51
755	... ...	9·0	1	15	41	7·97	104	22	9·7	1	0·39
756	W. B. K. XV. 886 ...	8·0	1	15	44	0·90	104	27	34·5	1	0·43
757	36 Serpentis b	5·2	...	15	44	20·20	92	41	8·3	1	0·82
758	46 Libræ θ	4·8	...	15	46	15·24	108	20	13·6	1	0·36
759	O. A. S. 14996	8·9	1	15	46	41·14	105	16	1·1	1	0·40
760	R. P. L. 115	6·9	...	15	48	8·96	4	44	29·8	2	0·40
761	O. A. S. 15058	8·0	2	15	49	12·88	106	26	29·7	2	0·50
762	W. B. K. XV. 928 ...	9·8	1	15	49	23·16	104	56	37·5	1	0·51
763	O. A. S. 15146	9·5	1	15	54	57·92	107	29	20·8	1	0·39
764	O. A. S. 15148	8·9	1	15	54	59·87	107	47	53·6	1	0·46
765	W. B. K. XV. 1044 ...	7·7	2	15	56	59·51	94	27	43·4	2	0·46
766	1 Scorpii β[1]	2·9	.	15	57	42·38	109	26	20·2	7	0·30
767	Lalande 29306	8·0	1	15	59	56·56	107	34	29·4	1	0·49
768	14 Scorpii ν	4·2	...	16	4	16·72	109	6	46·7	4	0·49
769	1 Ophiuchi δ	2·8	.	16	7	22·62	93	20	50·5	2	0·50
770	Lalande 29610 ..	8·0	.	16	8	20·46	105	32	50·9	1	0·83

737.—Groombridge 2258.
742.—Comparison star for Comet 2, 1867.
740.—763.—Comparison stars for Comet 2, 1862.
747.—750.—755.—756.—759.—761.—762.—Comparison stars for Asia in 1861.
751.—767.—770.—Comparison stars for Donati's Comet in 1858.
754.—758.—764.—767.—Comparison stars for Sylvia in 1866.
760.—Carrington 2380

Observed with the Madras Meridian Circle in that Year.

Number.	Star.	In Right Ascension. Annual Precession.	In Right Ascension. Secular Variation.	In Right Ascension. Proper Motion.	In Polar Distance. Annual Precession.	In Polar Distance. Secular Variation.	In Polar Distance. Proper Motion.	Number in B A C
		s	s	s	"	"	"	
736	32 Libræ ξ¹ . ..	+ 3·3718	+ 0·0148	− 0·002	+ 12·840	− 0·354	+ 0·05	5080
737	R. P. L. 114 ...	− 22·9852	+ 7·7277	..	+ 12·901	+ 2·575	.	5140
738	. ... .	+ 4·9872	+ 0·0892		+ 12·700	− 0·567		
739	.	+ 3·9613	+ 0·0827		+ 12·570	− 0·454		
740		+ 3·7430	+ 0·0252		+ 12·540	− 0·431		
741	Lacaille 6421 ...	+ 3·7440	+ 0·0252	..	+ 12·496	− 0·433		
742	38 Libræ γ ..	+ 3·3417	+ 0·0180	+ 0·002	+ 12·340	− 0·389	− 0·02	5134
743	5 Cor. Bor. α (*Alphecca*.)	+ 2·5806	+ 0·0088	+ 0·000	+ 12·274	− 0·297	+ 0·07	5143
744	.. . .	+ 3·6740	+ 0·0224	...	+ 12·386	− 0·429	.	
745	. ..	+ 3·6755	+ 0·0234	...	+ 12·340	− 0·428		.
746	Lalande 28530	+ 2·0916	+ 0·0025		+ 12·073	− 0·240		
747	Taylor 7300 ..	+ 3·3812	+ 0·0131		+ 12·040	− 0·398		..
748	... . ..	+ 3·6697	+ 0·0278	.	+ 11·944	− 0·453	.	..
749	24 Serpentis α ...	+ 2·9415	+ 0·0062	+ 0·009	+ 11·866	− 0·354	− 0·06	5196
750	O. A. S. 14874 ..	+ 3·3616	+ 0·0133	...	+ 11·590	− 0·406	...	.
751	Lalande 28757 ..	+ 3·1266	+ 0·0088		+ 11·341	− 0·381		..
752	.. ..	+ 3·3860	+ 0·0130		+ 11·386	− 0·409		.
753	. .. .	+ 2·4796	+ 0·0027		+ 11·311	− 0·304		.
754	O. A. S. 14984 ...	+ 3·4287	+ 0·0145		+ 11·287	− 0·415		..
755	.	+ 3·3862	+ 0·0130		+ 11·205	− 0·411		..
756	W. B. E. XV. 898 .	+ 3·3830	+ 0·0130	...	+ 11·208	− 0·411	...	
757	36 Serpentis b	+ 3·1242	+ 0·0087	− 0·004	+ 11·180	− 0·385	+ 0·02	5205
758	46 Libræ θ	+ 3·3992	+ 0·0136	+ 0·009	+ 11·080	− 0·416	− 0·12	5237
759	O. A. S. 14996 .	+ 3·3770	+ 0·0131	...	+ 11·010	− 0·416	...	.
760	R. P. L. 115 ..	− 10·4483	+ 1·5488	...	+ 10·912	+ 1·374	..	...
761	O. A. S. 15053 ..	+ 3·3830	+ 0·0131		+ 10·894	− 0·480		.
762	W. B. E. XV. 983. ..	+ 3·3720	+ 0·0128		+ 10·815	− 0·410		.
763	O. A. S. 15146	+ 3·4822	+ 0·0135		+ 10·407	− 0·482		.
764	O. A. S. 15148	+ 3·4860	+ 0·0137		+ 10·404	− 0·488		
765	W. B. E. XV. 1044..	+ 3·1517	+ 0·0092		+ 10·380	− 0·402		
766	8 Scorpii β¹ .	+ 3·4758	+ 0·0142	− 0·002	+ 10·300	− 0·411	+ 0·02	5389
767	Lalande 29306	+ 3·4285	+ 0·0132	...	+ 10·097	− 0·486	.	.
768	14 Scorpii ν	+ 3·4770	+ 0·0136	− 0·002	+ 9·762	− 0·408	+ 0·08	5392
769	1 Ophiuchi δ	+ 3·1411	+ 0·0081	− 0·005	+ 9·484	− 0·408	+ 0·13	5414
770	Lalande 29610 ..	+ 3·4010	+ 0·0119	.	+ 9·285	− 0·402	.	

736.—739.—Proper motions from "*Greenwich Catalogue* 1872."
757.—Proper motions from "*Greenwich Catalogue* 1864."

Mean Positions of Stars for 1867 January 1st.

Number.	Star.	Magnitude.	Estimations.	Mean Right Ascension.			Mean Polar Distance.			Observations.	Fraction of Year.
				h.	m.	s.	°	′	″		
771	O. A. S. 16070 ...	8·0	1	16	9	10·13	112	35	20·5	1	0·41
772	R Scorpii Var. 1	9·5	2	16	9	40·69	112	36	40·8	2	0·51
773	S Scorpii Var. 2 ..	10·0	4	16	9	44·87	112	33	44·2	4	0·42
774	...	8·7	1	16	9	53·80	112	34	4·8	1	0·51
775	...	8·0	1	16	14	27·36	146	11	32·2	1	0·50
776	4 Ophiuchi ψ .	4·6	...	16	16	19·36	109	43	25·2	2	0·45
777	O. A. S. 15613 .	7·0	1	16	17	25·04	113	0	2·8	1	0·49
778	21 Scorpii α (*Antares*)	1·1	.	16	21	15·98	116	9	1·4	1	0·52
779	40 Herculis ζ	3·1	..	16	36	16·28	58	9	17·4	3	0·51
780	Lacaille 6984	8·7	1	16	39	51·74	120	57	45·6	1	0·50
781	.. ..	9·8	1	16	44	47·60	130	18	30·3	1	0·51
782	... ...	10·8	1	16	46	1·08	76	17	3·9	1	0·51
783	Taylor 7816 .. .	7·5	1	16	46	42·30	130	18	18·2	1	0·51
784	S Herculis Var. 3 ...	6·9	1	16	46	50·51	74	40	57·9	1	0·39
785	46 Herculis	7·0	1	16	46	1·33	74	48	3·2	1	0·50
786	Taylor 7882 ...	8·0	1	16	47	41·26	130	17	40·5	1	0·51
787	Taylor 7942 ...	6·8	...	16	48	21·87	106	36	20·0	1	0·53
788	... ...	8·0	...	16	49	8·74	125	31	36·6	1	0·86
789	27 Ophiuchi κ ...	3·4	...	16	51	22·35	80	24	57·6	3	0·48
790	29 Ophiuchi ..	6·8	.	16	54	4·72	108	41	13·9	1	0·90
791	O. A. S. 16283	8·0	1	16	54	9·45	110	23	30·8	1	0·38
792	... .	8·8	1	16	55	27·43	109	56	40·5	1	0·36
793	O. A. S. 16288	7·7	1	16	56	39·03	119	30	34·5	1	0·61
794	22 Ursæ Minoris ε ...	4·5	...	16	50	42·41	7	44	59·5	2	0·07
795	R Ophiuchi Var. 2 ...	7·0	3	17	0	7·86	106	54	46·4	3	0·48
796	35 Ophiuchi η	2·6	.	17	2	44·06	105	33	36·5	2	0·40
797	. .. ..	8·0	1	17	5	56·72	130	30	30·0	1	0·61
798	... .	9·0	1	17	6	19·28	130	54	14·7	1	0·88
799	64 Herculis α Var. 1 ...	3·3	..	17	8	34·96	75	27	21·3	3	0·52
800	Taylor 8017	6·9	1	17	16	38·30	114	46	8·3	1	0·48
801	42 Ophiuchi θ	3·4	.	17	13	30·36	114	51	49·7	4	0·82
802	.. .. .	8·5	2	17	21	16·71	130	48	48·4	2	0·58
803		9·3	1	17	21	36·74	130	46	59·2	1	0·85
804	... . ..	9·7	1	17	27	14·31	130	36	46·1	1	0·88
805	23 Draconis β . .	3·0	...	17	27	36·61	37	35	37·1	1	0·80

771.—774.—Stars observed for map of R Scorpii Var. 1.
772.—R Scorpii Var. 1.—Period 223 days.—Range, 9th to below 14th magnitude.
773.—S Scorpii Var. 2.—Period 177 days.—Range, 9th to below 14th magnitude.
777.—Comparison star for Angelina in 1866.
784.—S Herculis Var. 3.—Period 309 days.—Range, 6th to 12th magnitude.
791.—792.—Stars observed for map of T Serpentis Var. 4.
795.—R Ophiuchi Var. 2.—Period 303 days.—Range, 7·5 to below 14th magnitude.
799.—α Herculis Var. 1.—Changes irregularly between 3rd and 4th magnitude.

Observed with the Madras Meridian Circle in that Year.

Number.	Star.	In Right Ascension.			In Polar Distance.			Number in B. A. C.
		Annual Precession.	Secular Variation.	Proper Motion.	Annual Precession.	Secular Variation.	Proper Motion.	
		s	s	s	"	"	"	
771	O. A. S. 15470	+ 3·5648	+ 0·0147		+ 9·313	− 0·464	..	.
772	R Scorpii Var. 1.	+ 3·5086	+ 0·0147		+ 9·282	− 0·466	...	...
773	S Scorpii Var. 2.	+ 3·5646	+ 0·0146		+ 9·280	− 0·466	..	...
774	.. ..	+ 3·5649	+ 0·0147		+ 9·267	− 0·465	...	...
775	... ...	+ 4·8608	+ 0·0192		+ 8·912	− 0·669	...	...
776	4 Ophiuchi ψ . .	+ 3·5082	+ 0·0139	− 0·004	+ 8·780	− 0·464	+ 0·06	5497
777	O. A. S. 15613 .	+ 3·6375	+ 0·0141	...	+ 8·679	− 0·476	...	...
778	21 Scorpii α (*Antares*)	+ 3·6681	+ 0·0150	− 0·001	+ 8·370	− 0·491	+ 0·03	5496
779	40 Herculis ζ	+ 2·2964	+ 0·0088	− 0·034	+ 7·166	− 0·316	− 0·45	[illegible]
780	Lacaille 6964 ..	+ 3·8259	+ 0·0149	...	+ 6·867	− 0·598	...	[illegible]
781	...	+ 4·1463	+ 0·0192	...	+ 6·465	− 0·375	...	...
782	... ..	+ 2·7306	+ 0·0039	...	+ 6·448	− 0·361	...	...
783	Taylor 7815	+ 4·1471	+ 0·0191	...	+ 6·396	− 0·570	...	...
784	S Herculis Var. 3	+ 2·7535	+ 0·0039	...	+ 6·370	− 0·360	...	...
785	49 Herculis ...	+ 2·7276	+ 0·0040	− 0·001	+ 6·301	− 0·361	− 0·02	5674
786	Taylor 7832	+ 4·1499	+ 0·0166	...	+ 6·223	− 0·576	...	...
787	Taylor 7842 ..	+ 3·4612	+ 0·0089	...	+ 6·168	− 0·482	...	[illegible]
788	... ...	+ 3·9815	+ 0·0156	...	+ 6·106	− 0·566	...	...
789	27 Ophiuchi κ ..	+ 2·8863	+ 0·0044	− 0·028	+ 5·916	− 0·492	− 0·02	5708
790	20 Ophiuchi	+ 3·5037	+ 0·0089	− 0·001	+ 5·800	− 0·492	− 0·01	5728
791	O. A. S. [illegible] .	+ 3·6459	+ 0·0093	...	+ 5·693	− 0·488	...	...
792	... .	+ 3·5358	+ 0·0091	...	+ 5·674	− 0·496	...	...
793	O. A. S. [illegible] ..	+ 3·3100	+ 0·0119	...	+ 5·478	− 0·387	...	...
794	22 Ursæ Minoris ε ..	− 6·4154	+ 0·2046	+ 0·009	+ 5·215	+ 0·901	− 0·01	5780
795	R Ophiuchi Var. 2	+ 3·4404	+ 0·0077	...	+ 5·179	− 0·487	...	...
796	36 Ophiuchi ν	+ 3·4827	+ 0·0073	+ 0·001	+ 4·958	− 0·467	− 0·12	[illegible]
797	.	+ 4·1960	+ 0·0148	..	+ 4·690	− 0·597	. .	...
798	. ...	+ 4·1957	+ 0·0148	...	+ 4·686	− 0·599	...	...
799	64 Herculis α Var. 1	+ 2·7389	+ 0·0085	− 0·008	+ 4·461	− 0·391	− 0·01	[illegible]
800	Taylor 8017	+ 3·0764	+ 0·0080	...	+ 4·087	− 0·527	..	[illegible]
801	42 Ophiuchi θ . .	+ 3·0791	+ 0·0090	− 0·008	+ 4·012	− 0·595	− 0·02	[illegible]
802	.. ..	+ 4·2069	+ 0·0111	..	+ 3·372	− 0·606	...	..
803	. . .	+ 4·2034	+ 0·0111	..	+ 3·350	− 0·606	...	. .
804	... .	+ 5·4303	+ 0·0219	...	+ 2·986	− 0·768	...	...
805	23 Draconis β	+ 1·3551	+ 0·0052	− 0·003	+ 2·841	− 0·197	0·00	[illegible]

789.—Proper motions from "*Greenwich Catalogue* 1872."
790.—Proper motions from "*Greenwich Catalogue* 1864."

Mean Positions of Stars for 1867 January 1st.

Number.	Star.	Magnitude.	Estimations.	Mean Right Ascension. h.	m.	s.	Mean Polar Distance. °	′	″	Observations.	Fraction of Year.
806	55 Ophiuchi α ...	8·2	...	17	28	45·61	77	30	38·0	3	0·55
807	Taylor 8141 ...	6·0	1	17	30	45·66	111	40	52·0	1	0·30
808	... ...	9·0	2	17	34	46·35	138	36	30·0	2	0·61
809	... ...	9·9	1	17	34	46·51	136	16	7·4	1	0·56
810	... ...	9·2	1	17	36	15·96	150	36	11·9	1	0·51
811	58 Ophiuchi ...	5·0	...	17	36	27·98	111	36	55·7	1	0·30
812	... ...	9·4	1	17	36	39·44	150	37	13·4	1	0·51
813	... ...	9·0	1	17	37	32·05	136	29	34·4	1	0·58
814	... ...	8·0	1	17	39	45·53	127	21	43·8	1	0·52
815	... ...	8·0	1	17	39	56·07	127	14	41·1	1	0·58
816	... ...	8·3	1	17	40	7·06	126	28	36·9	1	0·60
817	86 Herculis μ ...	3·5	...	17	41	15·21	62	11	50·4	6	0·54
818	Radcliffe 3765 ...	8·3	1	17	43	41·47	17	32	1·8	1	0·58
819	... ...	8·6	2	17	45	13·75	136	35	32·5	3	0·63
820	Lacaille 7499 ...	7·1	3	17	48	27·70	120	4	44·4	3	0·60
821	Taylor 8368	6·0	1	17	48	40·02	105	47	10·7	1	0·46
822	Lacaille 7504	6·9	1	17	48	44·40	129	6	58·1	1	0·30
823	4 Sagittarii b	4·6	...	17	51	40·31	113	46	1·8	1	0·64
824	... ...	9·0	2	17	59	36·77	130	36	10·2	2	0·57
825	13 Sagittarii μ¹	4·1	...	18	5	43·46	111	5	38·9	9	0·56
826	Lacaille 7644 ..	6·7	1	18	9	9·63	132	19	30·6	1	0·52
827	... ...	8·8	1	18	13	7·00	127	43	56·4	1	0·55
828	23 Ursæ Minoris δ ...	4·3	...	18	15	14·90	3	23	41·4	4	0·42
829	21 Sagittarii	4·9	...	18	17	36·61	110	36	36·7	1	0·32
830	Taylor 8509	4·7	..	18	21	36·94	104	36	32·2	2	0·56
831	.. ..	9·0	...	18	23	14·79	136	18	42·2	1	0·59
832	θ Coronæ Australis ...	6·0	1	18	23	50·97	132	24	17·6	1	0·59
833	... ...	10·3	1	18	30	30·44	136	56	30·1	1	0·51
834	... .	9·0	2	18	30	39·22	136	51	47·2	2	0·51
835	3 Lyræ α (*Vega*) .	0·2	...	18	32	26·01	51	20	19·3	5	0·50
836	... .	9·3	1	18	36	12·90	137	16	7·4	1	0·62
837	. .. .	9·3	1	18	36	2·46	137	10	52·1	1	0·36
838	... .	9·0	1	18	40	37·19	127	27	26·6	1	0·68
839	10 Lyræ β Var. 1 .	3·6	..	18	45	10·07	56	47	36·0	13	0·49
840	... .	8·3	1	18	46	2·94	125	40	44·2	1	0·31

809.—813.—814.—815.—819.—820.—822.—833.—834.—836.—837.—Comparison stars for Donati's Comet of 1858.

839.—β Lyræ Var. 1.—Period 12·91 days.—Range, 3·5 to 4·5 magnitude.

Observed with the Madras Meridian Circle in that Year.

Number.	Star.	In Right Ascension. Annual Precession.	Secular Variation.	Proper Motion.	In Polar Distance. Annual Precession.	Secular Variation.	Proper Motion.	Number in B.A.C.
		s	s	s	"	"	"	
806	55 Ophiuchi ε	+ 2·7746	+ 0·0030	+ 0·01	+ 2·726	− 0·408	+ 0·20	5941
807	Taylor 8141	+ 2·6088	+ 0·0054	...	+ 2·358	− 0·382	..	5954
808	.. ..	+ 4·1380	+ 0·0072	...	+ 2·268	− 0·600	..	...
809	... ...	+ 4·0467	+ 0·0069	...	+ 2·208	− 0·587	..	...
810	... ..	+ 5·4325	+ 0·0162	...	+ 2·068	− 0·797	...	...
811	58 Ophiuchi . ..	+ 3·3080	+ 0·0050	− 0·010	+ 2·148	− 0·488	− 0·04	5967
812	... .	+ 5·4340	+ 0·0155	...	+ 2·080	− 0·790	...	...
813	... ..	+ 4·0563	+ 0·0064	...	+ 1·904	− 0·590	...	.
814	. ..	+ 4·0880	+ 0·0060	...	+ 1·768	− 0·596	.	..
815	... ..	+ 4·0848	+ 0·0060	...	+ 1·758	− 0·594	...	...
816	... ...	+ 4·0608	+ 0·0057	...	+ 1·736	− 0·601	...	...
817	86 Herculis μ	+ 2·3095	+ 0·0025	− 0·026	+ 1·680	− 0·346	+ 0·74	6021
818	Radcliffe 3765	− 1·1469	+ 0·0163	...	+ 1·466	+ 0·166	...	...
819	... ...	+ 4·1368	+ 0·0040	...	+ 1·292	− 0·608	...	.
820	Lacaille 7490	+ 4·1365	+ 0·0042	...	+ 1·009	− 0·606	...	...
821	Taylor 8268	+ 3·4490	+ 0·0039	...	+ 0·991	− 0·508	...	6065
822	Lacaille 7604	+ 4·1560	+ 0·0042	...	+ 0·966	− 0·608	...	...
823	4 Sagittarii b ...	+ 3·6615	+ 0·0028	− 0·003	+ 0·720	− 0·534	+ 0·04	6077
824	... .	+ 5·4692	+ 0·0012	...	+ 0·084	− 0·792	...	.
825	13 Sagittarii μ[1]	+ 3·5976	+ 0·0009	− 0·004	− 0·508	− 0·528	+ 0·01	6168
826	Lacaille 7644	+ 4·2892	− 0·0010	...	− 0·801	− 0·626	...	..
827	... ...	+ 4·1081	− 0·0015	...	− 1·147	− 0·598	...	
828	23 Ursæ Minoris δ ..	− 19·4136	− 0·4457	+ 0·068	− 1·388	+ 2·881	− 0·08	6241
829	21 Sagittarii . ..	+ 3·5735	− 0·0004	− 0·008	− 1·548	− 0·519	+ 0·02	6247
830	Taylor 8809	+ 3·4200	− 0·0008	...	− 1·689	− 0·496	...	6279
831	... ..	+ 4·4145	− 0·0060	...	− 2·016	− 0·640	...	..
832	θ Coronæ Australis .	+ 4·2365	− 0·0049	0·000	− 2·086	− 0·620	− 0·03	6294
833	... ..	+ 4·4501	− 0·0062	...	− 2·576	− 0·646	...	
834	... ..	+ 4·1396	− 0·0060	...	− 2·838	− 0·641	.	..
835	3 Lyræ α (Vega)	+ 2·0181	+ 0·0016	+ 0·017	− 3·180	− 0·290	− 0·28	6355
836		+ 4·5024	− 0·0108	...	− 3·069	− 0·648	..	
837	.	+ 4·4072	− 0·0108	...	− 3·141	− 0·647	...	
838	. ..	+ 4·6604	− 0·0075	...	− 3·586	− 0·684	.	.
839	10 Lyræ β Var. 1	+ 2·2138	+ 0·0016	− 0·002	− 3·987	− 0·315	+ 0·06	6400
840	..	+ 4·0479	− 0·0091	.	− 4·063	− 0·576	.	.

832.—Proper motions from "*Stone's Cape Catalogue.*"

Mean Positions of Stars for 1867 January 1st.

Number.	Star.	Magnitude.	Estimations.	Mean Right Ascension.			Mean Polar Distance.			Observations.	Fraction of Year.
				h.	m.	s.	°	′	″		
841	Lacaille 7319	8·0	1	18	47	36·34	129	4	56·0	1	0·66
842	O. A. S. 18960	7·0	1	18	53	40·96	121	7	36·2	1	0·64
843	30 Sagittarii ε	3·9	...	18	56	42·66	111	50	0·0	1	0·46
844	17 Aquilæ ζ	3·1	...	18	59	17·71	76	19	54·0	12	0·59
845	21 Sagittarii π	3·1	...	19	1	51·16	111	13	56·5	1	0·46
846	... ...	9·3	2	19	5	21·12	126	28	0·4	2	0·66
847	... ...	9·0	...	19	7	18·60	129	47	26·3	1	0·56
848	... ...	9·7	...	19	8	30·66	129	48	46·5	1	0·56
849	... ...	8·3	1	19	10	19·98	146	13	40·1	1	0·66
850	36 Aquilæ ω ...	5·1	...	19	11	34·33	78	38	32·0	6	0·60
851	30 Aquilæ δ	3·5	...	19	18	47·43	87	8	58·1	7	0·57
852	... ...	9·3	1	19	19	21·59	128	39	7·2	1	0·66
853	... ...	9·0	1	19	26	52·14	127	46	53·8	1	0·66
854	52 Sagittarii h²	4·6	2	19	28	36·55	116	10	30·0	2	0·64
855	... ...	8·0	1	19	31	11·87	127	42	5·7	1	0·66
856	. ...	8·0	1	19	34	2·07	127	44	58·3	1	0·66
857	55 Sagittarii e¹	5·0	...	19	34	54·46	105	36	59·4	1	0·46
858	50 Aquilæ γ	2·8	..	19	39	50·05	70	42	31·5	6	0·63
859	O. A. S. 19806	9·5	1	19	42	27·53	108	11	37·3	1	0·64
860	53 Aquilæ α (*Altair.*) ..	1·0	..	19	44	17·51	81	29	30·1	4	0·61
861	55 Aquilæ η Var. 1 ...	3·9	...	19	46	41·71	89	20	1·2	1	[illegible]
862	60 Aquilæ β	4·0	...	19	48	46·71	83	55	34·0	7	0·65
863	λ Ursæ Minoris	6·5	..	19	57	13·66	1	5	24·7	6	0·64
864	... ...	9·3	1	19	59	27·77	129	10	55·1	1	0·66
865	O. A. N. 20040	9·5	1	20	2	49·30	32	38	1·4	1	0·65
866	. ... .	.	...	20	6	31·05	74	47	53·3	1	0·60
867	Lacaille 8870	7·6	1	20	7	13·75	152	18	42·3	1	0·75
868	O. A. S. 20086	9·0	1	20	8	31·97	110	36	36·6	1	0·62
869	6 Capricorni α¹ ...	3·8	..	20	10	40·30	102	57	17·9	5	0·65
870	. . ...	7·6	1	20	11	23·51	106	16	10·7	1	0·76
871	Lalande 39046 .	5·5	...	20	12	11·36	50	2	30·0	1	0·76
872	34 Cygni Var. 1	6·0	1	20	12	53·06	52	22	46·4	1	0·69
873	9 Capricorni β	3·4	...	20	13	33·00	106	11	56·7	2	0·46
874	Lalande 39136 .	8·0	1	20	15	41·33	106	12	46·0	1	0·76
875	.. ..	9·0	1	20	15	54·03	106	16	51·3	1	0·76

842.—Comparison star for Eurydice in 1866.
861.—η Aquilæ Var. 1.—Period 7·176 days.—Range, 3·5 to 4·7 magnitude.
865.—The s. f. companion of β Cygni Var. 4.
866.—The s. f. companion of S Aquilæ Var. 4.
868.—Comparison star for Parthenope in 1862.
870.—874.—875.—Comparison stars for Hestia in 1866.
872.—34 Cygni Var. 1.—Supposed to vary from 3rd to 6th magnitude in several years.

Observed with the Madras Meridian Circle in that Year.

Number.	Star.	In Right Ascension. Annual Precession.	In Right Ascension. Secular Variation.	In Right Ascension. Proper Motion.	In Polar Distance. Annual Precession.	In Polar Distance. Secular Variation.	In Polar Distance. Proper Motion.	Number in B. A. C.
		s	s	s	"	"	"	
841	Lacaille 7510	+ 4·1348	− 0·0098		− 4·188	− 0·596	..	.
842	O. A. S. 18960	+ 3·8672	− 0·0077	..	− 4·667	− 0·546	.	...
843	39 Sagittarii *o* ...	+ 3·5942	− 0·0053	+ 0·001	− 4·912	− 0·506	+ 0·06	6507
844	17 Aquilæ ζ . ..	+ 2·7578	+ 0·0003	− 0·006	− 5·131	− 3·347	+ 0·07	[illegible]
845	41 Sagittarii *π* ...	+ 3·5798	− 0·0057	− 0·004	− 5·347	− 0·500	+ 0·03	6548
846	.. .	+ 4·0203	− 0·0121		− 5·641	− 0·660		
847	.. .	+ 4·1380	− 0·0144		− 5·806	− 0·576		
848	... .	+ 4·1371	− 0·0140	...	− 5·907	− 0·574	.	.
849	. ...	+ 4·9767	− 0·0328	...	− 6·086	− 0·639	...	.
850	25 Aquilæ *ω*	+ 2·8165	− 0·0068	− 0·008	− 6·102	− 0·388	− 0·02	[illegible]
851	30 Aquilæ δ	+ 3·0096	− 0·0018	+ 0·014	− 6·760	− 0·410	− 0·10	6646
852	... .	+ 4·0773	− 0·0161	...	− 6·806	− 0·556	...	
853	.. ..	+ 4·0936	− 0·0167	...	− 7·289	− 0·546	.	.
854	52 Sagittarii *h*² ...	+ 3·8640	− 0·0102	+ 0·002	− 7·363	− 0·490	− 0·02	6700
855	.. . ..	+ 4·0948	− 0·0170	...	− 7·771	− 0·588	...	...
856	... .	+ 4·0213	− 0·0162	...	− 8·000	− 0·585	..	.
857	55 Sagittarii *e*¹ ..	+ 3·4332	− 0·0075	+ 0·001	− 8·080	− 0·456	− 0·02	6742
858	50 Aquilæ γ	+ 2·8680	− 0·0011	+ 0·001	− 8·470	− 0·378	0·00	6772
859	O. A. S. 19996 ..	+ 3·4063	− 0·0087	...	− 8·670	− 0·482	...	..
860	53 Aquilæ α (*Altair*)	+ 2·9221	− 0·0014	+ 0·036	− 8·815	− 0·374	− 0·36	[illegible]
861	55 Aquilæ η Var 1 ..	+ 3·0692	− 0·0031	− 0·001	− 8·994	− 0·396	+ 0·04	6811
862	60 Aquilæ β	+ 2·9466	− 0·0020	+ 0·003	− 9·166	− 0·375	+ 0·47	[illegible]
863	λ Ursæ Minoris ..	− 58·1026	− 20·8894	− 0·036	− 9·187	+ 7·406	− 0·01	[illegible]
864	... .	+ 4·0171	− 0·0098	...	− 9·967	− 0·504	...	..
865	O. A. N. 20040 ..	+ 1·9892	− 0·0074	...	− 10·288	− 0·184	..	.
866	... .	+ 2·7613	− 0·0004	...	− 10·443	− 0·369		...
867	Lacaille 8370 ..	+ 5·2872	− 0·0772	...	− 10·571	− 0·648		6046
868	O. A. S. 20846 ..	+ 3·4037	− 0·0116	...	− 10·667	− 0·427		..
869	6 Capricorni *α*² ..	+ 3·3810	− 0·0094	+ 0·001	− 10·886	− 0·408	0·00	6874
870	.	+ 3·3999	− 0·0098	...	− 10·880	− 0·412	...	...
871	Lalande 39046 ..	+ 2·1382	+ 0·0017	...	− 10·987	− 0·280	..	[illegible]
872	31 Cygni Var 1 ..	+ 2·2102	+ 0·0019	...	− 10·988	− 0·265	..	[illegible]
873	9 Capricorni β ..	+ 3·3754	− 0·0096	− 0·001	− 11·086	− 0·406	− 0·03	[illegible]
874	Lalande 39126 ..	+ 3·3947	− 0·0101	...	− 11·176	− 0·406	.	
875	. .. .	+ 3·3960	− 0·0101	..	− 11·207	− 0·406		

Mean Positions of Stars for 1867 *January* 1*st.*

Number.	Star.	Magnitude.	Estimations.	Mean Right Ascension.			Mean Polar Distance.			Observations.	Fraction of Year.
				h.	m.	s.	°	′	″		
876	... ...	9·8	1	20	17	1·92	121	11	13·7	1	0·86
877	11 Capricorni ρ ..	5·0	...	20	21	16·20	108	15	4·5	9	0·61
878	... .	8·6	2	20	23	10·13	125	57	58·1	2	0·73
879	... ...	8·7	1	20	23	27·08	124	56	27·3	1	0·85
880	Lalande 39685	7·0	1	20	26	5·04	86	1	51·2	1	0·75
881	...	9·0	...	20	26	32·29	121	12	4·1	1	0·64
882	...	9·0	1	20	27	31·47	121	5	21·7	1	0·62
883	...	9·5	1	20	28	54·11	121	5	54·5	1	0·62
884	...	0·0	1	20	29	59·00	143	51	26·0	1	0·73
885	...	9·8	1	20	31	42·69	124	40	12·0	1	0·55
886	Taylor 9618	7·1	2	20	32	6·37	106	26	26·8	2	0·66
887	... ...	9·0	2	20	34	52·00	128	13	6·3	2	0·74
888	... ...	8·6	1	20	35	58·17	124	50	44·1	1	0·75
889	... ...	8·5	1	20	36	46·46	148	22	57·3	2	0·72
890	50 Cygni α (*Deneb*) ...	1·5	...	20	36	53·52	46	11	36·8	4	0·62
891	... ...	10·4	1	20	39	28·25	74	4	40·4	1	0·73
892	O. A. S. 20941	7·9	3	20	39	23·58	116	53	58·9	3	0·76
893	2 Aquarii ε	3·8	..	20	40	26·15	99	58	30·5	1	0·64
894	... ...	9·8	1	20	43	48·10	124	57	31·1	1	0·73
895	6 Aquarii μ	4·8	...	20	45	25·84	99	28	50·9	1	0·72
896	32 Vulpeculæ	5·1	...	20	49	53·47	62	26	40·3	9	0·71
897	... ...	9·0	1	20	50	55·09	148	45	10·8	1	0·75
898	Lacaille 8680	7·0	1	20	51	36·61	126	38	31·4	1	0·74
899	... ...	0·6	2	20	57	56·21	120	24	37·2	2	0·74
900	8 Microscopii	6·8	3	20	57	59·04	120	39	4·2	3	0·76
901	23 Capricorni θ	4·3	...	20	58	26·27	107	45	34·9	1	0·77
902	... ...	9·7	1	20	59	53·59	129	0	47·2	2	0·63
903	... ...	9·0	2	21	0	44·06	130	4	3·4	3	0·77
904	61 Cygni ... ... 1st..	5·6	...	21	0	36·13	51	54	12·7	3	0·73
905	61 Cygni ... ... 2nd ..	6·3	...	21	0	37·40	51	54	15·0	1	0·72
906	.. ..	9·1	2	21	1	22·26	119	59	41·9	2	0·73
907	13 Aquarii ν	4·6		21	2	20·77	101	51	20·6	2	0·66
908	64 Cygni ζ .. .	3·5	..	21	7	16·80	60	19	3·2	6	0·67
909	.. . .	9·4	2	21	13	59·50	128	59	34·9	2	0·68
910	. ...	7·8	1	21	14	30·54	126	28	28·2	3	0·69

892.—Comparison star for Undina in 1867.
899.—900.—903.—905.—Comparison stars for Sylvia in 1867.

Observed with the Madras Meridian Circle in that Year.

Number.	Star.	In Right Ascension. Annual Precession.	In Right Ascension. Secular Variation.	In Right Ascension. Proper Motion.	In Polar Distance. Annual Precession.	In Polar Distance. Secular Variation.	In Polar Distance. Proper Motion.	Number in B. A. C.
		s	s	s	"	"	"	
876	.. ..	+ 3·7410	− 0·0191	...	− 11·[illegible]	− 0·447	..	...
877	11 Capricorni ρ .	+ 3·4918	− 0·0115	− 0·006	− 11·594	− 0·403	+ 0·01	7046
878	... .	+ 3·8890	− 0·0237	...	− 11·729	− 0·461	...	...
879	...	+ 3·8891	− 0·0239	...	− 11·749	− 0·447	...	...
880	Lalande 39623	+ 2·9078	− 0·0031		− 11·865	− 0·347	...	...
881	...	+ 3·7219	− 0·0200		− 11·980	− 0·481		...
882	...	+ 3·7172	− 0·0200		− 12·098	− 0·480		...
883	.	+ 3·7143	− 0·0201		− 12·132	− 0·467		...
884	...	+ 4·5245	− 0·0535		− 12·207	− 0·619		...
885	...	+ 3·8015	− 0·0234		− 12·327	− 0·492		...
886	Taylor 9518	+ 3·3630	− 0·0106	...	− 12·354	− 0·381	...	...
887	.. .	+ 3·8903	− 0·0275	...	− 12·544	− 0·486	...	...
888	... ...	+ 3·7897	− 0·0242	...	− 12·618	− 0·486	...	...
889	... ...	+ 4·7563	− 0·0384	...	− 12·674	− 0·588	...	...
890	50 Cygni α (*Deneb*)	+ 2·0433	+ 0·0021	− 0·002	− 12·682	− 0·286	0·00	7171
891	... .	+ 2·7798	0·0000	...	− 12·860	− 0·305	...	...
892	O. A. S. 20841 ...	+ 3·5028	− 0·0177	...	− 12·851	− 0·397	...	...
893	2 Aquarii ε	+ 3·2620	− 0·0084	− 0·001	− 12·928	− 0·336	+ 0·01	7196
894	.. ..	+ 3·7781	− 0·0248	...	− 13·144	− 0·410	...	...
895	6 Aquarii μ ..	+ 3·2806	− 0·0083	0·000	− 13·206	− 0·340	+ 0·04	[illegible]
896	32 Vulpeculæ	+ 2·5655	+ 0·0026	− 0·002	− 13·477	− 0·270	0·00	7295
897	... ...	+ 4·6800	− 0·0739	...	− 13·608	− 0·497	...	...
898	Lacaille 8680 ..	+ 3·8006	− 0·0272	...	− 13·651	− 0·400	...	...
899	... .	+ 3·6880	− 0·0215	...	− 14·063	− 0·372	...	...
900	8 Microscopii	+ 3·6372	− 0·0218	...	− 14·080	− 0·373	.	7310
901	28 Capricorni θ .	+ 3·3763	− 0·0126	+ 0·004	− 14·098	− 0·344	+ 0·06	7322
902	.. . ..	+ 3·8834	− 0·0306	...	− 14·175	− 0·391	...	...
903	.. ... .	+ 3·6177	− 0·0215	...	− 14·226	− 0·366	...	...
904	61 Cygni ... 1st	+ 2·3338	+ 0·0044	+ 0·339	− 14·241	− 0·288	− 3·22	7336
905	61 Cygni ... 2nd ..	+ 2·3340	+ 0·0044	+ 0·330	− 14·244	− 0·288	− 3·22	7337
906	..	+ 3·6146	− 0·0214	...	− 14·280	− 0·368	...	...
907	13 Aquarii ν . ..	+ 3·2996	− 0·0068	+ 0·001	− 14·326	− 0·336	+ 0·01	7344
[illegible]	64 Cygni ζ ...	+ 2·5606	+ 0·0023	+ 0·003	− 14·684	− 0·346	+ 0·07	7369
909		+ 3·7594	− 0·0215	...	− 15·082	− 0·360	...	..
910	... .	+ 3·7019	− 0·0274	...	− 15·160	− 0·351	...	...

Mean Positions of Stars for 1867 January 1st.

Number.	Star.	Magnitude.	Estimations.	Mean Right Ascension.			Mean Polar Distance.			Observations.	Fraction of Year.
				h	m	s	°	′	″		
911	32 Capricorni ι	4·4	...	21	14	50·15	107	28	57·5	8	0·62
912	... ...	9·1	2	21	18	48·44	153	51	49·9	3	0·69
913	.. ..	7·9	1	21	20	10·50	150	47	4·9	1	0·75
914	... ...	8·0	2	21	21	21·72	138	55	30·2	2	0·69
915	Lacaille 8829	7·2	2	21	23	46·50	127	7	46·8	2	0·67
916	... ...	9·0	1	21	23	7·08	110	6	41·8	1	0·70
917	22 Aquarii β	3·1	...	21	24	38·84	96	9	17·6	7	0·71
918	... ...	9·5	...	21	27	16·97	132	37	30·0	1	0·78
919	... ...	7·7	1	21	30	54·66	127	46	35·1	1	0·65
920	40 Capricorni γ	3·8	...	21	32	42·07	107	15	42·2	2	0·59
921	Taylor 10068 ...	7·2	2	21	34	32·46	134	5	55·3	2	0·74
922	Taylor 10065 ...	6·6	1	21	34	40·69	146	6	17·8	1	0·68
923	... ...	8·3	1	21	34	56·33	138	59	40·1	1	0·75
924	8 Pegasi ε	2·4	...	21	37	30·11	80	44	0·3	8	0·73
925	... ...	9·2	1	21	37	51·09	127	47	22·8	1	0·65
926	45 Capricorni λ	5·4	...	21	39	22·37	101	58	40·4	2	0·77
927	49 Capricorni δ	3·0	...	21	39	41·68	106	48	46·2	2	0·62
928	... ...	9·6	2	21	41	6·30	127	46	38·4	2	0·69
929	Taylor 10126	7·0	1	21	41	11·11	137	13	36·7	1	0·76
930	... ...	8·8	1	21	43	4·19	132	30	37·6	1	0·75
931	... ...	7·5	2	21	46	13·02	127	31	6·9	2	0·61
932	51 Capricorni μ	5·2	...	21	46	2·53	104	10	35·8	3	0·72
933	16 Pegasi	5·0	...	21	47	0·60	64	41	59·4	4	0·69
934	... ...	9·4	2	21	50	52·80	127	36	36·6	2	0·75
935	... ...	9·1	3	21	52	56·59	127	29	45·0	3	0·64
936	... ...	8·7	3	21	52	58·40	129	31	50·3	3	0·74
937	ε Indi	5·0	2	21	58	10·12	147	19	50·0	2	0·70
938	Lacaille 9006 ..	7·3	2	21	56	30·34	129	31	2·1	2	0·75
939	34 Aquarii α .	3·2	...	21	58	37·05	90	57	51·2	4	0·74
940	33 Aquarii ι .	4·3	..	21	59	15·02	104	30	50·2	2	0·64
941	W. B. K. XXI. 1318 ..	9·0	1	22	1	35·24	78	8	55·6	1	0·73
942	.. ..	9·7	1	22	3	31·97	130	4	27·9	1	0·74
943	43 Aquarii θ .	4·3	..	22	9	48·74	98	36	40·9	8	0·74
944	.. . ..	8·2	1	22	12	14·21	139	25	48·2	1	0·78
945	... .	7·5	1	22	12	30·64	130	34	36·8	1	0·74

Observed with the Madras Meridian Circle in that Year.

Number.	Star.	In Right Ascension. Annual Precession.	In Right Ascension. Secular Variation.	In Right Ascension. Proper Motion.	In Polar Distance. Annual Precession.	In Polar Distance. Secular Variation.	In Polar Distance. Proper Motion.	Number in B. A. C.
		s	s	s	"	"	"	
911	32 Capricorni ι ...	+ 3·3486	− 0·0130	0·000	− 15·089	− 0·316	− 0·02	7497
912	... ...	+ 4·3944	− 0·1061	...	− 15·396	− 0·440		...
913	... ...	+ 4·6064	− 0·0871	...	− 15·391	− 0·425		...
914	... ...	+ 3·7613	− 0·0816	...	− 15·489	− 0·348		...
915	Lacaille 8829... ...	+ 3·7188	− 0·0896	...	− 15·518	− 0·336		...
916	... ...	+ 3·3617	− 0·0147	...	− 15·597	− 0·306	.	...
917	22 Aquarii β	+ 3·1686	− 0·0071	− 0·001	− 15·616	− 0·282	0·00	7478
918	... ...	+ 3·8829	− 0·0871	...	− 15·764	− 0·390	.	...
919	... ...	+ 3·7694	− 0·0806	...	− 15·906	− 0·332	...	...
920	40 Capricorni γ ...	+ 3·3211	− 0·0130	+ 0·018	− 16·064	− 0·288	+ 0·08	7585
921	Taylor 10068... ...	+ 3·8408	− 0·0896	...	− 16·130	− 0·325	...	7588
922	Taylor 10065... ...	+ 4·2078	− 0·0849	...	− 16·157	− 0·387	...	7640
923	... ...	+ 3·8560	− 0·0894	...	− 16·167	− 0·324	...	...
924	8 Pegasi ε	+ 2·9451	− 0·0005	+ 0·008	− 16·310	− 0·242	0·00	7661
925	... ...	+ 3·6747	− 0·0807	...	− 16·380	− 0·304	...	...
926	45 Capricorni λ ...	+ 3·2855	− 0·0101	− 0·002	− 16·397	− 0·285	+ 0·01	7677
927	49 Capricorni δ ...	+ 3·2030	− 0·0128	+ 0·014	− 16·410	− 0·270	+ 0·28	7680
928	... ...	+ 3·6694	− 0·0807	...	− 16·488	− 0·297	...	...
929	Taylor 10126 ...	+ 3·8940	− 0·0464	...	− 16·487	− 0·317	...	7691
930	... ...	+ 3·7616	− 0·0872	...	− 16·560	− 0·302	.	...
931	.. ..	+ 3·6416	− 0·0804	...	− 16·686	− 0·287	...	..
932	51 Capricorni μ ...	+ 3·2646	− 0·0112	+ 0·021	− 16·736	− 0·255	− 0·02	7619
933	16 Pegasi	− 2·7260	+ 0·0032	+ 0·001	− 16·772	− 0·210	+ 0·01	7687
934	... ...	+ 3·6195	− 0·0801	..	− 16·956	− 0·275	...	...
935	... ...	+ 3·6121	− 0·0804		− 17·048	− 0·270	...	...
936	... ...	+ 3·6823	− 0·0880	...	− 17·053	− 0·273	..	...
937	ε Indi...	+ 4·1681	− 0·0724	+ 0·480	− 17·061	− 0·313	+ 2·46	7686
938	Lacaille 9006 ..	+ 3·6857	− 0·0820	...	− 17·306	− 0·266	...	...
939	34 Aquarii α	+ 3·0894	− 0·0041	− 0·008	− 17·392	− 0·219	+ 0·02	7698
940	33 Aquarii ι	+ 3·2451	− 0·0113	− 0·001	− 17·386	− 0·231	+ 0·07	7691
941	W. B. K. XXI. 1418.	+ 2·9880	+ 0·0014	...	− 17·464	− 0·208	..	...
942	... ...	+ 3·6010	− 0·0812	...	− 17·514	− 0·280	..	...
943	43 Aquarii θ	+ 3·1698	− 0·0075	+ 0·006	− 17·762	− 0·206	+ 0·08	7773
944	... .	+ 3·5708	− 0·0825	...	− 17·678	− 0·289		.
945	... ...	+ 4·1460	− 0·0848	...	− 17·686	− 0·265	.	

937.—Proper motions from "*Stone's Cape Catalogue.*"

Mean Positions of Stars for 1867 January 1st.

Number.	Star.	Magnitude.	Estimations.	Mean Right Ascension. h.	m.	s.	Mean Polar Distance. °	′	″	Observations.	Fraction of Year.
946	... ...	0·1	2	22	14	54·76	130	25	30·7	2	0·77
947	... ...	0·2	2	22	19	2·14	140	44	50·7	2	0·76
948	R. P. L. 150 ...	5·5	...	22	23	27·51	4	38	47·1	8	0·43
949	57 Aquarii σ ...	4·8	...	22	23	36·32	101	21	27·3	4	0·70
950	... ...	8·2	1	22	23	52·02	130	30	34·2	1	0·81
951	27 Cephei δ Var. 1. ...	4·0	...	22	24	14·21	82	16	54·8	2	0·75
952	... ...	9·3	1	22	24	47·95	120	43	40·9	1	0·80
953	62 Aquarii η	4·2	...	22	28	31·23	90	48	8·8	4	0·70
954	R. P. L. 153	7·6	...	22	29	8·81	2	35	45·1	1	0·84
955	63 Aquarii κ	5·5	...	22	30	51·95	94	54	48·3	2	0·62
956	... ...	8·0	1	22	34	34·37	155	30	24·8	1	0·76
957	42 Pegasi ζ ...	3·6	...	22	34	40·60	79	51	43·8	11	0·79
958	67 Aquarii ...	6·2	...	22	36	17·88	97	39	32·0	1	0·55
959	... ...	8·2	1	22	36	40·89	130	26	2·3	1	0·77
960	... ...	9·5	1	22	43	37·86	130	35	42·7	1	0·80
961	... ...	8·0	1	22	44	52·01	135	40	30·0	1	0·84
962	73 Aquarii λ	3·8	...	22	46	40·81	98	17	12·5	3	0·70
963	... ...	9·5	1	22	47	38·83	128	56	54·4	1	0·81
964	... ...	9·3	2	22	48	18·74	152	34	14·7	2	0·74
965	O. A. S. 22600	8·3	2	22	49	10·41	110	18	56·5	2	0·76
966	S Aquarii Var. 2	10·2	1	22	49	58·80	111	3	9·8	1	0·73
967	24 Pis. Aus. α (*Fomalhaut*)	1·3	...	22	50	17·65	120	10	36·8	4	0·81
968	... ...	8·0	2	22	53	34·73	128	4	20·8	2	0·74
969	... ...	10·3	...	22	54	49·70	101	42	8·6	1	0·73
970	53 Pegasi β Var. 1 (*Scheat*).	2·6	...	22	57	19·30	62	29	17·4	1	0·66
971	... ...	9·5	1	22	57	30·82	87	11	9·0	1	0·88
972	... ...	9·0	1	22	57	23·47	140	37	3·2	1	0·80
973	54 Pegasi α (*Markab*) ...	2·6	...	22	58	8·17	75	30	30·2	5	0·70
974	R Pegasi Var. 2	8·5	1	22	59	58·30	80	10	25·2	1	0·82
975	... ...	9·2	1	23	3	16·30	127	20	6·2	1	0·83
976	90 Aquarii φ	4·2	..	23	7	26·08	96	46	58·3	2	0·68
977	. . ..	8·2	3	23	7	25·30	139	54	6·3	3	0·82
978	. ...	9·0	2	23	10	0·65	151	48	30·6	2	0·76
979	6 Piscium γ ...	3·8	...	23	10	16·23	87	26	38·6	10	0·81
980	93 Aquarii ψ¹	4·5	1	23	10	50·37	99	54	30·1	1	0·86

948.—Groombridge 3930.
951.—δ Cephei Var. 1.—Period 5·366 days.—Range, 3·7 to 4·8 magnitude.
954.—Carrington 3460.
955.—Comparison star for Calliope in 1866.
966.—S Aquarii Var. 2.—Period 279 days.— Range, 8th to below 12th magnitude.
970.—β Pegasi Var. 1.—(*Scheat*).—Period about 6 weeks.—Range, 2·0 to 2·5 magnitude.
974.—R Pegasi Var. 2.—Period 382 days.—Range, 7th to 13th magnitude.

Observed with the Madras Meridian Circle in that Year.

Number.	Star.	In Right Ascension. Annual Precession.	Secular Variation.	Proper Motion.	In Polar Distance. Annual Precession.	Secular Variation.	Proper Motion.	Number in B. A. C.
		s	s	s	"	"	"	
946		+ 3·5860	− 0·0294	...	− 17·989	− 0·222	.	.
947	. ..	+ 3·7689	− 0·0516	...	− 18·140	− 0·227		
948	B. P. L. 150 ...	− 3·7778	− 1·1818	+ 0·048	− 18·302	+ 0·238	− 0·05	7851
949	67 Aquarii e	+ 3·1519	− 0·0089	− 0·004	− 18·207	− 0·182	− 0·05	7840
950	... .	+ 3·5807	− 0·0337	...	− 18·317	− 0·202	.	..
951	27 Cephei δ Var. 1 ..	+ 2·2129	+ 0·0165	+ 0·002	− 18·389	− 0·128	+ 0·02	7846
952	.	+ 3·5806	− 0·0394	...	− 18·349	− 0·190	.	
953	62 Aquarii η	+ 3·0798	− 0·0081	+ 0·008	− 18·470	− 0·166	+ 0·06	7888
954	B. P. L. 153	− 8·3166	− 3·8868	...	− 18·497	+ 0·476	.	..
955	63 Aquarii κ .. .	+ 3·1157	− 0·0051	− 0·007	− 18·557	− 0·164	+ 0·11	7894
956	... .	+ 4·1429	− 0·1087	...	− 18·672	− 0·213	.	.
957	42 Pegasi ζ . ..	+ 2·9852	+ 0·0023	+ 0·001	− 18·686	− 0·140	0·00	7905
958	67 Aquarii ...	+ 3·1868	− 0·0068	...	− 18·782	− 0·156	.	7921
959	... .	+ 3·4772	− 0·0327	...	− 18·744	− 0·173	.	...
960	. ..	+ 3·4470	− 0·0285	...	− 18·962	− 0·157		.
961	.	+ 3·5126	− 0·0396	..	− 18·987	− 0·157	.	.
962	73 Aquarii λ	+ 3·1341	− 0·0063	− 0·006	− 19·009	− 0·137	− 0·08	7970
963	.	+ 3·4070	− 0·0300	...	− 19·064	− 0·146	.	.
964	.. ..	+ 3·8654	− 0·0854	...	− 19·080	− 0·166		...
965	O. A. S. 22500 ..	+ 3·3004	− 0·0204	...	− 19·104	− 0·139		.
966	S Aquarii Var. 2 ..	+ 3·2269	− 0·0140	...	− 19·126	− 0·134	...	.
967	24 Piscis. Australis α.	+ 3·3063	− 0·0210	+ 0·022	− 19·186	− 0·135	+ 0·15	7982
968	.. ..	+ 3·3714	− 0·0256	...	− 19·219	− 0·132	...	.
969	. ..	+ 3·1499	− 0·0073	...	− 19·260	− 0·121		...
970	53 Pegasi β Var. 1 .	+ 2·8653	+ 0·0117	+ 0·014	− 19·310	− 0·100	− 0·15	8002
971	... . ..	+ 2·8893	+ 0·0141	...	− 19·310	− 0·104		
972	... .	+ 3·6876	− 0·0706	...	− 19·311	− 0·138	.	
973	54 Pegasi α (Markab).	+ 2·9800	+ 0·0066	+ 0·008	− 19·365	− 0·107	+ 0·02	8034
974	R Pegasi Var. 2 ...	+ 3·0122	+ 0·0084	...	− 19·371	− 0·108	.	.
975	... ...	+ 3·3233	− 0·0271	...	− 19·444	− 0·110		
976	90 Aquarii φ .	+ 3·1082	− 0·0045	+ 0·001	− 19·529	− 0·096	+ 0·19	8066
977	. .. .	+ 3·3365	− 0·0394	..	− 19·622	− 0·102		
978	... ...	+ 3·6100	− 0·0742	...	− 19·680	− 0·105		
979	6 Piscium γ	+ 3·0892	+ 0·0096	+ 0·047	− 19·866	− 0·087	+ 0·01	8105
980	98 Aquarii ψ[1] .	+ 3·1217	− 0·0061	− 0·008	− 19·886	− 0·088	− 0·01	8109

970.—Proper motions from "*Greenwich Catalogue 1872.*"

Mean Positions of Stars for 1867 January 1st.

Number.	Star.	Magnitude.	Estimations.	Mean Right Ascension.			Mean Polar Distance.			Observations.	Fraction of Year.
				h.	m.	s.	°	′	″		
981	... ...	9·2	2	23	11	44·28	131	7	11·8	2	0·78
982	... ...	9·5	1	28	17	4·56	127	38	29·6	1	0·80
983	... ...	9·7	2	23	18	24·73	131	7	28·0	2	0·78
984	... ...	8·7	3	23	18	39·68	127	16	47·2	3	0·79
985	8 Piscium κ ...	5·0	...	23	20	6·89	89	28	20·4	9	0·80
986	... ...	7·5	1	23	20	56·78	88	50	30·5	1	0·84
987	Lacaille 9496 ...	8·0	2	23	23	28·84	127	41	27·5	2	0·81
988	... ...	7·9	2	23	24	27·84	127	1	17·1	2	0·73
989	Lacaille 9514 ...	9·0	1	23	26	2·66	131	34	46·9	1	0·82
990	Lacaille 9517 ...	7·1	3	23	26	38·57	126	59	57·7	3	0·76
991	... ...	9·4	2	23	30	50·27	130	6	8·0	2	0·80
992	... ...	7·8	3	23	29	51·66	127	32	9·0	3	0·88
993	17 Piscium ι ...	4·3	...	23	33	6·60	86	5	39·7	0	0·80
994	18 Piscium λ ...	4·7	...	23	35	15·40	88	57	8·1	1	0·77
995	... ...	8·4	2	23	36	21·85	143	41	57·6	2	0·84
996	... ...	9·4	2	23	36	8·30	128	8	40·8	2	0·78
997	R Aquarii Var. 1	9·7	2	23	36	56·42	106	1	18·2	2	0·80
998	Lacaille 9583	8·0	2	23	30	0·04	123	42	54·0	2	0·84
999	Lacaille 9607	6·0	3	23	40	32·45	126	38	44·4	3	0·83
1000	δ Sculptoris	4·6	...	23	41	59·55	118	51	57·4	4	0·78
1001	... ...	7·5	1	23	42	8·16	142	3	37·1	1	0·72
1002	21 Piscium ...	6·1	...	23	42	38·80	89	39	45·6	1	0·86
1003	... ...	9·3	2	23	42	56·48	129	42	48·4	2	0·82
1004	... ...	8·7	3	23	47	1·42	126	8	51·7	3	0·80
1005	Lacaille 9641 ...	7·8	2	23	48	11·48	128	6	17·4	2	0·86
1006	Lacaille 9680 ...	8·2	1	23	49	16·00	129	47	10·3	1	0·88
1007	... ...	9·0	1	23	51	54·98	152	10	40·5	1	0·73
1008	28 Piscium ω ..	4·2	...	23	52	28·08	96	52	23·1	3	0·82
1009	29 Piscium ...	5·1	...	23	55	0·52	93	46	5·7	1	0·86
1010	... ..	9·5	2	23	56	7·88	130	16	1·7	2	0·86
1011	... ...	9·3	1	23	56	36·70	126	42	22·8	1	0·80
1012	Taylor 10990	8·8	1	23	57	8·48	146	34	13·8	1	0·88
1013	Taylor 10997	9·0	1	33	58	12·65	126	46	30·6	1	0·74
1014	... ...	9·0	2	23	58	44·51	126	58	14·8	2	0·79
1015	Lacaille 9721	6·0	1	23	59	26·37	139	49	51·5	1	0·86
1016	Lacaille 9728	7·7	1	23	59	51·63	126	49	27·3	1	0·85

997.—R Aquarii Var. 1.—Period 388 days.—Range, 8th to 11th magnitude.

Observed with the Madras Meridian Circle in that Year.

Number.	Star.	In Right Ascension. Annual Precession.	Secular Variation.	Proper Motion.	In Polar Distance. Annual Precession.	Secular Variation.	Proper Motion.	Number in B A C
		s	s	s	"	"	"	
981	.. ...	+ 3·3161	− 0·0304	...	− 19·613	− 0·083	...	.
982	... ...	+ 3·1627	− 0·0280	...	− 19·704	− 0·081	...	.
983	... .	+ 3·2[illegible]	− 0·0304	...	− 19·72[illegible]	− 0·078	.	.
984	... ...	+ 3·2347	− 0·0255	...	− 19·730	− 0·077	...	.
985	8 Piscium κ	+ 3·0899	0·0000	+ 0·005	− 19·752	− 0·080	+ 0·12	8160
986	... .	+ 3·0670	+ 0·0003		− 19·763	− 0·084		.
987	Lacaille 9406 ..	+ 3·2362	− 0·0254		− 19·800	− 0·080		...
988	... . .	+ 3·2278	− 0·0247		− 19·814	− 0·083		...
989	Lacaille 9514	+ 3·2472	− 0·0302		− 19·885	− 0·061		...
990	Lacaille 9517 ..	+ 3·2182	− 0·0244		− 19·848	− 0·059		...
991	... .	+ 3·2199	− 0·0272	...	− 19·852	− 0·052	...	...
992	... ...	+ 3·2069	− 0·0240	...	− 19·882	− 0·062	..	...
993	17 Piscium ι	+ 3·0686	+ 0·0080	+ 0·025	− 19·913	− 0·042	+ 0·16	8088
994	18 Piscium λ	+ 3·0696	+ 0·0011	− 0·011	− 19·938	− 0·080	+ 0·17	8048
995	... .	+ 3·3062	− 0·0561	...	− 19·963	− 0·048	..	...
996	... .	+ 3·1810	− 0·0344	...	− 19·940	− 0·080	...	...
997	R Aquarii Var. 1 ...	+ 3·1097	− 0·0081	...	− 19·966	− 0·096	...	...
998	Lacaille 9583 ..	+ 3·1702	− 0·0288	...	− 19·971	− 0·094	...	...
999	Lacaille 9597... ...	+ 3·1636	− 0·0246	...	− 19·983	− 0·090	...	...
1000	δ Sculptoris	+ 3·1290	− 0·0161	+ 0·009	− 19·993	− 0·086	+ 0·10	8073
1001	... .	+ 3·3067	− 0·0409	...	− 19·994	− 0·083	..	.
1002	21 Piscium .. .	+ 3·0714	+ 0·0011	− 0·002	− 19·993	− 0·086	+ 0·04	8081
1003	... ...	+ 3·1547	− 0·0251	...	− 19·999	− 0·085	.	...
1004	... ...	+ 3·1315	− 0·0232	...	− 20·088	− 0·017	.	...
1005	Lacaille 9641 ...	+ 3·1262	− 0·0250	...	− 20·027	− 0·015	..	...
1006	Lacaille 9650 ...	+ 3·1348	− 0·0244	...	− 20·033	− 0·012	...	..
1007	. ... ..	+ 3·1621	− 0·0500	...	− 20·042	− 0·009	...	.
1008	33 Piscium ..	+ 3·0673	+ 0·0017	+ 0·010	− 20·046	− 0·005	+ 0·13	8081
1009	30 Piscium	+ 3·0738	− 0·0004	− 0·002	− 20·051	0·000	+ 0·01	8046
1010	... . .	+ 3·0913	− 0·0240	..	− 20·063	+ 0·001	..	.
1011	... . .	+ 3·0869	− 0·0208	...	− 20·063	+ 0·003		...
1012	Taylor 10890	+ 3·1003	− 0·0462	...	− 20·063	+ 0·003		.
1013	Taylor 10937	+ 3·0797	− 0·0206	..	− 20·061	+ 0·005		.
1014	.. ..	+ 3·0774	− 0·0198	...	− 20·064	+ 0·007		.
1015	Lacaille 9721 .	+ 3·0762	− 0·0208	...	− 20·065	+ 0·008		.
1016	Lacaille 9728 .	+ 3·0727	− 0·0205	...	− 20·066	+ 0·009	...	.

1009—Proper motions from "*Stone's Cape Catalogue.*"

DISTRIBUTION LIST OF INSTITUTIONS AND INDIVIDUALS

TO WHOM COPIES OF THE MADRAS ASTRONOMICAL OBSERVATIONS ARE PRESENTED

BY THE GOVERNMENT OF MADRAS.

ALGERIA (FRENCH.)

Algiers ... The Observatory.

ARGENTINE REPUBLIC (SOUTH AMERICA.)

Cordoba ... National Observatory
Dr J. M. Thome.

AUSTRALIA (SOUTH.)

Adelaide ... Government Observatory.
C. Todd, C.M.G.

AUSTRALIA (VICTORIA.)

Melbourne ... Government Observatory.
R. L. J. Ellery, F.R.S.

AUSTRALIA (NEW SOUTH WALES.)

Sydney ... Royal Society of New South Wales.
Government Observatory
H. C. Russel, B.A.

Windsor ... J. Tebbutt.

AUSTRIA.

Buda-pest ... The Observatory.
Herény ... E. von Gothard.
Kalocsa ... Dr. C. Braun.
Kiskartal ... Baron von Podmaniczky.
Krakau ... Prof. F. Karlinski.
Kremsmuenster ... The Observatory.
O. Gyalla ... Dr N. von Konkoly.
Pola ... The Observatory.
Prague ... Prof. L. Weinek.
Prof. Safarik.
Trieste ... Prof. A. Kunes.
Dr. F. Anton.
Vienna ... Imperial Academy of Sciences.
Imperial Observatory.
Prof. E. Weiss.
Dr. F. Bidschof.
Dr. J. Holetschek.
Dr. J. Palisa.

BELGIUM.

Brussels ... Royal Academy of Sciences.
Royal Observatory
Prof. F. Folie.

BRAZIL (SOUTH AMERICA.)

Rio Janeiro ... Imperial Observatory.
Dr L. Cruls.

CAPE OF GOOD HOPE.

Cape Town ... Royal Observatory
Dr. D. Gill, F.R.S.
W. H. Finlay, B.A.

CEYLON.

Colombo ... Surveyor General

CHILI (SOUTH AMERICA.)

Santiago ... National Observatory

CHINA

Hong Kong ... Dr. W. Doberck, Govt. Astron.

DENMARK.

Copenhagen ... Royal Academy of Sciences.
Royal Observatory.
Prof. T. N. Thiele.
C. F. Pechule.

FRANCE.

Besançon ... The Observatory.
Bordeaux ... The Observatory.
Cherbourg ... Soc. Nationale des Sc. Naturelles.
Lyons ... The Observatory.
Marseilles ... The Flammarion Sc. Society.
Dir. E. Stephan.
A. Borelly.
—Coggia.
Nice ... Dir. J. Perrotin.
A. Charlois.
Paris ... Institute of France.
Bureau des Longitudes.
Office de la Conn. des Temps
National Observatory.
A d'Abbadie
H. A. E. A. Faye.
Camille Flammarion
P. Henry.
P. J. C. Janssen
C. Loewy.
L'Amiral E. Mouchez.
L. Schulhof.
F Tisserand
Toulouse ... The Observatory

GERMANY.

Bamberg ... Dr. E. Hartwig.
Berlin ... Imperial Academy of Sciences.
Imperial Observatory.
Prof. A. Auwers. Geh. Rath.
Prof. W. Foerster. Geh. Rath.
Dr. V. Knorrie.
Prof. F. Tietjen.
Bonn ... Royal Observatory.
Prof. E. Schoenfeld. Geh. Rath.
Bothkamp ... Count von Bulow.
Breslau ... The Observatory.
Prof. J. G. Galle.
Carlsruhe ... The Observatory.
Dresden ... Baron B. von Engelhardt.
Dusseldorf ... Dr R. Luther.
Gotha ... The Observatory.
Gottingen ... The Observatory.
Prof. W. Schur.
Halle .. Prof. O. A. Rosenberger.
Hamburg ... The Observatory.
Prof. G. Rumker.
Jena ... W. Winkler.
Kiel ... The Observatory.
Prof. A. Krueger.
Dr. E. Lamp.
Koenigsburg ... Royal Observatory.
Prof. C. F. W. Peters.
Leipzig ... Astronomical Society.
Prof. H Bruns.
Dr. R. Engelman.
Dr. W. Feddersen.
Prof. F. Zollner.
Manheim ... The Observatory.
Munich ... Royal Academy of Sciences.
Royal Observatory.
Prof. H. Seeliger.
Prof L. Seidel.
Potsdam ... The Observatory.
Prof H. Vogel.
Strassburg .. The Observatory.
Prof. E. Becker.
Prof. F. A. J. Winnecke.
Thorn The Copernicus Verein.
Wilhelmshaven. . The Observatory.

GREECE.

Athens ... Royal Observatory.

INDIA.

Arconam ... C. K. Winter.
Bombay ... Government Observatory.

INDIA—(continued.)

Calcutta ... Surveyor General.
Meteorological Reporter to Govt
Asiatic Society.
Geological Survey of India
Dehra Dun ... G. T. Survey of India.
Col. G. Strahan, R E.
Madras ... Christian College Library
Civil Engineering College Library
Government Central Museum
Literary and Philosophical Society.
Presidency College Library.
Prof. C. Michie Smith B. Sc

ITALY.

Florence .. The Observatory (Arcetri)
W. Temple.
Lombardy ... Royal Institution.
Milan ... The Observatory (Brera.)
Prof. G. V. Schiaparelli.
Naples ... Royal Observatory.
Prof. A de Gasparis.
Padua ... The Observatory.
Palermo ... The Observatory.
Prof. G. Cacciatore.
Rome .. The Observatory (Capitol.)
The Observatory (Collegio Romano)
Prof. E. Millosevich.
Prof. L. Respighi.
Prof. P. Tacchini.
Turin ... Royal Academy of Sciences.
The Observatory (Moncalieri)

JAPAN.

Tokio ... The Observatory.

MAURITIUS.

Pamplemousses.. C. Meldrum, M A., F R s.

NATAL (AFRICA EAST)

Durban The Observatory.

NETHERLANDS (HOLLAND)

Leyden ... The Observatory.
Prof H. G. van de Sande Bakhuyzen.
Utrecht ... The Observatory.
Prof. J. A. C. Oudemans.

NETHERLANDS (INDIA)

Batavia ... Surveyor General.

NORWAY.

Bergen ... The Observatory
Christiania Royal Observatory.
Prof. C. Fearnley
O. A. L. Pihl.

PERU.

Lima .. The Observatory.

PORTUGAL.

Coimbra The Observatory.
Lisbon Royal Observatory.

RUSSIA.

Dorpat The Observatory.
Helsingfors The Observatory.
Kasan The Observatory.
Kharkoff The Observatory.
Kiev The Observatory.
Kronstadt ... The Observatory.
Moscow The Observatory.
Prof. Th. Bredechin.
Dr. W. Ceraski.
Nicolaiew . The Observatory.
Odessa ... The Observatory.
Plonsk ... The Observatory.
Pulkowa ... Central Imperial Observatory.
Prof. W. Dollen, Geh. Rath.
Prof. M. Nyren.
Dr. H. Struve.
Prof. O. W. vonStruve, Geh. Rath.
St. Petersburg ... Imperial Academy of Sciences.
Dr. J. O. Backlund.
Prof. S. von Glasenapp.
Taschkent ... The Observatory.
Warsaw ... The Observatory.
Wilna .. The Observatory.

SPAIN.

Madrid ... Royal Observatory.
San Fernando ... Marine Observatory.

STRAITS SETTLEMENTS.

Singapore . Surveyor General.

SWEDEN.

Lund .. The Observatory.
Dr. N. C. Duner.
Dr. F. Hagstrom.
Prof. A. Moller.
Stockholm Royal Academy of Sciences.
Prof. H. Gilden.
Upsala The Observatory.
Prof. H. Schultz.
Dr. H. Thalen.

SWITZERLAND.

Geneva The Observatory.
Prof. E. Gautier.
Neuchatel The Observatory.

SWITZERLAND—(continued.)

Vevey Prof. F. F. E. Brunnow.
Zurich .. The Observatory.
Prof. R. Wolf.

UNITED KINGDOM (*ENGLAND*).

Blackheath A. M. Downing, M.A.
E. Dunkin, F.R.S.
J. Glaisher, F.R.S.
W. Thynne Lynn, B.A.
Birkenhead .. Bidston Observatory.
Bocking E. B. Knobel.
Bristol .. W. F. Denning.
Cambridge ... The Observatory.
Prof. J. C. Adams, F.R.S.
Prof. A. Cayley, F.R.S.
J. W. L. Glaisher, F.R.S.
Prof. G. G. Stokes, F.R.S.
Chepstow ... E. J. Lowe, F.R.S.
Collingwood .. Lieut.-Col. J. Herschel, R.E., F.R.S.
Cuckfield ... G. Knott, LL.B.
Darlington T. E. Espin.
Durham .. The Observatory.
Ealing ... Lt.-Gen. Tennant R.E., C.I.E., F.R.S.
A. A. Common, F.R.S.
Eastbourne ... G. F. Chambers.
Gateshead ... R. S. Newall, F.R.S.
Greenwich ... Royal Observatory.
Sir G. B. Airy, K.C.B., F.R.S.
W. H. M. Christie F.R.S., Ast. Royal.
E. W. Maunder.
H. H. Turner, M.A.
Harrow .. Lt.-Col. G. L. Tupman, R.M.A.
Ipswich ... Col. Tomline.
Leyton .. J. G. Barclay.
Liverpool .. Astronomical Society.
London .. Royal Society.
Royal Asiatic Society.
Royal Astronomical Society.
Royal Geographical Society.
Royal Institution.
Royal Meteorological Society.
British Museum.
Meteorological Office.
Nautical Almanac Office.
Sc. & Art Dep. South Kensington.
R. Bryant, B.A.
Col. W. M. Campbell, R.E.
Dr. W. De La Rue, F.R.S.
Dr. W. Huggins, F.R.S.
Cuthbert E. Peek, M.A.

UNITED KINGDOM (ENGLAND) (continued)

London	E. B. Powell, C.S.I.
	A. C. Ranyard, M.A.
	Gen. R. Strachey, R.E., F.R.S.
	Gen. J. T. Walker, R.E., C.B., F.R.S.
Manchester	Literary and Philosophical Society.
	Owen's College.
	Prof. A. Schuster, F.R.S.
Maresfield	... Captain W. Noble.
NewcastleonTyne	Prof. A. S. Herschel.
Oxford	. Radcliffe Observatory
	University Observatory.
	Rev. C. Pritchard, F.R.S.
	E. J. Stone, M.A., F.R.S.
Richmond	.. Kew Observatory.
Rugby	. Temple Observatory.
Scarborough	... J. Wigglesworth.
Southampton	... Ordnance Survey Office.
Southport	... J. Baxendell.
Twickenham	... Dr J. R. Hind, F.R.S.
Westgate on Sea.	J. N. Lockyer, F.R.S.
Whalley.	... Stonyhurst College Observatory.
Witham	... Lord Rayleigh, F.R.S.

UNITED KINGDOM (SCOTLAND).

Aberdeen	... University Library.
Dun Echt	... Earl of Crawford & Balcarres, F.R.S.
	Dr. Ralph Copeland.
Edinburgh	... Royal Observatory.
	Royal Society of Edinburgh.
	University Library.
Glasgow	... The Observatory.
	Prof. R. Grant, F.R.S.
	Sir W. Thomson, F.R.S.

UNITED KINGDOM (IRELAND).

Armagh	... The Observatory.
	Dr. J. L. E. Dreyer.
Ballysodare	.. J. F. Gore.
Collooney	... Col. E. H. Cooper.
	A. Marth.
Dublin	. Royal Irish Academy.
	Royal Dublin Society.
	Royal Observatory, Dunsink.
	Sir R. S. Ball, F.R.S., Ast. Royal.
	H. Grubb, F.R.S.
	G. Johnston Stoney, F.R.S.
Parsonstown	.. The Earl of Rosse, F.R.S.

UNITED STATES AMERICA

Albany, N. Y.	Dudley Observatory
	Prof. L. Boss
Allegheny, Pen.	The Observatory.
Amherst, Mass	Lawrence Observatory
Ann Arbor, Mich.	The Observatory.
Baltimore	... The John Hopkins University.
Boston, Mass	... American Academy of Arts & Sc
Cambridge, Mass.	Harvard College Observatory.
	S. C. Chandler.
	Dr. B. A. Gould.
	Prof. E. C. Pickering.
	O. C. Wendell.
Cambridgeport ..	E. F. Sawyer
Chicago, Ill	Dearborn Observatory.
Cincinnati, Ohio.	Mount Lookout Observatory
Clinton, N. Y.	Prof. C. H. F. Peters.
Georgetown, Columbia.	The Observatory
Glasgow, Missouri.	Morrison Observatory.
Madison, Wis.	Washburn Observatory
Mt. Hamilton Cal.	Lick Observatory.
	E. E. Barnard.
	Prof. S. W. Burnham.
	Prof. & Dir. E. S. Holden.
	J. M. Schaeberle.
New Haven, Conn.	Academy of Arts and Sciences.
	Dr. W. Elkin.
	Prof. H. A. Newton.
Phelps, N. Y.	. W. R. Brooks, Red House Obs
Philadelphia	American Philosophical Society
Princeton, N. J.	. Prof. C. A. Young.
Rochester, N. Y.	Prof. L. Swift, Warner Observatory
San Francisco, Cal.	Prof. G. Davidson.
Virginia.	... The Leander Mc. Cormick Obs.
Washington.	... American Ephemeris Office.
	National Academy of Sciences.
	Signal Office, War Department
	Smithsonian Institution.
	U. S. Coast & Geo. Survey Office
	U. S. Naval Observatory.
	Commander C. H. Davis, U.S.N.
	Admiral S. R. Franklin, U.S.N.
	Prof. E. Frisby.
	Prof. Asaph Hall.
	Prof. S. P. Langley
	Lieut. S. C. Lemley, U.S.N.
	Prof. S. Newcomb.
	Prof. W. C. Winlock
Williamstown Mass.	Prof. T. H. Safford.

www.ingramcontent.com/pod-product-compliance
Lightning Source LLC
Chambersburg PA
CBHW021355210326
41599CB00011B/884
9783337711979